The Information Technology Revolution

Also by Tom Forester

The Labour Party and the Working Class (1976)
The Microelectronics Revolution (ed.) (1980)

The
Information Technology
Revolution

Edited and introduced by
TOM FORESTER

The MIT Press
Cambridge, Massachusetts

Third printing, 1986
First MIT Press edition, 1985

Printed and bound in the United States of America

Library of Congress Cataloguing in Publication Data
Main entry under title:

The Information technology revolution.

Bibliography: p.
Includes index.
1. Computers and civilization. 2. Computers.
I. Forester, Tom.
QA76.9.C66I53   1985   303.4'834   84–23422
ISBN 0–262–06095–7 (hard)
        0–262–56033–X (paper)

Contents

Preface

A lot has happened since my previous collection, *The Microelectronics Revolution*, first appeared in June, 1980. Even so, it is remarkable how much of the discussion in that volume is just as relevant today. So rather than attempt to revise it into a second edition, I have left it to stand in its own right and have instead produced a sequel volume, consisting entirely of new material published in the four years between June, 1980 and June, 1984.

In this way we have been able, I hope, to cover the key technological developments over the period and to track the many social changes that are following on with great rapidity. And with more case study rather than speculative material available, I think we have been able to achieve a much higher overall standard.

Of the 48 pieces here, 33 are of US origin and most of the rest are British. A total of 29 were first published in 1983 or 1984, 15 in 1982, and only 4 before that. I am particularly proud of the fact that 13 were scripted by working engineers and/or computer scientists. Once again, our aim has been to help bridge the gap that C. P. Snow described as the "gulf of mutual incomprehension" between the "two cultures" of scientists and liberal arts students.

The pattern of each chapter remains the same: an introductory article is usually followed by either case studies or pieces that explore some of the issues in greater depth. This format has proved useful for teaching purposes.

Again, a key feature of the book is the inclusion of comprehensive Guides to Further Reading at the end of each chapter (and at the end of the book), which are longer and better organized this time. Obviously, there is some overlap between these, as indeed there is between the chapters themselves; such is the nature of a work that seeks to represent the complex, evolving whole that is advanced industrial society today.

Tom Forester

Acknowledgments

The editor and the publisher are grateful to the following:

Addison-Wesley Publishing Company, for Edward Feigenbaum and Pamela M. Corduck, "Land of the Rising Fifth Generation," reprinted from *The Fifth Generation*, by permission of Addison-Wesley. Copyright © 1983 by Addison-Wesley Publishing Co., Reading, Massachusetts.

Association for Educational and Training Technology (AETT), for Christopher Dede, "Educational and Social Implications," reprinted from "Educational, Social and Ethical Implications of Technological Innovation," *Programmed Learning and Educational Technology*, November 1981. Reprinted by permission of AETT. Copyright © 1981 by AETT.

Basic Books, Inc., and Harvester Press Ltd, for Seymour Papert, "Computers for Children," reprinted from *Mindstorms: Children, Computers and Powerful Ideas*, © 1980 by Basic Books Inc., New York City 10022, and Harvester Press Ltd., Brighton, Sussex. Reprinted by permission of the publishers.

Margaret A. Boden, for "The Social Impact of Thinking Machines," published in *Futures*, February 1984. Reprinted by permission of the author.

British Association for the Advancement of Science (BAAS), for John Child et al., "The Quality of Employment in Services," Bernard Leach and John Shutt, "Chips and Crisps: Labor Faces a Crunch," and Chris Freeman, "Long Waves of Economic Development," all reprinted from P. K. Marstrand (ed.), *New Technology and the Future of Work and Skills*, Proceedings of the Annual Meeting of the BAAS, August 1983, published by Frances Pinter, London, 1984. Reprinted by permission.

David A. Buchanan, for "Using the New Technology," reprinted from the *European Management Journal*, vol. 1, no. 2, 1982, and published in David A. Buchanan and David Boddy, *Organisations in the Computer Age*, Gower Press, Aldershot, 1983.

ʿBusiness Week, for: "Software: the New Driving Force," reprinted from the February 27, 1984, issue of *Business Week*; "Telecommunications Liberalization," reprinted from "Telecommunications: the Global Battle," in the October 24, 1983, issue of *Business Week*; "Personal Computers Invade Offices," reprinted from "Computer Shock Hits the Office," in the August 8, 1983, issue of *Business Week*; "High Tech is Low on Jobs," reprinted from "America Rushes to High Technology for Growth," in the March 28, 1983, issue of *Business Week*. All reprinted by special permission, copyright © 1983 and 1984 by McGraw-Hill Inc., New York City 10020.

Council for Science and Society (CSS), for Howard Rosenbrock et al., "A New Industrial Revolution?" reprinted from *New Technology: Society, Employment and Skill*, published by the CSS in 1981. Reprinted by permission.

Fortune, for Martin Mayer, "The Videotex Revolution," reprinted from "Coming Fast: Services Through the TV set," *Fortune*, November 14, 1983; and Gene Bylinsky with Alicia Hills Moore, "Flexible Manufacturing Systems," reprinted from "The Race to the Automatic Factory," *Fortune*, February 21, 1983. Both reprinted by permission of Fortune, New York City 10020. Copyright © 1983.

Harvard Business Review, for Fred K. Foulkes and Jeffrey L. Hirsch, "Robots at Work," reprinted from "People Make Robots Work," in *Harvard Business Review*, January—February 1984, reprinted by permission of the *Harvard Business Review*.

ILO, for Diane Werneke, "Women: the Vulnerable Group," reprinted from *Microelectronics and Office Jobs: The Impact of the Chip on Women's Employment*, © 1983, International Labour Organization, Geneva, Switzerland.

A. Richard Immel, for "The Automated Office: Myth versus Reality," reprinted from *Popular Computing*, May 1983, by permission of the author.

Manpower Services Commission (MSC), for Tim Brady and Sonia Liff, "Job Losses Now, Maybe Some Later," reprinted from *Monitoring New Technology and Employment*, by permission of the MSC. Copyright © 1983, HMSO.

Newsweek, for William D. Marbach et al., "The Race to Build a Supercomputer," reprinted from the July 4, 1983, issue of *Newsweek*. Copyright © 1983 by Newsweek, Inc. All rights reserved. Reprinted by permission.

New York Review of Books, for Joe Weizenbaum, "The Myths of Artificial Intelligence," reprinted from "The Computer in your Future," the *New York Review of Books*, October 27, 1983. Reprinted with permission from the *New York Review of Books*, copyright © 1983 by Nyrev, Inc.

Policy Studies Institute (PSI) for John Marti and Anthony Zeilinger, "New Technology in Banking and Shopping," reprinted from their *Micros and Money* by permission of the PSI. Copyright © 1982 by Policy Studies Institute.

Sage Publications, for Sherry Turkle, "The Psychology of Personal Computers," reprinted from "The Subjective Computer: A Study in the Psychology of Personal Computation," *Social Studies of Science*, vol. 12, 1982. Reprinted by permission of Sage Publications, copyright © 1982.

Science, for: James D. Meindl, "Micros in Medicine," reprinted from "Microelectronics and Computers in Medicine," *Science*, vol. 215, pp. 792–7, February 12, 1982; C. A. Hudson, "Computers in Manufacturing," reprinted from *Science*, vol. 215, pp. 818–25, February 12, 1982; and John S. Mayo, "Evolution of the Intelligent Communications Network," reprinted from *Science*, vol. 215, pp. 831–7, February 12, 1982. All copyright © 1982 by the American Association for the Advancement of Science, Washington, DC.

Scientific American, for Vincent E. Giuliano, "The Mechanization of Office Work," reprinted from *Scientific American*, September 1982; and Martin L. Ernst, "Electronics in Commerce," reprinted from "The Mechanization of Commerce," *Scientific American*, September 1982. With permission. Copyright © 1982 by Scientific American, Inc. All rights reserved.

Technology Review, for Leslie D. Ball, "Computer Crime," reprinted from *Technology Review*, April 1982; Robert Ayres and Steve Miller, "Industrial Robots on the Line," reprinted from *Technology Review*, May–June 1982; Charles H. Ferguson, "Chips: the US versus Japan," reprinted from "The Microelectronics Industry in Distress," *Technology Review*, August–September 1983; Robert Kahn et al., "The Impact of Cable," reprinted from *Technology Review*, January 1983; Jack Nilles, "Teleworking from Home," reprinted from *Technology Review*, April 1982; William J. Perry and Cynthia A. Roberts, "'Smart' Weapons" reprinted from "Winning Through Sophistication," *Technology Review*, July 1982; Hoo-min D. Toong and Amar Gupta, "Personal Computers," reprinted from "The Computer Age Goes Personal," *Technology Review*, January 1983; Duane L. Huff, "The Magic of

Cellular Radio," reprinted from *Technology Review*, November–December 1983. All reprinted with permission from *Technology Review*, copyright © 1982 and 1983.

Time, for Frederic Golden, "Here Come the Microkids," reprinted from *Time Magazine*, May 3, 1982. Copyright © 1982 by Time Inc.

George Weidenfeld & Nicolson Ltd, for David Burnham, "Data Protection," reprinted from his *The Rise of the Computer State*, by permission of Weidenfeld (Publishers) Ltd.

Wisconsin Medical Journal, for John Halton, "The Anatomy of Computing," modified from a series of six articles on computerization appearing in vol. 81, April, May, June, and November 1982, and vol. 82, February and April 1983. Copyright © 1982, 1983, State Medical Society of Wisconsin.

World Future Society, for Michael Marien, "Some Questions for the Information Society," reprinted from the *WFS Bulletin*, September–October 1983; Yoneji Masuda, "Computopia," reprinted from his *The Information Society as Post-Industrial Society*, 1983; Richard M. Neustadt, "Electronic Politics," reprinted from "Politics and the New Media," in Howard F. Didsbury (ed.), *Communications and the Future*, 1982; William L. Renfro, "Second Thoughts on Moving the Office Home," reprinted from *The Futurist*, June 1982. Reprinted with permission. Copyright © 1982 and © 1983 by the World Future Society, Bethesda, MD.

Editor's Introduction

The first electronic digital computer built in the US, ENIAC, was unveiled at the University of Pennsylvania in 1946. It weighed 30 tons, filled the space of a two-car garage, and contained 18,000 vacuum tubes, which failed on average at the rate of one every seven minutes. It cost half a million dollars at 1946 prices.

Today, the same amount of computing power is contained in a pea-sized silicon chip. Almost any home computer costing as little as $100 can outperform ENIAC. Put another way, if the automobile and airplane businesses had developed like the computer business, a Rolls Royce would cost $2.75 and run for 3 million miles on one gallon of gas. And a Boeing 767 would cost just $500 and circle the globe in 20 minutes on five gallons of gas.

This dramatic reduction in the cost of computing power made possible by microelectronics has coincided with the conversion of telecommunications networks from analogue to digital signals—the same "stream of numbers" technique used in computing. The result has been the "convergence" of electronics, computing, and telecommunications and the unleashing of a tidal wave of technological innovation which scientists are now calling the "Information Technology Revolution." This revolution is not confined to the world of science and technology: it is bringing about dramatic changes in the way we live and work—and maybe even think.

Information technology in its strictest sense is the new science of collecting, storing, processing, and transmitting information. Information is the lifeblood of complex industrial societies and it is growing in importance. A recent study by the British prime minister's Information Technology Advisory Panel put the number of people employed in the "tradeable" information sector (publishing, consulting, on-line data services, etc.) at about 5 percent of the UK workforce. But on a wider definition of the information sector, taking in the whole of banking and insurance, central and local government, and education and training, it is clear that 40–45 percent of the

UK workforce is directly involved in the processing of information. And since we all use information at some time or other, there is no one who won't be touched by the information technology revolution.

In the home, microchips are already controlling our washing machines, food processors, and VCRs. Personal computers are no longer just being used for playing games: more and more people, especially professionals, are using them to actually work at home, sometimes on a permanent basis. At school, computers have been welcomed into the classroom and a new generation of "microkids" have grown up who know more about computers than their parents—and even their teachers.

The personal computer has marched into the office, where new information technology is replacing paperwork and improving productivity, customer service, and job satisfaction for some. The worlds of banking, retailing, and financial services are being transformed by information technology—because money, after all, is merely information. No type of commercial service or public utility has remained untouched by the new technology. Micros in medicine are providing new help for doctors and offering new hope to the deaf, the blind, and the disabled. In factories, computers are becoming more and more involved at every stage in the manufacturing process. The robots have arrived at last, along with new systems like CAD/CAM and FMS, which are making the workerless factory a reality.

Some skeptics might still look around and conclude that information technology has not yet changed things very dramatically. It's rather like those who saw the first automobiles, telephones, and TVs and concluded they'd never catch on. Or those who failed to spot the potential impact of mains electricity, which is still only one hundred years old. Of course, some 1950s (pre-chip) predictions about instant automation by writers like Diebold and Wiener have never been realized. Even today, with the chip, progress might seem slow, especially in industry and commerce, where lack of investment, poor management, skills shortages, and the very complexity of new systems are holding things back.

But the pace of change can be deceptive. The chip itself was invented only in 1971, and mass-produced personal computers appeared on the scene less than five years ago. Taking a longer perspective, it is difficult for some to remember that 70 percent of the US workforce was employed in agriculture 150 years ago; by 1900, only 40 percent were still on the farm, and today the figure is down to a mere 3 percent. Only 50 years ago, during the Great Depression, 50 percent of the US workforce was still employed in the production of goods: now it is down to 30 percent. Services—many of them "information"-related— now employ nearly 70 percent of the work-force, the same proportion that toiled on the land only a century and a half ago.

Since the publication of *The Microelectronics Revolution* in 1980, there have been four major developments which stand out and they have been duly reflected in this new volume.

First, computer software has gained in importance relative to hardware.

The currently booming software industry is not only making millionaires, it is actually setting the pace of the information technology revolution.

Second, the remarkable rise of the personal computer, coupled with IBM's entry into the market, has placed cheap computing power on desks in homes, offices, and schools. The personal computer has thus unexpectedly become the basic building block of the information technology revolution.

Third, the race is now on to build the next or "fifth" generation of computers, machines that exhibit artificial intelligence. As governments and corporations pour resources into fifth-generation projects, there is growing controversy about whether machines can and should be made to think.

Fourth, there has been an explosion of innovation in the whole area of telecommunications. Microelectronics has made possible the "intelligent" digital network, and the invention of new gadgetry has been spurred by the deregulation of long-established PTT monopolies. Satellite communications, cable TV, cellular radio, and videotex are revolutionizing the way we receive entertainment, keep up with the news, and even work, shop, and conduct financial transactions.

In all these areas—and especially in certain types of hardware, office equipment, and computerized factory systems—one is all too well aware of the growing importance of the Japanese. Japan is already on a par with the US in many areas and ahead in others. The rise of Japan in the information technology revolution has been a particular feature of the past few years.

While the Silicon Valley-based chip industry itself still has the unnerving habit of seeming to go from boom to bust with great rapidity—with this month's "glut" story being followed by next month's "shortage" scare—every year since 1980 has been marked by new buzzwords, with unlikely acronyms like CCD, EEPROM, CMOS, CAD/MAT, and FMS being tossed around like confetti. In 1981 all the talk was of personal computers and expert systems; 1982 was the year of cellular radio and satellites; 1983 saw the rise of the professional personal computer and there was much talk of parallel processing; 1984 seems to be more the year of integrated software than of Big Brother.

All kinds of interesting new microchip applications have emerged in the last few years. Apart from micro-TVs, twentieth-century civilization has come up with the wristwatch TV and the talking wristwatch. The Japanese have invented a programmable tennis ball service machine that can give you a taste of John McEnroe's backhand, while in California a microchip-controlled telephone dialling machine automatically phones the parents of truants from local schools and delivers a short, sharp recorded message.

New information technology is transforming hotel and theater bookings, news-gathering, meter-reading, and weather forecasting. In Italy a national obsession, the football pools, is now fully computerized. If they don't have a mobile phone, construction and warehouse workers can carry portable microelectronic notepads. And we can all now purchase a microchip-based geiger counter for monitoring radiation levels on the day after World War III.

In Britain, there are new devices on the market for automatically logging the weight and speed of heavy trucks. And if you've lost your job to a computer, you can visit your local Jobcentre and consult the computer which put many Jobcentre workers out of a job. If you plan to travel, be sure to bring your "smart" credit card, your machine-readable passport, and—if you're a soldier—your "intelligent" dog tag in case you get run over by one of those speeding trucks. For relaxation, try visiting Britain's computer graveyard in the Forest of Dean, Gloucestershire, where old micros are left to rust in peace.

Despite the gee-whizzery that greets each new development, the path of technological innovation has not always been smooth. Some inventions, like videotex, have been slow to catch on or have caught on in ways that weren't envisaged. Others, like videodisks, appear not to have caught on at all. Some much-talked-about scenarios, like the fully automated office of the future, seem further away than ever; while the personal computer, originally conceived as a hobbyist's toy, has had a far greater impact in the commerical world than anyone ever imagined. Technological advances and social obstacles interact to influence the process of change. As John Naisbitt put it in his recent book, *Megatrends*, "Technological innovation does not proceed in a straight line. It weaves and bobs and lurches and splutters."

The overall impact of new information technology on employment is just as hard to predict, because we don't know quite how the revolution is going to proceed. Certainly, it is difficult to see where the new jobs are going to come from to replace those lost in the traditional manufacturing industries—as recent British and American studies have shown. But if information technology can help Western economies to generate new economic growth, then jobs may be created in reasonable numbers, perhaps in the 1990s. Analysis of so-called Kondratiev "long waves" would seem to point to this. Yet it's a mistake to see recent industrial changes as a once-and-for-all shakeout; artificial intelligence, "expert systems," and new software could have just as devastating an impact on professional occupations like doctors and lawyers as industrial robots have on production-line workers. And there is still a role for the traditional or "sunset" industries—if they can remain competitive with newly industrialized nations in the Third World.

But we already know a great deal about what is going on in many workplaces, thanks to case studies now being reported in Britain and the US. Information technology offers managers the chance to enrich jobs and to develop new forms of work organization, thus transforming old-style labor relations. Yet the picture emerging is a varied one: deskilling and loss of control has undoubtedly occurred in some sectors, leading to fears that the craftsman of the future will be computer-degraded rather than computer-aided. In others, job satisfaction has clearly been improved. So much depends on how managers handle the changes and their responses to a series of essentially political problems. Although organized labor seems to have opted out, the future of work is still very much a matter of social choice.

Social choices present themselves, too, when we confront the new social problems being thrown up by the information technology revolution. The true extent of computer crime needs to be verified, and more resources devoted to its detection. The growth of databases is creating widespread fears about loss of privacy and increasing the need for adequate data protection laws. These could restrict the use of personal information by credit card firms, telephone companies, mail order operations, and that other kind of salesman, the politician.

At the global level, we must consider the impact of information technology on the Third World and the possible exacerbation of North-South differences. Some argue that the less developed nations now have the opportunity to leapfrog the developed countries, but more likely, information technology will simply increase the obsolescence of their industries, services, and development strategies. Microelectronics has also made possible a new generation of sophisticated "smart" weapons, precision-guided armaments that threaten to blow us all up in more sophisticated ways.

No doubt society will survive, but what kind of society is emerging is not clear. The debate on the future of industrial society has become somewhat sterile, with labels like "the post-industrial society," "the information society," and even "the technetronic society" being bandied around with little thought given to their true meaning or analytical value. But we are certainly headed somewhere, and the future shape of society is still, to some extent, negotiable. We all have the right to debate and shape the kind of society we want to live in.

Part One: The Computer Revolution

1 Introduction to Information Technology

The Anatomy of Computing

John Halton

In order to understand the information technology revolution, you need to know something about how computers work. In this introduction to computing, the author first describes the importance of information processing and then takes us through the hardware and software of the modern microcomputer. Halton, an Englishman who settled in the US in 1962, was formerly Professor of Computing Sciences at the University of Wisconsin, Madison. He is at present Principal Engineer, Advanced Technology Department, Harris Corporation, Melbourne, Florida. Taken from a series of articles in the Wisconsin Medical Journal, *April 1982–April 1983.*

The Second Industrial Revolution

The world is undergoing a major social and economic change, a Second Industrial Revolution, through the new information-processing technology of communications and computers. While mankind has developed a myriad ways of applying and controlling power to dominate and shape our environment, through the use of tools, weapons, machines, fuels, vehicles, instruments, clothing, buildings and roads, metals, plastics and drugs, agriculture, and electricity, the handling of information has lagged considerably, perhaps because the human brain is itself so remarkably powerful. Until recently, there have been only three major developments in this area: the invention of written (or painted or carved) language, some five or six thousand years ago; that of simple arithmetic operations, using what would now be called a digital representation of numbers, about a thousand years later; and that of printing, about five hundred years ago.

With written language, we get the capacity to make a permanent record of information and also to convey messages across space and time: *storage*, *retrieval*, and *communication*. With digital arithmetic, we get the ability to perform accurate, repeatable *manipulations* of *quantitative data*. With printing, we can make many identical copies of the same record and so *broadcast* a single message to a wide and continuing audience. Beyond these outstanding advances, until the last hundred years or so, the only progress has been in the engineering and increasingly plentiful production of more powerful and reliable and efficient, faster and cheaper devices to implement these concepts (such as styli, chalks, brushes, and pens; slates, wax tablets, papyrus, parchment, and paper; typewriters; abaci, cash registers, and calculating machines; movable type and typesetting machines; and printing presses of increasing speed and complexity).

In the last hundred years, we see the rapidly accelerating advent of a technology so powerful, novel, widespread, and influential that we may indeed call it the Second Industrial Revolution. Its basis is electromagnetic, in many interconnected forms: photography, photocopying, cinematography, and holography; telegraphy, telephony, radio communication, radar, sonar, and telemetry; sound and video recording and reproduction; vacuum tubes, transistors, printed circuits, masers, lasers, fiber optics, and (in rapid succession) integrated circuits (IC), large-scale integration (LSI), and very large-scale integration (VLSI) of circuitry on a tiny semi-conducting 'chip'; and, finally, the bewildering variety of electronic digital computers. All these devices are intimately interrelated, and any advance in one tends to generate advances in all of them.

The progress has been truly amazing. In only about 40 years, electronic communications and news media have become commonplace and indispensable; computers have proliferated, becoming increasingly fast, powerful, small, and cheap, so that now there is scarcely a human activity in which they are not to be found, bearing an increasing share of the burden of repetitive information processing, just as the machines of the First Industrial Revolution have taken over the majority of heavy and unpleasant physical labor (we may say, energy processing).

Now, information can not only be stored, retrieved, communicated, and broadcast in enormous quantities and at phenomenal speeds; but it can also be *rearranged, selected, marshalled,* and *transformed*. Until recently, these activities were the sole province of the human brain. While creative, judicious, moral, and esthetic choices are still best left to people, all the tedious and mechanical mental processes can now be relegated to the accurate, fast, and tireless machines. Any sequence of operations on information that can be precisely specified can be carried out without further human intervention or supervision.

At first, computers were the experimental toys of university researchers; then they became the tools of government establishments and giant corporations, huge, expensive, individually designed and manufactured, and beyond the reach of any but the wealthiest organizations. People thought of the future in terms of machines of ever-greater speed and capacity; centralized behemoths would hold all the world's information in gigantic data banks, whence major decisions would issue to be imposed upon the populations at their mercy. With the emergence of powerful, cheap, mass-produced computers-on-a-chip, the picture has changed radically. Now we see tiny computers everywhere: in wrist-watches, microwave ovens, electronic games, pocket calculators, cameras, typewriters, musical instruments, etc. What used to be done, with few options, by intricate mechanical devices is now performed, with great flexibility and convenience and at much less expense, by the ubiquitous preprogrammed microcomputer. The probable future has become one of millions of small yet powerful computers, controlling virtually every machine and appliance, distributed in every home, on every desk, in every workshop; many of them connected in a maze of small and large networks, much like the present telephone network (and perhaps replacing it), so that individual computers could communicate, sharing information in a gigantic distributed data-base, and gaining, through distributed processing, computational power whose extent is yet difficult to gauge; all this following the individual requirements and choices of the owner or operator of each machine.

Increasingly, we are confronted, not only with the results of the use of computers throughout industry, commerce, banking, advertising, science, the communications industry, newspapers, airlines, and hospitals; but with the realistic possibility of purchasing computer power for our own small enterprises, offices, and homes. This may be done in a variety of ways; but in all of them, the real cost of computation is constantly diminishing. It is probably fair to say that the question of computerization is not "whether," but "when" and "how." We must choose whether to lease equipment or to buy it; whether to install terminals, connected to a computerized "service bureau," or a complete local computer system; whether to get a package of standard programs directed towards our kind of work, to commission new programs tailored to our special needs, or to learn programming and write our own; whether to go it alone or to share a system with a number of similar users (especially if they are in the same building); how far to take our first efforts at computerization; what to include and what to continue doing by hand. Then, having made the broad choices, we must select, from the wide range of available products, those that will suit our situation best.

Computer programming is likely to become the literacy of the third millenium AD. Elementary schools may well be teaching it before long, and we might be well advised to gain at least a smattering of knowledge of computers and of programming competence, especially since computer languages and programming environments are becoming increasingly helpful and friendly to the uninitiated user.

The anatomy of a computer

A computer is a machine for the automatic processing of *information*. Historically, this information was numerical, and computers were machines for doing arithmetic. Unlike the simpler *calculating machines*, which can perform only one elementary arithmetic operation at a time, then need to be told what to do next (usually, by suitable button-pushes); computers can be given a list of operations to perform (often with branching and repetitions, depending on tests of sign or value included among the operations), and will then execute these in proper sequence without further intervention. This sequence of instructions is called a *program*.

A *digital computer* stores its information in the form of *words*, finite ordered sets of *digits*, each of which can have only one of a finite set of values. (This is analogous to the *decimal* representation of numbers, in which each digit takes one of the values 0, 1, 2, 3, 4, 5, 6, 7, 8, 9; or the written representation of English words, in which the "digits" are the 26 letters of the alphabet.) Considerations of simplicity, reliability, and economy dictate that electrical engineers should design computers to consist of a great number of similar pieces of circuitry, each of which can only be in one of two states, usually denoted by 0 and 1. Such *binary digits* (or *bits*) are the elements of which computer digital representation is built. A row of eight bits is called a *byte*, and the majority of computers have their storage of information organized in words of one, two, four, or eight bytes (8, 16, 32, or 64 bits). The number of bits in a word is termed its *length*; if this is k, then the number of possible distinct pieces of information that can be stored in such a word is 2^k. In particular, four bits together can have 16 different contents, and these are standard *binary representations* of the numbers 0–15:

$$0000 = 0 \quad 0001 = 1 \quad 0010 = 2 \quad 0011 = 3$$
$$0100 = 4 \quad 0101 = 5 \quad 0110 = 6 \quad 0111 = 7$$
$$1000 = 8 \quad 1001 = 9 \quad 1010 = A \quad 1011 = B$$
$$1100 = C \quad 1101 = D \quad 1110 = E \quad 1111 = F$$

with A = 10, B = 11, C = 12, D = 13, E = 14, and F = 15. These may now be viewed as the 16 possible digits of a representation (the *hexadecimal*, or *hex*), which is much more compact and humanly intelligible than a long string of zeros and ones. For example, the byte 10110010 becomes "B2," and the four-byte computer word 01001100011100101101000110001110 becomes the eight-digit hex word "4C72D18E."

We now turn to Figure 1.1, which is a diagram of the various parts of which any computer is composed. These are, essentially,

1 a *central processing unit* (CPU), which is the controlling and computing center of the machine;
2 a *memory*, possibly of different levels, in which both *data* and *instructions* are stored;
3 a variety of *input* and *output* (*I/O*) *devices*, through which the machine communicates with the world outside it.

The CPU consists of an *operation control unit* (OCU), an *arithmetic/logical unit* (ALU), and a relatively small, very-fast accessible *local memory* (LM). The OCU keeps track of the memory location of the next instruction to be executed, and analyzes the current instruction, so as to activate the proper operation of

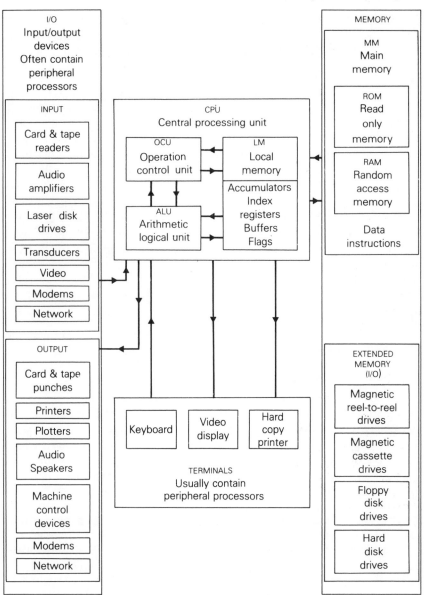

Figure 1.1   The anatomy of a computer.

a memory transfer, non-sequential jump (by appropriately changing the address of the next instruction), input or output of information, or computation (performed by the ALU), as is indicated by the instruction code. The ALU actually carries out the elementary arithmetic operations (addition, subtraction or negation, multiplication, division or reciprocation) and logical operations (these being bit-by-bit operations, without carry, such as **not,** **and,** and **xor**; eg., **not** 1010 = 0010, 1100 **and** 0101 = 0100, 0110 **xor** 1010 = 1100) on the data given to it by the OCU. The LM receives the operands called for by the OCU and also the results of the ALU's operations upon them. For example, the OCU may retrieve the factors of a multiplication from the main memory into a pair of LM registers and instruct the ALU to multiply them and place their product in another LM register. Such registers are usually called *accumulators*, and they are normally double-length (since the product of two k-bit numbers is a $2k$-bit number). Other LM registers are used for counting (eg. repetitions) and are called *index registers*; others hold intermediate values and are called *buffers*; and, finally, there are one-bit or two-bit registers which reflect the information on which tests are made by the OCU (for example, the occurrence of a carry, the vanishing or positivity of an answer, or the occurrence of an arithmetic overflow): these are termed *flags*.

Originally, the CPU was a sizable piece of electronics, hand-assembled and highly complex. With the advent of micro-miniaturization of circuitry, printing, and photographic techniques, and the mass production of components only the largest computers (*mainframes*) are built in the old way. Smaller systems generally have the entire CPU on a single chip. Among these, the name *microcomputer* is now applied to those with less than a million words of memory and a word length of one or two bytes; the name *minicomputer* applies to the larger machines, with two- to four-byte words and one to a hundred million words of memory. (The smallest micro is probably more powerful than the big computers used by universities and industry in the 1950s.)

The *main memory* (MM) consists of magnetic or electronic components which store the information (both data and instructions) needed by the computer. The individual words are directly *addressable* from the CPU by number (rather like houses in a street), and their contents are retrievable in very short times, of the order of the operation time of the CPU (ranging from fractions of a *nanosecond*, 10^{-9} or one billionth of a second, for the fastest main-frames to several *microseconds*, 10^{-6} or millionths of a second, for the slower micros). This is often referred to as *high-speed storage* or *random-access memory* (RAM). While most of the MM is erasable and may be changed at will, some memory is used to store constants and often-used utility programs and is not erasable by the CPU: such memory is called *read-only memory* (ROM). Sometimes this is optional and can be plugged into the computer: this is called *firmware*.

Note that a computer with one-byte-long addresses can have at most $2^8 = 256$ words of MM; a two-byte address can reach $256^2 = 65536$ words; a three-byte address can select any of 16777216 words of MM; and so on. In practice, most micros and minis have MM ranging in size from $2^{14} = 16384$ to $2^{18} = 262144$ words. It should be noted that $10^3 = 1000$ and $2^{10} = 1024$. Because computers are so heavily slanted towards base-2 representation, it has become almost universal computer parlance to use the prefix *kilo* or K (which usually denotes a thousand units) to denote 1024 and the prefix *mega* or M (which usually denotes a million units) to denote $2^{20} = 1024^2 = 1048576$. Thus, we write $16K$ for 2^{14} and $\frac{1}{4}M$ for 2^{18}. Perhaps the commonest size of MM consists of $64K = 65536$ words.

Almost all computer instructions comprise an *operation code* (usually one byte long, allowing 256 possible operations to be specified), followed by an *operand reference* (number, index, or address) of variable length (since some operations require more data than others; for instance, the STOP instruction needs no operand, so it is one byte long).

The *extended memory* (EM) is sometimes considered under I/O devices, both because it is often physically located outside the computer (while the CPU and the MM are usually in the same enclosure), and because its speed of access is much slower than the speed of operation of the CPU and is comparable with the range of speeds of I/O devices. Most read/write memory is magnetic (optical memory is read-only, and fast RAM is either magnetic or electronic), either in the form of *tape*, *drum*, or *disk*, coated with magnetic material, much like sound-recording equipment; it is similarly erased, read, and recorded upon by "heads" which contain electromagnetic sensors/polarizers. The cheapest (and most limited, in both speed and capacity) device is a common portable cassette recorder, with $\frac{1}{4}''$ tape cassettes. From this, one moves to specially engineered cassettes and recorders, and thence to high-speed $\frac{1}{2}''$ or $1''$ reel-to-reel drives carrying thousands of feet of tape at very high speeds. Access times can be quite good for *sequential* access, along the tape, but *random access* time is poor at best, running to seconds or even minutes. Economy and a virtually unlimited total storage capacity (on numerous cassettes or reels; but only as many units as one has on-line tape-drives are actually accessible without human intervention) are the only advantages.

When we wish for practically useful EM, combining large capacity with relative economy and speed of random access, we must turn to *drum* or *disk memory*; and, nowadays, the former have been practically replaced by the latter. Disk memory is of two types: *floppy disk* and *hard disk*, the first being the cheaper, slower, smaller-capacity option. Floppy disks are flexible, have diameters of $5\frac{1}{4}''$ or $8''$, generally, and are removable from the disk-drive, so allowing one to build up an unlimited *library* of stored data. The information is stored on concentric circular *tracks* (not on a single spiral track, as on a sound record), on one or both sides of the disk. The number of tracks and the number of bytes per track vary (the density increasing with precision of engineering, and so with cost of the drive), but the total capacity of a floppy disk is in the range of 50KB to 1MB. The disks rotate at, typically, 300 rpm, and access time is governed by the time required to place the *movable head* on the right track, a fraction of a second, plus the fifth of a second taken by the head to traverse the circumference of the track, in search of a record; thereafter, consecutive bytes are accessed at some thousands per second.

Hard disks are rigid and have larger diameters. There are drives with anything from one to a dozen disks, rotating at about ten times the speed of floppy-disk drives (and so diminishing the access time of records in a track), with one or several heads. *Fixed-head* drives naturally must have a head for each track (which costs more), but save head-movement time in random access. *Winchester* disks are movable-head drives with sealed-in disks, in which the heads ride very close to the disk, cushioned by the layer of air between. In floppy-disk drives, the head actually rides on the disk, eventually wearing it out. The capacity of hard-disk drive ranges from 10 MB to 100 MB in a single drive. Some movable-head hard-disk drives have removable disks or disk-packs, allowing for greater library storage.

The *I/O devices* are the computer's link with the outside world. In large mainframe computers, we see *paper card* (as in Hollerith or "IBM" cards) and *paper tape* readers and punches: the perforations in the paper carry the information. Increasingly in large computers, and almost universally in small ones, the main input is from the *keyboard* of a *terminal*. This is much like a typewriter keyboard, and depressing any key sends an 8-bit code to the computer. When the computer is waiting for input from this terminal, it reads the code and interprets it as a datum; when it is busy, either the message is lost or it is held in a "buffer" for subsequent input (this depends on how the connection is made and what the computer is made to do). It is quite common for the computer to be connected to several terminals, all competing for its attention. This is called *time-sharing*. The computer cycles around the terminals, looking for their several inputs and dividing its CPU time among them. The main output of the computer is to the display devices of the terminals; these are either *video displays* (cathode ray tubes, CRT, just like the screens of black-and-white or color TV sets; indeed, simple micros sometimes use ordinary television sets as display devices) or *printers* (in so-called *hard-copy terminals*). Of course, the computer may be connected to additional video displays and printers, of different qualities, as well as to *plotters*, a kind of printer for drawing graphs and diagrams. Many types and speeds of printers exist.

Other input devices, such as audio amplifiers (receiving signals from radio tuners, record players, etc), video receivers and recorders, laser disk drives, and a variety of scientific instruments, can all be classified as *transducers* (devices that transform physical quantities, such as position, conductivity, pressure, temperature, vibration frequency, or amplitude, into electro-magnetic impulses) linked to *digitizers* (which convert such impulses into sequences of zero/one pulses). Output from the computer can similarly follow the reverse process, yielding visible or audible results, or the control of mechanical or electrical equipment. Thus computers can draw pictures (often, moving pictures), make music and other sounds, and can control appliances, machinery, and whole manufacturing processes.

In order to connect *remote terminals* to a computer, use is often made of telephone lines; and, for this purpose one employs a device called a *modem* (for "MOdulator/DEModulator"), which converts the computer's digital signals to and from telephone audio signals. This has a cradle shaped to hold the ear and mouth pieces of a telephone's hand-set. It is also possible to connect several computers in this way. This is called the formation of a computer *network*. Of course, computers may also be connected by cable, fibre-optics, or microwave link.

It should be noted that terminals usually, and other i/o devices often, themselves contain computers of varying degrees of power and complexity; so that even a multi-user computer with a central CPU may still be seen as a kind of computer network in its own right. The computers in i/o devices such as terminals or printers are often referred to as *peripheral processors* (PP). Many networks do not have a central computer at all; but are simply a collection of independent computers linked for the sharing of information and, sometimes, computing capabilities. Often, they permit the exchange of messages (*computer mail*) and the pooling of data (*distributed data-base*). They may also share a common bank of memory, accessible to all. Finally, since the invention of computer networks, designers have been investigating the possibilities of computers made up of an array of CPUs (*multicomputers, parallel processors, distributed processing*). These new ideas are very powerful and far-reaching: they will probably revolutionize our ideas of computers, and of their applications, in the next few years.

I mentioned that the various peripheral devices are connected to the CPU without examining how. In fact, this may be done in several ways. We can simply have a separate connection (or *port*) for each device, but this limits rather severely the number of devices that may be connected to the CPU. Another way is to have a single *bus* (or connection) to which any number of devices may be attached. The information signal must then carry an appropriate *address*. The bus receives all signals, and individual devices (including the CPU) seek out and decode only those addressed to them. It is also possible to have a *switching device*, which receives addressed data and directs them to the appropriate recipient device, rather like a central post office. The decision on what communication arrangements to adopt is made on the basis of considerations of cost, capacity, and speed.

It is often the case that the several devices forming a computer or a computer network have their data coded in different ways. It is then the job of the CPU(s) and PPs to share the work of interpreting signals into appropriate codes for each machine. This is broadly termed the problem of *interfacing* devices. Sometimes, the solution is to have a standard code or structure for the communications device (one meets the s-100 bus, the RS-232 serial port, the ASCII character-code, and so on). Another interfacing problem arises from the difference in the rate at which different devices can send and receive information (this is measured by the *baud rate*, named after Baudot, the inventor of the first five-hole paper-tape code; one baud is one bit transferred per second; hence kilobaud, kb, and megabaud, Mb; typical rates range from 200 baud to 200 kb). One solution is to send each piece of information (usually one character at a time, which takes 8 to 10 bits) only when the last has been acknowledged (this is referred to as a *handshake*); this is sure, but slow. Another way is to use a storage *buffer* in which a large batch of information is accumulated for fast transmission, thus not wasting the time of the faster device. Yet another approach is to connect several slow devices to a fast one and weave all their communications together into a fast message for the rapid device to unravel (*multiplexing*: this is used in time-sharing a CPU among several terminals).

One last kind of choice must be mentioned: some channels of communication are *serial* (they transmit one bit at a time), while others are *parallel* (they can transmit a byte or a whole word at a time); the latter are obviously faster, more complex, and more expensive. When devices are connected by cable, the degree of parallel communication is exhibited in the width of a flat ribbon cable, carrying several wires, side by side, and in the number of pins in the plugs and sockets by which they are connected to the machines. Parallel transmission is a variation on multiplexing.

What we have described is generally referred to as the *hardware* of a computer. By contrast, the programs that make the computer work, the "soul" of the machine, as opposed to the hardware "body," are collectively called its *software*. Inevitably, there came to be programs that were *hard-wired* (in the now outdated phrase) into the computer, in the form of ROM. These are termed *firmware*. A computer without software is a helpless set of circuits, and the expertise required to create the basic software that will bring the machine to useful life is comparable to that required to design the machine itself. Indeed, these days, computers are designed in cooperation between *computer architects*, who design what the computer will do, *hardware engineers*, who design how it will be constructed to be able to do it, and *software engineers*, who design and program the *operating system* that will run the machine. Beyond this, the computer will also need *application software* of many kinds, to enable it to do a variety of jobs, such as file-handling, accounting, statistics, payrolls, inventories, complex graphic displays, games, and so on. Typically, the application software is written (*i.e.* programmed) by the end-users (if they are sophisticated enough) or firms of consultants and programmers and system analysts, often called *software houses*. Sometimes, a software house will produce part or all of an alternative operating system, or an addition to an operating system, to make a computer more flexible or more efficient than the manufacturer's own software system allows it to be.

Computer languages

The CPU of any computer is designed to accept and execute a specific set of *operation codes* (*op-codes*), ranging in number from a dozen or so to several hundred. Different makes and models of computers and microprocessors may have entirely dissimilar op-codes; but the operations that they represent are much more alike than different, both through functional necessity and historical development. The interpretation of the op-codes is built into the hardware of the OCU (though sometimes the details of interpretation may be modified by the user through what is called *microcoding*), and it is part of this interpretation that the complete instruction being decoded contains a certain amount of further information (such as parameters, indices, and one or more memory addresses). Thus, the "length" of a computer instruction (the number of consecutive memory words, or of bytes, occupied by it) may vary, but the OCU automatically adjusts to this. An *executable program* (sometimes called "object code") consists of a sequence of machine instructions, consecutively stored in the computer's memory and (with the exception of "jump" instructions) executed in the order in which they are stored. The aggregate of possible machine instructions is called the *machine language*. In the computer, a program consists of a long string of binary digits (bits), usually written as 0's and 1's; and, of course, the same program would be interpreted quite differently (usually as nonsense) by any computer for which it is not designed. Indeed, the slightest error in a program almost always leads to an error in its output (usually a fatal error!) This state of affairs is sometimes expressed by the computer adage, "Garbage in; garbage out," or just "GIGO." Errors in programs are called *bugs*; and the tedious, odious process of finding and correcting such errors is termed *debugging*. It is estimated that, in the production of a working program, the debugging time may be two to four times as long as the time it takes to plan and write the program initially.

To give the reader a feeling for the nature of machine language, we present a simplified, fictitious, but typical, machine language specification. Our computer has two 16-bit *accumulator* registers (acc), X and Y, which may be coupled into a single 32-bit acc XY, with X holding the more and Y the less significant digits; these are attached to the ALU; and a *program control* register (pc) Z, also of 16 bits, attached to the OCU, which contains the address of the next instruction to be executed. Instructions will contain a 2-bit *acc code a*, with $a=0$ referring to X, $a=1$ to Y, $a=2$ to XY, and $a=3$ to Z. The memory consists of $2^{16} = 65536$ 16-bit words, directly addressable with a 16-bit address, n. There are four *addressing modes*, denoted by a 2-bit *mode code d*, with $d=0$ referring to *absolute address* (address $n|0$ refers to the actual number "n"), $d=1$ to *direct address* (address $n|1$ refers to word n in the memory), $d=2$ to *indirect address* ($n|2$ refers to the memory address which is the content of the memory word with address n), and $d=3$ to *relative address* ($n|3$ refers to the memory address that is the sum of n and the content of the pc register Z, with any carry to the seventeenth bit ignored). We will write { $\{a\}$ } for the acc denoted by a, $\{n|d\}$ for the memory word with address n in mode d, and C[x] for the content of word or acc x (so, e.g., C[$\{n|2\}$]=C[C[n]], while C[$\{n|3\}$]=C[n+C[Z]). A *port code p* selects one of 8 *output ports* ($p=0$, 1, . . . ,7) and 8 *input ports* ($p=8, 9, . . . , 15$), these ports transmitting 16 bits at once and having to be reactivated before each function (i.e., each output instruction sends out one 16-bit number, and the machine must wait for the output to be acknowledged before making another output; while each input instruction reads one 16-bit number, if an input is available, and the machine must wait for a new input each time). Finally, a *bit code b* refers to each of the 16 bits in a word (bit 0 being the least significant—rightmost—and bit 15 the most significant—leftmost). In some cases, the codes a, d, p, and b are interpreted somewhat differently, depending on the particular op-code c. The instruction layout is as shown in figure 1.2. This 16-bit word is followed, in some cases, by one or two additional words. The op-code c has the effect given in figure 1.3.

15	14	13	12	11	10	9	8	7	6	5	4	3	2	1	0	Bit number
c_3	c_2	c_1	c_0	a_1	a_0	d_1	d_0	b_3	b_2	b_1	b_0	p_3	p_2	p_1	p_0	Bits in codes
c				a		d		b				p				Code
OPERATION				ACC		MODE		BIT				PORT				Name
4				2		2		4				4				Length (bits)

Figure 1.2   Typical instruction layout.

I should explain to the reader that what is important is not the actual set of specifications given in figures 1.2 and 1.3 (though it should be noted that a similar set of specifications, embodying most of the nine operations described there, will apply in any computer), but the kind and complexity of these specifications, and the very limited types of operations provided to program-mers who wish to or must write in machine language. They are compelled to walk down their chosen road in extremely short and humanly uncomfortable steps; and though great computer efficiency may thus be attained, and though the longest journey may be made in tiny steps, nevertheless, programmers soon found it unreasonable and intolerable to use machine language.

As an example of a very simple computer program, we consider the solution to the following computer problem. Our computer is to be fed a sequence of one thousand 16-bit numbers at input port #9 (these may be keyed in by hand or fed in by a digitizer connected to some experiment). They are to be stored in memory words with addresses 5000, 5001, 5002, . . . , 5999. Their sum is to be computed and stored in address 6000 and output to a printer through output port #2. The program is to be stored beginning at address 0 in the memory. A possible solution is given in figure 1.4.

Again, what is important about this example is not its detailed form, but the difficulty of its interpretation, and therefore also the difficulty of verification and debugging. The programmer must deal with a mass of details that are of a purely mechanical nature and have no relevance to the problem being solved. "Higher-level languages" are attempts at making the computer understand the programmer's way of thinking, rather than forcing the programmer to think like a machine.

OP CODE	EFFECT
0	I/O with {{a}} — p_3=1 for input, p_3=0 for output. a≠2. b=d=0. No n.
1	COPY into {{a}} — Copy {n/d}. If a=2 & d=0, use double length number m, n. b=p=0.
2	COPY from {{a}} — Copy into {n/d}. If a=2 & d≠0, copy XY into 32 bits {n/d}, {n/d} + 1. If d=0; if a=0, copy X into Y, if a=1, copy Y into X, if a=2, copy X into Z, if a=3, copy Z into X, and no n. b=p=0.
3	CHANGE ACC {{a}}. — CLEAR (i.e., insert zero) if d=0; NEGATE (change sign) if d=1; COMPLEMENT (**not:** replace 0 by 1 and 1 by 0) if d=2; if d=3, a=2, ROUND-OFF XY into X and clear Y. b=p=0. No n.
4	ARITHMETIC & LOGICAL OPERATIONS on {{a}}. — Arithmetic: ADD C[{n/d}] to C[{{a}}] and put answer in {{a}}, if p=0; SUBTRACT C[{n/d}] from C [{{a}}] and put answer in {{a}}, if p=1. Logical ops (if corresponding bits of C [{{a}}] and C [{n/d}] are α and β, respectively, put γ in corresponding bit position in {{a}}; repeat for all bits): **and** (γ=1 if α=β=1, else γ=0) if p=2; **or** (γ=0 if α=β=0, else γ=1) if p=4; **xor** (γ=1 if α≠β, else γ=0) if p=4. If a =2; if p<5; operate on Y with carry into X; if 4<p<10, operate with XY and 32 bits {n/d}, {n/d} + 1, and if d=0, use double length number m, n (i.e., instruction is 3 words long; see op-code 1), and operation corresponds to p-5 (i.e., ADD if p=5, . . ., **xor** if p= 9). b =0.
5	SHIFT {{a}}. — Shift to the LEFT if d=0 or 2; shift to the RIGHT if d=1 or 3. CIRCULAR SHIFT (bits ejected from one end of acc are reinserted at other end, in same order) if d=0 or 1; ARITHMETIC SHIFT (ejected bits are lost; on left shift, 0's are inserted from right; on right shift, copies of leftmost bit are inserted from left) if d=2 or 3. Shift by p places. b=0. No n.
6	JUMP — Each instruction is executed when C [Z] is equal to its address. On every instruction, the OCU adds to Z the length (1, 2 or 3) of the current instruction, so that Z will contain the address of the next instruction listed in the program (this happens before execution of the current instruction). In a jump instruction, C[{n/d}] is then put into Z, overwriting the address there. (Case p=0: equivalent to c=1, a=3, b=p=0). If p>0, we have a CONDITIONAL JUMP: C[{n/d}] is put into Z if (and only if) condition is satisfied; condition C[{{a}}] = 0 if p=1, C[{{a}}]≧0 if p=2, C[{{a}}]≠0 if p=3, C[{{a}}] >0 if p=4, and bit b of {{a}} = 0 if p=5, bit b of {{a}} = 1 if p=6. If p<5, b=0.
7	BIT operations — Force bit b of {{a}} to be: 0 if d=0, 1 if d=1, or complemented if d=2. d≠3. b=p=0. No n.
8	STOP — Terminates execution.
9	STOP if an INTERRUPT is flagged.

Figure 1.3 Typical op-code specification.

Higher-level languages

In the last section we examined in some detail a fictitious, but representative, sample of *computer language* in all its stark simplicity. After some introductory description of an imaginary computer, and especially of its CPU, and the establishment of some essential notation, a typical *instruction layout* was

ADDRESS	INSTRUCTION		c a d b p address	EXPLANATION
0	0011000000000000		3 0 0 0 0	X ← 0.
1	0010000100000000	0001011101110000	2 0 1 0 0 6000	C [X] → 6000.
3	0001000000000000	0001001110001000	1 0 0 0 0 5000	X ← 5000.
5	0010000100000000	0000000000011100	2 0 1 0 0 28	C [X] → 28.
7	0100000000000001	0001011101110000	4 0 0 0 1 6000	X ← C [X]–6000.
9	0110000100000001	0000000000011000	6 0 1 0 1 24	Jump to 24 if C [X] = 0.
11	0000000000001001		0 0 0 0 9	Input to X from port ≠ 9.
12	0010001000000000	0000000000011100	2 0 2 0 0 28	C [X] → C [28].
14	0100000100000000	0001011101110000	4 0 1 0 0 6000	X ← C [X] + C [6000].
16	0010000100000000	0001011101110000	2 0 1 0 0 6000	C [X] → 6000.
18	0001000100000000	0000000000011100	1 0 1 0 0 28	X ← C [28].
20	0100000000000000	0000000000000001	4 0 0 0 0 1	X ← C [X] + 1.
22	0110000100000000	0000000000000101	6 0 1 0 0 5	Jump to 5.
24	0001000100000000	0001011101110000	1 0 1 0 0 6000	X ← C [6000].
26	0000000000000010		0 0 0 0 2	Output from X to port ≠ 2.
27	1000000000000000		8 0 0 0 0	Stop.
28	*Arbitrary: this location is used as work-space for the program.*			

Figure 1.4   A simple program.

shown in figure 1.2 and a typical *op-code specification* in figure 1.3. The overwhelmed reader may balk at my use of the epithet "simplicity," but I shall cleave to it. Presented with a set of arithmetic and other transformations required by potential users of a proposed new computer, the electronic and logical design engineers seek the simplest circuitry that will execute operations sufficient to generate all the required transformations. Circuits must be *simple* to be *fast, efficient, reliable*, and *cheap*. However, when an instruction has 16 or more bits, most of which should, for the sake of efficiency, have some significant effect; the exact and complete specification of the action induced by it may well be somewhat forbiddingly intricate! And indeed, simplicity of circuitry does not usually lead to simplicity of use. An example of a *program* in this machine language, which takes 16 instructions and 28 16-bit words of memory storage, to read in one thousand numbers, store them in consecutive memory locations, sum them, and print out the sum, is given in figure 1.4. This illustrates the—from the programmer's point of view, unnecessary—difficulty of machine language to interpret the user's requirements. However, please note that the *English description* of what the program does (though much easier to say and comprehend) itself requires 103 characters, each equivalent to 8 bits; so that it is almost twice as long as the machine-language program. Indeed, the difficulty of the latter lies not in its *length*, but in its alien and opaque *form of expression* of what it does.

The first improvement on machine language was *assembly language*. This is not very different from machine language, but removes the most glaring and trivial irritations. First, *instructions* may be *labelled*, so that one no longer needs to count lines to address jump instructions; *one jumps to the label*. This is also a great help when one wishes to insert additional instructions, either to correct or modify a program (every jump need not be changed). Second, *memory* may be labelled, so that numbers may be stored in *symbolic addresses* and retrieved therefrom, leaving it to a computer program to assign actual storage. Third, the entire program becomes *relocatable* anywhere in the computer memory. Finally, and perhaps more trivially, the op-codes may be replaced by abbreviated mnemonics, and the instruction layout may be relaxed, using punctuation marks.

Of course, a program written in assembly language is no longer directly intelligible to the computer or executable by it. It becomes a piece of textual input, to a "translator program" called the *assembler*. Naturally, the program may be stored in the computer's memory, like any other piece of text; but, before it can be executed, it must be "assembled." The assembler finds appropriate space for the object code (i.e. the machine language equivalent of the input program) and for its needed storage space (often called *variables*). The assembly language instructions correspond, one-to-one, to machine-language instructions.

For example, we may denote memory locations by one, two, or three lower-case letters (both for labelling instructions and naming variables). A numerical address will still be allowed. Absolute addresses (i.e. actual numbers; $d=0$) will correspond to quoted numbers (e.g. "6000"); direct addresses ($d=1$) to variables named as stated; indirect addresses ($d=2$) to variable names placed in parentheses; and relative addresses will not be used (with relocatable programs and symbolic addresses, they are no longer useful). Accumulators (X, Y, XY, or Z) will not be referred to symbolically, but only by these names. Instruction labels will precede the instruction and will be terminated by a colon (:). Spaces will be ignored. Op-codes will be denoted by strings of capital letters and terminated by a comma (,), which will also separate multiple arguments. Instructions will be terminated by a semicolon (;) or by the ending of a line ("carriage-return" character); several instructions may appear on one line. A sample specification of assembly language op-codes is shown in figure 1.5. In the assembly language notation, *A* denotes X ($a=0$), Y ($a=1$), XY ($a=2$), or Z ($a=3$). Italic letters (p, x, or b) denote numbers (possibly in quotes, to denote absolutes) or variables names (possibly in parentheses, to denote indirect addressing), and are interpreted according to the op-code (e.g. *A* cannot be XY in IN and OUT instructions; if *A* is XY in instructions with $c=1$, 2, and 4 (with $p \geqslant 5$), the operand must be 32 bits long—either a double-length number, or an address referring to the first of two consecutive 16-bit memory words. Again, the reader must remember that it is not the particular details of our specifications that are important or memorable but the kind and degree of detail occurring.

ASSEMBLY LANGUAGE MNEMONIC	MACHINE LANGUAGE INSTRUCTION
IN **A**, (p-8)	$c = 0, a \neq 2, p \geqq 8$
OUT **A**, p	$c = 0, a \neq 2, p \leqq 7$
TO **A**, x	$c = 1$
FROM **A**, x	$c = 2, d \geqq 1$
FROM **A**, **B**	$c = 2, d = 0$
CL, **A**	$c = 3, d = 0$
NEG **A**	$c = 3, d = 1$
NOT **A**	$c = 3, d = 2$
RNDX	$c = 3, d = 3, a = 2$
ADD **A**, x	$c = 4, p = 0$
SUB **A**, x	$c = 4, p = 1$
AND **A**, x	$c = 4, p = 2$
OR **A**, x	$c = 4, p = 3$
XOR **A**, x	$c = 4, p = 4$
LADD, x	$c = 4, p = 5$
LSUB, x	$c = 4, p = 6$
LAND, x	$c = 4, p = 7$
LOR, x	$c = 4, p = 8$
LXOR, x	$c = 4, p = 9$
CSHL **A**, p	$c = 5, d = 0$
CSHR **A**, p	$c = 5, d = 1$
ASHL **A**, p	$c = 5, d = 2$
ASHR **A**, p	$c = 5, d = 3$
JMP, x	$c = 6, p = 0$
JPCZ **A**, x	$c = 6, p = 1$
JPCNN **A**, x	$c = 6, p = 2$
JPCNZ **A**, x	$c = 6, p = 3$
JPCP **A**, x	$c = 6, p = 4$
JPCB **A**, 0, b, x	$c = 6, p = 5$
JPCB **A**, 1, b, x	$c = 6, p = 6$
BIT **A**, 0, b	$c = 7, d = 0$
BIT **A**, 1, b	$c = 7, d = 1$
BITNOT **A**, b	$c = 7, d = 2$
STOP	$c = 8$
INTER	$c = 9$

Figure 1.5   Typical assembly language mnemonics.

Again, to illustrate our assembly language we return to the program given in machine language in figure 1.4. The same program in assembly language would take the form given in figure 1.6. Here, numbers are given in decimal notation (another convenience of assembly language; one can also give numbers in hexadecimal notation by prefixing them with "H," in octal with "O," and in binary with "B"). Note the tremendous improvement in direct legibility of the program in comparison with the binary strings of figure 1.4. The assembler is left to figure out where to store the sum (the variable "sum" which the machine language program put into address 6000) and the current address of the listed numbers (the variable "w," which the machine language

```
            CL X; FROM X, sum; TO X, '5000'
    ret:    FROM X, W; SUB X, '6000'; JPCZ X, out
            IN X, 1; FROM X, (W); ADD X, sum
            FROM X, sum
            TO X, W; ADD X, '1'; JMP, ret
    out:    TO X, sum; OUT X, 2; STOP
```

Figure 1.6   The summation program in assembly language.

program puts at location 28 if the program begins at location 0, necessitating counting the words occupied by the program; if the program were to be modified or relocated, this would have to be changed, and would be an unnecessary source of programming errors); and similarly, there is no need to count words to determine the addresses of "ret" and "out" (placed by the machine language program at 5 and 24, as it happens). As an exercise, the readers may wish to try to modify the program in figure 1.4 so as to allow for the possibility that the sum of the 1000 numbers might occupy more than 16 bits. In machine language this requires no less than 12 changes and seven shifts of memory location. In assembly language only the substantive changes (from the programmer's point of view) need be made. The amended program is given in figure 1.7. Here we note that only seven changes are needed; but observe that "sum" becomes a double-length address—just another piece of bookkeeping handled automatically by the assembler.

Of course, a program in assembly language still represents, instruction-by-instruction, a program in machine language, and therefore still has the defects of "walking the road in extremely short steps," as we put it earlier. All that shuffling of numbers in and out of the accumulators, counting and restricted testing, is not natural to us, and its mechanical nature suggests that an improvement is still possible. Thus it was not long before programmers devised much more humanly natural languages, which are generically called *higher-level* (or *algebraic*) *languages*. Some of the more common ones are ADA, ALGOL, APL, BASIC, C, COBOL, FORTRAN, LISP, PASCAL, PL/I, RPG-II, SNOBOL, SPITBOL, WATFOR, and WATFIV. There are more; and all come with a variety of versions and dialects.

```
            CL XY; FROM XY, sum; TO X, '5000'
    ret:    FROM X, W; SUB X, '6000'; JPCZ X, out IN X, 1; FROM X, (W); TO XY, sum ADD XY, (W);
            FROM XY, sum TO X,W; ADD X, '1'; JMP, ret
    out:    TO XY, sum; OUT X, 2; OUT Y,2; STOP
```
(Underlined material represents changes from previous program.) Alternative version would have third line replaced by:
```
            CL X; IN Y, 1, FROM Y, (W); LADD, sum FROM XY, sum
```

Figure 1.7   The summation program modified for long sums.

The characteristic that puts a language into this class is that one higher-level instruction translates into several assembly or machine language instructions. Beyond this, the languages differ according to the kind of program they are intended for. While most languages will perform most tasks, FORTRAN was clearly intended to do scientific and engineering computations, COBOL to perform business data-processing, LISP to manipulate lists, and SNOBOL, strings of text. As a final illustration of the power and intelligibility of such languages, we present higher-level versions of our summation program in BASIC, WATFIV, and PASCAL, in figure 1.8. Once more, note that the details of the individual languages do not matter to us at this point (though, if the reader ever intends to program, then *some* language will have to be thoroughly understood; but I assume that many of my readers have no such intention); it is their general flavor and appearance that is noteworthy; and the casual readability of the samples presented is their salient characteristic.

Of all the hundreds of programming languages devised, it is safe to say that more lines of program have been written in FORTRAN or COBOL than in all others combined, though this reflects their age rather than their desirability. The fast-gaining runner-up must surely be BASIC, which comes with every micro- and most mini-computers; though some would say that BASIC, which perpetuates the style of FORTRAN, spoils any chances of producing really good programs and should be replaced by PASCAL. Fashions come and go in programming languages, not always fully based on rational arguments, and personal preferences vigorously touted.

Given a higher-level language, it is necessary to have a *translator program* to turn it into machine (or more frequently assembly) language. Such translators are of two kinds. An *interpreter* operates at *execution* time: beginning with the first instruction of the stored higher-level program, it translates it into one or more machine language instructions, which it proceeds to execute; then it reads and translates the next higher-level instruction and executes that; and so on. If a jump is encountered in the *source program* (as the higher-level language program is often called), the interpreter goes to the next instruction in accordance with the jump and proceeds to read, translate, and execute it, without taking into account whether it has already encountered it. Thus, in a program such as those in figure 1.8, the same source instructions will have to be parsed and translated a thousand times (a very time-consuming and wasteful procedure). But the (usually much longer) object program need not be stored, and interpreters themselves tend to be shorter and simpler programs. This is why a BASIC interpreter is a natural adjunct to a microcomputer, which tends to be a little cramped for space; especially since many microcomputer users are not interested in extremely lengthy computations.

By contrast, a *compiler* is a program that translates a source program into an object program, the former being in higher-level language and the latter in machine or assembly language. The program is translated as it sits in

1 BASIC

```
100 DIMENSION A (1000): LET S = 0
200 FOR I = 1 TO 1000: INPUT A(I): LET S = S + A(I): NEXT I
300 PRINT ≠2, S: STOP: END
```

Note: The "DIMENSION" statement sets up the "array " of indexed variables A(I), corresponding to our memory locations 5000–5999. A statement such as "LET S= S+A(I)" computes the right-hand side of the "equation" and stores the result in the address named on the left. The statement "NEXT I" transfers control to the instruction following the "FOR" statement, until all the I-values (1 through 1000) have been dealt with; then execution moves to the instruction following the "NEXT" statement. BASIC is a terminal-oriented language and the "INPUT" instruction requires the user to type in the thousand numbers one-by-one as they are called for; but there are other ways of effecting the input   Note, too, that (typically of a higher-level language) no accumulator is mentioned. BASIC is rather unusual in requiring lines (but not individual instructions, which are separated by colons) to be numbered.

2 WATFIV

```
INTEGER*4 A (1000), S; READ A; S = 0; I = 1
WHILE (I.LE. 1000); S = S + A(I); I = I + 1;
    END WHILE
PRINT, S; STOP; END
```

Note: The "INTEGER"*4" statement specifies both the "dimension" of the array A and that each variable represents four bytes (32 bits) of memory and is treated as an integer. The "READ" statement will then input all 1000 numbers into consecutive array locations. The instructions between "WHILE (I.LE. 1000)" and "END WHILE" will be repeatedly executed, so long as the condition in parentheses is true. (".LE." denotes the less-than-or-equal-to sign ≦. Here, instructions are separated by semicolons. In both programs above, "STOP" denotes the end of execution and "END" the end of the program listing. (Actually, the former may be omitted, if it comes, as it does here, just before the latter.)

3 PASCAL

```
program SUM (INPUT, OUTPUT);
var       A : array [1..1000] of integer ;
          S , I : integer ;
begin
          S: = 0; I: = 1;
          repeat
              read (A[I]) ; S : = S + A[I] ; I : = I + 1
          until I = 1000 ;
          write (S)
end .
```

Note: PASCAL is rather more formal about specifying variables, and lends itself to "structured programming." The symbol ":=" replaces the misleading "=" of BASIC and WATFIV. The "repeat . . . until" may be replaced by a "while . . . begin do . . . end" construct, rather as in WATFIV.

Figure 1.8   Three higher-level language versions of the summation program.

memory, not during execution; so that the problem of repeated translation is avoided. Indeed, there are many so-called "optimizing compilers" which pass through the compiled object code several times, eliminating redundancies and inefficiencies to produce faster object programs (something that would be impossible for an interpreter to do). Speed is greatly increased, in general, at the expense of space in the computer memory. Of course, once a program has been compiled and proven to be free of bugs (debugging is far easier in source code than in object code), the source program may be stored for future use and only the object program kept in main memory. Compilers are available for almost all higher-level languages on almost all machines, and are essential for extensive applications.

Computer software

The programs, or *software*, available for computers fall into two classes, *system software* and *applications software*. The distinction is a fuzzy one; roughly, the system software comprises those programs that are considered indispensable to the general operation of a given computer system, forming what is often termed the *operating system* of the computer, and generally are supplied by the manufacturer, though alternative operating systems are sometimes available from software houses; while the applications software includes all those programs that are needed by one computer user but not by another, even if such programs are widely required (for example, an accounting program), and while some manufacturers will sell applications software (as optional additions to their systems), more often this is obtained from software specialists.

We begin with the *operating system*. This may be divided into a *kernel* of absolutely indispensible programs and a *shell* of almost indispensable, so-called *utility* programs. The kernel (or *nucleus*) is also sometimes termed the *monitor*.

To understand why this kernel program is needed, we consider what happens when we type a character (say an "A") on our keyboard, as part of a message to the computer. Because our timescale recognizes perhaps 1/20 second, while the computer may deal in units of 1/1,000,000 second, if we want to type an "A", we may find ourselves actually transmitting some 50,000 A's to the computer, before our finger leaves the key! If the keyboard mechanism were to transmit the A for only one microsecond, on the other hand, the computer might well be busy elsewhere during this particular μs and miss it altogether. Therefore, it is necessary to establish *a protocol* (in the usual diplomatic sense, a formal structure for the orderly passage of information) or "handshaking" procedure. This is often done by sending the A on one line (or lines) and simultaneously sending a pulse along an auxiliary line, which sets a "flag" bit in the CPU. When the computer is ready to receive a new character, it checks the appropriate flag bit over and over, until it detects that it has been set. Then it reads the character, resets the flag to its "null" state, and sends a signal pulse back to the keyboard (or its controlling circuitry), telling it that the character has been received. The flag is always open, so the transmit pulse need only be sent once by the keyboard; but the character continues to be sent until it is acknowledged by the computer. Thus a keyboard character will neither be missed nor read repeatedly. In addition, either the CPU or the keyboard controller should send the "A" to whatever device displays the typed characters.

If the particular keyboard were to be the only input device ever to be connected to the CPU, it would be best to incorporate all this in the circuitry of the CPU; but since there may be several different inputs to the CPU, and several types of terminal may be used, coming from different manufacturers, it is found preferable to have the *I/O procedures* made part of the operating system. In addition to the keyboard input already described, there will be short machine language programs for input from any other devices attached to the computer and corresponding programs for output to display screens, printers, and so on.

The orderly operation of a computer is subject to interruptions (or *interrupts*, as they are known in Computerese), either because a program contains an instruction that at some point, is not executable (such as a division by zero or a reference to a nonexistent memory location), or because someone needs to terminate execution for some overriding reason (such as the decision that a program is in error and is generating garbage or is in an infinite loop). *Handling errors and interrupts* is another function of the kernel of the operating system, as is the *start-up* (or "bootstrap") *procedure* which initializes operation.

If, as is often the case, several users are connected to the computer, then the operating system must handle the tasks of *job-scheduling* (and *job accounting* and *billing*, if this is appropriate), and the *allocation* of *storage*, in main and extended memory, and of other *resources* (such as printers or communication lines), and the management of *time-sharing* (as between several terminals).

Another task that is handled by the operating system is the management (including allocation, proper formatting, and appropriate handling) of *user files*, and their transfer between main memory and extended memory (such as disk). Indeed, one sees frequent reference to *disk operating systems*, such is the importance of this function of the operating system.

Another function of the kernel is to provide *protection and security* to users and to itself, both from authorized users encroaching on forbidden territory and from unauthorized users attempting to use the computer. Ideally, each user should, on giving the correct access code (such as a "password")—this feature did not exist on the first computers, and is still absent in personal microcomputers with only one user/owner—have access strictly limited to the parts of the computer allocated to him or her by the operating system, but in such a way that the impression is maintained that he or she is the only user present. Another very congenial aspect of this concept is that some operating systems conduct their memory management function in such a way that the user need make no distinction between main and extended memory; this is referred to as a *virtual memory* system.

The kernel program has to be able to *display, move, modify,* and *search,* at least the main memory, and to *initiate execution* at any given address, or activate any of the peripheral devices. Using these functions, by means of appropriate *commands* in the "operating system language," together with its file-handling capabilities, the system can *load, compile,* or *execute* any program stored in its memory (main or extended) written in machine, assembly, or higher-level language, provided that a suitable compiler or other translator is available to it.

We now turn to the *utilities* provided by most operating systems. First, we have an *assembler* (and perhaps also a *disassembler,* which translates a program written in machine language into the more intelligible assembly language; as well as a *macroassembler,* which allows the user to define his own *macro-*instructions in machine or assembly language), together with a selection of *translation programs,* either *interpreters* or *compilers,* for the higher-level languages that the user wishes to employ. The smaller microcomputers will provide an interpreter for some dialect of BASIC, since this a relatively simple language to learn and to interpret into machine or assembly language; beyond this, one must pay for additional languages. There may also be a variety of TRACING and DEBUGGING utilities, according to the cost, size, and sophistication of the computer and its operating system. There will be facilities for *linking* or *chaining* programs together.

While the kernel will contain the rudiments of a *file-handling* system (to create, destroy, list, locate, and transfer files), there will also be utilities for further management of these. Such programs will *sort, find* (among other files), *search* (in a given file, for information specified), *transform, edit,* and *combine* files. Indeed, a good *editor* program can enormously facilitate the rapid and painless creation and modification of files, which include both *text* and *programs.*

The line separating the kernel from the shell of utilities is purely conceptual and far from sharp. Similarly, the boundary of the entire operating system is far from definite. What is available beyond the essentials mentioned above will be called part of any decent system by some, part of a compiled language by others, and just applications software by still others, depending on their point of view.

The ability to handle a variety of *data structures* (such as *arrays, lists, strings* of characters, *trees, queues,* and *stacks*) may come from the use of a suitable higher-level language or from an extended "system development" utility package. Similarly, languages intended for scientific and engineering applications usually handle *floating-point* and *multiple-precision arithmetic,* though this may be provided even in computer hardware. The same applies to routines for computing, for example, sines, cosines, logarithms, etc., and pseudo-random numbers (the last beloved of computer games inventors!)

A *data-base management system* (DBMS) may be just a glorified file-handling utility; or may be an elaborate program, cross-indexed and relational, with its own language of special commands, for answering any conceivable question about a large amount of intricately structured data. As is often the case, you gets what you pays for, both in money and in memory space; and vendors' claims have to be carefully scrutinized and verified. It is advisable not to overbuy one's capabilities beyond one's needs.

Similarly, a *graphics package* may simply allow one to produce passable graphs, histograms (i.e. bar-graphs), pie-charts, and perhaps games, with a resolution of some 200 by 300 delightfully named *pixels*; or it may allow you to resolve perhaps 2000 by 3000 pixels, in a variety of colors, with the ability to draw complex three-dimensional shapes, properly shaded, illuminated, and textured, and move them by commands in a special language. The latter systems are a lot of fun to operate, and can be most helpful to draughtsmen, animators, film designers, and simulator-trainer designers; but they cost a bundle, and they require quite powerful computers to hold and run them.

Another offshoot of the file handler is a cluster of programs for generating *reports, journal ledgers*, and *accounts* of all kinds, as well as *forms* and *mailing lists* (the latter leading to a kind of DBMS in which mailing lists are matched to interests and characteristics of the individuals itemized).

Again, there are *statistical packages* of various degrees of sophistication, from a mean–variance–covariance calculator, to systems able to analyse very complex sets of data by elaborate techniques, using a whole statistical computer language. (In my opinion, these last, in the hands of the uninitiated, may be as dangerous as a loaded machine-gun!)

What has been described as "the most popular program ever written" is usually given the generic name of an *electronic worksheet*, though the trade name of the first such program, Visi-Calc, like Kleenex and Vaseline, has become almost generic. A table is presented on the video screen and each entry is allocated either a numerical value or a formula relating it to other entries. When the data are sufficient, the resulting numbers are displayed. When an entry is changed, it and all entries depending on it are altered accordingly at once. This provides a representation of a given situation surpassed (I think) only by a graph in its impact, and a first-class planning aid.

Finally, among the borderline system/applications software utilities, I would put what is usually called a *word-processing package* (though I would prefer either character or text processor). This is an extension of an editing utility, in which the text may by "scrolled" up and down on the screen, edited in the usual ways, but also more specifically for producing letters, articles, reports, and other written copy. Margins may be set and the text right, left, or double justified, or set up in multi-column pages, with page numbering, indexing, and even the use of different sizes and styles of typeface, in the most sophisticated systems. Here again, cost rises steeply, and one should buy only what one will need.

Beyond these programs, there lies an endless variety of unquestionable applications programs. There are programs for ballistics, boat design, analysis of molecular structure from x-ray diffraction data, tabulation of Bessel functions, simulation of naval battles and economic cycles, etc. Most of these are not on the market and only work on one machine; but there are very many programs available on the open market, produced or distributed by software houses, with various levels of efficiency, sophistication, and reliability.

Programming is a fascinating, intricate, rewarding, but unforgiving and at times infuriating occupation. I would encourage all of you who have the time to try it. If you have a computer at your disposal, it would be a shame not to learn a simple language, such as BASIC or PASCAL and try your hand at writing a simple program or two. The sense of achievement when you have a working program is great; perhaps because the process is addictive and consumes much more time than you would believe possible. The programmer is an eternal optimist, and every bug really seems to be the last!

Software: the New Driving Force

Business Week

Software—the set of instructions that tell a computer what to do—is becoming more important than the computer itself. The booming software industry is therefore setting the pace of the information technology revolution—and the pace is frantic. This article first appeared in Business Week, *February 27, 1984.*

Raw power. That is what sold computers in the early days. Salesmen would fire off strings of statistics—how many millions of instructions per second a machine could handle, how many bytes of data it could store, how many bits it could process at a time.

But hardware is no longer where the action is. Computers are becoming remarkably similar—in many cases they are turning into off-the-shelf commodity products. Now the computer wars are being fought on a new battleground: software—the instructions that tell computers how to do everything from processing payrolls to playing video games. "Hardware is getting less and less distinguished," says Jon A. Shirley, president of Microsoft Corp., a Washington State company that writes software for personal computers. "Software is what's leading the industry."

A key reason for the change in emphasis is an overwhelming demand from customers for packaged software that will let them apply computer power to a broad range of new tasks. Increasingly, corporations are finding they do not have the resources to write the programs they need. As a result, most companies have stopped writing their own software and are instead buying standard software packages.

So far, producers have been unable to keep up with the need, and there has been a severe shortage of software able to take advantage of the power of the latest machines. "We're getting the hardware [we need from the industry], but software has not moved out at the same rate," says Robert J. Metzler, vice-president of First Computer Services, the data processing arm of First

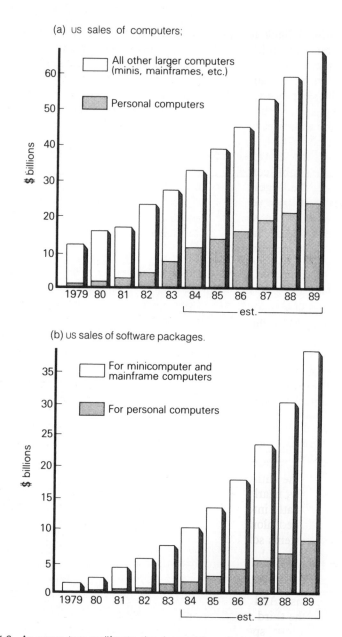

Figure 1.9 As computers proliferate, the demand for software soars.

Sources: (a) International Data Corporation, BW estimates; (b) Input.

Union National Bank in Charlotte, NC. "There's a tremendous cry in the industry [for software]." The flood of personal computers pouring into small businesses and homes has created an even faster-growing new market.

The booming demand for new and better programs has quickly turned software into big business. As recently as three years ago, software was still a cottage industry, with sales totaling just $2.7 billion annually. This year—according to estimates by Input, a California market research firm—sales are expected to top $10 billion.

With the industry growing that fast, several once-small software companies have become sizable corporations. Management Science America Inc. (MSA), for example, the largest independent software supplier, quadrupled in size during the past three years, topping $145 million in sales in 1983. Trying to capture some of the soaring market, thousands of new companies have entered the software business in recent years; by one count there are more than 3000 software companies now.

While over the long term a shakeout is probably inevitable, the near-term outlook for most of these companies appears extremely bright. "The upside potential has barely been tapped," says Robert M. Freeman, senior analyst at Input. He expects the market to keep growing by a dizzying 32 percent a year, topping $30 billion in 1988. Software sales of that magnitude would amount to half of the $60 billion hardware business expected for the same year; today, revenue from software is equal to only 27 percent of the value of all the computer hardware sold (figure 1.9).

Sales of software for personal computers—the fastest-growing part of the software industry—should rise by an astounding 44 percent annually over the same five years. "We're finally at the point where software applications are going to be a big moneymaker," says Jack M. Scanlon, vice-president of the Computer Systems Div. of AT&T Technologies Inc. But with the promise of such rapid market growth has come a feverish competition that is beginning to restructure the entire industry.

Until recently, nearly all software companies concentrated on a particular niche and fit neatly into one of three distinct market segments (figure 1.10) that were divided along the same lines for both mainframe and personal computers. Systems software, which handles basic housekeeping operations such as controlling the printer and memory, was supplied primarily by the computer makers. They were joined by independent software companies in providing utility software, which, among other things, helps programmers write programs. And a host of other independent software companies competed in the third market; applications packages, which tell a computer how to carry out specific tasks such as accounting, payroll, or word processing.

Suppliers for large computers
Software companies: American Management Systems, American Software, Anacomp, ASK Computer Systems, Comserv, Hogan Systems, Informatics Information Science, Integrated Software Systems, Management Science America, McCormack & Dodge (Dun & Bradstreet) Policy Management Systems, Shared Medical Systems, Technicon, Timberline Systems, University Computing (Wyly), Walker Interactive Products

Hardware makers: IBM, Sperry
Software companies: Applied Data Research, Artificial Intelligence, Cincom Systems, Computer Associates, Comshare, Cullinet Software, Informatics General, Information Builders, Johnson Systems, Mathematica, Pansophic Systems, Software AG, University Computing (Wyly)

Others: AT&T

Hardware makers: Burroughs, Digital Equipment, Honeywell, IBM, NCR, Sperry

Software companies: Applied Data Research, Computer Associates, Pansophic Systems

Others: AT&T

APPLICATIONS SOFTWARE
These programs turn the computer into something useful because they set up the system to handle a specific task. They can be for general-purpose applications, usable by a wide variety of companies, performing such jobs as general ledger accounting, payroll, and word processing. Other packages tailor the computer to the needs of a specific industry, such as banking, insurance, hospitals, manufacturing, or retailing

UTILITY SOFTWARE
These products are the middle-men between operating systems and applications software, and ensure that the applications programs are written and run efficiently. These include: compilers which turn programs into code that computers understand; data base management systems that act as electronic librarians to keep track of data; and, for now, only on personal computers, "windows" and integrated "environments", that permit users to do more than one task at a time

SYSTEMS CONTROL SOFTWARE
These are the "house-keeping" programs that manage the operation of the computer's various components, such as printers and memories, so that they work smoothly together as the applications software performs its tasks. These products include operating systems, communications monitors, and network control programs

---

Suppliers for personal computers
Hardware makers: Apple, IBM
Software companies: Computer Task Group, Digital Research, Eduware, Information Unlimited Software (Computer Associates), Informatics General, Lotus Development, MicroPro, Peachtree Software (MSA), Safeguard, SEI, Sorcim, Software Publishing, Spinnaker, VisiCorp
Publishers: CBS, Dow Jones, Dun & Bradstreet, McGraw-Hill

Hardware makers: Apple, IBM
Software companies: Ashton-Tate, Condor Computing, Digital Research, Microsoft, Quarterdeck, Software Publishing, Stoneware, VisiCorp

Others: AT&T

Hardware makers: Apple, IBM
Software companies: Digital Research, Microsoft, Softech Microsystems

Others: AT&T

Figure 1.10   A quick guide to software—and who supplies it.

But those distinctions are blurring:

Consolidation
International Business Machines Corp., to move faster into applications software, in November placed all its software efforts into a single, entre-preneurial unit—the same type of organization it used to launch the highly successful Personal Computer.

Acquisitions
To provide a full complement of software for their computers, makers such as Hewlett-Packard, Burroughs, and Prime Computer are rushing to buy applications software companies. Other manufacturers, including Honey-well, Sperry, and Digital Equipment, are setting up joint ventures with software suppliers. Says Roger T. Hobbs, vice-president for software products and services at Burroughs Corp.: "The demands for software are increasing so rapidly that it is impossible for the manufacturer to keep up."

Expanding product lines
Software companies are expanding product lines to maintain their competi-tive edge. Those that supply programs mostly for large mainframes—MSA, Cullinet Software, and Computer Associates International, for example—are snapping up personal computer software houses. And companies that specialize in systems software for personal computers—among them, Ashton-Tate, Digital Research, and Microsoft—are adding general-purpose applica-tions software to their product lines. "Today you have to have a broad line," says Terry L. Opdendyk, president of VisiCorp. Otherwise, "the vendor is increasingly vulnerable to competition."

New players
To grab a piece of the action, publishers and other communications companies outside the computer business—including CBS, Dow Jones, Dun & Bradstreet, McGraw-Hill, and Simon & Schuster—are licensing programs they then sell through their own distribution channels. "It's a very fragmented industry, but the potential is absolutely huge," says Richard W. Young, president of Houghton Mifflin Co.

Japan's drive
To catch up with their US rivals, Japanese computer makers are launching major software development efforts. Perhaps the best evidence of just how important software has become is the drive by Japan's Ministry of Inter-national Trade & Industry to change the law to boost Japan's fledgling soft-ware industry.

One result of many of these trends is that mergers and acquisitions in the industry are at an all-time high. Last year there were 146 acquisitions, valued at more than $1 billion—up 130 percent from 1982—reports Broadview

Associates, a New Jersey company handling such transactions. And this frantic pace of activity is turning the software business into a hot investment area.

To complete successfully amid all the turmoil, software suppliers are struggling to formulate new strategies. One common focus is advanced design—especially the development of software aimed at making computers easier to use (for details of these trends, see below).

But just as important to the success of new software as advanced design is marketing. "You need a great product and great marketing" insists David S. Wagman, chairman of Softsel Computer Products Inc., an Inglewood (Calif.) software distributor. "For a while it was either/or in this business, but now you absolutely must have both."

Nowhere is the new attention to marketing more noticeable than in the hotly competitive personal computer business. To reach the millions of personal computer users, software companies are spending huge amounts to introduce and advertise their products. Industry watchers have dubbed this obsession with splashy promotion "the Lotus syndrome"—a reference to the more than $1 million that Lotus Development Corp. spent over a three-month period in 1983 to launch its first product, the highly successful 1-2-3 package. "Lotus advertised so much that companies are going to be forced to step up their advertising just to be heard above the noise," says David E. Gold, a San Jose (Calif.) computer consultant.

Advertising, in fact, has grown so important to software success that it has become one of the biggest barriers keeping out new companies. "The cost of technology development is dwarfed by the marketing cost; the ante has really been upped," says Rodney N. Turner, vice-president for sales at Ashton-Tate, a Culver City (Calif.) software company best known for its dBase II package. It takes as much as $8 million to launch a new software product today, he estimates. Ashton-Tate, by contrast, was founded on shoestring— $7,500.

Any personal computer program must be carefully packaged and promoted. "It has to look good and has to be well supported by national advertising," maintains retailer Gregg E. Olson, a salesman at Mr Software in Boulder, Colo. "A company that wants to sell to us has to have all their marketing elements in place to establish credibility with us and our customers."

Even companies that write software for the large minicomputers and mainframes are plowing more money into marketing. While programs for the larger systems, unlike those for personal computers, do not require heavy consumer advertising, says Martin A. Goetz, senior vice-president at Applied Data Research Inc., his company still boosted its 1984 advertising budget by 60 percent over last year, to about $2 million.

This emphasis on marketing comes chiefly from the need to reach a different, far broader group of potential customers. "Ten years ago, [our customer] was the data processing department, and if you had a product, you sold it to the technicians," says Robert D. Baskerville, group vice-president for product management at Computer Sciences Corp. in El Segundo, Calif. Today, he notes, a software company has "to sell to end users, and you have to emphasize more than the technical capability—you have to really sell the benefit."

Software marketing now means beginning with detailed product planning, so that a program will better meet the needs of customers. Before Wyly Corp. even began a $20 million program to diversify into applications software, for example, it brought in all its salesmen from the field to tell programmers what customers wanted—something most software companies had never done. "With modern software development techniques, you can produce almost perfect code, but if you don't understand the market, it could still be useless," says Ron W. Brittian, Wyly's vice-president for research and development.

Critical, too, is a reputation for good service and customer support. "To grow from a startup into a large software company is not so easy anymore; it's a question of distribution and of educating the customer in how to use the product," says Anthony W. Wang, executive vice-president at Computer Associates International Inc. Toward this end, his company recently installed a $250,000 television studio—costing $200,000 or more a year to run—to make training videotapes for its software customers.

As competition heightens and customers become more demanding, companies selling personal computer software are providing a level of customer support far greater than anything they offered even a year ago. MicroPro International Corp., for example, has plowed $2.5 million into developing computer-aided instruction and retail-support aids for its software, which includes the best-selling WordStar. The company is also hiring journalists and other nontechnical writers to produce more readable instructions for today's less sophisticated customers. "A product is not a product until you have computer-aided instruction, it's not a product until you have video instruction, it's not a product until you have honest-to-God understandable language [in the manual]," says H. Glen Haney, president of MicroPro.

Establishing a marketing presence is especially critical to companies that provide personal computer software, because they must win space on already crowded retail shelves. Computer retailers are reluctant to take on new software without feeling confident it is a winner. "Unless we're absolutely convinced that it's a great program, we'll wait to see how other channels of distribution do with it," says John H. Rollins, national manager for Sears Business Systems Centers, the computer retailing arm of Sears, Roebuck & Co.

One group trying hard to enter the software business may have an edge: the traditional book publishers. Because they already sell books to computer stores, the publishers "provide ready-made distribution channels," says Input's Freeman. Agrees William M. Graves, president of MSA: "There is a tremendous similarity in the microcomputer business and publishing."

No matter how they distribute their products, software companies are finding that they must offer a broader range of products than ever before. "The strategy is to offer a complete solution," asserts Robert N. Goldman, president of Cullinet. "If a customer ends up with eight different vendors, none of the software works together." Cullinet, for instance, is augmenting its data base management software with such applications as general ledger accounting and manufacturing control programs.

Broadening a software line also helps a supplier leverage the large amounts of money spent on establishing a reputation and brand recognition. VisiCorp, which gained market fame with its hit program VisiCalc, is trying to expand its applications software with its family of Visi On integrated software. Similarly, Microsoft is plunging into such applications software as word processing and financial analysis to build on its reputation as a supplier of personal computer operating systems.

The trend toward comprehensive offerings will force small companies to target well-defined market niches. There will continue to be demand for specialized packages designed for the needs of a particular industry or profession. "There will be several very large software companies," contends John P. Imlay Jr., chairman of MSA, "but there will be literally hundreds of small companies, with under $100 million [in sales], that have specialized market niches."

Perhaps the most fundamental change in the software industry is the blurring distinctions among the suppliers. No longer can they be neatly divided into companies that make basic systems software and those that write programs for specific applications.

Moreover, the top mainframe software companies are rushing to market with software for personal computers. And vendors serving only the personal computer market are joining with suppliers of mainframe software. VisiCorp, for example, recently teamed up with Informatics General Corp. to offer Visi-Answer, a program that allows a user of an IBM Personal Computer to retrieve information from an IBM mainframe data base. These changes resulted in large part from the growing number of customers linking personal computers to large mainframe systems.

As the emphasis in the data processing industry shifts to software—and as software companies strengthen their sales, service, and distribution—the big-system makers, too, are scrambling to do more to provide their customers with software. "In the old days, our customer wrote his own application [software]," notes Jon Tempas, vice-president for software products at Sperry Corp.'s Computer Systems operation. Today, he says, "there's an increased expectation for hardware suppliers to provide the complete solution." That means the equipment makers will need to provide more of their own software. Sperry, for example, now writes 95 percent of the software it sells for its computer line.

Most of the big-system companies, however, are turning to software specialists for help, since much of the demand is for applications software finely tuned to specific industries. "I don't think there's any hardware manufacturer that can—or should—provide all the software," says John E. Steuri, general manager of Information Services Business Unit, IBM's new independent software group. "We'll be more and more dependent over time on software developed outside the company." For example, in January IBM began marketing with Comshare Inc. specialized software to help executives make decisions.

The pressure to team up with independent software companies is especially strong among makers of personal computers. "Without an adequate software base, a microcomputer dies," says Eugene W. Helms, vice-president for business development at Texas Instruments Inc.'s Data Systems Group. So TI is recruiting software suppliers to adapt their best-selling programs for its Professional Computer. Perhaps the most extensive effort to sign up independent software companies was made recently by Apple Computer Inc., which courted more than 100 companies to write software for its new Macintosh computer. The California company was successful in signing up more than 80 of them.

Companies that do not move quickly to develop a broad line of software will have trouble keeping up. Consider Tymshare Inc., a computer time-sharing company. As computer prices began to drop precipitously, more and more of Tymshare's customers stopped renting computer time and purchased their own machines. Tymshare did not have any software packages to sell to its former clients. "We used to look at software as simply an in-house tool that we needed to offer time-sharing," says one Tymshare executive. That attitude, he admits, "came back and bit us in the rear end." In 1983 the California company lost $1.7 million while sales fell 3 percent to $288 million. Now it is engaged in a major effort to expand into applications software, through internal development and by licensing packages developed by other companies.

With most software companies trying hard to move in the same direction, everyone will face stiffer competition. Already, prices for personal computer software are sliding wherever similar programs have proliferated—in electronic spreadsheets and word processors, for example. And a big battle is under way among suppliers of the various "windowing" software packages—software "environments" that permit the users of personal computers to display several tasks at once. VisiCorp has been forced to slash the price of its Visi On environment package from $495 to $95. "The consumer is going to force the prices down simply by demand," asserts Alvin B. Reuben, executive vice-president at Simon & Schuster Inc.'s electronic publishing division.

As prices fall, the opportunity for a newcomer to jump in and grab quick and easy profits will all but disappear. The cost of developing and marketing new programs is mushrooming just when margins are shrinking. VisiCorp successfully launched VisiCalc in 1978 with a $500 budget. But the California company has spent more than $10 million developing its latest product, the Visi On environment. Product life is also getting shorter as new products come out faster. As a result, says Softsel's Wagman, "the stakes have gotten higher and much riskier from a development point of view."

For many companies, the risks will ultimately prove fatal. "There are companies out there that are already feeling the pinch of increased costs, new product announcements, and a changing marketplace," warns consultant Gold. Industry watchers predict that some big-name failures will occur within the next 18 months. By the end of the decade, many experts expect the software industry to have consolidated its thousands of suppliers into a few major players.

For those that make it, the future is promising. The demand by the growing army of computer users for easier-to-use software to handle an exploding variety of tasks will drive the industry to create software with far more capabilities than anything available today. "The only limit," says Stuart A. Walker, vice-president of marketing at Knoware Inc., "is the limit of new ideas."

Software in the office

One of the main reasons software has become so highly visible is that the computer has moved from the back room into the office. Programs of the past, buried inside corporate data processing centers, primarily handled clerical and accounting tasks, such as turning out payrolls and keeping track of accounts receivable.

Today's computer handles a rapidly increasing variety of management tasks—employing information as a competitive marketing weapon, for example. That trend, coupled with the explosion in the use of personal computers by executives, has given new prominence and importance to software in US corporations. "The programs that we install to do our company's business are analogous to the tools an auto maker puts in place to manufacture auto parts," says Jeffry A. Alperin, an assistant vice-president in the Information Systems Support Dept at Aetna Life & Casualty Co. "Our plant is our data processing system."

Since the same computer hardware is available to everyone, it is the software that often gives a company a competitive edge. At oil companies, "software is now helping drive the search for oil and gas," says Michael C. Balay, general manager of information technology at Gulf Oil Corp. "You can have the same [seismic] information, but a unique software package may help you better interpret the data."

Nowhere is this increasing importance more obvious than in corporate data processing budgets. Balay estimates that Gulf spends about half of its annual data processing budget to write its own software and to buy packages from independent suppliers.

At Aetna, half of the 4000 people on the data processing staff are programmers. And Alperin says that 4 percent of the $237 million that Aetna spent on information processing last year went to purchase packaged software. Moreover, individual owners of personal computers are no different from large corporations: over the life of a computer, its owner will spend $2 on software for every $1 spent on hardware, according to industry estimates.

But users are finding it increasingly difficult to produce their own programs. Writing software is still a time-consuming process, and users find that, while they want to employ their computers to handle more applications, they do not have enough programmers, time, or money to write all of the necessary software in-house. In fact, at many companies users must now wait as long as 18 months for new programs to be written. "The end users are demanding faster development of solutions," declares Richard R. Douglas, group vice-president for the US Marketing & Services Group of Honeywell Inc.

To come up with this software faster, computer users are compromising— buying more ready-to-run software packages than ever before. "There has been a turnaround from five or six years ago," says James T. Manion, sales vice-president at ASK Computer Systems Inc. Few computer users write all their own programs anymore, he says.

But while users sacrifice some uniqueness by buying a software package, that disadvantage is often outweighed by the time and money saved. A package can cut the time it takes to get a system up and running by more than half, figures Mayford L. Roark, executive director of systems at Ford Motor Co. He also estimates that buying a program from an outside vendor can cut development costs by 30 to 75 percent. "It's almost impossible to think of a function you might want to do on a computer that you can't find several software packages for," says Roark.

Where a custom solution to a problem is required, companies often find they can save time by modifying a standard software package for their own use. That is what Aetna did when it needed software to offer a new type of life insurance called universal life, which features flexible premiums. That allowed the company to match its competitors' universal life offerings quickly.

With such a variety of software available, companies are finding that writing their own programs from scratch makes about as much sense as drawing their own road maps for their salesmen. "In the past there was very little software available, and people almost routinely designed a customized solution," says John M. Hammitt, vice-president for corporate information management at Pillsbury Co. Now, "companies are forced to do a better job of evaluating the make-or-buy decision."

US corporations brought 50 percent more software packages last year than in the previous year, says International Data Corp., a market researcher. Such purchases, along with software services, now account for more than 8 percent of the average data processing budget, up from 6 percent in 1980, IDC says. Similarly, purchases of personal computer software packages soared 74 percent last year to top $1 billion, according to market researchers at California's Input.

The customer's new attitude toward software is beginning to alter fundamentally the way that computer systems are sold. "We're selling to a much more sophisticated consumer base than in the past," points out Elizabeth M. R. Hall, product manager at Information Science Inc., a Montvale (NJ) software company. From now on, adds Honeywell's Douglas, "Software looms as a much more important factor in [a user's purchase] decision than hardware."

Easy-to-use software

Computing used to be the solitary domain of a priesthood of programmers. The arcane languages that they employed to command their giant mainframe systems were shrouded in such complexity that few laymen could understand, much less control, these behemoths. But the inexpensive yet powerful personal computer is changing all that.

Millions of nontechnical users are now running computers. These new operators, however, refuse even to read an instruction manual, let alone memorize the cryptic commands of the priesthood. Their aversion, coupled with the dramatic slide in the cost of computer power, has created a revolution in the way software is written. Today, making a program simple to control and easy to use is just as important as what the program actually does. Says Jeanne M. Baccash, a software engineer at American Telephone & Telegraph Co.'s Bell Laboratories: "There's a whole new thrust to reach a market that doesn't know or care what it means to 'boot a system' [start a computer]."

These efforts to make software simpler are crucial if the information processing industry is to continue its fast growth. "Rapidly advancing technology has left consumers, trainers, and computer salesmen behind," says Terry L. Opdendyk, president of VisiCorp. "How to spread the word of how to use the products is the key limiting factor [to the industry's growth]."

To make computers easier for novices to use, the industry is concentrating on development work in three key areas:

Man-machine communications
To reduce the number of commands that a user must memorize and type into a computer to get it to work, the latest software enables users to communicate with the machine in new ways. Some of these techniques are relatively simple, such as establishing menus to show users what commands are

available to choose from. The most elaborate methods, used on such machines as Apple Computer Inc.'s new Macintosh, replace commands with an array of tiny icons or pictures—for example, a file folder to indicate filing. To tell the computer what to do, users point to the appropriate picture on the screen by moving a pointing device called a "mouse" over a desktop.

Data exchange
Early programs were built to accomplish tasks, such as calculating financial forecasts, with only the data located within the user's own computer. But users have grown more sophisticated and now want to get data stored in other machines—say, the corporation's central mainframe system. This ability is starting to show up in programs such as Informatics General Corp.'s Answer/DB, which extracts selected data from a mainframe computer in a spreadsheet format for immediate use.

Artificial intelligence
This embryonic method of programming, which enables software to mimic human thought more closely, might one day be the easiest approach to use. It is starting to show up in programs such as Artificial Intelligence Corp.'s Intellect and Microrim Corp.'s CLIO, which let a user ask a computer for data with English sentences rather than esoteric commands.

Most programmers are ill-equipped to figure out how office workers and other nontechnical users best handle computers. So companies are bringing in professionals from disciplines as far afield as education and psychology to help in the design of man-to-machine communications. AT&T, for example, had a staff of psychologists survey about 400 computers users to help it decide how to add commands to its Unix operating system.

Designing easier-to-use software also requires new development techniques. For instance, International Business Machines Corp. now tests its new software in "usability labs." In these labs, volunteers try to use IBM software while researchers with video cameras watch from behind one-way mirrors. "In the long haul, those that have easy-to-understand software will be the successful companies," says John E. Steuri, general manager of IBM's Information Services Business Unit.

A lot of discussion is going on in research circles over how best to use menus, pictograms, and other techniques to help both computer novices and experts. The type of coaching that a neophyte computer user requires becomes annoying once the user has learned to use the software. So designers want "to make sure the user is not presented with extraneous information when he has to decide what to do next," says Brian K. Reid, a Stanford University electrical engineering professor. Micropro International Corp., for one, is trying to solve this problem by using a time-delay activator. If a user types in a command less than two seconds after being asked for it, the program assumes the user is proficient and skips over the menu listing of options. Users who take longer are given the benefit of directions from a menu.

Researchers are also struggling with the problem of how to standardize commands. Each computer program now uses different commands to accomplish the same function. An early attempt to solve this problem is the "environment" or "windowing" software packages for personal computers that are now coming to market. These packages—products such as Microsoft Corp's Windows and VisiCorp's Visi On—divide the computer screen into segments, each of which shows a different task. Common functions are performed the same way in each window.

The large amount of this kind of innovative software now being developed to make personal computers easier to use is also forcing the mainframe software companies to follow suit. "Mainframe vendors have been forced to make software easier, because users have been spoiled by micros," says David Ferris, an industry consultant.

For example, Information Builders Inc. has developed a version of its Focus data base management software that can run on an IBM Personal Computer. But the PC version, called PC-Focus, "has more features than the mainframe product because after using a personal computer customers have come to expect more," says Gerald D. Cohen, president of the New York company. One of the PC-Focus features is called TableTalk. Each time a user moves from one step to another in the course of retrieving information from a data base, the computer automatically presents him with a menu of the appropriate choices.

Another software challenge is to find simpler ways to exchange data between programs. In most cases, swapping data between a large mainframe and a personal computer is difficult because the two units use different formats for storing their data and different commands for retrieving them. The problem is similar to that of a person who is trying to communicate with 10 people, each of whom speaks a different language, explains Robert J. Spinrad, director of systems technology at Xerox Corp. "I could either learn all nine other languages," he says, "or we could all learn a common one—say, Latin."

But software vendors are not waiting until everyone in the industry agrees on what common language all computers should learn. In the past year a host of vendors, including Cullinet Software, Informatics General, and Cincom Systems, have announced products that link personal computers to large mainframes. Such capabilities will be a requisite for any software in the marketplace, says President Frank H. Dodge of McCormack & Dodge Corp. "If mainframe software doesn't allow that link," he adds, "it's going to be pushed aside pretty rapidly."

Some experts think the ultimate mechanism for exchanging data is the use of sophisticated file-management software, often called data base management systems. These large programs index information and then store it in such a way that it can be retrieved using a variety of names—much the way a library card catalog lists the same book by author, subject, and title. By providing this uniform filing structure, a data base system simplifies the exchange of data.

What may soon make computers even easier to use is artificial intelligence (AI) software. Some companies are already working on AI programs that will eventually be able to remember an individual's habits in using the computer. The first AI applications from Microsoft—expected within the next year—will be likely to use these rudimentary pattern-recognition techniques in tutorial programs that teach novices how to operate software. These programs will adjust the level of tutorial difficulty by determining the proficiency of the student running the program. Says Microsoft Chairman William H. Gates: "Just as humans take actions based on past experience without having to be told again and again, so will software."

The Japanese push

Japanese computer makers, virtually unknown a decade ago, are now household names—Hitachi, Fujitsu, NEC, Toshiba, and Mitsubishi. As the second-largest group of computer companies in the world—after the US—they exported $2.7 billion worth of computers last year. But in their single-minded determination to build and export hardware that is faster and less expensive, the Japanese have given short shrift to software development. "Hardware manufacturers have been lazy about developing software," acknowledges Hisao Ishihara, managing director of the Japan Software Industry Assn. But now, he points out, "that is suddenly changing."

Faced with the twin problems of rapidly falling hardware prices and the growing percentage of computer budgets being spent on software, Japanese hardware makers are scrambling to shift their resources into software development. Three years ago, for example, Hitachi Ltd's Computer Division spent just 10 percent of its research and development budget on software. This year, even though software will account for about 10 percent of its computer revenues, Hitachi will spend 30 percent of its R&D money on software. "These days, more and more of the value added in information processing comes from the software," says Toshimitsu Kaihatsu, head of Toshiba Corp.'s Software Management Dept.

The Japanese are also keenly aware that to continue to be competitive internationally in computers they must now start developing world-class software. "We have to sell more overseas if we want to recover our software development costs," comments Shoichi Ninomiya, general manager of Fujitsu Ltd.'s Information Processing Group. His company now spends close to $100 million a year—more than one-third of its R&D budget—to develop software for its mainframe computers. But almost all of that software is now sold to Japanese customers.

Until now, software has been the biggest handicap the Japanese have had in selling their equipment abroad—especially in the office and personal computer markets. This point was driven home in 1982 when International Business Machines Corp. sued Hitachi, one of the largest Japanese exporters of computers, charging the company with copying IBM software and reselling it with Hitachi's machines. Hitachi agreed in an out-of-court settlement to pay IBM between $2 million and $4 million a month in software license fees and agreed to let the US computer giant inspect all new Hitachi products before they go on the market to ensure that they do not infringe any IBM copyrights. To avoid a similar lawsuit, Fujitsu has also agreed to pay IBM many millions of dollars and promised not to copy the US company's software, although it would not disclose the details of its agreement.

Because of language and cultural differences, software written in Japan is obviously difficult to export. Not only must instruction manuals be translated, but often the programs have to be completely rewritten. Japanese accounting rules, for example, are different from those in the US, so accounting software written in Japan is useless in the US. This shortage of programs has severely limited the export potential of Japanese home and personal computers.

Japan's software industry is also held back by the preference of Japanese buyers for custom software rather than the standard software products that US customers are increasingly buying. As a result, an independent Japanese software industry has been slow to develop. And the computer manufacturers' efforts to develop all the software themselves have been crimped by the large sums they must invest to improve methods of getting information into the computer and processing it, using Japanese *kanji* characters.

But it would be dangerous to write off the Japanese as competitors in world-class software. "The Japanese are as dedicated to computers as they were to autos and shipbuilding," says John P. Imlay Jr, chairman of Management Science America Inc. "If US companies do not innovate and supply quality products and customer support, the Japanese could make inroads."

Already, Japan has a big chunk of the US market for the simplest type of software, that written for video games. Cultural differences are not a barrier to the export of such hit games as Pac-Man. The fast-growing software markets for engineering and scientific applications—which manipulate schematic drawings and numbers rather than words—are other areas where Japanese software could be sold overseas with little modification.

The Japanese have also demonstrated their technical ability by creating powerful supercomputer software and sophisticated banking and airline reservation systems. "People are misreading the capability of the Japanese when they say Japanese can't build good software," maintains Joseph C. Berston, president of Comstute Inc., a software consulting firm based in Japan.

Indeed, Japanese companies may actually have some advantages over their US rivals. Because of the legendary thoroughness of Japanese workers, "the finished product here is better, more reliable, and easier to maintain," says consultant Berston. Labor costs are also lower for Japanese software makers. Well-educated, highly disciplined Japanese programmers are paid an average of about $10,000 a year. Their US counterparts can expect a starting salary twice that—from $19,000 to $25,000 according to a survey by Robert Half International Inc., a personnel agency.

In addition to that salary differential, the Japanese claim their programmers are 10 to 15 percent more productive than their US counterparts because of Japanese investments in program development aids. To widen that margin, they are now building software factories that give their programmers access to even more sophisticated tools. Toshiba recently completed a factory that employs 3000 software engineers to develop industrial software. Now the company is building a second software factory that will employ 2000 more programmers. "To overcome Japan's language problem and compete with the US," says Toshiba's Kaihatsu, "we have to have productivity double that of the US."

NEC Corp., which already spends a significant portion of its annual $400 million software budget on productivity tools, says it will use its productivity and quality advantages to crack the US market. It has hired US software engineers to analyze the needs of US computer users. The resulting lists of requirements are then fed into computers at NEC's Japanese software factories, where programmers write the software. In this fashion, NEC has already started developing commonly used business applications, says Yukio Mizuno, an NEC vice-president.

Perhaps the best illustration of the importance the Japanese now attach to software is the move by the Ministry of International Trade & Industry (MITI) to back the industry. MITI has set up several research laboratories to work on software, including a lab that is developing the so-called fifth-generation computer and software. That lab is budgeted to receive $23 million in MITI support this year. The ministry is also giving low-interest loans and tax breaks to software developers.

MITI hopes to boost its fledgling software industry further with a proposed revision in the copyright law. The ministry is pushing to allow Japanese companies to save money and programming time by making it legal for them to copy portions of existing software products without the permission of the original developers. Industry observers say the law, if passed, would help the Japanese leapfrog US software companies by enabling them to copy popular US programs and incorporate them into Japanese software products.

But even with MITI's help, it will be a long time before Japan's software industry catches up with its US competitors—if it ever does. In 1982, the latest year for which figures are available, software sales in Japan were only $1.4 billion—just one-quarter of the total US software sales that year. Although these sales are growing by 25 percent a year, Japan's software companies complain that their domestic clients are still far more willing to spend money on new hardware than to invest in software. But while their progress may be slow, Japanese computer companies are confident they will be a strong future force in the booming worldwide software business. "We don't want to copy IBM," says Ishihara of the Software Industry Assn. "We want to beat IBM."

Chips: the US versus Japan

Charles H. Ferguson

While the Japanese may not have caught up with the US in software, they are ahead on key items of hardware. The author, who is with IBM, analyses the reasons why Japan now rivals the US in chip production, arguing that the US microelectronics industry is plagued by an outmoded structure. Urgent action is necessary. From Technology Review, *August–September 1983.*

Suggestions of trouble in the US microelectronics industry have been widely reported. For example, Japan leads the US in manufacturing and selling 64K RAMs (random-access memories)—silicon chips used in computers to store approximately 64,000 pieces of information. And, while four Japanese manufacturers already have prototypes of the more advanced 256K RAMs and two more are close to having prototypes, only two firms in the US have progressed this far. Japanese companies have begun to sell advanced supercomputers, and Japan's Fifth Generation Computer Project, seeking radical technological advances, has no true US counterpart. Efforts such as the Microelectronics and Computer Technology Corp. (MCC), a consortium of US firms, are responding only slowly to these challenges. But what is at stake is more than the profits of a few semiconductor firms; it is the future health of the US economy.

As always, when quantum changes occur in an industry, their significance is hard to appreciate at the time. Developments in electronics have been further obscured by that sector's volatile and complex history. An earlier period of technological change beginning in the 1950s overtook the leading firms of that day, the vacuum-tube manufacuturers. This early transformation of the electronics industry gained momentum in 1956, when AT&T agreed with the Justice Department in a consent decree to license its patents on the transistor in exchange for keeping its telephone monopoly. It was already clear that the transistor could do anything the vacuum tube could, namely amplify electrical signals, such as in radios, and perform mathematical and logical operations in computing devices. But the transistor's long-run potential was not so clear.

Made of "semiconductor" materials that may either conduct or not conduct a current, transistors are far more compact and durable than vacuum tubes, use far less electricity, and produce far less heat. Between 1956 and 1962 new firms such as Shockley, Fairchild, and Texas Instruments developed economical ways to manufacture integrated circuits, in which many transistors are cheaply combined on one silicon chip. "Systems" firms such as IBM, Honeywell, Control Data, and Sperry Univac began to combine large numbers of integrated circuits to make computers. In marked contrast to roomsfull of vacuum tubes, computational devices using integrated circuits could be made inexpensive enough for widespread commercial and military use. A new industry was born, in which the vacuum-tube firms played little part.

Spurred in the beginning by military purchasing policies, the semiconductor industry rapidly gained momentum. Small-scale integration (SSI), with up to about 10 transistors on a single chip, gave way to medium-scale integration (MSI), with up to 100 transistors. By the 1970s large-scale integration (LSI) technology, with thousands of transistors on a chip, made possible the mass production of memory chips to store data and logic chips to perform calculations.

By this time, the US microelectronics industry gained structural stability as US "merchant firms," semiconductor manufacturers selling mass-produced chips, continued to grow and dominate world markets. Their largest customers were the computer systems firms. Others included manufacturers and defense contractors who assembled the standard chips into electronic control devices for robots, telecommunications equipment, and other electronic products.

At the same time, LSI technology made it possible to fabricate powerful custom chips to perform specific functions. Systems firms had a particular need for custom chips, which were often vital to the commercial success of their products. Thus, to protect proprietary information, these firms set up their own research, design, and production operations for custom integrated circuits. For their part, the merchant semiconductor firms found it impractical to develop the special technology required to design and produce short runs of custom circuits. A few tried selling custom LSI circuits but had little success, so these firms concentrated largely on mass production of standard memories, microprocessors, and logic chips.

This structure of the industry did not seem overwhelmingly important at the time. However, it is proving to be poorly suited to producing very large-scale integration (VLSI), developed in the late 1970s, in which hundreds of thousands of transistors are embedded on a single silicon chip. VLSI is not merely another incremental improvement in technology. Rather, with VLSI, single chips produced by semiconductor manufacturers will soon compete in performance and speed with large systems of chips assembled by computer firms. This upsets the traditional division between semiconductor and systems companies. In the future they will increasingly employ the same techniques and compete for the same markets.

For example, the Hewlett-Packard 9000 microcomputer has as its central processing unit (CPU)—the core of a computer—a single VLSI "microprocessor" chip containing 450,000 transistors. Though the HP 9000 costs around $40,000 and sits on a desktop, it competes with larger "minicomputers" that have dozens of electronic chips in their CPUs, are the size of refrigerators, and cost hundreds of thousands of dollars.

That is only the beginning. The HP 9000 and other similar computers can perform 1 million instructions per second and have capabilities that border on those of a large mainframe computer. It is widely conceded that within a few years microcomputers—computers based on single-chip microprocessors—will push further into the mainframe category, performing 3 million calculations per second. Such "micromainframes" will compete with standard mainframes having hundreds or even thousands of individual chips in their CPUs.

The trade-offs possible in designing any computational device such as a computer or machine-tool control mechanism will become increasingly significant. To do a given job, a single custom chip may be designed, a piece of hardware may be constructed of many simpler chips, or a general-purpose microprocessor on a single chip may be specially programmed. Firmware—programming embedded in a chip—provides a compromise between pure software and pure hardware. Furthermore, to be effective, diverse pieces of equipment must be designed to communicate with one another. This requires standardization and compatibility among many products.

Given the division in the US between semiconductor and systems firms, no one firm is in a position to make a whole spectrum of standardized products and to take optimal advantage of these trade-offs. The sole possible exception is IBM, the largest US firm, which manufactures computers, semiconductors, and a wide variety of electronic products. While its size and scope give IBM unique opportunities to exploit advances in microelectronics, the company has proceeded slowly in some important areas. Committed to mainframe technology, it long ignored microprocessors: it uses a microprocessor manufactured by Intel in its Personal Computer. IBM has recently tried to rectify this shortcoming by purchasing 12 percent of Intel and launching a cooperative project with Carnegie-Mellon University involving the design of a 32-bit microprocessor. However, the result of these efforts remains in doubt. Equally important, no other US firm even approaches IBM's size and technical strength.

By contrast, Japanese computer manufacturers such as Hitachi, Fujitsu, and Nippon Electric Co. (NEC) are in a far better position to take advantage of advances in microelectronics. These firms belong to *kieratsu*—large families of companies that include robotics, semiconductor, and other high-technology firms. *Kieratsu* have ready access to relatively inexpensive, long-term financing, frequently through special relationships with banks, so member firms are able to plan large purchases of capital equipment. The *kieratsu* are better prepared than US producers to make optimal decisions regarding technology and strategy, such as whether to sell a given product—a chip, a system, a software package—on the open market or to reserve it for internal use.

While a US firm is producing the world's first micromainframe, the Japanese are well positioned for future competition in this market, and in three to five years US firms may find themselves in difficult straits.

From memory to logic

Gaining a better understanding of how the Japanese have exploited VLSI technology requires taking a closer look at the manufacturing process for microelectronics, especially memory chips. Semiconductor memories first appeared in 1971 with LSI technology. Though they are generally the simplest chips to design, they are used in the largest numbers and represent a very large market. Consequently, equipment for producing and testing memory chips is usually the most automated and advanced. Furthermore, memory circuits are a good proving ground for manufacturing: mass production of simple devices permits firms statistically to analyze defects and correct errors. Design rules—for example, the distance in microns by which transistors must be separated on a chip—are established. A firm that has fine-tuned the design, production, and testing of memory chips will therefore be able more efficiently to manufacture other, more sophisticated logic circuits.

With VLSI there is even more to learn from manufacturing memory chips. At lower densities, circuits could be economically produced without automated equipment. Indeed, fully 80 percent of the circuits sold by US manufacturers are assembled and tested in Third World countries, usually in low-technology plants using semi-skilled, poorly paid workers. Using such manual techniques and cheap labor once conferred a cost advantage on US firms—albeit one not passed along in great measure to the Third World countries that provided the labor—but with VLSI technology, that advantage will no longer hold.

For example, it will no longer be possible manually to produce or test the next generation of memories—the 256K RAMs. One problem is that they are virtually impossible to make defect-free, so manufacturers make them (like some 64K RAMs) with "redundant," or extra, memory cells. All memory cells must be tested and the defective ones removed from the circuit. People alone are simply incapable of testing hundreds of millions of cells a day; this will have to be done with highly automated, computer-controlled equipment.

Thus, VLSI production will require increasingly automated chip-fabricating equipment, robotized machinery to assemble the chips, and automatic testing devices. Ever more advanced computer-aided design equipment will be needed, primarily for more sophisticated logic chips. US semiconductor merchants are relatively inexperienced with this type of equipment and frequently lack sufficient capital to develop or purchase it. Conversely, US manufacturing equipment producers will need increasingly advanced custom circuits with which they are unfamiliar. Both types of firms will need robotics technologies in which the Japanese are currently world leaders. Finally, advanced computer-aided design systems require both systems expertise and detailed knowledge of semiconductor technologies. World-class systems and microelectronics firms must have expertise in all of these areas—systems design, chip making, programming, and capital equipment—as well as sufficient financing. In most of these areas, the *kieratsu* are in a better position than US merchants.

The Japanese appear to be ahead in automating their semiconductor industry, with the probable exception of computer-aided design systems. Japanese robotics are widely agreed to be the best in the world. Already 90 percent of Japanese circuits are assembled with automated equipment. Japanese purchases of such equipment are increasing rapidly: capital expenditures of the 10 leading semiconductor producers rose 86 percent from 1978 to 1979 and 49 percent from 1979 to 1980, the last year for which statistics are publicly available. Furthermore, the increase has been greatest in the most advanced equipment. By 1980 Japan bought 23 percent of the world's VLSI automatic-test equipment and 32 percent of the world's memory automatic-test equipment. And Japan's share of the world sales of automatic-test equipment for memory chips rose from 18 percent in 1975 to 25 percent in 1980.

To be sure, US manufacturers still purchase and sell over half the world's automatic-test equipment, but their share is declining, especially in advanced sectors. Furthermore, those figures take into account only purchases and sales on the open market, and US firms rely much more on the open market than Japanese firms, which are largely supplied from within their *kieratsu*. Japan's success in the market is increasing, despite the fact that the most advanced production technology is frequently not for sale, and certainly not for export. Thus, Japan's technical strength is even greater than its market share indicates.

Not surprisingly, the Japanese are well ahead in memory-chip markets. By 1978 Japan achieved technical parity with US semiconductor manufacturers in 16K RAMs (a later LSI technology), and Japanese manufacturers captured 40 percent of the world market. By 1982 the Japanese had achieved technical superiority in manufacturing memory chips and captured 70 percent of the world market for 64K RAMs, an early VLSI technology. That year they increased their open-market semiconductor sales by 25 percent, while US sales declined by 5 percent. Texas Instruments responded by transferring its memory production operations to Japan; Intel ceased production of 64K RAMs completely.

The next generation of memories, the 256K RAMS, are now out, and the

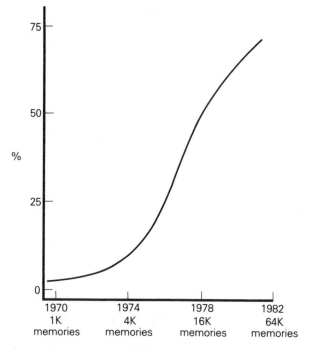

Figure 1.11   Growth in Japanese share of world market for RAMs.

Japanese are expected to dominate the market. As mentioned, four Japanese companies already have prototypes, and two others will have them soon. In the US only Western Electric and Motorola are reportedly in the running. And regardless of the outcome of the 256K RAM competition, Japanese dominance of 64K memory markets has already reduced economies of scale and manufacturing experience in the US industry.

So far, Japanese penetration of world markets has been restricted to commodity circuits that do not require advanced computer-aided design—particularly RAMs—but this is a very large market. Furthermore, there is every reason to believe the Japanese will try to build on their manufacturing experience to sell more sophisticated chips. They have already entered other growing markets, particularly for 16-bit microprocessors, logic circuits of intermediate capability, and gate arrays, mass-produced chips that can be customized at the end of the manufacturing process.

In late 1981, a large US systems firm estimated that Japan possessed a two-to-four-year advantage in so-called "CMOS" (Complementary Metal–Oxide Semiconductor) circuits at VLSI densities. CMOS is one of several technologies available for making integrated circuits; CMOS circuits consume little power and produce little heat. Cooling computers is expensive: mainframes are often fancily packaged and water-cooled, while supercomputers must be cooled by liquid nitrogen. Since CMOS circuits can be air-cooled, they are particularly well suited to inexpensive microelectronics applications. The Japanese lead in CMOS appears to be especially strong in two areas: 32-bit microprocessors used in micromainframes, and 64K "static" RAMs—memories that, unlike simpler "dynamic" RAMs, retain information even without a power supply. Indeed, the Japanese are expected to dominate the 64K static RAM market within about a year.

The advantage of planning

Japan has succeeded in microelectronics because it planned to do so and acted accordingly. The Japanese industry started from behind, producing unsophisticated devices for the consumer electronics market. Ten years ago, Japan exported only modest quantities of semiconductors, mainly to other Asian nations. But government protectionism, combined with Japan's overall economic growth, created sizable domestic markets. The growing importance of microelectronics in computer systems, and in industrial performance generally, led the Japanese to initiate ambitious research, development, and investment programs.

In the mid-1970s the government, along with the *kieratsu*, began preparing for VLSI. Beginning in 1976, the Ministry of International Trade and Industry (MITI) sponsored research and development on VLSI technology to be shared by major Japanese semiconductor firms. Systems and electronics firms such as Fujitsu, Hitachi, and NEC invested heavily and entered foreign LSI markets for the first time. As they began manufacturing LSI circuits such as 16K RAMs, they undertook major VLSI research and development programs. The Ministry of Finance (MOF) encouraged lending to VLSI manufacturers; MITI provided direct financing and coordination.

Nippon Telephone and Telegraph (NTT), the Japanese counterpart of AT&T, also contributed to the effort. Beginning in the late 1970s, NTT made a strategic decision to provide digital communications services for improved computer communications and to give Japanese semiconductor manufacturers a market for VLSI circuits. NTT's usual procedure is to perform research itself, share its results with several manufacturers, and contract out production. AT&T, by contrast, does not allow other US semiconductor firms to use its technology—indeed, Bell has refused merchant semiconductor manufacturers access to its computer-aided design systems.

Already coordinating efforts to excell in coming technologies, major Japanese systems firms and the government have launched the Fifth Generation Computer Project. While its announced goals are probably overly ambitious, the effort is serious and may give Japan world leadership in future high-performance computers. The immediate goals are twofold. First, supercomputers based on Josephson junction circuits, which more rapidly perform the same functions that ordinary integrated circuits do, may be extremely fast machines, though it appears that they will have conventional "architectures," or logic designs. Second, "data-flow" machines may prove able efficiently to employ tens or even thousands of processors. (So far, no one has figured out how to do this, but if it becomes possible tremendous power can be achieved extremely cheaply.) Japan has also begun large-scale work on software, especially artificial intelligence.

While government guidance has aided the Japanese industry, the once comparable force to abet electronics and development in the US—the military—has become a negative factor. Military purchasing helped the microelectronics industry during the 1960s, because when civilian markets were small the military was obliged to finance substantial basic research and development to produce the semiconductors it needed. However, during the 1970s, not only did the military become a much smaller share of the market, but the basic computational power it needed was more widely available. It mainly had to pay for its specialized needs, such as chips more resistant to radiation and temperature extremes. Although the Defense Advanced Research Agency finances some basic VLSI research, a large fraction of military-sponsored research directly on semiconductor circuitry is commercially useless.

Consider the Very High Speed Integrated Circuit (VHSIC) program, through which the Defense Department is allocating $40 million a year to develop advanced semiconductors for miliary applications. Such circuits typically have very different requirements from those of most commercial applications. For example, many military circuits employ CMOS technology, but require sapphire backings for additional speed. Such circuits cost too much for widespread commercial use. The firms doing this VHSIC research are defense contractors; commercial semiconductor manufacturers could benefit little from the program.

US microelectronics has also been damaged by the division between semiconductor and systems firms. One might expect IBM to possess the same strengths as the Japanese *kieratsu* and therefore to be able to bolster the position of US microelectronics, but the matter is not so clear. Unlike the Japanese *kieratsu*, IBM is alone in the US industry—more than eight times the size of its nearest competitor. As already noted, it has been biased towards conservative technologies, principally commercial general-purpose mainframes. And by producing its most advanced circuits itself but purchasing relatively outdated circuits from merchant firms, IBM has tended to reduce the technical sophistication and financial stability of open-market semiconductor manufacturers. IBM itself produces the complex microelectronic circuits and packaging most critical for its mainframe computers. By contrast, memory chips are important mainly because they are needed in large numbers. IBM produces a minimum of these and makes irregular purchases on the open market to meet peaks in demand.

Several systems firms are clearly interested in increasing their semiconductor capacity so they can make the full spectrum of choices and trade-offs available to IBM and the large Japanese companies. Digital Equipment Corp.

(DEC) and several others have substantial computer-aided VLSI design programs underway. Hewlett-Packard has developed some advanced production equipment and is making its 32-bit microprocessor for the HP 9000. Honeywell owns a semiconductor firm, Synertek, with which it has recently undertaken substantial joint efforts. And the Microelectronics and Computer Technology Corp (MCC) research collaborative includes both systems and semiconductor firms.

Nonetheless, the future of the major systems firms other than IBM is not bright. Their recent financial performance has been poor. None possesses semiconductor production operations more than one-twentieth the size of IBM's, and they generally lack the capital and expertise to start making VLSI devices on an efficient scale. Moreover, there are few likely semiconductor firms for them to buy. Fairchild has already been purchased by Schlumberger, a diversified French firm; Exxon owns Zilog; IBM has in effect staked a claim on Intel. Motorola and National Semiconductor remain independent, but they are basically too big for systems firms other than IBM to buy. Texas Instruments, which will lose hundreds of millions of dollars this year, can already be counted a victim of the VLSI revolution.

Furthermore, systems firms are wary of depending on independent semiconductor merchants for custom circuits critical to their systems. The chip maker's interest lies in its spectrum of customers, and no individual systems firm can assume it is the most important. Systems firms concerned about disclosing proprietary information prefer to design and produce their own chips. (Indeed, the security issue may significantly reduce Japanese penetration of US custom logic markets: several Japanese firms are widely and correctly regarded as unacceptable security risks by domestic systems firms.)

Thus, systems firms in need of a custom chip generally have two choices. They can make the chip themselves in relatively inefficient facilities or simply forgo the ideal custom design and use standard commercial chips.

For their part, US semiconductor firms lack the financial resources for adequate research, development, and investment. Part of the problem is that the cost of capital equipment has been rising precipitously. Ten years ago a semiconductor factory could be set up for $1 million: today that price buys only one electron-beam etching machine, and a state-of-the-art fabrication plant costs $20 million or more. Through their *kieratsu*, the Japanese semiconductor firms have access to the requisite amounts of capital at relatively stable, low interest rates. US interest rates are higher and less predictable. Furthermore, although US semiconductor merchants grew quite large during the 1970s, their profits have been squeezed by the Japanese, so that today they have difficulty securing capital to invest at the requisite scale. And US producers of semiconductor manufacturing equipment are in a similar plight.

The need for government assistance

Current US efforts to cope with the VLSI revolution are inadequate. A number of universities—principally MIT, Stanford, Berkeley, Carnegie-Mellon, Cornell, Caltech, and North Carolina—have research programs in VSLI systems financed by government and industry. These are of high quality, but have not been adequately financed.

The Semiconductor Research Cooperative, a facility established primarily by semiconductor firms, plans to spend several million dollars a year on research—far below what is necessary for competitive success. MCC, a larger research collaborative including both systems and semiconductor firms (but not IBM), will have a budget of $50 million per year, but its results will be available only to members, and it has yet to begin operations.

A systematic effort is required to understand and act upon the enormous implications of VLSI technology for the future of the US economy. Policies for the microelectronics industry must be coordinated by the government yet avoid the rigidity often introduced by bureaucracy. A small executive agency, representing all segments of the industry, universities, and labor, could be responsible for overall planning and allocating money. But detailed implementation should be carried out in a decentralized fashion by other government agencies, the educational system, and the private sector.

The following broad measures are clearly necessary:

1 The cost of efficient manufacturing operations appears to double every three to five years, and the industry's own resources will probably prove inadequate to meet these costs. The government should provide tax incentives for research and capital spending and should stimulate demand for advanced products—through purchasing for its own offices, for example.

2 Research and development for advanced microelectronics—apart from military technology—in universities, government laboratories, and industry needs additional financing. The government should award grants of perhaps $10 million a year to each of 20 or 25 universities and laboratories to strengthen existing microelectronics research groups and establish new ones. The government should also consider establishing national microelectronics laboratories that could pursue long-term research unattractive to other groups because of costs or uncertainty.

3 The government should encourage more joint ventures among companies—not just for basic research, as MCC is doing, but for product development. For 25 separate firms to develop the same generation of memory chips is hardly efficient. A few joint ventures, each composed of a number of individual companies, would achieve the same results more effectively and with less total cost.

US antitrust law poses a difficulty not encountered by Japanese firms, which routinely enter such joint research and development ventures. The Justice Department has liberalized regulations for joint ventures somewhat, but this effort should be carried further. For example, enterprises should be given unambiguous information at the start as to whether a joint venture is legal. In some cases, the government should promote these ventures by giving them research and development grants. To ensure that the entire industry benefits, the government might require the group to license technology to other firms.

4 Government must also assist in regulation and standardization. Today one manufacturer's products frequently cannot "talk" with another's. This menagerie of incompatible equipment wastes efforts and curtails competition: once a customer has bought one manufacturer's system, adding another company's products is hard. Establishing joint research ventures would be a good start toward making electronic products compatible, but the government should also set standards—not an easy task.

5 Education requires direct government attention. While universities in nations such as Japan and Germany are graduating increasing numbers of electrical engineers, the US has a notorious shortfall. The government must provide substantial financial support for faculty and students working on microelectronics theory, technology, and policy.

6 At the same time that there is a shortage of engineers, many blue-collar and clerical employees are losing their jobs. The government must address this tremendous waste of human resources. Clerical and industrial workers whose jobs are automated will require employment security and retraining. The government should also increase financial support for electronics-related career training in secondary schools and colleges.

These measures will not come easily. The field of microelectronics is complex, and the US has little experience in implementing a coherent technology policy comparable to that of MITI. Powerful groups such as the Department of Defense may find their interests threatened and may therefore lobby against an effective policy; even commercial firms may balk at incurring the costs of fundamental change. But the economy will suffer if the US does not make a serious attempt to confront these problems. VLSI is here to stay.

Guide to Further Reading

Hardware

Robert M. White, "Disk-Storage Technology," *Scientific American*, August, 1980.
Stephen E. Levinson and Mark Y. Liberman, "Speech Recognition by Computer," *Scientific American*, April, 1981.
John S. Mayo, "The Power of Microelectronics," *Technology Review*, January, 1981. On VLSI.
Gene Bylinsky, "And Now, Chips That Can See," *Fortune*, August 10, 1981. On CCDs or imager chips.
T. M. Cannon and B. R. Hunt, "Image Processing by Computer," *Scientific American*, October, 1981.
Ronald D. Levine, "Supercomputers," *Scientific American*, January, 1982.
Gene Bylinsky, "Intel's Biggest Shrinking Job Yet," *Fortune*, May 3, 1982. On the Intel 432, a so-called micromainframe.
Gene Bylinsky, "The Chip That Erases Repair Bills," *Fortune*, October 18, 1982. On memory chips known as EEPROMs (electrically erasable, programmable, read-only memory).
Eitan Abraham, Colin T. Seaton and S. Desmond Smith, "The Optical Computer," *Scientific American*, February, 1983. On computers based on beams of light rather than electric currents.
Jeff Hecht, "Processing Signals the Optical Way," *High Technology*, October 1983.
David A. Patterson, "Microprogramming," *Scientific American*, March, 1983. On how chips can be made with a low-level set of instructions built into them—so-called "firmware."
Paul Kinnucan, "Machines That See," *High Technology*, April 1983. On vision systems.
James B. Angell, Stephen C. Terry and Philip W. Barth, "Silicon Micromechanical Devices," *Scientific American*, April, 1983.
Sydney B. Newell, Aart J. de Geus and Ronald A. Rohrer, "Design Automation for Integrated Circuits," *Science*, April 29, 1983.
Gene Bylinsky, "The Next Battle in Memory Chips," *Fortune*, May 16, 1983. On the 256K RAM.

"The Technology That Will Create Tomorrow's Superchip," *Business Week*, May 23, 1983. On CMOS (complementary metal-oxide semiconductor) technology, the chip production process.

Albert J. Blodgett Jr, "Microelectronic Packaging," *Scientific American*, July, 1983. On new methods of housing and cooling chips.

Amar Gupta and Hoo-min D. Toong (eds), *Advanced Microprocessors* (IEEE Press, NY, 1983).

Cary Lu, "Microcomputers: the Second Wave," *High Technology*, September–October, 1983.

Gene Bylinsky, "The Comeback of the Chip of the Future," *Fortune*, October 17, 1983. A better account of CMOS.

Dwight B. Davis, "Supermicros Muscle Into Mini Markets," *High Technology*, December, 1983. On supermicrocomputers, the intermediate computers with powerful, standard chips at their heart which can use a variety of software.

"Machines That Can See," *Business Week*, January 9, 1984. On CCDs or vision systems.

Tom Alexander, "Reinventing the Computer," *Fortune*, March 5, 1984. On the faster, smarter technology of parallel computing used in big supercomputers.

Tom Alexander, "Computing with Light at Lightning Speeds," *Fortune*, July 23, 1984. On optical processing.

Electronics had a section on parallel processing techniques on June 16, 1983.

Jonathan B. Tucker, "Computer Graphics Achieves New Realism," *High Technology*, June 1984.

Speculative pieces on so-called "biochips" include:

Stephanie Yanchinski, "And Now—the Biochip," *New Scientist*, January 14 and 28, 1982.

Andrew Pollack, "Computers from Bacteria," *New York Times*, February 18, 1982.

"'Biochips' May Be the Brains of Tomorrow's Computers," *Business Week*, November 28, 1983.

Jonathan B. Tucker, "Biochips: Can Molecules Compute?" *High Technology*, February, 1984.

Software

Bro Uttal, "Famous Victories in Personal Software," *Fortune*, May 2, 1983.

"A Software Whiz Kid Goes Retail," *Business Week*, May 9, 1983.

"A Software Hard Sell," *Time*, June 20, 1983.

"How 'Searchers' Help Find the Friendliest Software," *Business Week*, June 27, 1983.

"Capturing the World of Software," *Time*, July 18, 1983.

"Mass Marketers Are Reprogramming the Software Business," *Business Week*, August 29, 1983.

"A Fierce Battle Brews Over the Simplest Software Yet," *Business Week*, November 21, 1983.

"Windows on the World," *Time*, November 28, 1983.

"Software for All Seasons," *Time*, December 19, 1983.

Bro Uttal, "The Best Software for Executives," *Fortune*, December 26, 1983.

Stratford P. Sherman, "Microsoft's Drive to Dominate Software," *Fortune*, January 23, 1984.

"The Wizard Inside the Machines," *Time* cover story, April 16, 1984.
The *Financial Times* (London) had useful features/surveys on software on September 21 and 30 and October 10, 1983, and on January 10 and October 31, 1984.

Software piracy is covered in:
David A. Blumenthal, "Lifeforms, Computer Programs and the Pursuit of a Patent," *Technology Review*, February–March, 1983.
"Software Rentals: Piracy is the Hot New Issue," *Business Week*, August 7, 1983.

The Microelectronics Industry

Ernest Braun and Stuart Macdonald, *Revolution in Miniature: the History and Impact of Semiconductor Electronics* (Cambridge University Press, Cambridge, England and New York, 2nd edn, 1982).
Tracy Kidder, *The Soul of a New Machine* (Little, Brown & Co, NY, 1981 and Penguin, Harmondsworth, 1982). Account of the race to build a new minicomputer at Data General's Westborough, MA, plant.
Katherine Davis Fishman, *The Computer Establishment* (Harper & Row, New York, 1981). How IBM came to dominate the computer industry. (*Time* also had a cover story on IBM, "The Colossus That Works", July 11, 1983. See also "Striking It Rich", *Time*, February 15, 1982 (cover story on Silicon Valley).
Everett M. Rogers and Judith K. Larsen, *Silicon Valley Fever* (Basic Books, New York, 1984). Account of Valley culture and lives of electronic engineers.
Simon Caulkin, "The Silicon Valley multiplier," *Management Today*, October, 1982.
Bro Uttal, "The Man Who Markets Silicon Valley," *Fortune*, December 13, 1982. Profile of Regis McKenna, PR man to Silicon Valley companies.
Laura Mazur, "Boston's High-Tech Highway," *Management Today*, January, 1984. About the renaissance of Route 128.
Bro Uttal, "The Coming Glut of Semiconductors," *Fortune*, March 19, 1984.
Joseph LaDon, "The Not-So-Clean Business of Making Chips," *Technology Review*, May/June, 1984. Major exposé piece on chemical hazards in chip manufacturing.
"Reshaping the Computer Industry," *Business Week* Special Report, July 16, 1984.

The Silicon Valley "spy" scandals are discussed in:
Christopher Joyce, Ian Anderson and John Lamb, "The Night They Raided Silicon Valley," *New Scientist*, July 1, 1982.
David B. Tinnin, "How IBM Stung Hitachi," *Fortune*, March 7, 1983.

The Japanese

Gene Bylinsky, "The Japanese Chip Challenge", *Fortune*, March 23, 1981.
Justin L. Bloom and Shinsuke Asano, "Tsukuba Science City: Japan Tries Innovation," *Science*, June 12, 1981.
Gene Bylinsky, "Japan's Ominous Chip Victory," *Fortune*, December 14, 1981.
John Walsh, "Japan–US Competition: Semiconductors Are the Key," *Science*, February 12, 1982.
Gene Bylinsky, "US Chip Makers Are Back in the Race," *Fortune*, June 28, 1982.
Richard Casement, "Tomorrow's Leaders: A Survey of Japanese Technology," *The Economist*, June 19, 1982.

"A Big Fight Over Tiny Chips", *Time*, August 9, 1982.

Gene Gregory, "Japan Challenges the Computer Giant," *New Scientist*, January 6, 1983.

John Thackray, "The Microchip Mayhem," *Management Today*, May, 1983.

"Chip Wars: the Japanese Threat," *Business Week*, May 23, 1983.

Lewis M. Branscomb, "Improving R&D Productivity: The Federal Role," *Science*, October 14, 1983.

Bro Uttal, "Japan's Latest Assault on Chipmaking," *Fortune*, September 3, 1984.

Britain and Europe

Nicholas Newman, "Britain's Silicon Gulch," *Management Today*, December, 1982. The M4 corridor.

Richard Lambert, "Scotland: a Painful Rebirth," *Financial Times*, March 6, 1984. On Silicon Glen.

Barry Fox, "Ireland Moves From Spuds to Chips," *New Scientist*, October 20, 1983.

Luc Soete and Giovanni Dosi, *Technology and Employment in the Electronics Industry* (Frances Pinter, London, and Dover, NH, 1983).

The *Financial Times* had pieces on ACORN (August 24, 1983); Ferranti (November 2, 1983); the Cambridge area (November 30, 1982); the Thames Valley/M4 corridor (July 19, 1982); and special surveys on "Electronics in Europe" (March 28, 1984) and "The Semiconductor Industry" (September 4, 1984).

The UK government's role in promoting information technology through "Information Technology Year" in 1982 and so on is best dealt with by the former Permanent Secretary at the Department of Industry, Sir Peter Carey, in his speech to the Royal Society of Arts, on February 25, 1981 (*Journal of the RSA*, June, 1981) and by John Lamb, "IT 82 — a Critical Year for Britain," *New Scientist*, January 28, 1982. Also very useful are:

Sir Ieuan Maddock, *Report on the Commercial Exploitation of Defence Technology*, to the Electronics Industry Economic Development Committee of the National Economic Development Office, September, 1982.

A Policy for the UK Information Technology Industry, National Economic Development Office, February 1983.

Reports on information technology from the European Commission, the EC's Esprit project and the French effort in Télématique are dealt with in *Business Week*, May 30, 1983 and July 4, 1983; and the *Financial Times*, January 12, 1982; March 12, 1982; April 30, 1982; November 22, 1982; and January 25, 1983, as well as by K. Hartley, *The Implications of National and European Commercial Policies for the Development of Europe's Technological Industries* (Centre for European Policy Studies, Brussels, 1984).

2 Artificial Intelligence and the Fifth Generation

The Race to Build a Supercomputer

William D. Marbach et al.

Although microchip-based computers are less than ten years old, Japan and the US are now rushing to produce the next or "fifth" generation of computers—machines that exhibit "artificial intelligence" and the so-called "supercomputers." Whoever wins the race will increase their control over the information revolution and thus their geo-political influence. This scene-setting article first appeared in Newsweek, July 4, 1983.

One day in 1981, Michael L. Dertouzos, the director of the renowned Computer Laboratory at the Massachusetts Institute of Technology, received from a colleague just back from Japan a draft of a research paper. It outlined Japanese proposals for long-range research projects in advanced computer science—plans to build revolutionary artificial intelligence computers and supercomputers a thousand times faster than today's machines. To Dertouzos's ears, the document had a familiar ring. "I looked at it and I started panicking, panicking, panicking," he recalls. "I said, 'My God, this is the research charter for my laboratory. These guys have stolen it!'"

The Japanese had in fact taken nothing but the initiative, yet Dertouzos immediately saw the threat: Japan's JIPDEC plan, as it was then called, was a carefully conceived blueprint of the research and engineering needed to leapfrog the US computer industry and destroy its world supremacy. Worse, even though American universities had produced the basic research the Japanese would rely on, American companies were as serenely unaware of danger as the battleships that swung at anchor in Pearl Harbor more than 40 years ago. "My good friends in the US corporations were deeply asleep," Dertouzos says.

Today the battle lines are drawn. Armed with fresh commitments of money and manpower, the US is taking on Japan for control of the advanced technologies that will dominate computing in the late 1980s and the 1990s.

Nor is this simply a struggle for an industry: entire economies will be reshaped by the coming radical changes in information processing. Supercomputer speed is already being used commercially for aircraft design, oil, and mineral exploration, weather forecasting and computer circuit design—all of which require vast amounts of calculation. Supercomputers may soon be put to work in the automobile and shipbuilding industries, and pressed into the service of genetic engineers and economic forecasters.

The supersecret National Security Agency, the cryptographers' "Puzzle Palace," is already a heavy user of today's supercomputers. And the Pentagon's futuristic laser weapons systems based in space will depend on supercomputers. "This assault is far more serious to our future than the automobiles sold from Japan, because the computer is at the root of every major future change," Dertouzos warns. "The Japanese recognize that whoever controls the information revolution has, in effect, some form of increased geopolitical control."

Until now the US has dominated advanced computer technologies: the world's supercomputers have all been American-made. There are 74 in operation, and they are very powerful machines, capable of performing several hundred million operations per second. They are so fast and their electronic circuitry is so dense that giant refrigeration units must pump a freon-gas coolant through the machines just to keep them from melting down.

Yet the current supercomputers are only at the threshold of what computer designers think can be achieved; the next generation of advanced supercomputers will make today's machines look like handheld calculators. "We have problems that would take 500 to 1,000 hours to solve [on today's supercomputers]," says David Nowak, division leader for computational physics at Lawrence Livermore National Laboratory, where a cluster of seven supercomputers—known as "Octopus"—is used for nuclear weapons research. Before the end of the century, computer scientists hope to develop machines that not only crunch numbers at high speed but also exhibit artificial intelligence—computers that can think and reason somewhat like human beings and that can understand information conveyed by sight, speech, and motion.

The question is, which nation's scientists will get there first? The Japanese have announced a two-pronged plan to build advanced computer technologies. One project is the $100 million, eight-year National Superspeed Computer project, which aims at producing machines 1000 times faster than the existing Cray-1 supercomputer built by Cray Research of Minneapolis. The other, the $500 million, 10-year Fifth Generation Computer project, is focussing on artificial intelligence. Both are now being countered by American efforts, including a Pentagon request for up to $1 billion over the next five years for superspeed and artificial intelligence (AI) technologies. Although behind, Great Britain and France have also launched national supercomputer projects.

The great danger for the losers in the race—and the opportunity for the winners—is that whoever builds the next generation of computers will have a huge technological and commercial advantage: these computers will be used for computer and microelectronics design—to build even smarter and more powerful machines. They won't be self-replicating machines, but they will be close. "It takes you a long time to catch up," says computer scientist Raj Reddy of Carnegie-Mellon University, one of the top US computer-research centers. "In some of these areas, that is the difference between a first-rate power and a second-rate power—from an economic point of view and from a security point of view." The losers in the race will fall farther and farther behind.

The leading edge of computer science is still a black art; there are no fixed laws, and the field is highly experimental. That is what worries US scientists about Japan's approach—some success is inevitable. "Because the field is experimental, [the Japanese] will come out with something," says Dertouzos. "It may not be what they wanted, but they'll come up with new architectures, new insights, new design techniques.

To build the computers that will dominate the 1990s, both Japan and the US are depending on the onrushing technological advances in microelectronics. Japan's Fifth Generation project will use faster, denser circuitry to create a new class of superintelligent computers. The 24 projects in the Fifth Generation concentrate on artificial intelligence, a goal of American computer scientists for more than a quarter century. "We are trying to catch up to *you*, and not the other way around," says Tokyo University Professor Tohru Moto-oka, who organized the project for the Japanese government.

Although the Japanese are highly regarded as superb engineers, Japanese computer scientists have often been faulted for failing to develop innovative computer software. But in a break with tradition, Kazuhiro Fuchi, the Fifth Generation project director, has deliberately assembled a young team: "The question was who would adapt most easily to this research," says Fuchi. "Young people have fewer fixed ideas." The project, headquartered in a downtown Tokyo skyscraper, will focus on computer architectures, software, and the symbolic logic necessary to build thinking computers.

In the US, meanwhile, three huge new programs in electronics and computing are getting under way.

1 The Microelectronics and Computer Technology Corp. (MCC)
Last year William Norris, the founder and chairman of Control Data Corp., convened a meeting of top computer and semiconductor industry executives at the Grenelefe Golf and Tennis resort in Orlando, Florida, to discuss setting up a hugh research cooperative. The companies agreed to form a nonprofit joint venture so that they could pool their resources and share the cost of doing long-range research. Twelve major US corporations joined the new organization, including Honeywell, Motorola, RCA, and Control Data.

MCC is a bold departure from the way research is usually done in American universities and corporations, and it will probe the limits of the nation's antitrust laws. The venture partly follows the Japanese model: the companies will donate scientists and researchers to MCC, loaning them for up to four years. The corporate co-owners will also put up the money to fund MCC's research, in return for the rights to use the results. Whether the scheme will work is an open question. The 12 companies in MCC are competitors in fast-moving, high-technology markets, and ordinarily they jealously guard any technological edge they gain. In fact, many top US firms—Cray Research, Texas Instruments, Intel, and others—chose to stay out of MCC. "That's not our style," says John A. Rollwagen, chairman of Cray Research, an 11-year old company proud of its entrepreneurial creed. "We don't want to participate." The biggest market force of all—IBM—reportedly stayed out of MCC because it feared antitrust action against it if it joined.

So far, however, the creation of MCC has not provoked any such suits. In January San Francisco antitrust lawyer Joseph M. Alioto did write to the chief executives of the companies that were about to form MCC: "In my opinion, your contemplated conduct is an unequivocal combination in violation of the antitrust laws of the United States." But the threat did not deter MCC's co-owners and, for the time being at least, the Justice Department has allowed the MCC plan to stand.

To run the new corporation, MCC's directors chose retired Admiral Bobby Ray Inman, former director of the National Security Agency and former deputy director of the CIA. Inman is widely respected for his managerial abilities and is an adept politician besides. "The day they picked Bob Inman to head MCC," says George W. Keywork II, Ronald Reagan's top science adviser, "any concern about its success diminished in my mind."

Over the past five months, Inman orchestrated a competition among 57 cities for the MCC headquarters; the winner was Austin, Texas, after private donors, the state, and universities put together a generous package of incentives.

The consortium will have a budget of about $75 million a year and a staff of 250. Its first projects include programs in semiconductor packaging and interconnect technology, advanced software engineering and computer-aided design and manufacturing (CAD/CAM) for the electronics and computer industries. Most ambitious is a 10-year program aimed at breakthroughs in computer architecture, software, and artificial intelligence. MCC will own the licenses and patents to the technologies; the manufacturing and marketing will be left to the companies that sponsor the projects. MCC will give them a competitive edge on the market—they will have exclusive rights for three years before the research is published and other firms are allowed to buy licenses.

2 The Semiconductor Research Corp.

Over the past three years Japan has captured a vital segment of the world semiconductor industry, the market for so-called RAM (random access memory) chips, a technology invented in America. Japan now supplies 70 percent of all 64K RAMs sold, and it appears that, as the next generation of memory chips, the 256K RAMs, is being readied for market, the Japanese semiconductor companies are threatening to take a big share of sales. It makes for a grim reminder: a decade ago, before an all-out government project to build up the industry, Japanese semiconductor firms lagged far behind American and European chip companies.

Last year 13 US chip manufacturers and computer companies banded together to form a nonprofit research consortium, the Semiconductor Research Corp. (SRC), to share the spiraling costs of advanced research and development. SRC's founders include Control Data Corp., Digital Equipment Corp., Hewlett-Packard, IBM, Intel and Motorola. Unlike MCC, however, SRC, which is headquartered at Research Triangle Park in North Carolina, does not carry out its own research. Instead, it sponsors research at universities. SRC is spending $12 million this year and has allocated $30 million next year. The goal: "To assure long-term survival in the market," says Larry Sumney, SRC's executive director.

3 DARPA

The Pentagon's Defense Advanced Research Projects Agency (DARPA) is, more than any other single agency in the world, responsible for the shape of advanced computer science today—and for many technologies now in widespread commercial use. Over the past 20 years, DARPA has poured half a billion dollars into computer research, in the process virtually creating the science of artificial intelligence. The first supercomputer, built in 1964, was a DARPA project. Computer time sharing, a fundamental advance, came out of work sponsored by DARPA; so did packet-switched networks, the workhorses of today's telecommunications data networks. And computer graphics—now used on desktop computers and video-arcade screens as well as in F-16 cockpits—is a DARPA-sponsored invention.

DARPA's next priority is a push for advanced supercomputing and artificial intelligence technologies that may cost as much as $1 billion. DARPA plans to do everything the Japanese have set out to accomplish—and more. Earlier this year DARPA proposed a "Strategic Computing and Survivability" project, which it hopes will lead to a variety of new machines. "We want some architectures that are good for building semantic memories, memories that can hold knowledge," says DARPA's computer director, Robert Kahn. "Other kinds of systems are good for logic processing. We want architectures that can do very rapid signal processing [and] structures that can handle very, very large amounts of data in communications."

Once they are in place, these technologies will make possible an astonishing new breed of weapons and military hardware. Smart robot weapons—drone aircraft, unmanned submarines, and land vehicles—that combine artificial intelligence and high-powered computing can be sent off to do jobs that now involve human risk. "This is a very sexy area to the military, because you can imagine all kinds of neat, interesting things you could send off on their own little missions around the world or even in local combat," says Kahn. The Pentagon will also use the technologies to create artificial intelligence machines that can be used as battlefield advisers and superintelligent computers to coordinate complex weapons systems. An intelligent missile guidance system would have to bring together different technologies—real-time signal processing, numerical calculations, and symbolic processing, all at unimaginably high speeds—in order to make decisions and give advice to human commanders.

While the national security needs are driving supercomputing technologies, there is a growing market for commercial spinoffs. The same technology that can be used to simulate an antitank missile smashing into a heavily armored tank can also be put to work on less martial arts. In Los Angeles, Digital Productions, Inc., is using a Cray supercomputer to produce television commercials for Mattel, rock video segments for Turner Broadcasting, and special effects for Lorimar's "Star Fighter" space epic, which will be released next summer. Instead of shooting a commercial the conventional way—a costly photo session—Digital can create the "pictures" it needs in detail so precise that it's impossible to distinguish between the supercomputer's graphic image and the real photo. For inanimate objects, that is: the Cray-1/S can simulate a car right down to the glint of sunlight on the windshield, but not a human being.

As the commercial market grows, the Japanese believe that their strategy of building supercomputers for the general-purpose market will pay off. During the past year Fujitsu, Hitachi, and NEC have each announced supercomputers faster than the most powerful American machines now on the market, the Cray x-mp and Control Data's Cyber 205. Moreover, all three are designed to use the standard Fortran language, lifted from ordinary mainframe computers, but at much faster speeds. "Although supercomputers right now are mainly being used by specialists," says Takamitsu Tsuchimoto, Fujitsu Corp.'s development manager for supercomputers, "we believe that in the next 5 to 10 years they will be used by a lot of ordinary people, so we wanted to design a machine [they could use] without making great efforts."

Designing a supercomputer is no easy task. America's premier supercomputer designer is Seymour Cray, 57, who designed the Cray-1 and the basic architecture of Control Data's Cyber 205. The Cray x-mp contains a dense pack of 240,000 silicon chips arranged to shorten the distances the electrical signals must travel, thereby decreasing the time it takes to perform an operation; the new Cray x-mps will run at 400 million operations per second. And Japan's superspeed goal is 10 billion operations per second. (An Apple IIe computer contains 31 chips and can execute 500,000 operations per second.) Because of the cost of supercomputers—Cray's x-mp will sell for $11 million—the market will remain limited. Japan's Fifth Generation project, however, plans to build "super personal computers."

These huge increases in raw computer power are just the first step. The most profound changes brought by the new technologies will be the development of reasoning computers that will use superspeed symbol manipulation to simulate human thought. "We are about to see the next explosion, which is the application of computers to reasoning," says Stanford University computer scientist Edward A. Feigenbaum, who is a founder of two artificial intelligence companies, Teknowledge Inc. and IntelliGenetics, and coauthor of a new book on Japan's challenge, *The Fifth Generation*. The field is not new: scientists in the US have been studying artificial intelligence for 25 years, struggling to understand the nature of knowledge and how to represent it in forms adaptable to computer usage.

Limited forms of artificial intelligence are already enjoying commercial success. Digital Equipment Corp. the nation's second largest computer firm, uses an AI program called X-CON to make custom designs of its computer systems. Using a set of more than 2500 rules programmed into the system, X-CON examines a customer's specifications, determines whether all the necessary components are included, then draws a set of diagrams showing the proper spatial relationships among the components.

Scientists studying artificial intelligence have been aided by advances in computer hardware, too. In 1980 a group left MIT's Artificial Intelligence Laboratory to found a company, Symbolics Inc., of Cambridge, Mass., to build computers specially designed to run LISP, a language used to develop artificial intelligence programs; and Xerox has begun to sell similar computers. Symbolics has been selling its machines, the Symbolics 3600, to a broad range of university and industrial research labs. The company has another customer as well: the managers running Japan's Fifth Generation project have bought 10 Symbolics machines—and they have 15 more on order.

In exploring the brave new world of artificial intelligence, computer scientists are concentrating on several important problems. "Knowledge engineers" are building so-called expert systems that can mimic human expertise in a narrowly defined area. The firm of Teknowledge, in Palo Alto, California, for example, has built an expert system for the French Elf Acquitaine Oil Co., a system that will give advice on one of the industry's most costly technical problems—what to do when a drill bit gets stuck thousands of feet below the earth's surface. To build the system, Teknowledge engineers interviewed Elf Acquitaine's top troubleshooter, Jacques-Marie Courté, and then programmed his answers into a computer. The program is, in effect, a computer replica of Courté's expertise; the computer will ask the drilling-rig foremen questions, just as Courté would. Once it gathers the information it needs, the computer will make recommendations by drawing images on the screen and giving suggestions on how to retrieve the bit. Because daily drilling costs are high, Elf Acquitaine may well recover the program's development costs the first time it is used successfully.

General Electric is building a software program that will provide expert advice on repairing locomotives. The Pentagon would like to build artificial intelligence programs that could serve as a pilot's assistant in the cockpit. Stockbrokers and insurance agents may also soon get help: "Some people are beginning to see a gold mine in [building artificial intelligence programs for] financial services," says Patrick Henry Winston, chairman of MIT's Artificial Intelligence Laboratory. And savvy software designers for the personal computer industry are beginning to look at artificial intelligence as the next big wave that will sweep the market.

Despite the successes, there are problems ahead as researchers attempt to move beyond the building of narrowly defined "experts." "There's a lot of hard stuff out there we just don't have the answers to," says Roger Schank, director of Yale University's Artificial Intelligence Laboratory. Computers have always been maddeningly literal machines, subject to the absolute tyranny of the binary codes they use to do their calculations. (The switches are either on or off, simulating ones or zeros, nothing else.) That literal-mindedness can transform a problem that would seem trivial to humans into a nightmare. Consider the simple statement "Mary had a little lamb." For a computer to translate the text into another language—a function scientists are now trying to develop—it would have to sort through what is, by one count, 28 possible interpretations (Mary owned the lamb, Mary ate the lamb, Mary had sexual relations with the lamb, Mary gave birth to the lamb, and on and on).

The kind of understanding humans experience as a "flash of recognition" is also difficult to instill in a computer. The statement "Ronald Reagan is president" carries a number of immediate meanings to a flesh-and-blood American, but a computer would have to rummage through its silicon memory chips in search of dozens of facts—what the word "president" means and biographical facts and details that tell who Ronald Reagan is. And the biggest challenge of all—teaching computers to *learn* to acquire knowledge on their own—is nowhere near being solved.

Mankind has long been enchanted—and frightened—by the prospect of creating machines that think. "I don't see any limitations to artificial intelligence," says Nobel laureate Herbert Simon, professor of computer science and psychology at Carnegie-Mellon. "All the mechanisms for human intelligence are present. Already machines can think just like people—in a limited sense. Man isn't unique in that respect." Moreover, many machines may soon possess sight, touch, hearing (in the form of voice recognition), and speech, thus imitating humans' sensory capacities along with their intellectual ones. But whole areas of the human thought process—volition, emotion, the creative uses of error—still lie well outside a computer's experience. And scientists doing research on artificial intelligence are far from their ultimate goal—a computer-based analog of the human brain. Given the malign intelligence of the supercomputer Hal in Arthur C. Clarke's fantasy *2001: A Space Odyssey*, that may be just as well.

Still, the race to build superintelligent computers—Japan's challenge to the US—will almost certainly push the technology to new levels. "While at the beginning I was angry at Japan because I felt that they were swiping our best ideas, really, on second thoughts, I couldn't blame them," says Dertouzos of MIT. "And I don't blame them today at all. I think they're doing exactly what they should." And the Japanese challenge has at last spurred the US into action. "If we really wake up, I'm very optimistic," says Dertouzos. "We could beat the daylights out of them." No one knows where the competition will ultimately lead. But if the Merlins who are plotting the next computer revolution succeed in building superintelligent computers, Hal or, more welcome, '*Star Wars*' R2-D2, won't seem so farfetched after all.

Inside the Supercomputer

In the last 35 years, as computers have grown immensely more powerful, the basic scheme of their operation—their "architecture," as engineers call it—has never changed. In the late 1940s, John von Neumann, a Hungarian-born mathematician, conceived what was first called the stored-program computer: a central-processing unit—the brain of the computer—that executed its calculations one step at a time, storing each result in its memory before moving on to the next calculation. The primitive vacuum-tube models worked that way, as did the next generation of transistorized computers. With its almost unimaginable speed, the Cray-1 supercomputer still depends on serial processing, its signals shuttling back and forth through the dense mass of 350,000 silicon chips. Even as computer designers use faster and faster microelectronic circuits in the quest to build ever-faster machines, however, the von Neumann architecture has become the "von Neumann bottleneck," a traffic jam that limits the speeds existing computers can attain.

In the race to build the next generation of supercomputers, scientists are experimenting with a variety of designs that will break the von Neumann bottleneck between the processor and memory. These "non-von Neumann architectures" range from machines that will have two or more processors and shared memories to extensive parallel architectures with hundreds of local memories and processors, all executing instructions simultaneously. "Given the state of the art today and the way the physics are formulated, we think parallel processing is clearly the wave of future," says John A. Rollwagen, chairman of Cray Research of Minneapolis, one of the world's leading supercomputer manufacturers. "We and Control Data and every-one else who wants to play the game will have to have a completely different architecture." Where the Cray-1 had only one processor, the Cray

X-MP now coming onto the market has two processors and the Cray-2 due in late 1984 will have four processors. The next machine, the Cray-3, will probably have 16 processors; Seymour Cray, the master designer of supercomputers, is working out final designs for it now. One machine already in use, the HEP supercomputer built by Denelcor, Inc. of Aurora, Colorado, uses four processors to reach speeds up to 40 million instructions per second.

Road map
The more radical solutions to the von Neumann bottleneck involve networks of many more processors and their liberation from the tyranny of a central memory. At the University of Texas at Austin, James Browne has built a small prototype of a parallel-processing machine. The Texas Reconfigurable Array Computer (TRAC), as it is called, has four processors and nine memories and works more like a telephone network than a traditional von Neumann computer. "Instead of processing in a straight line, it looks like a road map where the cities are processors and the roads are the communications links between them, says Browne. "If one processor wants to talk to a certain memory, it can, in effect, dial it up." The scheme allows for much faster speeds, just as the phone network could handle more traffic when direct dial replaced human operators. The trick is to organize and synchronize the communications between the processors and memories.

For their Fifth Generation Computer project, the Japanese are considering a radical departure from the von Neumann architecture, the so-called "dataflow" computer championed for the past 15 years by Jack Dennis at MIT. Dataflow computers will have huge numbers of processors, each with its own memory; and, as in simpler parallel schemes, the computer will have a routing network so that the processors and memories can communicate with each other.

But the dataflow computer will go even further: "The rules about when instructions are executed are different," says Dennis. Conventional computers process a stream of instructions, one after another, in the order the program tells the computer to follow. In dataflow machines, on the other hand, the processing units don't have to go looking for data in memory; they simply do whatever calculation is necessary when a "data packet" arrives. Dennis is now planning to build a dataflow computer with 256 processors and memories, and scientists at the University of Manchester in England are working on a similar machine.

Tree leaves
There is another way to break the von Neumann bottleneck, one that doesn't depend on dataflow concepts. At the University of North Carolina at Chapel Hill, a team of scientists led by Gyula Magó has designed a "binary tree" computer, multiple processors arranged like leaves on a tree with the branches carrying information to and from the processors. IBM scientist John Backus, who invented FORTRAN, the most widely used programming language on mainframe computers, is now working on "function-level programming" languages that can run on computers like Magó's machine. One advantage of Backus's is that the programming instructions and the data are intermingled, making programming easier and providing the mathematical tools that will enable the program to run more quickly.

Before computer manufacturers abandon von Neumann architecture in favour of these more radical designs, however, many problems will have to be solved. Designers must prove that the prototype machines will indeed outperform conventional computers. "People are now dazzled by the prospect that they can have 1,000 chips, 10,000 chips or 1 million chips," says IBM scientist Herbert Schorr. "But the question of how to organize 1 million chips to do anything effectively is still very open." Despite the promise of the radical designs, von Neumann's imprint on the computer world has yet to be erased.

Land of the Rising Fifth Generation

Edward Feigenbaum and Pamela McCorduck

This more detailed description of the enormous Japanese artificial intelligence effort also analyzes its strengths and its weaknesses. It is excerpted from the authors' book, The Fifth Generation: Artificial Intelligence and Japan's Computer Challenge to the World *(Addison-Wesley, Reading MA, 1983) and first appeared in* High Technology, *June 1983. Feigenbaum is Professor of Computer Science at Stanford University; Pamela McCorduck is a journalist.*

The American computer industry has been innovative, vital, and successful. Today we dominate the world's ideas and markets in this most important of modern technologies. But what about tomorrow?

Japanese planners view the computer industry as vital to their nation's economic future and have audaciously made it a national goal to become number one in this industry by the latter half of the 1990s. They aim not only to dominate the traditional forms of the computer industry, but to establish a "knowledge industry" in which knowledge itself will be a saleable commodity like food and oil.

To implement this vision the Japanese have both strategy and tactics. Their strategy is simple and wise: to avoid a head-on confrontation in the marketplace with the currently dominant American firms; instead to look out into the 1990s to find an arena of great economic potential that is currently being overlooked by the more short-sighted and perhaps complacent American firms; to move rapidly now to build major strength in that arena. The tactics are set forth in a major and impressive national plan of the Ministry of International Trade and Industry (MITI) called Fifth Generation Computer Systems.

The Japanese plan is bold and dramatically forward-looking. It is unlikely to be completely successful in the ten-year period. But to view it therefore as "a lot of smoke," as some American industry leaders have done, is a serious mistake. Even partially realized concepts that are superbly engineered can have great economic value, pre-empt the market, and give the Japanese the dominant position they seek.

If we focus our efforts, we should have little trouble dominating the second computer age as we dominate the first. We have a two- or three-year lead; that's large in the world of high technology. But we are squandering our lead at the rate of one day per day.

America needs a national plan of action, a kind of Space Shuttle program for the knowledge systems of the future. The stakes are high. In the trade wars, this may be the crucial challenge.

Here we will describe the Japanese program, followed by some comments on its likely success and problems.

Maybe $1 billion budget

MITI's budget for the Fifth Generation project has been announced as $450 million, over a ten-year period. The first three-year phase allots $45 million for basic research and will be funded entirely by MITI. After this initial phase, heavy budgeting is planned for the years of expensive developmental engineering. In the second and third phases, MITI expects its funding will be matched by the participating companies, bringing the total project budget to approximately $850 million. It's very possible that, if the project is meeting its intermediate targets at the end of the first phase, and if the Japanese economy is strong, the total budget could well escalate to more than $1 billion.

In 1978 MITI assigned the Japanese National Electromechanical Laboratory the task of developing a computer system for the 1990s. To meet this long-range charge, a new *ad hoc* group, the Institute for New Generation Computer Technology, ICOT, was formed in April 1982. This "instant institute" drew together 40 researchers from participating firms including the eight firms that make up the consortium backing ICOT: Fujitsu, Hitachi, Nippon Electric Corp. (NEC), Mitsubishi, Matsushita, Oki, Sharp, and Toshiba. As ICOT's director, Kazuhiro Fuchi, realizes, MITI's plan intends a revolution, and revolutions are not made by the elderly. In his mid-40s, he is the oldest member of the team. The emphasis in selecting the researchers therefore was on potential and on youth—all are under 35 and, in the director's words, "young and excellent." The team was assembled in two weeks, most of them hand-picked by Fuchi, once the project got a go-ahead.

As head of this pioneering effort, Fuchi has already become the subject of minor legends. One story has it that as a young man Fuchi became so perturbed with the way things were being run at his laboratory that in fury and desperation he stalked out and stayed away for a month, coming back only after his supervisor came to his house and pleaded with him to return. Unlike the classical Japanese technological manager who, as he climbs the ladder of authority, gradually loses touch with the technology he manages, Fuchi commands the admiration of his staff for his deep involvement in projects. An energetic, intense, awesomely informed man, he seems, with his 40 modern samurai, set on shattering the copycat stereotype of the Japanese. A former senior member of the Electromechanical Laboratory, he resigned that secure position to place all his chips into the Fifth Generation ring. The enthusiasm he commands stirs his researchers to forget what will be for many a hiatus in their corporate careers. As he said to them at the ICOT's opening, "These will be great years for you. We will all work very hard. If the project fails, I will take the entire responsibility. But, of course, we won't fail."

In parallel with the formation of ICOT, closely allied R&D groups in the company labs have been targeted to track the progress at ICOT and absorb it for proprietary use. To ensure this transfer, researchers at ICOT will rotate back to their companies after three or four years. They will be sent back to their firms routinely, perhaps once a week, to report on progress. With the researchers at the outside labs, some 200 technologists will be directly involved in the project. Many others will serve on numerous steering committees.

Technology targets

The Japanese Fifth Generation project aims to design and produce computer hardware and software for knowledge engineering in a wide range of applications—including expert systems, natural language understanding by machines, and robotics. To accomplish these goals, the Japanese must improve present computing capabilities dramatically, but they must also make major innovations in existing technology that will enable Fifth Generation computers to support very large knowledge bases, allow very fast associative retrievals, perform logical inference operations as fast as current computers perform arithmetic operations, utilize parallelism in program structure and hardware to achieve high speed, and develop a machine–user interface that allows significant use of natural speech and images. In his speech at a Fifth Generation conference, Fuchi coined the term *knowledge information processing* to mean "an extended form of knowledge engineering." He adds, "This, it is thought, will represent the form of information processing in the 90s."

All experts systems built by knowledge engineers to date consist of three main parts. First is the subsystem that "manages" the knowledge base needed for problem solving and understanding. Second is the problem-solving and inference subsystem, which discovers what knowledge is useful and relevant to the problem at hand, and with it constructs—step by step—a line of reasoning leading to the problem solution, the plausible interpretation, or the best hypothesis. Third are the methods of interaction between human and machine, in modes and languages that are "natural" and comfortable for the user. Ordinary human natural language is often preferred, but the stylized notations of some fields like chemistry are also desirable for specific groups of users.

Knowledge-based management, problem solving and inference, and human interaction—these have all been approached in our present expert systems via software innovations, innovations that have pressed traditional von Neumann hardware architectures to their limits. The Fifth Generation plan organizes its work around these three subsystems, but with a critical added dimension: for each component subsystem there is a hardware level and a software level. And between the levels the Japanese designers must define a "language" with which the software and hardware interact.

The knowledge in the knowledge base must first be represented in symbolic form, and in memory structures, that can be used efficiently by the problem-solving and inference subsystem. This representation can take many forms. One of the most common is the *object*, a cluster of attributes that describe a thing. An object is usually associated with other objects by symbolic references (*links*) in the memory. A typical kind of associative network is the taxonomy, known as "The . . . . . . is—a (*hierarchy*)." For example, "The sparrow is—a kind of bird." In this case, both sparrow and bird are objects within the knowledge base. If the knowledge base is informed that "The bird is—a kind of animal that can fly," the knowledge-base management system must automatically propagate the deduction that sparrows can fly. It must also be able to handle the exceptions it is told about, such as flightless birds like ostriches, penguins, and kiwis, and it must also make the distinction that the dodo is both flightless and extinct, but nonetheless has its valid place in this taxonomic knowledge base as a bird.

Another common and useful representation is the *rule*. A rule consists of a collection of statements called the "if" part, and a conclusion or action to be taken called the "then" part. For example, "IF the fog ceiling is below 700 feet and the official weather forecast calls for no clearing within the hour, THEN landing is dangerous, will violate air traffic regulations, and diversion to a neighboring airfield is recommended." To find out if a rule is relevant to the reasoning task at hand, the problem-solving program must scan over the store of "ifs" in the knowledge base. That search can be immense in the size of knowledge base the Japanese plan to make possible. Here again, the knowledge-base management subsystem will be designed to organize the memory in ways that will reduce the amount of processing to be done. Parallel processing capabilities in both the software and hardware levels of the system will also speed associative retrievals.

In the Fifth Generation plan, knowledge will be stored electronically in a large file known as a *relational data base*. The job of automatically updating the knowledge in the file, and of organizing appropriate searches for relevant knowledge, will be performed by the knowledge-base management software. The interaction between the hardware file and the software file manager will be handled by a logical language called a relational algebra.

The Fifth Generation prototype knowledge-base subsystem will handle a modest knowledge base—thousands of rules and thousands of objects—about the size needed for current expert system applications. Each object will be allotted 1000 characters of file storage space (so that, in addition to storing the knowledge that sparrows are birds that can fly, the memory will also be able to hold knowledge pertaining to their size, weight, color, diet, range and habitat, breeding patterns, migratory routes, and so on).

Within the ten-year trajectory of their plan, the Japanese goal is to develop knowledge-base capacity in their systems that will be able to handle tens of thousands of interference rules and one hundred million objects! What could so much knowledge encompass? An American firm interested in the possibilities of representing large bodies of knowledge in computer knowledge bases has estimated that such a memory file could store the entire *Encyclopaedia Britannica*.

Adding reasoning

Knowledge serves as the basis for reasoning by a knowledge information processing system, but it is not sufficient in itself to discover and use lines of reasoning. Piecing together an appropriate line of reasoning that leads to the solution of a problem or the formulation of a body of consultative advice is the job of the inference process and the problem-solving strategy that employs it. Inference processes can be very much of the common-sense sort in which relevant knowledge is simply chained. A syllogism (IF X implies Y and IF Y implies Z, THEN X implies Z) is an example of such an inference process. Inference processes have been studied by logicians and mathematicians for centuries, and many different procedures for inference are known. From this logician's tool kit, artificial intelligence uses routinely only a few devices. Some of these methods allow for reasoning "inexactly" from knowledge that is uncertain. One, a favorite of AI, is resolution, constructed on a foundation of mathematical logic formulated in the 1960s by the logician Allan Robinson. Resolution is subtle, nonintuitive, and especially suited for computer processing.

An inference process is the tool of some problem-solving strategy. For example, the strategy of one kind of problem-solving might be goal-directed backward chaining. One works backwards from a desired set of end results through all the steps that must be taken along the way to ensure that all objectives are met.

The Japanese are betting on resolution as the best inference method for which to target their logic processing hardware. This is an approach that has great credibility in Europe, where a computer programming language called PROLOG (for "PROgramming in LOGic") that overlays resolution-based inference was developed (invented in France, polished in England). The Japanese have chosen PROLOG as the language of interaction between the logic processing hardware and software that implements the various problem-solving strategies (in other words, PROLOG is the machine language of the logic processor).

The Fifth Generation project envisions computer hardware engineered for logic processing, analogous to the well-engineered capabilities of earlier generations of computers to handle arithmetic processing. At present, we speak of computer capabilities in terms of millions of arithmetic operations per second. The initial milestone in the Fifth Generation plan is a one-user PROLOG work station capable of performing one million logical inferences per second (LIPS), where one logical inference equals one step in a syllogistic, or IF/THEN, sequence of reasoning. This machine is intended to be both a prototype for further development and an intermediate product that may be on the market by 1985. The prototype would give an order of magnitude improvement over software-based PROLOG implementations on today's common mainframe computers like the DEC 2060. The final target for this subsystem is extraordinarily ambitious, however. It aims for an inference supercomputer that can perform one hundred million to one billion LIPS. Such incredible speed can be achieved only by the insightful use of a great deal of parallel processing in the computing hardware—a major advance over today's von Neumann architecture.

Most knowledge-based systems are intended to be of assistance to human endeavor; they are almost never intended to be autonomous agents. A human–machine interaction subsystem is therefore a necessity in the Fifth Generation design. The Japanese intend to make this interaction as natural as possible for users in both language and mode of interaction. This means language understanding—the ability to speak directly to the machine—as well as image understanding—the ability to show it pictures.

To realize these objectives across the spectrum of human knowledge and images is one of the most difficult of the long-term goals of artificial intelligence research. But if constraints are applied to the amount of vocabulary and areas of subject matter the subsystem is expected to handle, the problem becomes tractable, though still very difficult. The Japanese have recognized this. Effective processing of the electrical signals that represent speech and pictures, first of all, requires specialized hardware to determine the most basic features of the words and images. But that's only the beginning. Software capable of inducing an understanding of the language being spoken or the image being shown must be developed, and it must be able to use the knowledge base efficiently to create a correct context. (It's much easier to understand what's being said or seen if you know something about the subject matter.)

That's the essence of the Japanese plan: hardware and software for each of three subsystems—knowledge base, problem-solving and inference, and human–machine interaction.

The realization of visionary engineering goals usually requires much time and much money. The Japanese are accustomed to investing both in their major technology projects. The Fifth Generation project is structured over a ten-year period. The first three-year phase is targeted for building the research teams and laboratories, learning the state of the art, forming the concepts that will be needed in the later work, and building hardware and software tools for the later phases. The single-user sequential PROLOG work station is one of these tools. The work station itself will be a prototype of later machines, as will be its problem-solving software. Early expert system prototype applications will be written. Three will be selected from a variety of areas such as medical diagnosis, equipment failure diagnosis and repair, intelligent computer-aided design (CAD) for integrated circuit designers, intelligent CAD for mechanical equipment, and intelligent software production aids.

The second phase, four years, is one of engineering experimentation, prototyping, continuing experiments on significant applications, and the initial experiments on systems integration (making the subsystems work together smoothly). The first thrust at the major problem of parallel processing will be done in these years.

The final phase of three years will be devoted to advanced engineering, building final major engineering prototypes, and further systems integration work. The earlier work on CAD for VSLI will be used at this stage to assist the design of hardware. Experiments with some difficult applications will be attempted in this period. Finally, in the third phase, the results of the R&D will be distilled into a set of production specifications for commercial products.

Advancing allied technologies

Since the Fifth Generation is so far-reaching, it demands dramatic improvements in other technologies that support the main-line knowledge information processing systems (KIPS) goals. Essential to the future of the enterprise, for example, are extremely high-speed processors, capable of processing by orders of magnitude faster than anything now available.

Artificial intelligence made its debut on first-generation machines and has been implemented subsequently on second- and third-generation machines, but not yet on fourth-generation supercomputers.

Some computer scientists argue that this hasn't been necessary because AI programs have been designed to behave like non-von Neumann machines would behave anyway, illustrating the lag between computer structures—the actual machines themselves—and computer concepts—the way the machines are put to use. However, to design a program for a von Neumann machine and intend it to behave in a non-von Neumann fashion seems unnecessarily awkward and eventually limiting to the Japanese.

The Japanese aim for chips with 10 million transistors. Chips in current production carry a few hundred thousand transistors at most. Such processors are being developed in the course of another MITI effort, the SuperSpeed Computing Project, and will be adapted into the Fifth Generation machines. In addition, the Fifth Generation depends on access to knowledge bases in many locations, so its technology will ultimately be fused with the most advanced communications technologies.

The whole area of intelligent interfaces—the ability the machines will have to listen, see, understand, and reply to human users—will require extensive R&D in natural language processing in speech understanding, and in graphics and image understanding. All these have been concerns of artificial intelligence research from virtually its beginning some 25 years ago, and basic research in each of these fields has made reasonable progress. Still, state of the art in each is primitive compared with what the Japanese have in mind.

Because nonexperts will be the largest group of users, natural language processing is one of the most important research goals of the Fifth Generation. Research here will cover speech wave analysis, phonetic and syntactic analysis, semantic analysis, and pragmatic analysis, which derives understanding by extracting themes or foci in a given sentence, detecting focus shifts, and so on. For speech output, sentence generation will also be studied. Text analysis is also considered a part of natural language processing by the Japanese, although they are quite aware that the techniques used for large-scale text analysis are different from the techniques needed to smooth the way for an individual user to talk to his machine.

The Japanese see knowledge systems as a potential solution to the steadily increasing amount of text and documents that must be handled by computer. "Sooner or later, as the problem of extracting useful information becomes more severe, we will have to turn to computing power in order to process these huge amounts of documents at reasonable speed. Our research on intelligent man-machine interface will help to solve this problem," reported a group of Japanese scientists at a Fifth Generation project meeting. Present artificial intelligence research suggests this can be done—in a prototype system, intelligent automatic analysis has been successfully applied to a wire news service in the US, for instance—but the sheer scale of the automatic analysis planned by the Japanese dwarfs any existing systems.

Natural language processing will also be put to use in the development of a highly ambitious machine translation program (initially between English and Japanese) with a vocabulary of 100,000 words. The goal is 90 percent accuracy (the remaining 10 percent to be processed by humans).

All this research in natural language processing will proceed in three stages, beginning with an experimental system, followed by a pilot model implementation stage that is connected with the inference and knowledge-base machines, and concluding with prototype implementations. At that point the machines will be expected to understand continuous human speech with a vocabulary of 50,000 words from a few hundred or more speakers with 95 percent accuracy. The speech understanding system is also expected to be capable of running a voice-activated typewriter and of conducting a dialogue with users by means of synthesized speech in Japanese or English. The machine's capacity to respond intelligently to users, known as its question-answering system, will first be designed to handle queries in the computing field, but it is expected to be a prototype for such systems in many professional fields: in addition to the query system's 5000 or more words of vocabulary, it will have 10,000 or more inference rules.

Interpreting images

Picture and image processing are considered almost as important as language processing, especially as they contribute to computer-aided design and manufacture (CAD/CAM) and to the effective analysis of aerial and satellite images, medical images, and the like. Here again the research will take place in three phases, beginning with an experimental phase to tackle such topics as the hardware architecture of "feature extractors"—for example, to distinguish the boundaries of objects—display generators, and the image data base. The second phase will produce a pilot model, and in the third and final phase a prototype will be built, integrated into the Fifth Generation machine, and applications will be studied. One obvious use is in constructing robots that can see, understand, and act under novel circumstances. The bulk of robotics R&D, however, will be done in a Robotics National Project. Eventually, the image understanding system is expected to store about 100,000 images. In this, as in voice recognition, the Japanese are building upon superb R&D done in the 1970s during their Pattern Information Processing Systems (PIPS) National Project.

In short, the Japanese have studied the results of a quarter-century of AI research and have concluded that many of its areas are ripe for serious methodical—and ultimately sensational—development. They are confident it can be done, and that they're the ones who can do it.

What's wrong?

There are both positive and negative sides the the Japanese project. First, its flaws. The science upon which the audacious Fifth Generation plans are laid lies at the outermost edge and beyond what computer science presently knows. The plan is risky; it contains several "scheduled breakthroughs." There are major scientific and engineering challenges in every aspect, from artificial intelligence through parallel architectures and distributed functions to VLSI design and fabrication.

The project demands early successes to maintain its momentum and funding, and that could be a problem. Conversely, meeting or exceeding the goals of the first three-year period might well propel the Japanese ahead of their timetable, bringing increased support from the participating companies.

Central to the success of the project are the Japanese managers, both governmental and industrial. Generally conservative and risk-averse, they are now to be charged with managing a very ambitious, high-risk project based on technology they hardly understand.

Most of the breakthroughs the Fifth Generation project must achieve are basically innovations in software concepts. The key ideas in the approach to knowledge information processing systems came out of the software world—ideas about the creation, maintenance, and modification of large and complex symbolic data structures in computer memories and the discovery of symbolic lines of reasoning. A quick fix to the problem is to work on the intermediate territory of the so-called firmware—intricate and detailed "programming" of the hardware–switching functions that sit at the bottom of the computing process. This is not a desirable final solution, however, since interpreting and executing the "firmware program" consumes time and slows down the machine. Japanese computer specialists and managers are not, and never have been, comfortable with software—it's intangible to them, and its production is notoriously difficult to manage "on schedule and on budget."

The Japanese lack the experience base in knowledge engineering and expert systems from which to draw as they begin to work out the details of what to build. In addition, the Japanese lack a large corps of university educated computer scientists.

Finally, from the AI viewpoint, two elements of the plan are questionable. First, the priority given to the highspeed logic processor. Are all those millions of LIPS needed? Second, is PROLOG the best choice for the machine language of the logic processor?

In the American engineering experience, few applications have been limited by the number of inference steps per second that could be performed. Rather, limitations in performance arise from limitations in the quantity and quality of knowledge available to the machine (too little and not well refined); the facility with which it can be managed and updated; and the speed with which it can be searched and accessed. Thus, the early focus on the inference subsystem in the Japanese plan, rather than on the knowledge-base subsystem, is puzzling.

The PROLOG language has good features as well as flaws. One is a logical calculus, called first-order predicate calculus, which has certain elegant and universal properties for the representation of knowledge. Its flaw is that the knowledge so represented is often opaque, incomprehensible, and arcane. A second plus for PROLOG is that it solves problems by proving theorems in first-order predicate calculus using computationally fast methods. The user never has to be concerned with the details of the problem-solving process. But PROLOG detractors see this as a serious flaw. The major successes of AI have come from mastering the methods by which knowledge can be used to control the search for solutions in complex problems. The last thing a knowledge engineer wants to do is abdicate control to an "automatic" theorem-proving process that conducts massive searches without step-by-step control exerted by the knowledge base.

Such uncontrolled searches can be extremely time-consuming. The parallelism that can be brought to bear is a mere palliative, because the searches become exponentially more time-consuming as the problem complexity increases. One can't keep up with exponential growth simply by lining up a hundred or a thousand more parallel processors.

What's right

But there is much that is right about the Japanese project also. Their strategy has been preceded by a thorough planning effort. It focuses attention on the right set of issues, properly structured.

Creating the knowledge industry, with hardware, software, and knowledge system applications, is a great bet. Indeed, it is one of the few great bets sitting out there now in the information processing industry, ready for a major push toward exploitation. Of course, the traditional modes of numerical calculation and data processing will continue to develop and prosper. But these will see steady incremental growth, not explosive growth. The exponential growth will be seen in symbolic computation and knowledge-based reasoning by computer.

MITI's key economic insight is correct. For an island trading nation, exports create wealth and in the knowledge industry the value of exports is enhanced by indigenous resources—the intelligence, education, and skill of people. Further, knowledge information processing systems will significantly enhance the productivity of many other industries, thereby indirectly contributing to the value added.

The creating of ICOT, the pooling of talent in a cooperative endeavor, plus the well-coordinated transfer of technology between ICOT and the parallel labs of the firms, seem inspired.

The ten-year planning horizon is excellent. Ten years is a long time in the information processing industry. Ten years ago pocket calculators cost hundreds of dollars, video games were primitive laboratory toys, and the Japanese had yet to produce their first viable microelectronic chip. As we live through it, we tend to underestimate the speed of technological change.

MITI's concern for nurturing the innovative talents of Japanese computer scientists appears well placed. The US and the advanced European nations have become wary of providing the leading technologies upon which Japanese technical achievements have hitherto relied. Trade wars are under way, and blockades are inevitable.

Though solutions to the technological problems posed by the Fifth Generation plan may be hard to achieve, paths to possible solutions abound. The Japanese are rich with excellent engineering talent and have an adequate supply of cutting-edge computer scientists. The mix of talents enables (but does not guarantee) a good chance of success.

What's real?

The Japanese undoubtedly will have a partial success. Managers of the Fifth Generation project have said that it would not disturb them if only 10 percent of the project goals were achieved; others have remarked that the ten-year planning horizon should not be taken too seriously; that the project goals are so important that an extension over another half or full decade would not be unreasonable.

Partially realized concepts that are superbly engineered can have great utility and be of great economic benefit. At the very least, a partial success can pre-empt the area and make it not worthwhile for others to enter playing catch up. The first 20 percent of the technical achievement may skim off 80 percent of the potential economic gain. If true, firms in the American industry might never find it in their economic interest to enter the arena. Being late might put them out of the contest. Consider this: though videotaping was invented in the US, the lengthy and expensive R&D process for the consumer-oriented video cassette recorder led to an all-or-none market share result, with American industry getting the "none."

No matter how partial its success may be, the Fifth Generation project will provide a decade-long learning experience for a new generation of Japanese computer scientists. They will be called upon to confront and perhaps solve the most challenging problems facing the future of information processing, rather than re-engineering traditional systems. They will be learning advanced software concepts in a way that has never been done before in Japan and has never been widely done in the US or Europe.

The Fifth Generation project, in its short life, has emplaced the technology transfer mechanisms necessary for Japanese industry to move effectively to bring its developments to market. Right now, the US has a substantial lead in virtually every area of Fifth Generation work. But *Fortune's* article on the Fifth Generation project concludes with this observation: "Even if the US retains its lead in AI research, there is no guarantee that the laboratory work will end up in products. Computer research tends to seep into the American marketplace slowly except when companies perceive a competitive threat. Assuming that ICOT can do even a fraction of what it intends, the results will show up quickly in Japanese computer products."

One of author Feigenbaum's experiences with Japanese researchers demonstrates their ability to advance rapidly in new fields.

A few years ago, engineers from Hitachi called on Feigenbaum with a list of possible candidates for expert systems treatment. There might have been 35 items on the list, rather vaguely explained. But all the visitors wanted from him was advice on whether each project was likely or unlikely (some "warmer–colder" judgments). A year later, they returned with their list pared down to six, those problems beautifully analyzed. One in particular intrigued him: the debugging of an integrated circuit fabrication line. Here the problem was slightly different from the other candidates that had been put forward for expert systems treatment, where the issue was replication of human experience. In the IC fabrication line problem there was no single human expert who held—or could hold—all the expertise necessary to make this complex industrial process function with high yield. Thus the problem was one of integrating expertise from many different experts. Rumor has it that Hewlett-Packard is working on a similar expert system. But it was the Japanese who identified the problem as a splendid one to work on, and it was a good indication of their growing sophistication in knowledge engineering.

The Myths of Artificial Intelligence

Joe Weizenbaum

Readers of my earlier volume, The Microelectronics Revolution, *will be familiar with Joe Weizenbaum's assaults on unrestrained computer enthusiasts and the artificial intelligence lobby in particular. In this savage, wide-ranging critique of Feigenbaum and McCorduck's book, the Professor of Computer Science at* MIT *ridicules the claim of the authors that revolutionary developments are just around the corner. From the* New York Review *October 27, 1983.*

A little quiz: who spoke the following lines, and on what occasion?

... no plausible claim to intellectuality can possibly be made in the near future without an intimate dependence upon XXX. Those intellectuals who persist in their indifference, not to say snobbery, will find themselves stranded in a quaint museum of the intellect, forced to live petulantly, and rather irrelevantly, on the charity of those who understand the real dimensions of the revolution and can deal with the new world it will bring about.

Of course, the answer depends on what one substitutes for "XXX." Readers who filled in, say, "national socialism" could have supposed that these words were spoken, as a warning to German intellectuals who had not yet appreciated the glory of the Nazi revolution, by Josef Goebbels on the occasion of the book burning in Berlin on May 10, 1933. Readers could substitute "the ideas of the great leader and teacher" for "XXX" and leave open what particular revolution is being talked about. Leaders who come to mind, and whose names would render the quoted paragraph plausible, are, to name just a few: Karl Marx, General Pinochet, Stalin. The Germans, by the way, had a word for what intellectuals are here being warned to do: *Gleichschaltung*, which is translated as "bringing into line" or "coordination."

But, implausible as it may seem at first glance, "XXX" in the quoted passage stands for "this new instrument," meaning the computer. The authors of *The Fifth Generation* maintain that intellectuality, the creative use of the mind engaged in study and in reflection, will soon become inevitably and necessarily dependent on the computer. They are astounded that American intellectuals aren't rushing to enlist in their revolution. They would expect, they say,

that American intellectuals (in particular those who still talk so reverently about the values of a liberal education, the sharing of the common culture, and so on, and so on) are eager to mold this new technology to serve the best human ends it possibly can.

Unfortunately, they're not. Most of them haven't the faintest idea what's happening . . . they live in a dream world, irresponsible and whimsical, served by faithful old retainers (in the form of periodicals that are high in brow, even higher in self-importance, but low in circulation) that shamelessly pander to their illusions.

In this book Edward Feigenbaum, a professor of computer science at Stanford University and a cofounder of two commercial companies that market artificial intelligence software systems, and Pamela McCorduck, a science writer, give the reader an idea of what's happening in the world of computers. They make the following claims.

First, certain American computer scientists have discovered that, if computers are expected to intervene in some activity in the real world, then it would help, to say the least, if they had some knowledge of the domain of the activity in question. For example, computer systems designed to help make medical diagnoses had better know about diseases and their signs and symptoms.

Second, other American computer scientists have described designs of computers, "computer architectures," that depart radically from the industry's traditional design principles originally laid down by the pioneer computer scientist, John von Neumann. In orthodox, so-called von Neumann, machines, long chains of computations are organized as sequences of very small computational steps which are then executed serially, that is, one step after another. The new architectures allow computational chains to be decomposed into steps which can be executed as soon as the data for executing them are ready. They then don't have to wait their turn, so to speak. Indeed, many steps can be executed simultaneously. Computation time is in a sense "folded" in such machines, which are consequently very much faster than their orthodox predecessors.

Third, still other computer scientists, mainly French and British, have created a computer language which they believe to be well suited for representing knowledge in computers in a form that lends itself to powerful logical manipulation. The Japanese hope that the conjunction of this computer language with the new architecture will allow ultrarapid computation of "inferences" from masses of stored knowledge.

Fourth, these developments have taken place at a time of continuing dramatic progress in making computers physically smaller, functionally faster, and with increasing storage capacities. Computer hardware becomes constantly cheaper.

Finally, the Japanese, who already dominate the world market in consumer electronics, have seized on the resulting opportunity and decided to create entirely new and enormously powerful computer systems, the "Fifth Generation," based on the developments described above. These systems, as the book jacket puts it, will be "artificially intelligent machines that can reason, draw conclusions, make judgments, and even understand the written and spoken word."

This appears to be a very ambitious claim. But not to seasoned observers of the computer scene, who have long since learned to penetrate the foggy language of the computer enthusiasts. What have they boasted of before and how were such boasts justified in reality? One example will do: in 1958, a quarter of a century ago, Herbert Simon and Allen Newell, both pioneer computer scientists and founding members of the Artificial Intelligence (AI) movement within computer science, wrote that

there are now in the world machines that think, that learn, and that create. Moreover, their ability to do these things is going to increase rapidly until—in the visible future—the range of problems they can handle will be coextensive with the range to which the human mind has been applied.[1]

In other words, the most recent ambitions of the Japanese were already close to being realized according to leaders of the American artificial intelligence community a quarter of a century ago! All that remained to be done—and it would be done within the "visible future"—was to extend the range of the problems such machines would solve to the whole range of the problems to which the human mind has been applied.

That ambition remains as absurd today as it was twenty-five years ago. In the meanwhile, however, much progress has been made in getting computers to "understand" the written word and even some words spoken in very highly controlled contexts. Is the Japanese project then really not very ambitious? The answer depends on the standards of intelligent computer performance one adopts: by those Simon and Newell evidently held 25 years ago, not much remains to be done. If, on the other hand, words like "judgment," "reason," and "understanding" are to be comprehended in their usual meanings, then the prospects for anything like full success for the Japanese project are very dim. I believe the Japanese will build some remarkable hardware in the coming decade, but nothing radically in advance of American designs. However, in order to do all they intend to do in a single decade, the Japanese have organized a huge effort involving the close cooperation of Japan's Ministry of International Trade and Industry (MITI) and the major Japanese firms in the electronics industry.

That, briefly, is "what's happening." But Feigenbaum and McCorduck do more than merely deliver the latest bulletins from the high-technology front. They also present their own vision of the world that is within our grasp, if only we would reach for it. Here are just a few of their speculative glimpses into that future.

Loneliness will have been done away with—at least for old people. There will be friendly and helpful robots to keep them company.

The geriatric robot is wonderful. It isn't hanging about in the hopes [*sic*] of inheriting your money—nor of course will it slip you a little something to speed the inevitable. It isn't hanging about because it can't find work elsewhere. It's there because it's yours. It doesn't just bathe you and feed you and wheel you out into the sun when you crave fresh air and a change of scene, though of course it does all those things. The very best thing about the geriatric robot is that it *listens* [emphasis in the original]. "Tell me again," it says, "about how wonderful/dreadful your children are to you. Tell me again that fascinating tale of the coup of '63. Tell me again . . ." And it means it. It never gets tired of hearing those stories, just as you never get tired of telling them. It knows your favorites, and those are its favorites too. Never mind that this all ought to be done by human caretakers; humans grow bored, get greedy, want variety.

Then there will be the "mechanical" doctor. But

if the idea of a mechanical doctor repels you, consider that not everyone feels that way. Studies in England showed that many humans were much more comfortable (and candid) with an examination by a computer terminal than with a human physician, whom they perceived as somehow disapproving of them.

Another helpful device that awaits us is the

intelligent newspaper [which] will know the way you feel and behave accordingly.

It will know because you have trained it yourself. In a none-too-arduous process, you will have informed your intelligent newsgathering system about the topics that are of special interest to you. Editorial decisions will be made by you, and your system will be able to get upon them thereafter. . . . It will understand (because you have told it) which news sources you trust most, which dissenting opinions you wish to be exposed to, and when not to bother you at all.

You could let your intelligent system infer your interests indirectly by watching you as you browse. What makes you laugh? It will remember and gather bits of fantasia to amuse you. What makes you steam? It may gather information about that, too, and then give you names of groups that are organized for or against that particular outrage.

We may suppose, by the way, such a system would be just as ready to help other agencies, say the police, gather information and obtain names of groups that may be for or against "outrages" that interest *them*.

But, much more importantly, what Feigenbaum and McCorduck describe here is a world in which it will hardly be necessary for people to meet one another directly. Not that this is a consequence they hadn't foreseen and from which they would recoil. They have seen that aspect of the new world, and they greet it as a welcome advance for mankind:

Despite the gray warnings about how the computer would inevitably dehumanize us, it has not. We are just as obstreperously human as ever, seizing this new medium to do better one of the things we've always liked to do best, which is to create, pursue, and exchange knowledge with our fellow creatures. Now we are allowed to do it with greater ease—faster, better, more engagingly, and *without the prejudices that often attend face-to-face interaction* [emphasis added].

A geriatric robot that frees old people from the murderous instincts of their children and is programmed to lie to them systematically, telling them that it understands their petty stories and enjoys "listening" to them. Mechanical doctors we can be utterly candid with and which won't disapprove of us as human doctors often do. Technical devices in our own homes that gather information about us and determine to what group we ought to belong and which ones we should hate. Technical systems that permit us to exchange knowledge "engagingly" with our fellow creatures while avoiding the horror of having to look at them or be looked at let alone *touched*. This is the world Feigenbaum and McCorduck are recommending.

In fact, they haven't told us the whole story. Professor Tohru Moto-oka of Tokyo University and titular head of the Japanese Fifth Generation project promises even more:

. . . first, [Fifth Generation computers] will take the place of man in the area of physical labor, and, through the intellectualization of these advanced computers, totally new applied fields will be developed, social productivity will be increased, and *distortions in values will be eliminated* [emphasis added].

In existing totalitarian societies, "distortions in values" are eliminated by very unpleasant methods indeed. Professor Moto-oka promises a future in which computers will do that without anyone's noticing, let alone feeling pain. This would be truly an advance for human civilization. Can it be done? It is a sign of the times that people will think this is a question about the technical power of computing systems. It is, however, a question of the willingness of populations to surrender themselves to the "conveniences" offered by technical devices of all sorts, particularly by information-handling machines. Will they be eager to make the Faustian bargain by which they will have the things and entertainments they want in exchange for their right and responsibility to determine their own values? Does that exchange sound so preposterous? Think of the enthusiasm with which millions of parents turn their children over to the television set in order to escape having to bother with them. Nor do most such parents exhibit any worries about what values their children might learn while watching hundreds of murders annually.

We are well on our way to the kind of world sketched here. Already, we are told, and it is undoubtedly true, many people prefer to "interact" with computers. Schoolchildren prefer them to teachers, and many patients to doctors. No one seems to ask what it may be about today's doctors and teachers, or with the situations in which they work, that causes them to come off second-best in competition with computers. Perhaps it would help to learn in what way and why such interpersonal encounters fail, and then— remedy whatever difficulties come to light. Perhaps the computer ought not to be wheeled into the classroom as a solution before the problems plaguing the schools, not merely their manifestations, have been identified.

The computer has long been a solution looking for problems—the ultimate technological fix which insulates us from having to look at problems. Our schools, for example, tend to produce students with mediocre abilities to read, write, and reason; the main thing we are doing about that is to sit kids down at computer consoles in the classrooms. Perhaps they'll manage to become "computer–literate"—whatever that means—even if, in their mother tongue, they remain functionally illiterate. We now have factories so highly computerized that they can operate virtually unmanned. Devices that shield us from having to come in contact with fellow human beings are rapidly taking over much of our daily lives. Voices synthesized by computers tell us what to do next when we place calls on the telephone.

The same voices thank us when we have done what they ask. But what does "the same voices" mean in this context? Individuality, identity, everything that has to do with the uniqueness of persons—or of anything else!—simply disappears. No wonder the architects of, and apologists for, worlds in which work becomes better and more engaging to the extent that it can be carried out without face-to-face interaction, and in which people prefer machines that listen to them to people, come to the conclusion that intellectuals are irrelevant figures. There is no place in their scheme of things for creative minds, for independent study and reflection, for independent anything.

These people see the technical apparatus underlying Orwell's *1984* and, like children on seeing the beach, they run for it. I wish it were their private excursion, but they demand that we all come along. Indeed, they tell us we have no choice, unless, that is, we are prepared to witness "the end, the wimpish end, of the American century," the conversion of the United States of America to an agrarian nation. Perhaps we ought to consider that and other alternatives to the Faustian bargain I mentioned. Increasing computerization may well allow us to increase the productivity of labor indefinitely—but to produce what? More video games and fancier television sets along with "smarter" weapons? And, with people's right to feed their families and themselves largely conditional on their "working," how do we provide for those whose work has been taken from them by machines? The vision of production with hardly any human effort, of the consumption of every product imaginable, may excite the greed of the society whose appetites are fixed on things. It may be good that, in our part of the globe, people need no longer sort bank checks or mail by hand, or retype articles like this one. But how far ought we to extrapolate such "good" things? At what price? Who stands to gain and who must finally pay? Such considerations ought at least to be part of a debate. Are there really no choices other than that "we" win or lose?

Aside from whatever advantages are to be gained by living our lives in an electronic isolation ward, and aside also from the loss to "them" of a market "we" now dominate, what good reasons are there for mounting our own Fifth generation project on a scale, as Feigenbaum and McCorduck recommend, of the program that landed *our* man on the moon? "If you can think of a good defense application," Feigenbaum and McCorduck quote "one Pentagon official" as saying, "we'll fund an American Fifth Generation Project." "But," Feigenbaum and McCorduck insist, "there are other compelling reasons for doing it too." Forty pages of text, not to mention appendices, indexes, and so on, follow this remark, but no "other" reasons are given— other, that is, than "defense" reasons.

One chapter of the book is devoted to "AI and the National Defense." However, the chapter is only six pages long, and is mainly a song of praise— perhaps gratitude is a more apt word—for the Pentagon's "enlightened scientific leadership." Laid on thick, the praise tends to betray its own absurdity:

Since the Pentagon is often perceived as the national villain, especially by intellectuals, it's a pleasure to report that in one enlightened corner of it, human beings were betting taxpayers' money on projects that would have major benefits for the whole human race.

Notwithstanding the characteristic swipe at intellectuals, the principal author of this book is a university professor. The corner mentioned is the Defense Department's Advance Research Projects Agency, DARPA, also often called just ARPA. Well, there is reason for workers in AI to be grateful to this agency. It has spent in the order of $500 million on computer research, and that certainly benefited the AI community—the artificial intelligentsia—enormously. But "major benefits for the whole human race"?

The military, however, has good reason to continue, just as generously as ever, to provide funds for work on the fifth generation, as Feigenbaum and McCorduck make clear:

The so-called smart weapons of 1982, for all their sophisticated modern electronics, are really just extremely complex wind-up toys compared to the weapon systems that will be possible in a decade if intelligent information processing systems are applied to the defense problems of the 1990s.

The authors make five points. First, they believe we should look "with awe at the peculiar nature of modern electronic warfare," particularly at the fact that the Israelis recently shot down 79 Syrian airplanes with no losses to themselves. "This amazing result was achieved," they write, "largely by intelligent human electronic battle management. In the future, it can and will be done by computer."

Second, we cannot afford to allow the "technology of the intelligent computer systems of the future . . . to slip away to the Japanese or to anyone else. . . . Japan, as a nation, has a longstanding casual attitude toward secrecy when it comes to technological matters." Third, with the ever-increasing cost of military hardware, "the economic impact of an intelligent armaments system that can strike targets with extreme precision should be apparent . . .—fewer weapons used selectively for maximum strike capability."

Fourth, "it is essential that the newest technological developments be made available to the Defense Department." Finally, "the Defense Department needs the ability to shape technology to conform to its needs in military systems. Here, perhaps more than in any other argument of the book, we are close to what it's "all about."

A much shorter, and better, account of the Japanese Fifth Generation project appeared recently as a cover story of *Newsweek* magazine (pages 60–70). The authors of that story reached pretty much the same conclusions about the reasons the US should invest heavily in "these technologies":

Once they are in place, these technologies will make possible an astonishing new breed of weapons and military hardware. Smart robot weapons—drone aircraft, unmanned submarines, and land vehicles—that combine aritificial intelligence and high-powered computing can be sent off to do jobs that now involve human risk. [To the people on our side. It is the intention, of course, to expose the others to considerable risk.] "This is a very sexy area to the military, because you can imagine all kinds of neat, interesting things you could send off on their own little missions around the world or even in local combat," says [the head of ARPA's information processing research office]. The Pentagon will also use the technologies to create artificial intelligence machines that can be used as battlefield advisors and superintelligent computers to coordinate complex weapons systems. An intelligent missile guidance system would have to bring together different technologies—real-time signal processing, numerical calculations, and symbolic processing, all at unimaginably high speeds—in order to make decisions and give advice to human commanders.

People generally should know the end use of their labor. Students coming to study at the artificial intelligence laboratories of MIT, my university, or Stanford, Edward Feigenbaum's, or the other such laboratories in the US should decide what they want to do with their talents without being befuddled by euphemisms. They should be clear that, upon graduation, most of the companies they will work for, and especially those that will recruit them more energetically, are the most deeply engaged in feverish activity to find still faster, more reliable ways to kill ever more people—Feigenbaum and McCorduck speak of the objective of creating smart weapons systems with *"zero probability of error"* (their emphasis). Whatever euphemisms are used to describe students' AI laboratory projects, the probability is overwhelming that the end of their research will serve this or similar military objectives.

The tone of the book, moreover, is as doubtful as its approach to the machines, projects, and people it seeks to promote. The authors, for example, speak of each other in the third person. If the first-person singular were substituted, some of the things they say would be absurd. For example: "'I know,' Feigenbaum said affably. . . ." Suppose they had written, "'I know,' I said affably." Nor is this a trivial point. The device is used in the service of reconstructing history. Even moderately well-informed computer professionals could be forgiven if, after reading this book, they came to a number of conclusions that are implausible to say the least: that Feigenbaum had single-handedly brought about a revolution in computer science; that the whole of the Japanese Fifth Generation program was designed by Japanese scientists informed largely by visits to Feigenbaum's laboratory; and that IBM missed its chance to mount a similar, but American, project because they didn't listen to Feigenbaum when, at their invitation, he told them the way things were.

Such effects are produced through promotional rhetoric. McCorduck, for example, remembers a lecture Feigenbaum had given at Carnegie-Mellon University. Herbert Simon, who had directed his thesis, was there, and

beside Simon was Allen Newell, another artificial intelligence great, and scattered about the room were some of the best and brightest in computer science in general and artificial intelligence in particular. But the Carnegie mood was skeptical.

Feigenbaum . . . threw out a challenge. "You people are working on toy problems," he said. . . . "Get out into the real world and solve real-world problems."

. . . It is sound science strategy to choose a simplified problem and explore it in depth to grasp principles and mechanisms that are otherwise obscured by details that don't really matter. But Feigenbaum was arguing to the contrary, that here *the details not only mattered; they make all the difference* [emphasis in the original].

There were murmurings among the graduate students. Maybe Feigenbaum was right. . . . And not immediately, but later, Carnegie-Mellon came around. . . .

Could that have been written with "I" substituted for "Feigenbaum"? Not without making the reader giggle. As it is, the account suggests the young Newton lecturing the Royal Society whose members are a little slow, but eventually come around—an impression conveyed throughout the book. Early on, the authors talk of the insight that

"computer" implies only counting and calculating, whereas [the computer] was, in principle, *capable of manipulating any sort of symbol* [emphasis in the original].

Though younger men eagerly pointed it out to them, that insight was simply unacceptable to many computer pioneers. John von Neumann, for example, who is widely acknowledged as a giant in computing, left as his last piece of published writing a long argument that computers would never exhibit intelligence.

In fact, John von Neumann was one of the first—if not the first—to recognize that commands to computers are merely symbols that could be manipulated by computers in the same way that numbers—or any other symbols—could be manipulated. He is responsible for a computer architecture, based on that insight, which is still a world-wide industry standard. But here von Neumann is cited as "an example" of a computer pioneer who didn't accept his own insight when "young men," if we are to believe the authors, "pointed it out" to him. As for von Neumann having written "a long argument that computers would never exhibit intelligence," no such published document exists so far as I know. Von Neumann's last published work is, in von Neumann's own words, "an approach toward the understanding of the nervous system from a mathematician's point of view." It simply has nothing to do with "computers exhibiting intelligence."[2]

Actually, von Neumann had a standard answer for anyone who asked him whether computers could think, or be intelligent, and so on. He argued that, if his questioner were to present him with a *precise* description of what he wanted the computer to do, someone could program the computer to behave in the required manner. Whether he thought there were some things in the human experience that could not satisfy his criterion, I simply don't know. The position that every aspect of nature, most importantly of human existence, must be precisely describable, and its corollary that all human knowledge is sayable in words, is central to the credo to which all true believers in the limitless scope of artificial intelligence must hold.

As a computer scientist, I agree with Ionesco, who wrote, "Not everything is unsayable in words, only the living truth." The Japanese, if one is to believe this book, appear to be true believers; von Neumann, perhaps, was not. It is unfortunate that Feigenbaum and McCorduck don't supply a reference for their claim about him. Indeed, whenever criticism of one of Feigenbaum's beliefs is alluded to, it remains anonymous. We are told, for example, that arguments against machine intelligence fall into four broad categories. These are briefly sketched, without mention, let alone identification, of serious proponents of such arguments or any hint where they might be found. Von Neumann's "long argument" must, I suppose, fall into at least one of the four mentioned categories. Which one? We would suppose that an argument by such a giant in the discipline in question might well be decisive—or that Feigenbaum has developed a powerful counter-argument. If he has, it remains well hidden.

It is as if someone trying to build a vehicle that is to travel at twice the speed of light categorizes the arguments against the possibility of success— but only in a caricature that makes them appear ridiculous on their face. And then remarks casually that Albert Einstein had a "long argument" for the impossibility of the proposed project, but gives no hint of the nature of the argument or where it may be found, and fails to supply the refutation that someone investing money in the project—say the taxpayer—might wish for.

Henry Kissinger, commenting on the Watergate scandal, once spoke of "the awfulness of events and the tragedy that has befallen so many people." The use of the passive voice here renders invisible the various actors in the drama and their degrees of responsibility; it denies the role of human will in human affairs. Feigenbaum and McCorduck use similar devices to shield themselves from having to quote their potential critics or to give citations that interested readers might follow up. "Cogent arguments have been made that . . . Others have argued that. . . ." Such phrases recur throughout the book. The reader has no way of knowing the authors' own standards of cogency. If the arguments matter, they ought to be stated, or we should be told where to find them.

To underline the importance of the computer as the "main artifact of the age of information," Feigenbaum and McCorduck instruct us just wherein the importance of the computer lies:

[The computer's] purpose is certainly to process information—to transform, amplify, distribute, and otherwise modify it. But more important, the computer *produces* information. The essence of the computer revolution is that the burden of producing the future knowledge of the world will be transferred from human heads to machine artifacts [emphasis in the original].

How are we to understand the assertion that the computer "*produces* information"? In the same way, presumably, as the statement that a coalfired electric power station produces energy. But that would be a simple and naive falsehood. Coalfired power stations transform energy, they do not produce it. Computers similarly transform information, generally using information-*losing* operations. For example, when a computer executes an instruction to add 2 and 5, it computes 7. But one cannot infer from "7" that it was the result of an addition, let alone what two numbers were involved. That information, though it may have been preserved elsewhere, is lost in the performance of the computation. Perhaps a more generous interpretation of what the authors are trying to say is that a computer produces information in the way a Polaroid camera produces—what? Information, or a picture? But how much information is there in a photograph that wasn't already in the world to be photographed? To be sure, the photographer selects from the world and composes aspects of it in order to create what will ultimately be encoded on film. But it is the photographer, the artist, who produces the art. The camera transforms images, but only those that were placed in front of it. I think Professor Feigenbaum has to explain just what information a computer produces and how.

But more important: how can the authors' jump from the idea of producing information—never mind whether that is nonsense or not—to that of producing knowledge, indeed *the* "future knowledge of the world," be justified? The knowledge that appears to be least well understood by Edward Feigenbaum and Pamela McCorduck is that of the differences between information, knowledge, and wisdom; between calculating, reasoning, and thinking, and finally of the differences between a society centered on human beings and one centered on machines.

Notes

1 H.A. Simon and A. Newell, "Heuristic Problem Solving: The Next Advance in Operations Research," *Operations Research*, January–February, 1958, p. 8.
2 *The Computer and the Brain* (Yale University Press, 1958).

The Social Impact of Thinking Machines

Margaret A. Boden

After the charges and counter-charges of the previous debate, a sober assessment of the key research areas in artificial intelligence and their prospects by the author of Artificial Intelligence and Natural Man *(1977). Here, she discusses the social implications of the use of expert systems in medical diagnosis, legal financial advice and education, challenging some of the more sensationalist claims. Boden is Professor of Philosophy and Psychology at the University of Sussex, England. Taken from a paper prepared for a workshop at the International Institute for Applied Systems Analysis, Laxenburg and published in* Futures, *February 1984.*

Artificial intelligence (AI) was conceived in the mid-1940s, born in the early 1950s, and christened in 1956 (at a small meeting of computer scientists, psychologists, and physiologists). Since then, it has had some notable successes, enabling computers to perform—albeit in a very limited way—some of the tasks normally done by our minds. Some AI workers see AI as a way of helping us understand human psychology; they try to write programs that tackle their tasks in the sort of way in which we do. Others see it as an approach to a theory of intelligence in general, human (and animal) intelligence being a special case. Still others simply want to write programs to *do* something (to understand language, to describe visible objects, or to solve problems of various kinds), irrespective of how we do it. And most of these hope that what their programs do will be not only interesting, but useful.

The technological aspects of AI have suddenly become more visible. Public interest in AI, and media coverage of it, have increased enormously over the last two years. More and more people view it as an incipient technology of great potential power and social significance. The public interest dates from the announcement in 1981 of Japan's ten-year national plan for developing "Fifth Generation" computers. These are defined as incorporating AI techniques (as well as large-scale parallel processing). Since then, large sums of money for AI-research have been made available also by governments and industry in the Western industrialized nations.

This paper discusses some of the prospects for AI applications and basic research, in the next decade and in the longer term. I have tried to identify the main areas of promise and the most recalcitrant intellectual problems, and also to highlight some significant social implications.

Developments in the next ten years

Core research areas and likely results
Several core research areas are likely to make solid progress within the next decade. Each of these is already being worked on in various countries, and progress does not depend upon the success of Japan's ambitious "Fifth Generation" project (though it might be accelerated by associated hardware and software developments).

One is low-level vision, based on techniques using parallel hardware and cooperative processing. Current "connectionist" research in this area differs in its approach from work on two-dimensional pattern recognition by "property lists," and from-the-top-down "scene analysis" of three-dimensional scenes. Based on detailed studies of image formation, it is able to extract from the ambient light information about three-dimensional features (such as shape, depth, texture, and surface orientation) which in previous approaches could have been computed only, if at all, by way of high-level knowledge of the expected scene. Some of this work is being done in the context of human psychology and neurophysiology, some in a more technological context. Dedicated (massively parallel) machines are being designed for this research, and major advances depend upon such hardware.

A second area in which we can expect significant progress is robotics. This includes problems of movement control, trajectory planning, and visuomotor coordination (and will take advantage of advances in low-level vision). As in the case of vision, some projects will rely on "artificial" means to ensure success (such as light stripes for automatic welding machines, capable of recognizing different sorts of weld joint and guiding the welder accordingly), while others will relate more closely to psychophysiological theories of motor control and visuomotor coordination in living organisms.

Knowledge-based "expert" systems will multiply enormously in the next decade, not least because there is considerable commercial interest in them. Different domains of human expertise may require different approaches to knowledge engineering. In domains less fully covered by an explicit scientific theory, it may be easier to extract knowledge from human experts who are competent but who have not yet achieved the "intuitive" mastery of the domain that topflight experts enjoy. The latter give the right answer more often, but cannot easily introspect their reasoning processes, which happen very fast and are not consciously accessible. The former take time to come to a decision, after consciously weighing distinct considerations against each other and verbally identifying areas of unclarity. Domains (such as medical radiology) that depend on the comparison and interpretation of complex visual images are especially difficult to automate, since low-level visual processes are not open to voluntary inspection or control. Indeed, experts often give highly misleading advice about how they may be carrying out the relevant comparisons. (Eye movement studies show, for instance, that expert radiologists do not scan x-ray photographs in the way that they say that they do.) In tandem with the increasing experience of AI-trained knowledge engineers, further psychological studies of the organization of knowledge in different domains should be useful.

Research on expert systems will also focus on the computational architecture required to deal with large, complex knowledge bases. Current systems are relatively simple and inflexible, and are restricted to very narrow domains. They can be incrementally improved, but only up to a point. Eventually, the interactions between the increasing number of independently added rules become too difficult to control, and the system's reliability and intelligibility are jeopardized. Current systems have no access to higher-level representations of the knowledge domain and their own problem-solving activity (see below). Special problems arise if a system has to work in real-time, where unexpected events can require quick switching from the current activity to some other. The next ten years will see some general work on powerful (IKBS) architectures (as well as the production of more examples of specific commercially useful systems), including parallel-processing devices.

Progress can be expected also in natural language processing, both of individual sentences and of texts. Key issues include syntactic parsing, the integration of syntax with semantics, and the understanding of connected text. Machine translation could in principle benefit from advances both in single-sentence parsing and in text analysis.

Current work on parsing is motivated both by theoretical (linguistic) interests and by the hope of improving the man–machine interface so as to make it possible for nonspecialist users to communicate with programs in (some reasonable subset of) natural language. Where a program is used for some specific purpose, semantic factors can be more readily used to help in the parsing and disambiguation of queries and instructions input by the user. Verbal interchanges about lunar geology, or about airline reservations, are already reasonably "natural" because of the exploitation of semantic constraints, and further domain-specific semantics will be developed over the next decade. More generally applicable (theoretical) research will continue into the best point at which to use semantics in parsing: from the beginning of the sentence, or spreading out from the middle, or only after an initial parse of the entire sentence?

Text-analysis programs can already give a precis of most short news stories about specific topics (such as earthquakes, hijackings, and road accidents). But they rely on rigid, preprogrammed schemata, which provide the semantic skeleton of the types of stories concerned. Some recent research is aimed at enabling a text-analysis program to learn new schemata for itself, to integrate one schema with another so as to understand a story combining both, and to use a given schema to reason analogically in an unfamiliar context. A high degree of success cannot be expected within the next ten years, but our understanding of the relevant problems should be advanced.

A variety of educational applications is already receiving attention. Some are focused on particular curricular subjects, and require both a model of the theory of that subject and a model of the student's knowledge of it (which varies in level and in organization, from person to person and from time to time). Others are less specific, and aim to use AI-based techniques to improve the pupil's attitude to intelligence in general. There is some evidence that both normal and handicapped students can attain greater self-confidence and intellectual achievement by experience with these specially designed programming environments. Controlled research into the classroom effects of AI-based systems has recently been initiated, and this can be expected to bear fruit within the next decade.

An extremely important area, which is increasingly being studied because of recent hardware developments, concerns the computational properties of large parallel systems. At present, we understand very little of the potential and limitations of such systems. Some of the connectionist work mentioned above suggests that cooperative processing may have some highly surprising properties. For example, the number of individual processors required to make the "human" range of visual shape discriminations appears to be markedly less than one would naturally assume. Again, making a connectionist system stochastic rather than deterministic *improves* its chance of finding an optimal solution. The computational properties of parallel machines will not be well understood for a long time, but experience with these new systems in the near future will doubtless lead to some advance.

Impacts on other sciences and technologies

The impacts of AI on other technologies will include many different examples of applications to individual problems. For example, an olfactory chip is being designed using AI techniques of pattern recognition. Given advances in very large-scale integration (VLSI), instruments and products of many different kinds will come to include chips whose design makes use of AI methods. Any commercial–industrial task that could benefit from even a limited degree of intelligence could in principle be performed better with the help of AI, so the technological applications of AI will be extremely diverse.

AI will influence other sciences in their general philosophical approach as well as their specific theoretical content. Indeed, psychology and (to a lesser degree) biology have already been affected by computational ideas. And, contrary to what most people assume, AI has had a humanizing effect in psychology. The behaviourists had outlawed reference to "mind" and "mental processes" as unscientific and mystifying, but AI—based as it is on the concept of *representation*—has made these concepts theoretically respectable again.

AI's influence will be especially strong in the psychology of vision and language, and, as noted above, it is likely that robotics will engage with the psychophysiology of movement. Psychological research will feed back into AI; for example, insofar as psychologists arrive at a better understanding of the organization of knowledge, their work may be useful in designing computerized expert systems. Cooperative interdisciplinary research should be encouraged: the institutional separation of empirical psychology and AI or computer science has hindered fruitful collaboration between these groups.

Social implications

Social impacts will be of various types. First, there will be effects on individuals and institutions brought about by specific applications of AI, such as expert systems for medical diagnosis, legal and financial advice, or educational help.

These programs will not merely provide a service (whose adequacy should be very carefully monitored), but will very likely change the social relations of the profession or institution concerned. For example, if general practitioners, or nurses, can use an AI program to aid in various aspects of patient care, the social image of the specialist physician may be profoundly affected. (And legal responsibilities for medical decisions may be assigned in a way very different from today.) Likewise, legal programs may undermine the status of lawyers, and alter the nature of their work. In both cases, while the mystique of the human experts may be lessened, their opportunity for exercising their specifically human powers may be increased.

The general public might come to be less dependent on human experts than they are today. Reducing the power of professionals such as doctors, lawyers, and teachers would certainly have advantages. But replacing human professional advice by computer programs is dangerous to the extent that AI systems in public use are inadequate—and/or ill-understood. Systems that have taken several man-years to develop (and whose original programmers may be retired, or dead) are often very difficult to evaluate or alter, because even computer scientists do not fully understand how they work. (It follows that attention should be given to methods of perspicuous documentation, to help make clear what it is that a given program is actually doing, and how.)

A second type of social impact concerns general social trends brought about by applications of AI and information technology (IT). These include changes in the proportion of the workforce in service and leisure industries, changes in the division of labour and sexual roles, and changes in general lifestyles and patterns of interaction.

For example, males will be increasingly freed to take up jobs in the "caring" professions (such as nursing, education, and social welfare). This could change the general evaluation of emotionality in the masculine role, an effect that could also be encouraged by men's having increased leisure time to spend with family and friends. Such an effect could be liberating and humanizing, leading to a more convivial society than we have today.

But other potential consequences of AI point in the opposite direction. The widespread use of home terminals, for instance, threatens to have an isolating influence even more powerful than that of television. If people are encouraged to work, and to shop, from their sitting-rooms, there may be unfortunate psychological effects in terms of personal stress and loneliness. Community computer centres could offset these effects to some extent, providing a social meeting place outside the confines of the home and nuclear family. Some writers even predict that commercially available (and highly profitable) AI systems will be heavily used not only in task-oriented ways, but as surrogates for human contact. On this view, the strong tendency to anthropomorphism that most of us share will result in patterns of interaction being skewed away from human beings, and towards quasi-human computer systems (with naturalistic "voices," and sometimes even "bodies"). Although such forecasts grossly underestimate the technological difficulties involved in building programmed "friends-off-the-shelf," they do suggest that human interactions could be impoverished to some degree in the future.

These contrasting examples show that widespread application of AI will have subtle, and varying, influences in society. Moreover, AI could foster a general view of humanity as either "mechanistic" or "non-mechanistic," depending on how it is interpreted by the public. The commonest interpretation is that AI presents us as "mere machines," with no free choice or moral responsibility. Since this image of man could have socially pernicious effects, people should be helped to understand that it is fundamentally mistaken. The education for computer literacy discussed below could help here. More generally, we should start thinking now about what the optimal social arrangements might be for a post-industrial society.

Economic aspects

The economic impacts will be far-reaching. Traditional manufacturing and clerical–administrative jobs will be decimated. But new jobs will be created: some directly connected with new technology (like computer engineers and programmers), others made possible because people are freed to devote their time to services (caring professions, education, leisure). Whether there will be enough new jobs to compensate for the loss of old ones (as has always happened in the past, at least eventually) is however unclear, for AI can apply potentially to all jobs where personal human contact is not essential. New methods of work-sharing and income distribution will have to be worked out (with income not necessarily being closely linked to jobs). Radical structural changes in society are likely, and the transition phase will not be easy.

Long-range AI research

There will be "more of the same," in that the areas mentioned above will provide perplexing problems for many years to come. Especially hard problems include learning, high-level vision, naive physics, and abstract work in computational logic.

I referred earlier to short-term research on IKBS architecture. But the deep problems involved in the organization and control of large knowledge bases will not be solved within a decade. This is so quite independently of the fact that parallel machines may support forms of inference radically different from those implemented today.

For instance, expert systems are at present unable to explain their reasoning except by "backwards-chaining": giving a resumé of the chain of inferences (rules) that led up to their conclusion. They cannot relate their conclusion to the domain in general, or rely on an overview of the problem to assess the relative theoretical reliability of different hypotheses. Nor can they monitor and adjust the structure of their own problem-solving, for they have no high-level representation of it. They are unable, too, to integrate different knowledge domains, and to use concepts and patterns of inference taken from one domain to reason (analogically) in another. Nor can current systems explain their conclusions differently to different users, taking account of the specific user's knowledge. The user can ask for a "deeper" explanation (a more detailed inference resumé), but the program has no user-model in terms of which to adjust its explanations to the human's particular range and level of knowledge. For this reason also, the pattern of interaction between user and system is at present very limited. The user cannot offer his own conclusions for comment and criticism, for example, as students can do with human teachers.

All of these abilities that current programs lack will need a richer understanding of the structure and content of different knowledge domains. Some of the projects mentioned elsewhere (such as research on naive physics) are highly relevant to many domains, and psychological research into human reasoning processes could be useful also. This is just one illustration of the fact that theoretical and empirical research may be needed for radical improvements in technological applications.

A special case of human knowledge is "naive physics," one's everyday knowledge of the properties and behaviour of different sorts of physical substances, and the nature of the causal relations between them. This knowledge enters into vision and motor control, and also into natural language. For example, a language-using program would have to understand the differences in meaning between verbs such as *pour, flow, spill, drop*, and the like if it were to give instructions or understand texts about activities dealing with liquids. Similarly, a robot capable of seeing that a container was just about to spill its contents onto the object below, and of adjusting its movements accordingly, would need some representation of the behaviour of fluids. Very little work has been done on these issues so far, and they are likely to provide a challenge for many years.

"Computer-aided design" is typically thought of as involving the graphical display of precise three-dimensional specifications of various products (from machine tools through cars to buildings), taking into account a wide range of values of many parameters. But a recent form of computer-aided design involves suggestion rather than specification, in the sense that the design program originates novel ideas—ideas that are not merely quantitatively different from previous specifications. For example, heuristic programs are already being used to suggest novel experiments (described at the intra-molecular level) in genetic engineering, or to help design new sorts of three-dimensional silicon chips. These programs were developed in tandem with a similar system that originates interesting mathematical ideas from a basis of elementary set theory.

The potential of systems like these should be further explored. The computer modelling of creative thinking will require long-term research, especially with respect to domains whose crucial concepts cannot be so readily defined as the concepts of molecular biology, chip circuitry, or set theory. We need a better understanding of how conceptual structures (and the inference patterns associated with them) can be explored, represented (on varying levels of abstraction), compared, and indexed.

Machine learning is a pressing problem for the future. If a program cannot learn for itself, its development is limited by the time and ability of the programmer to provide it with new information and ideas. The system should be able to induce regularities for itself. Some progress is being made in enabling computer systems to learn about specific aspects of particular domains. But "open-ended" learning, where what is to be learnt is not defined beforehand, is especially intractable.

In general, the properties of parallel computation will be a focus of research in the long term as well as in the next few years. It remains to be seen whether the Japanese hopes concerning VLSI and PROLOG (the computer language thought to be especially suited to a parallel architecture) will be achieved. But massively parallel hardware will become increasingly available, and will enable AI to progress in ways that are impossible using traditional types of machine.

Sensationalism feeds on ignorance, and many descriptions of artificial intelligence in the media, and in popular books about the subject, are sensationalist in nature. Whether proclaiming the "wonders" or the "dangers" of AI, they are not only uninformative but highly misleading—and socially dangerous to boot. They suggest that things can be done, or will be done, tomorrow which in fact will be feasible only (if ever) after decades of research.

Unfortunately, these sensational reports are sometimes encouraged by ill-judged remarks from the AI community itself. A recent hour-long BBC-TV science programme began and ended with a quote from a senior computer scientist at MIT, gleefully forecasting that the intelligent machines of the future would worry about all the really important problems for us (*for* us, not *with* us). As he put it (with apparent satisfaction), if we ever managed to teach chimps to speak, we wouldn't talk to them for long—for they would want to talk only about bananas; super-intelligent machines will be similarly bored by people, for we won't be capable of understanding the thoughts of the machines. His conclusion was that the super-intelligent AI systems will justifiably ignore us, leaving us simply to play among ourselves.

Humanity has of course been advised before to neglect the difficult moral and philosophical questions, to live life on the principle that "*Il faut cultiver son jardin.*" But that was said in a rather more ironic spirit. Enthusiasts evaluating AI's contribution to society would do well to emulate the common sense, if not the scepticism, of Voltaire.

Guide to Further Reading

Articles

"Machines That Think," *Newsweek* (cover story), June 30, 1980.

"Artificial Intelligence: The Second Computer Age Begins," *Business Week*, March 8, 1982.

"Artificial Intelligence is Here," *Business Week* (cover story), July 9, 1984.

Tom Alexander, "Thinking Machines," a three-part series in *Fortune*, May 17, May 31, and June 14, 1982.

Tom Alexander, "Why Computers Can't Outthink the Experts," *Fortune*, August 20, 1984. Says expert systems aren't so smart.

Peter Marsh, "The Race for the Thinking Machine," *New Scientist*, July 8, 1982.

David L. Waltz, "Artificial Intelligence," *Scientific American*, October 1982.

Richard O. Duda and Edward H. Shortliffe, "Expert Systems Research," *Science*, April 15, 1983.

James Dray and Joseph A. Menosky, "Computers and a New World Order," *Technology Review*, May–June 1983.

William B. Gevarter, "Expert Systems: Limited But Powerful," *IEEE Spectrum*, August 1983.

"Tomorrow's Computers", special issue of *IEEE Spectrum* edited by Edward A. Torrero, November 1983.

Marvin Minksy, "Why People Think Computers Can't," *Technology Review*, November–December 1983. Interesting discussion by the Donner Professor of Science at MIT and the founder of MIT's Artificial Intelligence Laboratory. In the same issue of *Technology Review* may be found Joel N. Shurkin, "Expert Systems: The Practical Face of Artificial Intelligence," which is well written and contains lots of real examples.

Paul Kinnucan, "Computers That Think Like Experts," *High Technology*, January 1984. On expert systems.

The *Proceedings of the IEEE* had a special issue on supercomputers and their impact on science and technology in January 1984.

Dwight B. Davis, "Supercomputers: A Strategic Imperative?" *High Technology*, May 1984.

Britain's Fifth Generation effort, heralded by the Alvey Report (September 1982) and launched in November 1983, was featured in the *Financial Times*, September 6, 1982, April 29, 1983, and May 9, 1983; *New Scientist*, October 14, 1982, and the *Sunday Times*, May 1, 1983.

Apart from *Newsweek* and Feigenbaum and McCorduck, the Japanese are discussed in:

Peter Marsh, "Japan Plans New Generation of Computers," *New Scientist*, July 16, 1981.

Philip C. Treleaven and Isabel Gouveia Lima, "Japan's Fifth Generation Computer Systems," *Computer*, August 1982.

B. L. Buzbee, R. H. Ewald, and W. J. Worlton, "Japanese Supercomputer Technology," *Science*, December 17, 1982.

Rex Malik, "Japan's Fifth Generation Computer Project," *Futures*, June 1983.

Pamela McCorduck, "Introduction to the Fifth Generation," and Ehud Y. Shapiro, "The Fifth Generation—A Trip Report," *Communications of the ACM*, September 1983.

Books

Joseph Weizenbaum, *Computer Power and Human Reason: From Judgement to Calculation* (W. H. Freeman, San Francisco, 1976, and Penguin, Harmondsworth, England, 1984).

Margaret A. Boden, *Artificial Intelligence and Natural Man* (Harvester, Brighton, 1977).

Patrick Henry Winston, *Artificial Intelligence* (Addison-Wesley, Reading, MA, 1977).

Pamela McCorduck, *Machines Who Think: A Personal Inquiry into the History and Prospects of Artificial Intelligence* (W. H. Freeman, San Francisco, 1979).

Herbert A. Simon, *The Sciences of the Artificial*, 2nd ed (MIT Press, Cambridge, MA, 1981).

A. Barr, P. R. Cohen and E. A. Feigenbaum, *The Handbook of Artificial Intelligence*, 3 vols (Kaufmann, Los Altos, CA, 1981).

J. E. Hayes, D. Michie and Y.-H. Pao (eds), *Machine Intelligence 10* (John Wiley, New York, 1982).

John Kirk (ed.), *When Machines Replace People* (Society for Social Responsibility in Science, Canberra, Australia, 1982).

J. E. Hayes and D. Michie (eds), *Intelligent Systems: The Unprecedented Opportunity* (Ellis Horwood, Chichester, England, 1983).

Geoff Simons, *Are Computers Alive? Evolution and New Life Forms* (Harvester, Brighton, England, 1983).

Igor Aleksander and Piers Burnett, *Reinventing Man: The Robot Becomes Reality* (Kogan Page, London, 1983; Penguin, 1984).

3 The Telecommunications Explosion

Evolution of the Intelligent Network

John S. Mayo

This outline of the evolution of the modern telephone network from 1876 to 1951 provides essential background to the new revolution in telecommunications. The author shows how microelectronics has since made possible the "intelligent" digital network, which supports a wide variety of new services and has enormous potential. Mayo is an executive vice-president with Bell Laboratories, New Jersey, and this article first appeared in Science, *February 12, 1982.*

Anyone, at any time and virtually anywhere in the US, can pick up a telephone and be connected almost instantly to any of about 175 million other telephones in the country and many more abroad. Placing a telephone call—something Americans do more than three-quarters of a billion times per day—is a simple task. But the nationwide telecommunications network that makes this possible is the largest and most complex mechanism on earth, capable of interconnecting both people and machines and of carrying not only telephone calls but also a wide variety of other signals such as data communications, television, facsimile, and teletype. It is an "intelligent" network, and it is our society's key link to the information age.

This article focuses on the evolution of the nationwide telecommunications network, whose facilities are owned by about 1500 independent telephone companies and other common carriers as well as the Bell System. This network has become the world's most sophisticated information-handling system by steadily incorporating state-of-the-art technology based on fundamental scientific advances. Increasingly, information in the network is handled by high-speed, digital transmission systems connected to electronic switching systems that are controlled by digital computers.[1] The network is called intelligent because these computers perform extremely complex tasks and can be reprogrammed to perform new tasks and provide new services without rewiring or physical modification. Increasing intelligence is making the network more versatile and flexible, enabling it to provide a wide variety of information-age telecommunications services.

The facilities of the nationwide network include 22,000 switching centers and over 1 billion miles of transmission paths. Its trillions of interrelated parts, all designed to interact compatibly and reliably on user command, can promptly complete any of 6×10^{15} possible connections. Overall, the network presents a capital investment of about $170 billion—about £1000 worth of equipment behind every telephone.

The network's ability to adapt rapidly to changing conditions saves money by making the most efficient use of all the equipment in the network. It also permits quick and effective communication during disasters and emergencies. During the 1979 nuclear accident at Three Mile Island, for instance, emergency calls poured into and out of the accident area. To handle the surge and avoid blockages, calls were rerouted according to the coordinated, preplanned network management strategy that allowed 1,745,000 calls—almost double the normal traffic volume—to flow smoothly into and out of the area on a single day. In 1975, when a major fire destroyed a New York City telephone switching office, many callers in the New York area never realized that they were speaking to one another via connections set up through Los Angeles. And during the New York City blackout of 1965, the telephones continued to work.

Telecommunications is such an essential part of our nation's social and economic fabric that the network's computers must be even more reliable than the best general-purpose computers. The computer that controls an electronic switching system, for example, includes redundant hardware and self-diagnostic software designed to keep the system operating trouble-free for all but two hours of its 40-year lifetime. The network is never shut down for maintenance, checkups, repairs, or improvements, so stringent reliability standards apply to other network elements as well. Continuous modification and improvement—from the installation of a 100,000-circuit switching system to the nationwide introduction of Touch-Tone signaling—are accomplished without any interruption in service.

Today's network is an accomplishment largely of the last few decades, but it was made possible by a stream of improvements over the past century. All have been driven by the nation's communications needs and made possible by intensive research and development in many areas of science and technology. In addition, all improvements have been based on planning for the network as a whole—a hallmark of telephony from its beginning. Advances in telecommunications technologies gave birth to the solid-state and data-processing industries. As a result, the US not only has enjoyed the world's finest telecommunications service, but also is leading the world into an age in which most of the workforce will create, process, or disseminate information.

From telephony to telecommunications

During three-quarters of a century, from Alexander Graham Bell's invention of the telephone in 1876 to the introduction of direct distance dialing in 1951, the goal of telephony was the achievement of "universal service"—providing telephone service to anyone who wanted it at reasonable cost. This goal, proclaimed by AT&T president Theodore Vail in 1907, was later codified by Congress in the Communications Act of 1934. But several difficult technical challenges had to be overcome before universal service could be achieved.

The first was the need to find efficient ways to interconnect telephones. Even before the turn of the century, the telephone instrument itself had become fairly rugged and reliable. As the number of telephones grew arithmetically, however, the number of potential connections among them grew geometrically, and routing calls quickly became a problem. The concept of a central office—to which a large number of local telephones were commonly linked—enabled operators to route calls through plug-in switchboards. In 1889 A. B. Strowger invented a dial telephone system that paved the way for users themselves to place calls through the central switching office. Such advances spurred the rapid growth of local urban telephone networks.

The next challenge, interconnecting local networks into a national network, required systematic planning and establishment of network-wide standards for performance and design. It also required great improvements in the quality of voice transmission over long distances. Some of the best scientific and engineering minds of the day tackled this problem. M. I. Pupin's invention of the loading coil boosted signal strength sufficiently for calls to be made between cities, but it was Lee De Forest's 1906 invention of the Audion, made practical as a vacuum-tube amplifier by H. D. Arnold, that allowed operators to set up the first long-distance calls from coast to coast in 1914.

The wave filter and H. Black's 1927 invention of the negative feedback amplifier further improved voice quality. In the 1930s and 1940s, development of carrier modulation techniques led to multichannel coaxial cable and microwave radio transmission systems that dramatically lowered the cost of long-distance calling.

Once local networks were interconnected in a reliable national network, the volume of telephone traffic increased so rapidly that switching equipment became strained, and more and more operators were needed to handle administrative functions. This posed a third challenge—the development of mechanized equipment. Rudimentary intelligence began to appear in automatic systems that recorded billing information in a form suitable for machine processing. In the 1930s plans were made for a hierarchy of switching systems, which would be necessary to handle nationwide traffic automatically.

Development of the crossbar switching system, which used electromechanical relays to perform intelligently, greatly increased the network's capabilities. In earlier "step-by-step" switches, the equipment used to set up connections was also required to maintain them. But in crossbar switches, the equipment used to establish connections was then released to handle other calls. "Common" control equipment also could "remember" each called number, select a route through the switching system, and seek alternative routes if necessary.

The culmination of these advances was the introduction of direct distance dialing in 1951. Nationwide direct dialing was the result of a systems engineering effort that included establishment of a network hierarchy and routing plan, a standard nationwide numbering plan, and systems that could handle automatic billing and perform all the necessary control functions. With direct distance dialing, the telecommunications network could automatically interconnect tens of millions of users through thousands of switching centers across the country.

Science and technology were also transforming telephony into telecommunications as the network began to interconnect machines as well as people and to carry pictures and data as well as voice. As early as 1924, a method for transmitting pictures over telephone lines, developed by H. E. Ives, was used at the national political conventions. Telephoto was soon followed by work on television, and in 1927 the first long-distance transmission of television in the US was achieved. Teletypewriter service was introduced in 1931 and was followed by improvements in systems and devices for transmitting digital data over telephone circuits. By the 1950s this work resulted in dataphone service, which made it possible for users to transmit information from outlying points to a central computer installation and then to distribute the results. In addition, by the 1950s the telecommunications network was distributing nationwide radio and television programming.[2,3]

Over its first 75 years, the network had become a masterfully construed machine capable of providing virtually everyone with low-cost, high-quality telephone service. It was on its way to becoming a versatile system capable of handling information in a variety of forms, and on the horizon lay several powerful technologies that clearly marked the course for the future.

Key technologies

Prior to World War II, people of foresight had recognized the limitations of the vacuum tube. In 1939 M. J. Kelley of Bell Laboratories established a research program with the explicit goal of funding a smaller, more reliable amplifying device. The search paid off magnificently in 1947, when J. Bardeen, W. H. Brattain, and W. Shockley invented the transistor. The transistor made possible solid-state integrated circuits and modern telecommunications. With rapid dissemination of transistor technology through publications, symposia, and demonstrations, the transistor's invention put the US far ahead in technnology, created a solid-state electronics industry with over $10 billion in current sales, and revolutionized data processing and consumer electronics.[4]

Even before the transistor was invented, A. H. Reeves, working with vacuum tubes, suggested in 1937 a feasible means of using on–off signals to transmit voice. The approach, called pulse code modulation, measures the amplitude of the speech wave form about 8000 times a second and converts each measurement into a binary number. The sequence of binary numbers is transmitted in the form of electrical pulses, and the speech wave form is reconstructed at the receiving end. There were early applications of pulse code modulation in military systems. But vacuum tubes were unreliable, power-hungry, and costly. The availability of the transistor made pulse code modulation practical for commercial applications.

The transistor also had a great impact on digital computers. In 1937, G. R. Stibitz at Bell Laboratories used relays and other telephone equipment to construct the first electrical digital computer. Soon afterward, he demonstrated that a remote terminal could be linked to a host computer via telephone lines.[5] During World War II the needs of the military—particularly for fast, accurate fire-control systems—hastened progress in the computer art. Harvard University's Mark I computer, the University of Pennsylvania's ENIAC, and J. von Neumann's conception of a stored-program, general-purpose computer demonstrated the potential of machine intelligence for a wide range of applications. Electrical computers soon left their mechanical predecessors far behind. In 1958 J. H. Felker and his associates at Bell Laboratories built the first fully transistorized digital computer. The transistor proved to be an ideal device for use in digital systems, and the digital computer revolution quickly followed.

At about the same time, transistor technology meshed neatly with rapidly evolving space technology, leading to the success of J. R. Pierce's proposal for communications satellites. After several experiments with passive reflecting satellites, the Telstar satellite—the first radio repeater to be placed in earth orbit—was launched in 1962. With Hughes Aircraft's development in 1963 of the Syncom synchronous satellite, which remained "fixed" over one point on the earth's surface, satellites opened the door to instant communications throughout the world. Although satellite technology has had a modest impact within the nationwide network, it is now being used extensively in point-to-point communications and distribution of many cable television channels, and is being developed to provide extensive digital communications capability.

Another breakthrough occurred in 1957, when C. H. Townes and A. Schawlow proposed the laser. This technology soon evolved into continuously operating gas lasers and tiny, reliable solid-state lasers. The extremely intense, highly directed beam of coherent light from solid-state lasers has enormous information-carrying capacity.[6] But an efficient transmission medium for light had to be found. The ideal solution came with the emergence of transparent glass fibres which "conduct" photons even more efficiently than copper wires conduct electrons.[7,8] Both lasers and fibre lightguides are ideally suited to high-speed digital transmission. The laser is simply turned on and off by the digital pulse stream. Glass fibers are now becoming economically competitive with copper cable,[9] and lightwave (or fiber optic) technology, already well established in the network, is expected to permeate all transmission elements.

Putting the key technologies together

The potential impact of the transistor on the telephone network was recognized immediately, and Bell Laboratories began planning for the introduction of solid-state digital technology. Unlike analog technology, which represents information as continuously varying electrical currents, digital technology represents information by the presence or absence of electrical pulses. Although digital technology requires greater transmission bandwidths than analog, digital signals are more robust and digital techniques are extremely well suited to solid-state electronics and to computer-generated information.

Early solid-state circuits, though costly, were promptly used in two digital systems: the T-1 carrier system, which placed 24 voice channels on a single pair of wires in ordinary telephone cable, and the No. 1 electronic switching system (EES),[10] which used the techniques of a digital computer to control the switching of telephone calls. Both applications of solid-state digital technology proved remarkably successful. The T-1 system went into service in 1962,[11] and the Bell System now has 130 million circuit-miles of digital transmission facilities—much more than exist in the rest of the world. The No. 1 EES went into service in 1965, and today thousands of computer-controlled electronic switching systems offer a wide variety of advanced services to both residential and business users.

The potential of a largely digital telephone network also was quickly recognized. The T-1 digital carrier system interconnected local switches and coupled them with the large toll switches that handled long-distance calls. In the 1960s the growth of T-1 systems, along with the rapidly declining cost of digital integrated circuits, suggested that, rather than switching analog voice signals, large toll switches might switch the digital pulse streams carried by T-1. Digital switching could greatly simplify the transmission-to-switching interface. In addition, it would allow time-division switching, in which some switching paths are time-shared among many calls. In certain cases, this would offer cost and operational advantages over space-division switching, in which a path through the switch is dedicated to a single call for its duration.

Following a massive research and development program, which cost $400 million and took 2500 person-years of effort, the No. 4 ESS "superswitcher" began commercial service in 1976. This machine, which is capable of handling more than ½ million calls per hour, is based entirely on the technology of the information age. It is a solid-state, time-division switch for digital signals, controlled by an ultra reliable digital computer containing several million words of programmed instructions. In addition, the No. 4 ESS offers great flexibility through software changes in the network that controls it. It also provides a convenient digital access point in the heart of the network, facilitating attachment of equipment frames for new network services. And it marks a giant step towards a largely digital nationwide network.[14]. In addition, existing space-division electronic switching systems are being programmed to switch digital pulse streams as well as analog signals.

Similarly, advancing technology is extending digital transmission to extremely high rates and is making it economic on long as well as short routes. For example, many small, digital lightwave transmission systems are now in service, and some large systems are being planned and built. The Bell System is installing a high-capacity lightwave system over a 600-mile route connecting Boston with Washington, DC. Another large lightwave system is being installed on the West Coast, and the first lightwave systems, microwave radio and coaxial cable routes, along with satellites, are being used to transmit both digital and analog signals.

A largely digital network requires a means for delivering high-speed digital signals directly to homes and businesses. This is now becoming possible, mainly because of the falling cost of electronic circuits. For instance, a digital transmission system called SLC-96 (subscriber loop carrier 96) now employs ten pairs of copper wires to provide each of 96 users with 56,000 bits of information per second.

Two additional developments have helped bring the key technologies together into the systems and services of the intelligent network. One is the use of a dedicated data network to interconnect the computers that control call-handling functions. This interconnection, known as common channel interoffice signaling (CCIS), has added enormous flexibility to the network and has formed the world's largest packet-switched data system. The other development is the use of large databases in the network. These databases and CCIS, working together as the result of network-wide planning and systems engineering, have made possible new call-handling functions. For example, they allow a user to dial one telephone number, have the computer that controls the local switching machine query a remote database over CCIS, and have the call routed to another number stored in the database.

Thus, an enormous amount of research and development translated the key technologies—solid-state devices, pulse code modulation, digital computers, satellites, lasers, and lightguides—into real communications systems. At the same time, a comprehensive systems engineering effort at Bell Laboratories enabled both the independent telephone companies and the Bell System to merge new technology smoothly with the old. As a result, the network evolved in a constructive, efficient manner. This deliberate yet rapid evolution has opened the door to a largely digital, multipurpose intelligent network and to the myriad new service capabilities of the information age (see figure 3.1).

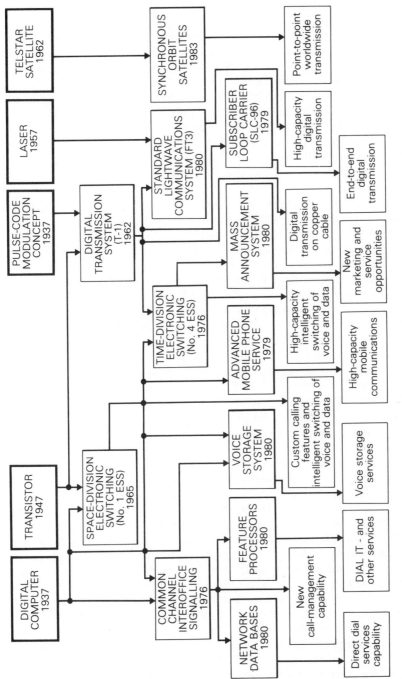

Figure 3.1 Evolution of the intelligent telecommunications network. Five key breakthroughs were the transistor, digital computer, pulse code modulation concept, laser and lightguide, and communications satellite. These have led to sophisticated service-providing systems.

Service today and tomorrow

The intelligent network supports a wide variety of new services today and offers enormous potential for tomorrow. For example, the ability to dial one telephone number and have the call automatically routed to another has numerous applications. Nationwide franchises could advertise a single number throughout the country. When that number is dialed, the call could be automatically routed to a preprogrammed destination—for example, to the nearest local franchise known to be open at that time of day. Also, a telephone number could be assigned to a person rather than to a telephone set. By dialing into the network database one's itinerary and the telephone numbers at which one can be reached, all calls could be automatically routed to an individual.

The intelligent network is now allowing more users to dial credit-card or bill-to-third-number calls without operator assistance. A range of call management services, including calling number display and special ringing for calls from selected phones, could give people more control over their communications services. For businesses, several new services based on a mass announcement system developed for No. 4 ESS will open up new marketing and service opportunities. DIAL-IT services let users dial a single nationwide number to receive recorded announcements and make possible instant national polls such as the one following the Carter–Reagan debate in 1980.[15]

While traditional telephone services help conquer distance, new services could also help conquer time. The computers within the network could interact with a digital system that stores voice messages. This would allow calls to be placed or received even at times when the caller or recipient is not at a telephone. For instance, users could record voice messages for delivery at specified times to specified telephone numbers. Or they could have their incoming calls automatically answered and recorded for later playback.

The intelligent network is also at the heart of the advanced mobile phone service. This mobile communications system, first demonstrated almost two decades ago, uses an intelligent switching system to "hand off" calls to another radio channel as a vehicle moves frrom one area to another. Thus scarce radio frequencies can be reused many times, and mobile service can be provided to hundreds of thousands of users in urban areas (see figure 3.2).

Other services now being planned and developed will take advantage of both the network's intelligence and its expanding digital capabilities. In the office of the future, for instance, new services will improve productivity. Flexible, software-based PBXs will make it easier for intelligent terminals to access host computers,[16] and advanced digital PBX controllers will provide integrated electronic mail, automated filing and retrieval, text editing, and voice and data communications. Widespread use of teleconferencing, now available through Picturephone Meeting Service in a dozen cities will eliminate needless travel and save time, money, and energy. And business systems such as the DIMENSION PBX, now used to control energy consumption in

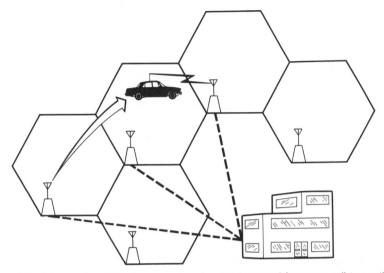

Figure 3.2 Advanced mobile phone service. As vehicles travel from one cell to another, electronic switching equipment automatically hands off calls, without interruption, from one radio transmitter to another.

hotels and motels, will find even broader use for monitoring and controlling energy consumption.

Other intelligent communications capabilities will have dramatic effects on office routines. For example, private communications services now use the network's intelligence to provide advanced voice and data services to large organizations, allowing users to configure and control their own service features. The packet-switched Advanced Communications Service, along with such existing packet-switched networks as Telenet and TYMNET,[17] will interconnect a wide variety of computer and display terminals. And an end-to-end digital communications capability will soon be added to the network. It will flexibly and economically support a variety of public needs, allowing people and machines to set up high-speed connections as easily as they now make telephone calls.

Ultimately, the expansion of digital technology and user demand for data-oriented services will lead to realization of the integrated services digital network (ISDN). This public, nationwide, end-to-end digital capability will let the user choose the type of terminal to be used, the rate at which information will be transmitted, and the size and density of messages—and will also allow simultaneous provision of a variety of voice and data services.[18]

In the home, new communications services based on the capabilities of the intelligent network will make life richer and more secure. "Smart" telephones, such as the Touch-a-matic series, are already widely available. In the future, "teleterminals" combining features of both the computer terminal and the telephone will raise even more exciting possibilities, such as the electronic home information systems already under trial in the US and abroad.

In Florida, for instance, AT&T and Knight-Ridder are providing 160 families with news, information, and in-home shopping services through a modified television set with keyboard and controller. And in North Carolina, a system now being tried uses the network to link users' homes with energy utilities. The system lets consumers control their energy consumption, while the utilities can read meters remotely, detect tampering, and shut off service when users move. Farther in the future, local transmission systems will allow bursts of digital data to travel side by side with voice signals on ordinary home telephone lines, improving the economic potential of home information systems. Sensors linked to such transmission systems might let people control their energy usage and enjoy improved home security.

Dealing with complexity

Today's network is so complex, and its parts so intimately interrelated, that efficient operations require the use of numerous computer-based support systems. The rapid growth of such systems has paralleled the growth of the network's intelligence. Over 100 different types of systems, running on over 4000 minicomputers and 300 maxicomputers, are now accessible over more than 100,000 computer terminals. These systems are deployed in such locations as PhoneCenter stores, business offices, central switching offices, repair centers, and remote equipment installations, as well as factories and laboratories.

For example, one set of operations systems gathers information every five minutes from each switching system in the country. This information, collected at 27 regional centers, is sent to the national Network Operations Center in Bedminster, New Jersey, where it is displayed on a large wall map. Although the computer-based systems coordinate the multitude of automatic instructions necessary to monitor network operations on such a grand scale, control of the network ultimately depends on the skill and judgment of human managers.

Other operations systems, such as CAROT (centralized automatic reporting on trunks), continually measure the performance of over 5 million transmission circuits and compile and analyse the resulting data. TIRKS (trunks integrated records keeping system) keeps track of over 25 billion bits of information concerning interoffice trunk circuits; PICS (plug-in inventory control system) controls inventories of 80 million plug-in circuit boards worth over $7 billion; and LMOS (loop maintenance operations system) keeps records on 60 million lines. Still other systems replace paper records, monitor the status of equipment, generate reports, mechanize testing of facilities, keep track of equipment needs, and carry out a variety of other routine but necessary tasks. Without these operations support systems, it would be extremely difficult, if not impossible, to keep today's intelligent network operating.

The challenges ahead

For over a century, steady progress has been made in network capabilities. But in the past few decades, the pace of change has quickened. As information-age services proliferate, the development of software technology has become a major challenge. It is largely the software—the detailed programs of instructions that tell the hardware what to do—that makes the network intelligent. In fact, the network can be viewed as an enormous software package where system architecture and software, rather than hardware, form the enduring framework. The intelligent network, like all other computer-based systems, has an enormous appetite for software. Yet software development remains more of an art than a science. Currently the principal way to get more software is to use more programmers, so improvements in programming productivity are sorely needed.[19]

Programming is being improved somewhat through development of better tools for programmers. For example, the UNIX family of computer operating systems, initiated by K. Thompson in 1969, includes a unified package of tools called Programmer's Workbench that helps Bell System programming teams create and modify large software systems. Ultimately, more dramatic increases in programming productivity might result from using more powerful hardware. The cost of digital integrated circuits has dropped by a factor of 1000 in the past 20 years, and hardware costs are expected to continue to fall at least until the end of this decade. Integrated circuits will soon contain one million components on a tiny silicon chip, which will cost about the same as present chips with fewer components. These highly complex chips should allow more modular and functional hardware to simplify and reduce the cost of software. Also, more powerful chips should make possible better and more economic computer aids to software design.

Another challenge is keeping telecommunications services simple to use. It is no accident that almost anyone, even a child, can use a telephone without special training. Human factors work has gone into every aspect of the telephone's operation, from the shape of the dialing pad to the nationwide numbering scheme. The same kind of "engineering for people" must be applied on an even wider scale to the design of increasingly complicated services and terminals.

For everyone to have access to the benefits of the information age, "friendly" interfaces with complex machines are needed. Present efforts to make machines friendlier include development of simplified computer operating systems and languages, and advances in display and sensing technologies. In addition, work on speech synthesis and voice recognition will permit people to communicate with machines in ordinary English.

Still another challenge is to make the terminals and devices attached to the ends of the network as reliable and trouble-free as the network itself. For instance, fire and burglar alarm systems that send warning signals through the telecommunications network have great potential for improving home security. But fire and police departments cannot tolerate a large percentage of false alarms, so alarm systems must be intelligent enough to distinguish real emergencies from false ones. The same kind of careful systems engineering and the same approach to reliability that helped build the intelligent network must be applied to the intelligent terminals and sensors that interact with it.

The capabilities of the intelligent network have evolved in close synchronization with advances in science and technology. But in recent years the merging of telecommunications and computer technologies has raised difficult regulatory issues with potentially profound effects on the evolution of the network. There are many points of view on these issues. But on at least one point there seems to be consensus: that the benefits of rapid technological progress—new products and services—must continue to become promptly and widely available.

Notes

1 The computers that control electronic switching systems differ from general-purpose computers principally in their added hardware and software features to achieve high reliability, the speed with which they execute some operations, and their ability to respond instantaneously to service demands. The computers referred to in the later discussion of operations support systems are general-purpose computers.

2 P. C. Mabon, *Mission Communications: The Story of Bell Laboratories* (Bell Telephone Laboratories, Inc., Murray Hill, NJ, 1975).

3 J. Brooks, *Telephone: The First Hundred Years* (Harper & Row, New York, 1975).

4 *Microelectronics, A Scientific American Book* (Freeman, San Francisco, 1977).

5 G. R. Stibitz, in *History of Computing in the Twentieth Century*, N. Metropolis, J. Howlett, G.-C. Rota (eds) (Academic Press, New York, 1980), pp. 479–783.

6 J. A. Giordmaine, *Appl. Opt.*, **11**, 2435 (1972).

7 K. C. Kao and G. A. Hockham, *Proc. IEEE*, **113**, 1151 (1966).

8 F. P. Kapron, D. B. Keck, R. D. Maurer, *Appl. Phys. Lett.*, **17**, 423 (1970).

9 J. B. Macchesney, P. B. O'Connor and H. M. Presby, *Proc. IEEE*, **62**, 1280 (1974).

10 "No. 1 Electronic Switching System" (special issue), *Bell Syst. Tech. J.*, **43**, 1831 (1964).

11 K. E. Fultz and D. B. Penick, *Bell Syst. Tech. J.*, **44**, 1405 (1965).

12 "No. 4 Electronic Switching System" (special issue), *Bell Syst. Tech. J.*, **56**, 1015 (1977); **60**, 1041 (1981).

13 H. E. Vaughan, *Bell Syst. Tech. J.*, **38**, 909 (1959).

14 A. E. Joel, Jr, *IEEE Trans. Commun*, COM-27, 948 (1979).

15 W. O. Fleckenstein, "Switching Technology and New Network Services," paper presented at the International Switching Symposium, Montreal, Canada, September 1981.

16 "SL-1 Digital PBX" (special issue). *Telesis* 4, (no. 3), 66 (1975).

17 J. Martin, *Networks and Distributed Processing* (Prentice-Hall, Englewood Cliffs, NJ, 1981).

18 I. Dorros, *IEEE Commun. Mag.*, **19** (No. 2), 16 (1981).
19 F. P. Brooks, Jr, *The Mythical Man-Month: Essays on Software Engineering* (Addison-Wesley, Reading, MA, 1975).

Telecommunications Liberalization

Business Week

The convergence of computing and telecommunications has brought about the liberalization of the telecommunications industry and the deregulation of long-established PTT monopolies. The resulting innovation explosion has in turn created a booming worldwide market for new telecommunications equipment, as described in this report which appeared in Business Week, *October 24, 1983.*

The emerging global battle

For nearly a century, telecommunications was a sleepy, provincial business. Each country had its own Postal Telephone & Telegraph (PTT) administration controlling everything related to communications, from the phone on the desk to the phone line overseas. And local equipment makers carved out cozy relationships with their PTTs.

That world is fast coming apart. Technology is changing so rapidly that equipment formerly enjoying a 30-year life cycle is becoming obsolete almost as soon as it is installed. As quickly as new technology creates new products, customers everywhere want a wider choice of telecommunications services.

The single greatest impact on world communications is being made by digital computers. As these machines multiply throughout business, telecommunications becomes an important strategic weapon to all companies. Instead of needing communications systems to transmit phone calls and telex messages, companies must now have them for such tasks as sending huge volumes of computer data at high speeds, transmitting facsimiles of blueprints, and holding video conferences. "Customers are demanding more of their communications systems, because, to be competitive in the world market, they have to automate more functions," says Robert E. La Blanc, a New Jersey consultant and former vice-chairman of what is now Continental Telecom Inc.

These powerful winds of change are forcing most governments to re-examine the regulations governing their PTTs. "One force that will drive them to a liberalization is the convergence of computers and telecommunications technology," declares Robert J. Gressens, president of GTE International Inc. "They are either going to have to expand their monopoly and regulation [to include computers] or else go in the other direction and liberalize."

Growing numbers of PTTs around the world are starting to loosen regulations and permit competition. "No monopoly can match all of the needs," says Björn Svedberg, president of L. M. Ericsson, the big Swedish maker of communications gear. "It is the marketplace that is driving the deregulation."

The stakes in this new global battle are enormous. This year nearly $59 billion will be spent on communications equipment alone, estimates Edgar A. Grabhorn, managing director of Arthur D. Little Inc.'s world telecommunications information program. By 1988, he predicts, the annual market will grow 50 percent, to $88.4 billion. The lion's share of equipment purchases is being made by the PTTs for switching equipment, cable, and other network gear.

But as deregulation opens up markets, selling directly to the users is becoming more important to makers. This year, users will buy nearly $19 billion worth of equipment ranging from simple telephones and telex machines to sophisticated private branch exchange (PBX) telephone systems. In addition, Little's Grabhorn estimates that expenditures for telephone calls, telex messages, and other transmission services will run three to four times the money spent on hardware.

American roots

The worldwide upheaval traces its roots to the US, the world's largest telecommunications market, which has been dramatically deregulated over the past 10 years culminating in the split-up of American Telephone & Telegraph Co, the closest thing to a PTT in the US.

Liberalization in the US has resulted in a plethora of new communications networks, services and products coming from such US suppliers as GTE, MCI Communications, and Rolm. At the same time, long-established foreign manufacturers—Britain's Plessey, West Germany's Siemens, and Japan's NEC, as well as Ericsson—began attacking the wide-open US market.

Faced with this new competition at home, the American companies, led by Western Electric Co., AT&T's giant manufacturing arm, began looking for their own export markets. "The US market will still be very strong in communications for the next 10 years," says John G. Puente, vice-chairman of M/A-Com Inc., a US communications equipment maker. But exports are essential, he says, because "all of our competitors are going global—you have the whole world really growing in communications."

To keep pace, many of the makers of switching equipment are looking outside their traditional markets and are lining up new international business arrangements. One of the best examples is the cooperative venture that was launched recently by AT&T and Philips. At the same time, the merging of telecommunications and data processing is prompting a new wave of joint efforts by such teams as Ericsson/Honeywell and IBM/Rolm to strengthen their assault on the user-equipment markets. "The style of the [telecommunications] business is changing," says Gerrit Jeelof, a member of the management board at Dutch-based Philips. "There is a completely new competitive environment."

Giant ITT, the world's second largest communications equipment maker, is also looking for business partners. The company is inviting more local ownership of some of its overseas subsidiaries: Last year, ITT reduced its ownership of Standard Telephones & Cables, its British subsidiary, from 75 to 35 percent. And it is establishing joint ventures, such as the one set up last July with National Postal & Telecommunications Industry Corp. of China to make switching equipment for the Chinese market. Such arrangements are necessary now, says John W. Guilfoyle, president of ITT Telecommunications Corp., because "no company can afford to develop a network switch without export markets."

Telecommunications has become the object of this mad scramble, because it has become so critical to the world's economy. The investment in communications plant and equipment by such industrialized nations as the US, France and Japan already represents as much as 9 percent of their total gross domestic product, according to the International Telecommunications Union, a UN agency.

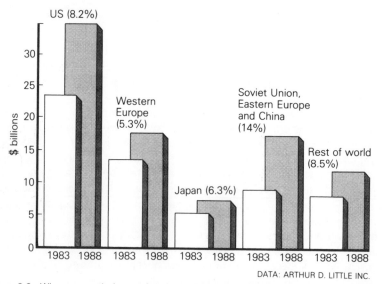

DATA: ARTHUR D. LITTLE INC.

Figure 3.3 Where growth is coming in world communications markets: some annual growth rates.
Source: Arthur D. Little Inc.

At the same time, a nation's telecommunications infrastructure is becoming essential in maintaining a good business climate. Without modern, flexible, and competitively priced communications facilities, a country runs the risk of slowing overall economic development, if not stopping it all together. "Telecommunications are as essential to business infrastructure as good highways," says Donald Smith, executive vice-president for business development at Canada's Mitel Corp., an equipment maker.

Digital revolution

Underlying the new uses for telecommunications is the technical revolution in the world's networks. Yesterday's copper wires and mechanical switching relays were designed to carry only a single type of information: telephone calls, telegrams, or telex messages. But today's high-capacity optical fiber and sophisticated digital computers can transmit everything—voice, data, or image—in the same way, by converting them into a stream of computer on–off pulses. "Before there was room for different systems to work at the same time," explains Grabhorn. Now, he adds, "systems have to work together."

So telephone companies around the world are racing to renovate their networks. Once the equipment is updated to handle everything in digital form, networks become multifunctional links that can carry anything from a simple telephone call to computerized airline reservations and television pictures. At the same time, new ways of moving digital information, such as microwave radio, satellites, and optical fiber cables, are dramatically lowering the cost of sending messages across vast distances.

These fundamental industrial and technical changes sweeping the US are creating enormous pressures for a similar wave of deregulation in Europe and Japan. The pressure is also coming from equipment makers that want to sell overseas to expand their markets so they can compete with US companies. Putting on even more heat are foreign phone customers who want the latest in technology and services. "There's no reason why the PTT shouldn't restrict itself to examining and approving the equipment and letting the client decide what to buy," says Wilhelm Hübner, managing director of Verband der Postbenutzer, a German PTT user association.

But deregulation is a sticky political problem for almost every government. Because the government-owned PTTs are such large employers and are so heavily unionized, any attempt to turn them into competitive private companies invites a nasty political backlash from vast numbers of civil servants. What is more, the PTTs generally contribute healthy profits to government coffers.

Even more threatening, however, is the damage that deregulation could do to domestic manufacturers that have been growing lazy and fat off protected home markets. Telecommunications is one of the few high-technology areas where European manufacturers have been successful. And the industry is a large employer. In Sweden, for example, communications accounts for 3 percent of manufacturing employment.

Any move toward deregulating the PTT monopoly and introducing competition for equipment or services would make it increasingly difficult for governments to nurse traditional suppliers. "The opening of markets is more a political issue than a technological one," admits the chief executive of one large European equipment maker.

European political leaders on both the left and right reject any comparisons between the US and their countries. European nations, they say, are too small to support such diversity and competition among their manufacturers. "The European view is that the US can afford deregulation because it will not reduce the overall capacity to undertake the large research and development investments needed to be at the forefront of technology," maintains Henry Ergas, a telecommunications specialist at the Paris-based Organization for Economic Cooperation & Development (OECD).

Many Europeans also argue that any eventual benefits from deregulation would come at the expense of the general public, especially in rural areas. Telecommunications is first and foremost a "public service," which should be operated by the government for the general good, maintains Louis Mexandeau, France's Socialist PTT minister. In a competitive environment, he fears, small communities where the market is tiny would suffer at the hands of large ones. "Lots of people could well be hurt," agrees Stephen C. Timms, a consultant at Logica Ltd. in London.

Status quo

As a result, except for the US—and, to a lesser extent, Canada and Britain—governments are vigorously trying to maintain the status quo for their huge PTTs. "We have been able to provide quite a sufficient telecommunications system in the past, and we hope this will be possible in the future as well," says Konrad Schmidt, head of marketing at the Deutsche Bundespost, West Germany's PTT. "So far, nobody has furnished any proof that one approach is better than the other; as long as it is only a philosophical question, there is no need to change anything."

And even in those countries, such as Japan, that have publicly supported the idea of deregulating communications, the PTT is not necessarily throwing its equipment market open to foreign suppliers. "If Japan doesn't change, we will be far behind," warns Hisashi Shinto, the 73-year-old president of Nippon Telegraph & Telephone Public Corp., which is Japan's PTT. The Japanese "have signed various agreements with the US promising to open purchasing," agrees GTE's Gressens. But, he says, "we have yet to see the results of either of those."

Nonetheless, in most countries the telecommunications industry leaders, corporate communications users, and—in private—even many of the PTT officials believe that technological progress makes gradual liberalization inevitable. "The changes in policy are driven very much by the [pace] of technology in the marketplace," says Sir George Jefferson, chairman of British Telecom, his country's PTT. "In the long run, each country will make some adaptation."

Jefferson speaks from first-hand experience. Two years ago, British Telecom was set up as a government corporation, and the market was opened to competition. Now the British government is going one step further: A bill expected to win parliamentary approval shortly will authorize the government to sell 49 percent of British Telecom to the public.

Despite the deregulation talk in Britain, some of Telecom's aspiring competitors charge that it is still trying to preserve its monopoly. For example, they claim that, to hobble the competitive suppliers of telephones and other equipment, the British PTT has done its best to stall the publication of technical specifications for attaching outside equipment to its network. As a result, in two years the British Standards Institution has approved the specifications for only eight attachments.

But overall, competition appears to be having the desired effect. British Telecom has transformed itself into a dynamic, market-oriented business that is slashing prices, offering advanced new services, and in general trying to please rather than irritate customers. For instance, waiting time for renting a new circuit in London's financial district has dropped from several months to just a few weeks. Competition "has been a tremendous spur to British Telecom, and it has responded far better than anyone thought it would," says Stephen Finch, chairman of the British Telecommunications Managers Assn.

British Telecom's traditional equipment suppliers—Plessey, British General Electric, and Standard Telephones & Cables—were hardly enthusiastic two years ago over the prospect of their sole phone customer losing its monopoly. But today they acknowledge that the change has had benefits. "The liberalization has stimulated us to bring ourselves up to date with the rest of the world," says Ian W. Maclean, chief executive of the office-systems subsidiary of Plessey Co., which still counts on Telecom for 49 percent of its total sales.

There are even glimmers of change in countries more attached to their government PTT monopolies than Britain is. In socialist France, which has nationalized its equipment makers, the PTT's communications arm, Direction Générale des Télécommunications, has moved many of its new services under the control of semi-independent subsidiary companies, some of which involve private capital. For example, PTT Minister Mexandeau notes that farmers' organizations in southwestern France are using videotex to set up their own agricultural data services. "The PTT is only serving as the intermediary between the customers and the service provided," says Mexandeau.

New challenge

Although the French PTT may be willing to permit others to operate information processing services, it insists on maintaining its monopoly on the nation-wide backbone network. But the PTT's communications head, director Jacques Dondoux, notes that the cable TV networks being installed in France are apt to provide a new challenge to the PTT monopolies on transmission. "These local cable operators may very well want to amortize their investments by offering telecommunications services over their networks," says Dondoux.

In the face of such budding competition, the PTTs in Europe are counting on technological progress to help them maintain their monopolies. They argue that, once phone networks have been fully converted to multifunction digital systems, they will operate like public highways, carrying any type of information. That helps to explain why telecommunications administrations, especially those in Europe, are working as fast as they can to develop what they call the Integrated Services Digital Network. "For all PTTs, including the Bell System in the US [such a digital network] is the great white hope," says Ergas of the OECD.

The prospect does not overjoy corporate users, who know that these very sophisticated networks will be financed through higher communications tariffs. "Do we need this elaborate pipeline?" asks Ernst Weiss, chairman of the International Telecommunications Users Group (Intug). "There is concern that [the digital network] will be a monster, with no possibilities of change."

West Germany's Bundespost recently struck terror in the hearts of many large European corporations by modifying its tariff structure in preparation for the digital network, which it plans to begin operating by 1987. Starting in 1985, the Bundespost will charge companies with leased lines according to the amount of traffic they send over those lines, instead of the flat rates they now pay.

As a result, costs for the largest users, especially for the multinational corporations, are expected to skyrocket. One corporate telecommunications director describes these leased lines as the "jugular vein" for many large companies. Many users "are scared to death by what is happening in Germany," says Intug's Weiss. The user group also fears that the Bundespost is breaking ground for a concerted tariff restructuring by all of the European PTTs, through a secretive body known as the Conference of European Post & Telecommunications Authorities. This conference is a closed forum where European PTTs coordinate services and policies.

The Bundespost says the tariff change is necessary to protect its investments in the public networks. Otherwise, it says, large companies will use private leased lines for the bulk of their traffic and send only the occasional overflow via public networks. This kind of "cream-skimming," says the Bundespost's

Schmidt, "would put the Deutsche Bundespost at a great disadvantage."

User groups, as well as the International Chamber of Commerce, argue that they will still need affordable private lines because the PTTs are ill-equipped to satisfy demands for new services. In the long run, the chamber says, a "competitive environment" would not only best serve users but also increase the volume of communications, which would provide the PTTs with more revenues from their networks.

Breaking the PTT monopolies, the users argue, is also necessary if they are to get the most up-to-date information processing equipment. "European suppliers suffer from an enormous technological lag, and the reason is their close links to their national postal and telecommunications agencies," says Intug's Weiss. It is as if the electric utilities decided they would supply not only the electricity but also everything that plugs into an electric socket, explains the OECD's Ergas. If that were true, he says, "we would probably never have washing machines."

As far as most observers are concerned, the question now is not *if* the PTTs will deregulate, but how long they will hold out. If the PTTs wait too long, they could be bypassed by increasingly restive users for a large part of their communications needs. The tools for circumventing the PTTs could be provided by the rapidly changing technology and by the stepped-up worldwide competition among communications equipment manufacturers. British Telecom's Jefferson, agreeing with this view, comments: "If you sit on a chair on the beach until the waves have come over your head, you are liable to drown."

Business turns to new technology

Developing the Escort, Ford Motor Co.'s "world car," forced major changes in the way the multinational company builds automobiles. Assembling world cars in several countries and selling them around the globe also pressured Ford into reconsidering its communications network.

To ensure that its marketers and engineers can better coordinate their efforts on future world cars, Ford has centralized control of its worldwide telecommunications network. Previously, each region—Europe, Latin America, North America, and Asia-Pacific—was responsible for its own communications. "We are looking at product design and engineering worldwide and decided we needed closer communications," says Gerald L. Decker, Ford's manager of telecommunications services at that time.

At many companies, such systems were long considered merely utility overhead, like electric power. But Ford and growing numbers of other corporations are finding that a well-managed communications system can be an important strategic weapon for competing effectively. "Telecommunications is beginning to be the thread that enables us to pull everything together and to reach out for new markets," says William R. Synnott, senior vice president of the Information Systems & Services Div. at First National Bank of Boston. "It's more economically feasible to do things we could never do before."

At a minimum, the latest communications and computer technology enables businesses to deliver information more quickly to the people who need it and at lower cost than ever before. For example, First National of Boston is installing a data communications network linking its offices in Boston, New York, London, and Hong Kong. "In the past, it's taken days to take information for a loan application in London, mail it to Boston for review, and then mail it back," says the bank's Synott. With the new network, he says, "we can transfer a document from London to Boston in a minute."

Ultimately, telecommunications technology holds the potential for bringing about fundamental changes in the way a company operates. For example, Atlantic Richfield Co. has just installed a $17 million video tele-conferencing system that will connect its Los Angeles headquarters with offices in Philadelphia, Denver, Houston, Dallas, and Washington, DC. Now, instead of traveling to meetings, executives can speak to—and watch—one another on a television linkup without leaving their office buildings. In addition to the $10 million in travel expenses that Arco estimates it will save each year, the new system will allow closer management of far-flung facilities, the company hopes.

Such benefits are being made possible by the wide assortment of new equipment coming out of the merger of fast-moving computer and communications technologies. "I look at the whole market as being very explosive because of the advent of sophisticated technology," says Gus V. Morck, manager of electronics and telecommunications at Arco. At the same time, the spreading competition among suppliers is making this technology available to all users. "The freedom to choose among vendors is a boon; we can come closer to meeting our needs," says James F. McAloon, group manager for voice and data communications at Continental Illinois National Bank & Trust Co. of Chicago.

Setting up an effective international communications network, however, is not easy. Unlike the US, which will completely open up its market to competition by splitting up American Telephone & Telegraph Co. on January 1, overseas PTT administrations still keep tight control over their markets. And because different countries set different technical standards, communications equipment from one may not work in another.

Many of the PTTs still do not offer most of the basic services, let alone the advanced features, that US customers take for granted. For instance, throughout most of Europe it is impossible to get an itemized monthly telephone bill. Residential and business users alike receive lump-sum quarterly phone bills that they have no way of investigating. "It's unrealistic in 1983 to say you can't have [billing] information that's been readily available in the States for years," declares Stephen Finch, chairman of Britain's Telecommunications Managers Assn.

Compounding the problem, whatever is available in Europe or Japan is usually more expensive than the same type of product in the US. "For instance, a phone call from the US [to Germany] is about a fifth of what they cost in the other direction," says Herbert M. Harrell, telecommunications specialist with Kienbaum Unternehmensberatung, a West German consulting firm.

To circumvent these limitations and high costs, US multinationals are locating their foreign offices in cities with the most accommodating PTTs. "Business gravitates to places where it is easiest to get things done," says T. Patrick French, director of PTT liaison for American Express Europe Ltd, a division of AmEx's Travel Related Services Co. For example, Ford has decided to set up a communications hub in England, because British regulations permit the auto marker to establish a private electronic mail system that can service the company's European operations.

Such independence on the part of users is generating competition among suppliers worldwide. Other nations watching a growing number of corporations move to relatively liberal nations such as Britain, may begin to relax their regulations and make it easier for users to get what they need.

"The users will put pressure on government and PTTs to increase competition and services," agrees David J. Markey, head of the Commerce Dept's National Telecommunications & Information Administration (NTIA), which advises the White House on telecommunications. To help users deal with foreign PTTs, Markey foresees the NTIA taking on the role of arbiter. In the long run, he believes, if the PTTs fail to offer new services, they will not reap the necessary revenue to build new networks. Says Markey: "If you try to close the borders, it's easy to fall behind in technology."

Doing nicely

To most people, American Express Co. is one of the nation's leading financial service companies. After all, nearly 15 million people never leave home without the famous green American Express card, and millions more buy traveler's checks or insurance or trade securities through the company. But B. Garland Cupp does not see his company quite that way. "The business we're in is really the communications and information-processing business," says Cupp, who is executive vice-president of systems and operations for American Express Travel Related Services Co. (TRS).

Few if any companies, in fact, demonstrate as well just how important telecommunications and computers can be as corporate strategic tools. American Express spends $300 million to $400 million annually—some 4 to 5 percent of its $8.1 billion revenues last year—on information processing. From its Phoenix computer center, the company approves 250,000 credit-card transactions a day from all over the world—in an average of five seconds or less.

Besides providing customers with speedy service, the AmEx network lowers the company's operating costs. Even though the number of cardholders has increased by more than 50 percent over the past five years, AmEx has been able to hold losses resulting from the fraudulent use of credit cards at nearly the same rate. More important, says Cupp, AmEx's losses "are far less than those of our competition."

The company now uses its network as a vehicle for delivering new services. This year, AmEx has made it possible for customers to obtain refunds for lost or stolen traveler's checks by phone from any of its 800 automatic traveler's check dispensers.

As communications becomes a basic corporate tool, American Express is finding that it must pay more attention than ever to government regulations. The company has become an outspoken critic of regulations by several European PTT administrations that outlaw the routing of data from a private network to a public one unless the data are processed in some way at the interconnection point. Such laws could handicap the overseas transmission of credit-card authorizations from Phoenix. "If you compare it with transportation facilities," says T. Patrick French, director of liaison for TRS's American Express Europe Ltd, "it is like saying that you have to shower and change your clothes before you can get from a train into your private car."

American Express has found that the key to utilizing communications effectively amid the tumultuous technical, industrial, and regulatory changes under way is simply more attention from top management. "The chairman pays an enormous amount of personal attention to communications," says Cupp, who meets frequently with AmEx Chairman James D. Robinson III and updates him on the latest happenings in communications.

The scramble to sell

When Japan Air Lines Co. decided earlier this year to replace the antiquated telephone reservation system at its office at Tokyo's Haneda Airport, it could easily have turned to one of the Japanese suppliers of private branch exchanges (PBXs). Instead, JAL purchased three of the electronic switchboards from Rolm Corp., the California company that already had supplied the airline with similar systems in the US.

Such a purchase would have been impossible as recently as two years ago. Like many countries, Japan was virtually closed to foreign makers of PBXs and other telephone equipment that subscribers use on their own premises. Now, however, increasing numbers of countries—including Japan, Italy, and Britain—are opening up their markets to outside producers of subscriber equipment. International trade in this private equipment, virtually nonexistent a few years ago, is "just starting to boom," says Robert F. Campbell, telecommunications analyst for Arthur D. Little Co., the Massachusetts consulting firm. "Once more barriers fall," he adds, "this could become a huge business."

Since the government-owned telephone agencies themselves are still reluctant to buy from foreign suppliers, equipment makers from around the world are focusing their attention on selling equipment directly to users. And in today's international market for subscriber products—which includes telephones, facsimile machines, teletypewriters, and computer terminals—sophisticated PBXs hold the spotlight. "It's a very dynamic market right now—in fact it's *the* telecommunications market," says Campbell.

Most of the action now is in the US. Since 1968, when the Federal Communications Commission ruled that non-telephone company equipment could be hooked up to the public telephone network, foreign manufacturers "have moved into the US by the horde," says consultant Robert E. La Blanc.

Two Canadian manufacturers—Northern Telecom Ltd and Mitel Corp.— will capture 28 percent of the $2.8 billion in PBX systems sold in the US this year, while Japanese and European companies will grab an additional 22 percent says Kevin J. Sara, industry analyst for Northern Business Information Inc., a New York market research company.

Until recently, besieged US equipment makers were content to keep a smaller piece of a growing pie—sales of PBXs have doubled since 1977. But market saturation threatens to slow domestic sales of switchboards. "The beauty of telecommunications systems is that everybody needs one. The problem is that they only need one," comments M. Kenneth Oshman, president of Rolm. One pessimistic forecast from Eastern Management Group, a Morris Plains (NJ) market researcher, predicts that US sales of PBX systems will reach a high of $3.6 billion in 1984, then slide back to $2 billion in 1987.

So American companies are commencing a "mad scramble for export outlets," says John F. Malone, president of Eastern Management. Because most developing nations are only starting to install telephone systems, demand in those countries for sophisticated PBXs is still limited. The significant switchboard markets, and the focus of most of the US efforts, are in Western Europe and Japan—which together account for one-third of the world's equipment markets.

International competition is also being given a nudge by the rapid pace of technology change. When the office switchboard merely routed phone calls, most users were content to buy or rent telephones from the local PTT. Modern digital PBXs can switch computer data, telex messages, and even facsimile images as easily as voice conversations. "The PBX has developed into a communications controller and is the logical, natural heart for any office system," says Rolf Eriksson, who has been named president of Ericsson Information Systems, the business communications subsidiary of Sweden's L. M. Ericsson. And customers are much more choosy about picking a system that is to become the hub of office communications.

Despite the growing cry for more competition, PBX makers face an uphill battle. To boost local employment, many countries still prohibit imports. For example, in France, where CIT-Alcatel, Thomson, and Jeumont-Schneider dominate the PBX market, the only foreign company able to sell switchboards is International Business Machines Corp.—and it is allowed only because it makes its PBXs in France. Even in Japan, despite the government's stated intention to open the local market, foreign companies still run into nontariff roadblocks and bureaucratic snafus. Japanese makers still capture more than 95 percent of their PBX market, estimates John E. P. Borden Jr., a Yankee Group analyst.

In nations where trade rules have been liberalized to allow foreign companies to sell products, most of the PTTs still require that those products contain some locally manufactured pieces. Rather than fight these local-content laws, many companies are opening foreign factories. "Up to now we tried to manufacture 100 percent in Japan," says Masaka Ogi, general manager of international operations at Fujitsu Ltd. But now, he says, "we will have to make 50 percent in Japan and 50 percent overseas, or we won't be able to do business."

Besides having to overcome trade barriers, switchboard makers must also solve the technical problems of readying a PBX for foreign markets. Technical standards, such as the operating voltage and the number of wires per cable, differ in every country. Users in different countries also demand different operating features. For example, when Britain's Plessey Co. first looked at the US market, it discovered that US customers wanted their PBX to route all outgoing long-distance phone calls automatically via the least expensive carrier. Plessey had not designed this feature into its equipment, because in Europe there is only one long-distance company in each country.

Customizing the PBX for a new market is an expensive and time-consuming undertaking. It took Rolm, for example, one year to adapt its electronic switchboard for the Japanese market and another 1½ years to modify it for Italy. Explains Ian W. Maclean, chief executive officer of Plessey's Office Systems subsidiary: "Redesigning a [semiconductor] chip takes a lot longer than redesigning [an electro-mechanical] relay."

Although all suppliers of subscriber equipment face the same challenges when exporting products, the European companies certainly have a head start. Many have been selling in the international arena for decades and understand the intricacies of dealing with the PTTs much better than do North American companies. L. M. Ericsson, for example, gets 95 percent of its sales outside of its own country and already has orders from 20 countries for its new MD-110 digital switchboard. By contrast, American Telephone & Telegraph Co. captures less than 5 percent of its equipment sales outside the US. "American companies in general are not as attuned to the international marketplace because we have such a good market in this country," explains John G. Puente, vice-chairman of M/A Com Inc.

However, if European suppliers are to take advantage of their greater experience, they will have to update their product lines. Because they have sold primarily in their protected home markets, they have been slower to respond to technical changes. "A sort of sclerosis has set in among some equipment manufacturers because of their traditional captive markets," says Pierre Chavance, executive director of France's CIT-Alcatel. To compete internationally, he says, "we must get accustomed to having foreign manufacturers in our home markets." Although Chavance is not talking about his own company, CIT-Alcatel still does not have a fully digital PBX on the market, despite the fact that it was the first company in the world to develop fully digital, public switching equipment.

What US and Canadian suppliers may lack in international savvy, they make up for in technology. Rolm and Northern Telecom both have state-of-the-art, fully digital PBXs on the world market, and top-quality smaller producers—such as InteCom Inc.—are eying overseas sales. It was superior technology that won Rolm its JAL contract. "The Japanese makers have little experience with digital equipment," says Yoshiaki Eto, head of JAL's Telecommunications & Network Development office.

Many US equipment makers are looking for overseas partners to guide them through the brambles and thickets of PTT politics and local market requirements. "Our partners need to be technically astute in order to assist us in adapting our equipment to the local country. And they need to be politically astute," says Thomas H. Aschenbrenner, senior vice-president and a founder of InteCom, which is now looking for foreign associates. Even giant AT&T, which has formed a joint venture with Dutch-based Philips for selling equipment to telephone companies, is looking for additional partners to help it sell its Dimension PBXs directly to users.

Similarly, overseas competitors are joining with American firms, both for access to their technology and to acquire instant marketing strength in the hotly competitive US market. Late last year, Plessey bought Stromberg-Carlson Corp. and is converting part of the US company's Florida plant to turn out Plessey switchboards. And L. M. Ericsson in September worked out an agreement with Honeywell Inc. to market PBXs in the US.

Obviously, the preparation needed to start selling in the international markets will not pay off immediately. The attitudes of governments "don't change too easily—even in this country there was a bit of resistance in the beginning to [competition]," notes Thomas L. Kelly Jr, president of TIE/Communications Inc., a maker of electronic telephones.

But, to succeed in the long run, companies must be prepared to compete around the world. "We have to position ourselves to be ready when [the foreign markets] do open up," says Wayne P. Glaser, vice-president for international marketing at Paradyne Corp. Companies do not have much time to prepare for the battle, warns Rolm's Oshman. "There is a window of opportunity," he says. "If a company is not at critical mass in terms of market share, worldwide, it will miss the boat."

International deregulation

No part of the world telecommunications industry is more topsy-turvy than the international common-carrier business. By setting domestic carriers free to compete for the overseas message business, the US has ended up exporting competition to reluctant authorities that hold communications monopolies in foreign countries.

"We'll see many more carriers coming into the [world] market," predicts Frank Urbany, special assistant for international affairs at the US Commerce Dept's National Telecommunications & Information Administration (NTIA). "We'll have our domestic carriers attempting to export overseas the varieties of services they offer here." This international slugfest for customers and routes, however, will undoubtedly knock out some carriers over the long run.

The battle actually began in 1981, when Congress and the Federal Communications Commission deregulated the business by permitting the traditional international record carriers (IRCs), previously restricted to carrying telex or low-speed computer data outside US borders only, to compete inside the country. At the same time, they unleashed Western Union Corp., a domestic carrier, on the international scene.

Deregulating the $2.6 billion international carrier business seemed the logical thing to do, at least from the US point of view. American customers had grown irritated about having to send a telex overseas via two separate carriers—across the US via Western Union, which handed the message to an IRC for the trip overseas. "You can't divorce the international from the domestic market any longer," asserts Eugene F. Murphy, chairman of RCA Communications Inc., which handles both US and overseas links. "Customers are looking for end-to-end service."

The FCC also recognized that it had become unnecessary to distinguish between voice, telex, and computer data transmissions. These technologies are coming together, so it is becoming one homogeneous market rather than three separate ones," explains Robert J. Braverman, senior vice-president for ITT Corp.'s Communications & Information Services group, a leading IRC. As a result, the FCC gave AT&T, which could previously carry only voice telephone calls overseas, the go-ahead to send data as well. And the IRCs were allowed to send voice.

The US-based carriers supply only half the international connection, however: they must link up with a foreign PTT to complete delivery of a message. And the PTTs are not eager to link up with any new competitors. "We would not wish to do anything here that would undermine the international agreements that have been built up over a number of years with other carriers and PTTs," says Sir George Jefferson, chairman of British Telecom.

To persuade the PTTs to go along, US carriers are increasingly using stronger tactics. Japan, for example, has refused to accept any messages from Western Union, so that company is routing its Japan-bound traffic through another country. By sending its messages through a middleman with a low-rate agreement with Japan, Western Union is able to ensure that the Japanese earn less money than if they accepted messages directly. The Japanese are already losing about $1 million a year, says Thomas Mathai, Western Union's vice-president for office message services.

One way to get around such problems is to buy a carrier that already has agreements with PTTs. That is what MCI Communications Corp. did. The long-distance phone company acquired Western Union International Inc. (no relation to Western Union Corp.) from Xerox Corp. "The barriers are too stiff in international communication for newcomers to get a handhold, except through acquisition," explains H. Brian Thomson, vice-president for corporate development at the MCI International subsidiary. The company now plans to expand WUI's existing telex agreements with the PTTs to include overseas telephone calls.

The carriers are also fighting an increasingly fierce battle for customers, because international communications is growing at about 15 percent annually, compared with a 10 percent yearly increase for domestic business. Last July, for instance, a major rate war was sparked by MCI International, which was offering customers a 20 percent discount if they increased their traffic through MCI. Its competitors—ITT, RCA, and TRT Communications—cut their rates even more, and by September discounts of as much as 50 percent were offered.

Price is only one field of battle. With so many competitors, plain-vanilla transmissions are becoming a commodity business. As a result, the success or failure of a carrier will depend on what extra services it offers. "Almost anyone can do the transmission; value-added services are the name of the game," says Richard A. Kozak, vice-president for finance at TRT Communications Corp., a United Brands Co. subsidiary that plans to add such services as electronic mail to its overseas offering.

While US companies move overseas, European carriers are launching forays into the unregulated US market. Britain's Cable & Wireless PLC, for example, owns TDX Systems Inc., a US discount long-distance phone company. And France Cables & Radio, the international subsidiary of the French PTT, in September invested in Argo Communications Corp., a new US intercity carrier.

The growing competition is even spreading to communications satellites. Two US companies plan to launch spacecraft over the Atlantic to compete with existing transatlantic cables. International Satellite Inc.—formed by a consortium that includes TRT—and Orion Satellite Corp. each plan to launch two satellites in 1985 or 1986. To avoid dealing with the restrictions imposed by the PTTs, the companies want to sell the radio relay gear on the satellite to other users and let them worry about getting their messages through.

These new satellite proposals are confounding Washington regulators. Because the companies are not common carriers, they would not technically violate the Intelsat agreement signed by the US and 105 other countries. That agreement gives Intelsat a monopoly on international satellite traffic. Yet law-makers worry that the new satellites may violate the spirit of the original Intelsat agreement and spur a worldwide free-for-all. The FCC is now awaiting comments from the State Dept and the NTIA before it rules on these licenses.

To pre-empt this potential satellite competition, as well as to bypass its terrestrial competitors, Intelsat on October 1 launched its first business service for customers in Europe, Africa, and the Americas. Instead of routing messages to the nearest Intelsat earth station through AT&T, Western Union, or another carrier, customers can now install their own antennas and beam messages directly to an Intelsat satellite.

The Magic of Cellular Radio

Duane L. Huff

Cellular radio now makes it possible to use public telephones on the move. Soon high-quality, hand-held mobile phones will be widely available for use in taxis, trains, buses, planes, and private cars. Industries like construction and fast-food will greatly benefit. The author is vice-president in charge of cellular development at Bell Laboratories and this piece first appeared in Technology Review, *November–December 1983. Ithiel de Sola Pool, author of the box article, is Professor of Political Science at* MIT.

Most of us regard a telephone on the dashboard or a "walkie-talkie" radio in the hand as a luxury or a status symbol rather than a necessity. But a new technology called cellular radio is now emerging to make high-quality, full-service mobile telephones and radio transceivers much more widely available. Mobile communications that have been for 20 years a curiosity will become commonplace—and soon enough a necessity for many.

Most market studies suggest that cellular radio technology will increase the number of mobile telephone users tenfold within the first few years. Projections for the use of hand-held, portable radiotelephones based on cellular technology are equally optimistic. The result is likely to be one of the greatest changes in communications patterns since the invention of the telephone.

After years of effort to select the most efficient way to meet increasing demand for high-quality mobile communications, the Federal Communications Commission (FCC) has accepted the cellular radio concept. Early this year the agency began taking applications to operate such systems in the 90 largest US markets, and it will shortly accept applications to serve other markets as well. No technical, legal, or regulatory uncertainties remain. Many other industrialized countries have already installed or are actively planning cellular radio systems. Third World countries are also planning to use this technology in the absence of conventional wire or optical transmission facilities.

Searching for the choicest cell

Less than one-tenth of 1 percent of the 150 million vehicles now in use in the US have mobile telephones. This is because, until now, mobile communications have been limited by a lack of radio channels: each mobile telephone has required exclusive use of one channel in the available spectrum of frequencies. Cellular systems conquer this limitation by using low-power transmitters of limited range, so that each channel can be used simultaneously in many different geographic "cells." Taken together, these cells make up the total service area of a system (see figures 3.4 and 3.2, p. 115).

This idea of frequency reuse is familiar. Hundreds of television stations across the country that are out of one another's ranges, and therefore cannot interfere with one another, reuse television channels 2 through 13 in the VHF band. Cellular telephone systems reuse frequencies on a much smaller geographic scale. Instead of covering an entire service area with one transmitter with high power and an elevated antenna—the technology used for conventional mobile communications—cellular service relies on transmitters of moderate power distributed throughout a service area.

Each transmitter is only powerful enough to communicate with the radiotelephones in its "cell"—the area surrounding it. Thus, other transmitters in distant cells can use the same frequencies at the same time to communicate with mobile telephones in *their* cells. The only requirement is that transmitters in adjacent cells avoid using the same frequencies to keep from interfering with one another. Cellular radio's more efficient use of the fre-

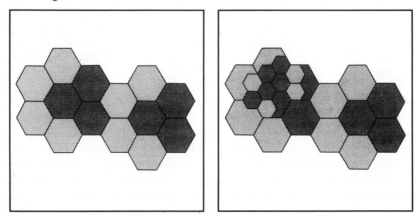

Figure 3.4   The advantage of cellular radio for mobile telephone systems is "frequency reuse": a limited spectrum of radio frequencies can be made to serve many users. The area to be served is divided into cells—in this case, 14. A low-power radio transmitter—powerful enough to reach any mobile unit in the cell but not powerful enough to interfere with transmissions in distant cells—is located in each cell. Cell-splitting is a simple way to keep up with growing demand for mobile telephone services in a cellular system. The right-hand diagram shows how 9 cells have been created out of parts of 4 cells.

quency spectrum was the reason the FCC chose it as the standard technology for mobile public communications.

Consider how a mobile telephone in a cellular system operates. As a vehicle in which a telephone call in progress on a particular channel moves from one cell to another, the call is automatically transferred to the neighboring cell. There, the call is conducted on different frequencies, to which both mobile and fixed transmitters and receivers are automatically assigned. This transfer should take place without the users' awareness; there is no break or perceptible difference in the communication. This function—the transfer of a conversation from one cell to another—is usually called "handoff."

Although the system is simple in concept, it is complex in execution, demanding specialized computers and intricate sensors and controls. For example, one popular way to perform the handoff function is to monitor the quality of the signal for each telephone call received at a cell site. When the signal quality starts to degrade, nearby cells are automatically requested to measure the quality of the signal to determine which cell could better serve the call. When a better candidate is chosen, an idle radio channel at the new cell is selected, and a digital message on the "old" radio channel instructs the mobile equipment to tune to the new channel in the "new" cell. Control techniques have now been developed so that conversations continue without interruption. (If no new channel is available, the handoff procedure can be briefly delayed, or the call can be transferred to an adjacent cell where transmission is adequate though not optimal.)

When a customer's mobile equipment is turned on but not in use, it periodically scans special control channels broadcasting at each cell site in the system, selecting the strongest signal to monitor for incoming calls. When an incoming call is detected, or when the user wants to make a call, the user's equipment rescans the cells to be sure it is still working with the strongest signal (usually the nearest cell).

Calls to a mobile terminal are initiated just like normal telephone calls. A seven-digit number (ten digits if the mobile party is in a different area code) is dialed, and the conventional telephone network routes the call to the central computer of the mobile party's cellular system. From here the number is sent to all cell sites in the system, and each transmits the number of the called terminal on its control channel. When the called mobile terminal detects the incoming call, it selects the best cell with which to establish the communication and sends its identification back to the system through this cell. The system then uses a digital message to designate a frequency for the mobile terminal to use, the receiver tunes to this frequency, and the user is alerted to the incoming call by ringing.

A similar sequence is involved when a mobile user originates a call. The user first dials the desired number at a register in the mobile unit. The unit then chooses the best cell site and transmits the called number over the control channel. The main computer designates the frequency to be used to complete the call, and the mobile unit is automatically tuned to this frequency.

All the control information between the terminal equipment and the system is redundantly coded and repeated to prevent errors. The entire exchange of control information takes place in a fraction of a second, and none is heard by the user. To the user, the sending and receiving of calls is routine—standard telephone procedure.

Cellular radio systems actually use two separate frequencies for each transmitter–receiver pair. This allows "full-duplex" service, which means that both parties in a conversation can talk at once. This constrasts with one-way systems such as citizens'-band ship-to-shore, and private radio. In these, the same frequency is used for transmission and reception, so transmission requires pressing a "push-to-talk" button that disables the receiver.

The computer control systems are the key to efficient operation of cellular radio communication. Computers at the mobile telephone switching office (MTSO) activate computers at the cell sites and, through these, the computers in the mobile units. Both cell sites and mobile units possess a degree of autonomy and a considerable amount of intelligence. In many cellular systems, the computer in the cell site performs most of the handoff functions, including measuring signal quality and comparing results with other cell sites. It also performs diagnostic tests and reports trouble in any equipment at the cell site. The microprocessors at these sites must cooperate with the processors of the MTSO to form an efficient system. The microprocessors in mobile units have important error detection functions, in addition to controlling some terminal hardware and keeping mobile units tuned to the strongest cell sites.

Special features for cellular systems are limited only by the ingenuity of the developers. These currently include options such as one or two-digit speed calling; call-forwarding, message-waiting, and three-way calling; locking devices to prevent unauthorized use; data transmission; and automatic callback. In short, mobile telephones based on cellular technology offer essentially all the services available on conventional telephones.

The same cellular concept can be applied to serve portable radiotelephones with built-in power supplies. Many market studies suggest that these radio links may be the most important early uses of the cellular concept. A briefcase-size radio weighing 10 to 20 pounds that can be used for a number of days without battery recharge is now available. Construction supervisors, for example, use such units to maintain contact with their own and subcontractors' offices while on the job. At the other end of the spectrum is a hand-held, lightweight radiotelephone unit that pedestrians—doctors, for example—can use where conventional telephones are not available. This unit requires new batteries (or recharge) after a few hours of use. We may not yet have achieved Dick Tracy's wristwatch-sized communications unit, but we are clearly making a close approach!

A service called "roaming" allows customers with mobile radio service in one cellular system to use their units in another system as they travel. Roaming works this way. When mobile equipment is taken into a new cellular system, the roaming feature identifies the equipment to the new system, which then informs the home system of the user's whereabouts. Calls coming into the home system for the user are rerouted to the correct "foreign" system. The roaming function includes computerized systems that exchange billing and payment information and permit operators to identify fraudulent users, illegal equipment, and bad credit risks. When cellular systems are fully developed, roaming—essentially complete mobile long-distance service—will be possible, with no special actions required by the user.

Though developers expect strong demand for cellular mobile services, startup facilities can be small, with orderly growth possible as mobile traffic increases. When the traffic in one cell becomes greater than that cell can handle, the cell can be divided into smaller cells. Transmission power is reduced to avoid radio interference among the new cells, each of which serves about as many customers as the original cell.

Most startup systems use cells with radii of 8 to 12 miles. These systems can be split three times if necessary, with the minimum radius being 1 to 1.5 miles. New cells can be created to meet demands for service outside the original area. In most systems, the maximum number of customers per cell, without excessive service delays, is 3000.

The demand to talk on the move

The potential market for mobile communications was recognized soon after the invention of radio. The first major use of this potential came quickly, as radio was employed to communicate with moving vessels at sea for navigation and safety. Today the use of mobile radio has spread dramatically to include land vehicles—public-safety and service vehicles and taxis, to name but three—as well as aircraft and ships. Indeed, some 8 million mobile units are licensed for private radio service in the US today. Over 8 million "citizens'-band" radio units are also in use in this country.

At the end of World War II, the demand for public mobile telephone service induced the Bell System to test-market the service in St Louis. Six frequencies were chosen, but the equipment was not sophisticated enough to prevent interference from adjacent channels. Thus, just three frequencies were actually used, so only a few customers could be served. One year later the Bell System tested a public mobile system along the highway between New York and Boston. This system also proved troublesome because of erratic radio propagation: short-range radio was blanked out in some areas, while interfering conversations were carried for long distances.

Despite these difficulties, Bell gradually extended mobile telephone service to other markets, and by the late 1940s there were long waiting lists in several metropolitan areas. Indeed, over 25,000 people are on waiting lists for mobile telephones today, with many more would-be users not even bothering to add their names. The number of channels assigned to a typical mobile telephone system are so few that subscribers frequently cannot make or receive calls when they wish, so new customers obviously cannot be accommodated.

Bell Laboratories planners were already looking forward to a more economical and efficient form of mobile telephone service when the first systems were being installed in the 1940s. The basic cellular concept of reusing frequencies was conceived as early as 1947. However, there was then no way rapidly to change the frequencies of transmitters and receivers and no techniques for managing the movement of calls among many cells.

Advances in electronic switching systems, low-cost frequency synthesizers, and high-capacity microprocessors stimulated the Bell System to suggest, late in 1971, how a cellular system might be developed and operated. The FCC responded in 1974 by allocating a substantial block of frequencies and soliciting proposals for tests that would prove the feasibility of cellular systems. A year later AT&T applied to operate a trial cellular system in Chicago, the FCC granted a license in March 1977, and the first such service began in late 1978, growing quickly to its limit of 2000 customers. Soon thereafter American Radio Telephone Service began a similar trial in the Washington–Baltimore area using a system developed by Motorola. Both projects have shown that the basic cellular concept is feasible and can win strong customer support. A third trial, by Millicom, Inc., was authorized in May 1981 and should be operating in the Raleigh–Durham, NC, area in 1984.

The Chicago system provided the first market test of this new technology. There were 136 voice channels spread across 10 cells, making possible "full-duplex" service—with few busy signals—to a 2100-square-mile metropolitan area. Potential customers, randomly selected from lists of businesses in the Chicago area, were reached by direct mail, telephone, and personal visits. One out of eight companies contacted subscribed to the service, and most requested more than one mobile unit. Construction companies were especially receptive to mobile telephone service. Least receptive were food retailers, whose customers generally come to them.

Almost all customers cited time saved and convenience as major benefits of cellular radio; many said that productivity increased 20 to 30 percent. An executive of a fast-food chain predicted his company could save 1 million gallons of gasoline a year if all service vehicles were equipped with cellular radios. A trucking company attributed a 30 percent increase in business and a 4000-gallon fuel saving to cellular service. A real estate executive claimed a 50 percent increase in productivity, and a representative of a waste disposal company reduced his automobile mileage by 20 percent. Many executives said they extended their working days by using their telephones while commuting or traveling.

The beginning of the cellular age

With testing complete, widespread use of cellular technology is beginning in the US. In June 1982, conventional telephone and radio communications companies filed 196 applications to operate cellular systems in the top 30 markets. Two competitors, one a telephone company and the other a radio communications company, will be licensed to serve each area. Competing carriers for each of two licenses in each market are agreeing to a number of joint ventures, but many disputes must still be resolved. The FCC has granted a Bell subsidiary, Advanced Mobile Phone Service (AMPS), permission to construct the first 22 commercial cellular systems; the first of these has just begun service in the Chicago area. The FCC has also awarded permits to other telephone carriers, as well as to a number of companies set up expressly to provide cellular radio. Meanwhile, nearly 400 applications to provide service in the 30 next-largest markets were filed in November 1982, and 567 applications to serve the sixtieth through ninetieth markets were filed in a third round.

Consumer rates for the new telephone service and for leasing the mobile equipment are being established by the marketplace. Prices will vary widely around the country, depending on individual companies' plans for recovering their large investments. However, the prices tested in Chicago in 1978–9 may typify what many companies' will start off with. Basic service (a telephone number and connection to the system) with 120 minutes of free usage cost $25 a month, with "overtime" usage charged at $0.25 per minute. Lease of the mobile terminal cost either $45 or $60 per month, depending on the type of set. Thus, the minimum bill was either $70 or $85 per month. Most customers used much more than the 120 minutes of free time, so that the average bill was double the minimum—over $150.

Most experts predict that minimum costs for cellular service will run about $150 per month during the first few years of operation. As the number of users increases, the cost of mobile equipment and services will decrease. If the FCC allocates more frequencies to cellular radio, construction of new cell sites can be delayed, decreasing costs further.

The FCC has reserved a spectrum of 40 mHz for cellular radio and will assign 209 mHz to each of two competing companies in a market. Each system will use 30 kHz of its assigned frequency for each path between cell and mobile unit—enough to prevent interference and allow all accessories to be used effectively. Thus, 333 channels can be operated in each band of 20 mHz assigned by the FCC. AMPS estimates that it may develop up to 100 cells in Chicago by 1986 to meet demand. Each cell might have an average of 30 radio channels, for a total of 3000 channels (which can support 3000 simultaneous conversations). The total number of customers in that market might approach 75,000 by 1986.

Cellular systems are also being developed overseas. The Nordic Mobile Telephone System, which began operating in 1981 in the four Scandinavian countries, had over 40,000 subscribers in the fall of 1983. The system was originally designed to accommodate only 96,000 subscribers, and expansion to a capacity of 200,000 is now being planned.

A cellular system designed by Nippon Telephone and Telegraph, in operation in Tokyo for a number of years, has recently been extended to Osaka and Kyoto. The system currently serves more than 10,000 subscribers, with over 100,000 projected by 1987. Smaller systems are in use in Mexico, Qatar, United Arab Emirates, Australia, Canada, Austria, and Singapore. Nippon Electric Co. equipment is used in Mexico, Singapore, and Australia; Matsushita equipment is used in Qatar and the United Arab Emirates; and Motorola equipment is used in Austria.

Meanwhile, even while cellular radio is being commercialized, the technology on which it is based is by no means standing still. Improvements such as digitized transmission will reduce the amount of the radio spectrum required for a single call, thus increasing the number of calls that can be processed in a cell. Mobile telephone equipment will shrink in size, and data terminals and printers will be available. Public telephones will be installed in trains, buses, and taxis; indeed, trials of public telephones in commercial airliners are already underway. Cellular telephone services will soon be coupled with answering and message services, encryption and scrambling for privacy, dictation services, data transmission, alarm calls, automatic callback, and all the other auxiliaries now available on conventional telephones.

Cellular services will be available in all major us markets within three to five years. Thus, this technology will form the basis of a competitive marketplace, with unlimited opportunity to provide truly personal telecommunications to users on the move.

Will mobile telephones move?
Ithiel de Sola Pool

Grey-haired readers will recall the flying heroes of their childhood comic books talking across the reaches of space on wristwatch radiotelephones. Now one small step on the road to the fantasy of omnipotent communication is becoming reality with the advent of cellular mobile telephone systems.

This long history of yearning is, however, not yet ended. We must realize that what we see in the next 10 or 15 years, though significant, may change a great deal.

First of all, we must disabuse ourselves of the notion that people want to talk all the time. The average person spends only a quarter of an hour a day talking on the telephone outside of the office. Of course, phone junkies use the telephone much more, and we can expect that they will install cellular phones in their cars as soon as possible and thereafter will feel lonely unless they are talking as they drive along. But many others will hate the very idea of this invasion of the precious solitude of the automobile.

The same uncertainty applies to travelers. We will not know until current market tests are completed how many business travelers will take advantage of the new plane-to-ground telephones. Nor will we know how many passengers will be less comfortable because they are no longer able to sit back in their airline seats, and feel cut off from the tension-loaded world they just left, with nothing to do but drink, eat, sleep, or read.

The hard facts of economics
But the desire for personal conversation is not the main test of the prospective significance of mobile telephones. Their role in increasing business productivity is likely to be far more important than their role in casual conversations.

About 16 million mobile units already serve the roughly 200 million American adults. Where are all of these? (Only 0.1 percent of our vehicles have two-way radios.) They are in the taxis we call, in the police cars and fire engines that serve us, in the repair trucks that come when needed, in the ships and planes that transport us and our goods. There is now an enormous backlog of demand for additional two-way mobile channels in our major metropolitan areas.

But is that perhaps an illusion? In 1977 the Federal Communications Commission (FCC) had made a study of the demand for radiotelephony in Chicago. Participants in the study, Robert Crandall and Robert Taggert of MIT's Research Program on Communications Policy, quickly recognized that there were many ways that distributed services such as taxis and delivery trucks could keep in touch. The choice of mobile radio was based on economics—the government was giving these frequencies away free. Once one's vehicles had a transceiver, mobile communication was a gift. Crandall and Taggart suggested that, even with a fairly modest charge for frequency use, congestion might disappear. Now, cellular radio services will have a genuine market test. In contrast to previous mobile radios, cellular service will be billed to customers on a services-rendered basis through a telephone company or other common carrier.

There is good reason to expect that there will be demand for the service. The old semifree congested service was not of very high quality. With the new cellular service, there will be no more "push-to-talk" and no serious interference. The new systems will bring their users the ability to call anywhere in the country and to add on, at a charge, many of the enhanced services that telecommunications companies will be promoting.

Two-way trucking
Those who have studied the possible uses of mobile telephones agree that truck operators—seeking to increase efficiency—will be a major market. When a garbage truck without two-way radio goes out, its managers can only guess at when it will be filled; with mobile telephone service, a second truck could be sent to the right point on the line. So, too, with delivery trucks: computer control and timely communication could vastly improve their efficiency. Traffic congestion could be markedly eased by communications that would tell drivers where there is clogging and how to select alternative routes. A form of jitney service called "dial-a-bus" that is halfway between taxis and mass transportation, based on routing optimized by computer, has long been discussed. Emergency services of all kinds could be improved.

There will be many other potential beneficiaries, too. In the accompanying article, Duane Huff tells us that in the AT&T Chicago experiment, building contractors signed up for the new service quickly but food vendors tended to be laggards. Although that may be a good short-run indication of demand, the long-term outlook should be different. The handling of perishable foods may well be improved by effective mobile communication. The food industry is today organized to minimize dependence on real-time communication. Early in the morning, food is delivered from warehouses to the refrigerators of supermarkets, there to wait for the capricious decisions of the day's (and the next day's) customers. The wastage and loss of flavor are enormous. There could be a system in which customers, as of old, are delivered the perishables that they now pick up themselves. With the day's demand recorded on computers, the various halfway storage systems could be eliminated.

One of the remarkable features of cellular radio systems is their propensity for graceful growth: cells can be split as traffic grows. Unlike other modern communications systems such as direct satellite broadcasting and videodiscs, cellular radio does not require a massive up-front investment into a system whose capacity must exceed demand for many years before it begins to repay its installation costs.

Making systems compatible
Cellular radio systems are not inherently monopolistic, and they are being set up to encourage competition. But to reach their full potential, communications systems have to be universally interconnectable, and that is not a trivial requirement. How will messages go easily from one system through connecting, competing systems to a system of yet another vendor? Can those services be rendered efficiently and economically? Even such mundane issues as cooperative billing systems may turn out to be costly and complex.

The Impact of Cable

Robert D. Kahn and Martin L. Ernst

Cable TV is bringing major changes in the way we receive entertainment and information. A fully-fledged cable network could revolutionize the way we work and conduct our affairs. The authors describe the cable revolution so far and take a look at what's happening to the industry in the US. From Technology Review, *January 1983. Robert Kahn is member of the board of directors of Davis (Calif.) Cable Cooperative; Martin Ernst is vice-president and senior staff consultant of Arthur D. Little Inc.*

The earliest automobiles were called "horseless carriages" because that was the only way people at the time understood the new technology. This is not very different from calling today's cable systems "television." And, just as the automobile went on to alter American life in important and lasting ways, so too may cable communications. Historically, cable has provided quality reception for remotely located consumers, but today it is the purveyor of myriad programming and services that redefine what we mean by, and expect from, TV.

Cable began to expand in the mid-1970s when several court decisions obliged the Federal Communications Commission (FCC) to relax some serious constraints. Until then, cable companies could not rely on national cable program sources but had to function on a local basis. In addition, space satellites were opened to commercial use, offering the cable industry an opportunity to establish its own national networks and to "cablecast" programming that broadcast TV—the established networks—either could not or would not offer.

Cable programming was linked to satellite transmission for the first time in 1975 when Time, Inc. established the first national network to distribute cable programs to local operators. And innovations such as all-movie channels—which showed current, uncut, commercial-free feature films—made a great impact on urban markets. In the three years from 1977 to 1980, cable subscribership went from 11 to 14 million households. Today that number stands at approximately 28 million, and the industry reports that 250,000 households per month are being wired for cable. Approximately 34 percent of US residences now receive cable, with half the homes having access to cable subscribing to it.

Cable differs from broadcast television in three fundamental ways.

Cable offers expanded channel capacity. Cable systems installed today promise upwards of 100 channels, offering local communities a varied, and unprecedented, communications menu. Not only do these channels offer today's mass-oriented broadcasting, but cable also offers "narrowcasting"—special programs going out to special-interest viewers.

Cable offers two-way communication. While the majority of cable systems are now limited to conventional one-way communication, newer systems also offer the option of consumer-originated messages. These may take the form of polling, emergency calls, or requests for information.

Cable can easily be coupled with other communications technologies. Cable is one medium that can carry virtually all others: newspapers, radio, film, books, even still pictures. While cable can't really substitute for a visit to the movies or a museum, it can distribute images widely and inexpensively. Linking computer technology with cable—in transmitting data, for example—is especially powerful.

One sign of cable's maturity is the profusion of new networks. A recent count by the National Cable Television Association identified 51 national satellite-distributed cable networks, including three all-new channels, two all-sports channels, four religious networks, three movie channels, a health network, a Spanish-language network, and numerous others. Soon cable systems will even carry an electronic-games channel.

Cable has begun to cut into the historic dominance of television programming by the nation's three broadcast networks. In some cable-serviced communities, the major networks combined attract only 70 percent and less of the audience. In the US overall, the three networks' share of prime-time ratings declined from 92 percent in 1979 to 81 percent in 1981.

Cable offers three kinds of programming: basic programming, pay programming, and two-way services. "Basic" is covered by the monthly subscription fee paid by the customer. The "first tier" closely resembles conventional television and must (under federal law) carry local broadcast TV stations. The first tier also frequently includes so-called "superstations" such as WGN-Chicago or WTBS-Atlanta—independent broadcast stations that specialize in sports and entertainment and are transmitted by satellite to hundreds of local cable systems. Basic services also include "community-access" programs generated within the cable-served community, under franchise terms negotiated between the operator and the local political jurisdiction.

"Pay TV" is the second tier: the system operator can offer subscribers numerous networks, in addition to the basic set, at additional cost. Moreover, special one-time programs such as concerts and boxing matches may be charged to subscribers on a "pay-for-view" basis.

Pay-TV programming accounts for much of cable's popularity and certainly contributes to its profitability; many subscribers appear willing to pay for the uninterrupted programs unavailable on commercial broadcast networks. Revenue from pay-cable channels has been rising steadily. According to Paul Kagan Associates, a leading cable analysis firm, approximately 19 million subscriptions to cable programs were sold in 1982. A conservative estimate of overall industry revenue for the year just ended is $4.36 billion. Kegan projects overall industry revenues to reach $16.6 billion by 1990, fueled largely by pay services.

But there are limits to pay cable's popularity. A 1981 study conducted by Benton and Bowles, Inc., a marketing research firm, determined that nearly half of those cable subscribers who presently reject pay TV would buy it if it were less expensive, even if this meant that advertising would have to accompany programming. In fact, cable viewers can expect to see more and more advertising over both basic and pay-cable channels. According to *Marketing News*, 70 percent of cable stations already seek advertising.

Two-way services constitute the third programming form that cable-TV systems make available. These services are now offered to only a fraction of the communities wired for cable, but they are becoming standard as new cable systems are installed. The offerings range from home banking, energy management, and security alarms to medical referrals and information-retrieval systems.

The best-known two-way system is Warner-Amex's QUBE, used by 38,000 residents of Columbus, Ohio. Using the five-button keyboard distributed to every household served by QUBE, consumers transmit elementary messages that are processed by a computer at the "headend" (the operator's central facility). For example, viewers may request a copy of a recipe demonstrated on a cooking program by flicking a switch on their home terminal; they will later receive a copy of the recipe by mail.

QUBE has polled viewers for their opinions on everything from politics to TV shows, and once had them collectively "quarterback" a local football game. Warner-Amex loses money on QUBE in Columbus, but the two-way channel seems to have paid off in the highly competitive "franchise wars"—where cities must choose among competing cabled companies before granting a monopoly contract. By breaking new ground with QUBE, Warner-Amex won a reputation for state-of-the-art technology, which likely helped it go on to win franchises in Cincinnatti, Houston, Dallas, Pittsburgh, and St Louis.

Not everyone is enthralled with cable's two-way capabilities: the potential for abuse is considerable. Accounts of the pay-TV programs individual subscribers watch, from special entertainment events to pornography, are monitored routinely. Operators of two-way systems are privy to polling results, purchase decisions, personal schedules, and whatever other household and commercial transactions subscribers make. Like the old-time village switchboard operator, today's cable operators could know more about us than we wish, with far more serious consequences.

In the fierce competition for local franchises, a cable company presents itself—in effect—as a "common carrier," available to serve much of the host community's communication needs. According to Michael Dann of ABC Video Enterprises, "Cable must be seen as a new utility, a utility people will use when they need and want it, not so much for entertainment but for information and services."

But cable does not operate as a utility—cable operators decide independently what programs to carry and what two-way services to provide. For example, if an entrepreneur develops an improved way to offer security and emergency services over cable, the operator that offers a security option of its own can either deny this competitor access to the system or charge prohibitive rates.

The issue of equal access to cable, and whether it should be defined as a common carrier, was actively debated last spring in the California legislature. Assemblyman Richard Robinson, a moderate Democrat from Orange County, introduced a bill to regulate the California cable industry—the nation's largest—as a public utility. Robinson argued that the installation and operation of a cable system should be separate from the control of its programming. The bill included a series of equal-access provisions to stimulate the use of cable by California's "knowledge industries," and to promote widespread reliance on cable as an alternative to the phone company. Robinson's measure was soundly defeated after industry opposition was mobilized. For now, California remains among the nation's 39 unregulated cable states.

"Information equity," like equal access, is another issue suggested by defining cable TV as an information utility. With the advent of some cable services, viewers pay for what used to be free. For example, a whole new set of children's programming, adult education offerings, and information services will be available only to those who can afford it. Les Brown, editor-in-chief of *Channels* magazine, quips that society may eventually be required to provide "information stamps," akin to today's food stamps, to establish minimum access to information.

Questions of information equity, equal access, and privacy safeguards arise from the cable industry's embryonic moves into the "information age." But the industry is approaching a watershed: it may soon have to choose between continuing exclusively in entertainment—eventually competing with the major broadcast networks—and evolving to provide information services such as data transmission and teleconferencing in competition with AT&T.

If cable firms continue to move in the latter direction—that is, if they function like utilities—the more they will be treated like utilities and face frequent political challenges. Such a cable industry must accept regulatory oversight and content itself with the bargain Americans have traditionally made with their utilities—equitable and affordable services in exchange for fair and protected profits. But so far, the industry has resisted even the thinnest local regulation, and Washington and the state houses have yet to assert themselves seriously.

R.K.

Economic impacts

After Gutenberg invented printing, massive amounts of information could be made widely available, often at little or no direct cost to users. Disseminating information no longer depended on the scarce resource of skilled and dedicated copyists; instead, information could be provided by a greater supply of far more productive labor. Fulfilling the print revolution required many other innovations, however. New industries had to supply paper, ink, and printing and binding equipment. Writers, publishers, and printers had to establish new business practices to work together effectively. New laws, such as copyright, were needed to provide economic incentives to create information. And new institutions, such as libraries and compulsory education, evolved as nations recognized the economic and social values of literacy and widespread knowledge. The result was that information was democratized—and, of course, copyists and parchment makers disappeared.

Cable systems and related technologies may well contribute to an equally significant set of changes. Large multichannel cable systems, low-power TV (permitting closer geographic use of the same broadcast frequencies), direct-broadcast services (whose authorization involves expanding the portion of the electromagnetic spectrum devoted to TV services), and, eventually, fiber-optic networks (that can provide multichannel capabilities far exceeding those of current coaxial cable) may convert a scarce resource—broadcast TV channels—into an abundant commodity. And coupling these systems with advances in electronics, such as in computers, videotape systems, and videodiscs, may yield enormous impacts.

Historically, the scarcity of local broadcast TV franchises, expecially those for the higher-quality VHF channels, made them very valuable business properties; because the choice of channels was limited, the audience was highly concentrated. Networks and local stations competed for this audience by appealing to the broadest possible segments. At least for prime viewing periods, they could afford to produce elaborate and expensive programming. Advertising, the system's means of financial support, similarly focused on goods and services with mass appeal. As the viewing population grew, advertising rates per broadcast minute also grew. Ultimately, television advertising rates became the standard, or "umbrella," for other media rates, including those of major newspapers and national magazines.

The early cable systems had little impact on the structure of the broadcast industry. Few cable channels were available, programming was restricted to locally produced materials, and the systems tended to be regarded simply as grandiose antennas for providing better reception.

But the large-capacity systems now being put in place, free from programming restrictions and offering 50 or more channels, are a totally different kind of proposition. Because of the large profits anticipated from pay TV and other new types of programming, local cable franchises have increased greatly in value. This has permitted local authorities to make far better bargains that usually involve lower basic connection charges for a larger set of free broadcast services and more channels for community use. Moreover, though early cable systems were priced to recover their costs entirely from basic connection charges, newer ones will likely require additional monthly revenues of at least $25 to $30 per connection from either viewers or advertising if they are to be profitable.

The differences between traditional broadcast operations and future cable systems may be quite striking. Cable system operators will have to recover greater investments but will have more direct relationships with viewers and better records of their interests. Most important, operators will be able to deliver programs at very low marginal cost per additional channel. Broadcasters can offer only one "page" (that is, one program) at a time, and must compete with other broadcasters, each of whom is also offering a "page" that usually has similar contents at the same time. Cable operators will be able to deliver 50 or more "pages" simultaneously—provided they can acquire the programming—and will face no direct competitors with equivalent capabilities. Where broadcasters had little choice but to cater primarily to mass audiences, and broadcast advertisers could justify promoting only those products with mass appeal, cable operators can provide highly specialized programming and seek advertising that will appeal to particular audience segments.

As a result, broadcast franchises are likely to decrease in value; they may lose viewers seeking more specialized entertainment and information. Advertising directed at mass audiences will bring in less revenue as these audiences diffuse. New forms of advertising directed toward specialized consumer segments, already being delivered by cable, will grow, incorporating far more details on products and how they are used. Another attraction to advertisers will be the ability to conduct market research at the same time.

Other likely losers in this realignment will be the traditional print media. Although newspapers and magazines have responded flexibly to new competition in the past, and although they have many advantages over electronic media in terms of portability, display quality, and durability, they are quite vulnerable to loss of advertising. This source provides more than 60 percent of the total revenues of most mass-audience print publications. The major newspapers may well remain the superior means for receiving in-depth news, but loss of advertising revenues may force them to raise prices and thus lose readers. This, in turn, may force owners to make major changes in the size, format, content, and even social function of newspapers.

If broadcasters and traditional print media turn out to be the losers, who will be the winners? What new industries will have to arise to support cable

TV and allow it to be fully used? This is an area of much controversy, but many of the forces at work are already visible and provide a basis for speculation. Here are a few possibilities.

1 The great increase in the number of channels will lead to major economic incentives for lower-cost production of high-quality programs. The costs for basic equipment, such as color TV cameras, have already dropped to a fraction of earlier levels. Prices for associated devices, such as computer-controlled editing and graphics equipment, are following the same trend. Particularly valuable would be the development of large videodisc archives (the "library" of the future?) indexed by content, visual details, and other characteristics. With access through computer, a wealth of quality material would be available.

2 Because more people will become participants in television as well as spectators, cable will help "democratize" the provision of entertainment and information, much as printing did their availability. Schools and training in program production and performing arts are likely to grow. Some cable systems already offer opportunities for local amateur groups to use community channels and even provide assistance in preparing material. Thus, the programs offered on such channels will gradually become more professional.

3 Because cable TV is in many ways more analogous to publishing than to film production, successful publishing formats will be applied to the field. Specialized "program-of-the-week clubs" will likely evolve, comparable to book-of-the-month clubs. Owners of video cassette recorders will be able to pick up such programs at their convenience. Packages combining video programs with supporting print materials could be developed. And the video equivalent of indexes and abstracts will be produced. The wealth of material that can be transmitted will tax the effectiveness of printed program schedules; these may have to be replaced by computerized schedules that offer brief samples of programs instead of written abstracts.

4 The tremendous demand for materials by both audiences and advertisers could lead to a variety of new forms of entertainment and recreation. Even the limited two-way features available on cable systems today are already being used to provide interactive games, and expansion of such activities is limited only by people's creativity.

5 Video education, whether basic or refresher courses, offers unusual opportunities. At first, cable systems may be used to present and upgrade correspondence-type courses. The next step could involve lectures transmitted to the home, "classroom questions" answered over the telephone, and local facilities established for doing laboratory and practice work. Eventually, fully accredited schools based on such systems could be established.

Business practices and relationships must evolve to support these kinds of services. But there are impediments. The first may be the cable operators themselves, who have complete control over the content of everything they deliver other than programs on community channels and the basic free services. While these operators have financial incentives to use their system's full capacity, restricting usage might increase profits in some situations. For example, if cable system owners develop some of their own programming, they may seek to limit competition with these products. In many cases, cable operators receive a percentage of program providers' income from advertisers and subscribers. The alternative approach—leasing channels to the providers at specific, time-based rates—is unpopular with operators because they lose control. But the revenue-sharing approach can prompt operators to discriminate against low-fee providers because audiences from higher-fee programs might be diverted.

Numerous other business practices will have to be changed or established before cable's potential can be realized. Means of payment for access to video archives will have to be developed. These mechanisms may be more complicated than those for print media. Growing participation in program development may blur the distinction between amateur and professional production, with important implications for stage, film, and television unions. Copyright practices are already a subject of much controversy because photocopiers and videotape recorders have become widely available; it is not yet clear whether the outcomes of these debates will expand or limit the range of programs available on cable.

But the most fundamental long-term regulatory problem relates to the roles of, and competition between, cable operators and the telephone network. Cable systems can deliver a great variety of materials to homes, but they have only limited capabilities in the reverse direction. The telephone system, when associated with home computers, offers clear advantages in transaction services such as the ordering of goods or the paying of bills. This capability, already being used, will probably grow rapidly in the next decade. As a result, two very different sets of sophisticated information-related services will soon be available to large numbers of households.

But questions will certainly arise over why it is necessary to have two sets of wires running into households—one a pair of low-capacity copper wires and the other a high-capacity coaxial or glass-fiber cable. The phone service could develop the switching necessary for direct video communications among households, but it won't have adequate incentive to make the necessary investments without the authority to offer video entertainment as well. Cable systems will have the video home-delivery capacity but will lack the switching authority and capability.

Current laws treat cable systems as an overlay on our existing "information infrastructure." And cable system operators are granted monopoly privileges to induce them to make major investments. But eventually, fundamental changes in the older institutional patterns will be necessary if the full benefits of cable are to be realized. Such conflicts may be difficult to resolve, since owners of both the phone and cable systems will have already made massive capital investments.

M.E.

The Videotex Revolution

Martin Mayer

Banks, retailers, and publishers have begun services that enable people to under-take financial transactions, purchase goods, and keep up with the news—all without leaving their homes or talking to a human. But progress with the videotex revolution will largely depend on whether businessmen can find ways of making money out of it, as journalist Martin Mayer makes clear in this article from Fortune, November 14, 1983.

The systems coming on line vary radically from one to another. Some are limited to plain-vanilla banking transactions, while others offer long menus of services. Some send information to a customer's home computer; others use special computer-like terminals that display words and pictures on TV sets. But all the systems share two features: they are expensive, and the companies offering them are unsure whether enough customers will pay enough money to make them profitable. As the head of home transaction services at one company puts it, "Everybody's nervous because nobody's paid for this yet."

The diversity of services reflects a broad range of visions about what consumers want. Nearly all the players foresee vast electronic networks linking business and households, but no one knows which features will catch on first or ultimately dominate. Banks and retailers are betting that most consumers will want systems that let them make purchases and other transactions. Publishers tend to see information as the central element customers will want. But each group believes it must also offer the other's service to survive.

The Knight-Ridder newspaper chain plans to begin selling the first full-blown transaction and information service to Miami-area households in November. Knight's system, called Viewtron, uses a special terminal made at AT&T to transform a television set and telephone line into a computerized communications marvel. The system vividly demonstrates the state of the art of home video services. It also highlights the cost. The AT&T terminal, called Sceptre, sells for a daunting $600, as much as a fairly complete home computer system.

What arrives in the home via Viewtron and Sceptre is a cornucopia of information: from national news via the Associated Press to local Little League scores from the Dade County Recreation Department; any page from a Grolier encyclopedia; the Scott Foresman reading series for the tykes and a cram course on the Scholastic Aptitude Test for teenagers; the Official Airline Guide, road maps, and nautical maps showing where the fish are biting. All in living color, 24 hours a day, for $12 a month plus about $1 in telephone charges for each hour of use.

Knight Ridder's service differs importantly from the data-base networks—like CompuServe and the Source—that have been available to owners of personal computers for several years. For one thing, Viewtron is local, with scads of news and information about Miami; CompuServe and the Source are national networks with only national and international news. Knight-Ridder and AT&T emphasize that Sceptre is much simpler to use than a computer and can reproduce pictures like the ones in video games.

The crucial difference, however, is that Viewtron users can *do* things. They can monitor their accounts at any of a dozen Florida banks, transfer funds from one account to another, and pay bills. And they can shop electronically at more than 100 south Florida retailers and travel agents. A Viewtron subscriber can flip through the pages of a store's electronic catalogue and simply press a series of buttons to order something. CompuServe, the Source, and Dow Jones News/Retrieval offer a similar feature through a catalogue discounter called Comp-U-Card, but they can't present sketches of the merchandise, and don't have local shopping.

Knight-Ridder has invested $26 million in Viewtron in hopes of becoming the main "system operator" for home transaction and information services in Miami and other cities. System operators control the computer link between households and the "service providers" who sell goods, services, and information, and they hope to profit by collecting from everyone involved.

Knight's objective is to sign up 5000 Viewtron subscribers the first year. Achieving even that modest goal could be hard given the $600 cost of a Sceptre terminal. Knight's subscriptions could get a boost from an unlikely helper, however. Dow Jones plans to offer its News/Retrieval service—which provides financial and general news, data on 8700 corporations, sports, weather, and other information—to Sceptre owners in the Miami area for a flat fee of just $10 a month. (More than 100,000 subscribers, many of them businesses, already pay stiff hourly rates, up to $72, for News/Retrieval.) It's hard to imagine that anyone would spend $600 on a Sceptre to get the News/Retrieval bargain and then turn down the extra goodies on Viewtron for only $12 month.

If Viewtron does well in Miami, Knight is set to sweep the country with it.

Plans are in place to introduce the system in five more cities where Knight owns newspapers (Detroit, Philadelphia, Charlotte, St Paul, San Jose) and in 12 others where local newspaper proprietors have signed marketing agreements.

Other major publishers are only steps behind in developing fancy systems of their own. Field Enterprises, owner of the Chicago *Sun-Times*, is part of a consortium that plans to offer a service called Keycom in the Windy City next April. Centel, a big Midwest telephone company, is the major partner and Honeywell is the third party in the venture. The $10- to $15-a-month service is quite similar to Viewtron, but it doesn't require a special terminal like Sceptre. Instead, Keycom will sell a $50 program that enables a home computer to receive video pictures. The owner also needs a $100 modem that allows the computer to make phone calls. Keycom believes that with the lower front-end charge it can sign up 20,000 Chicago-area subscribers in the first year.

The Times Mirror Co., publisher of the *Los Angeles Times*, will begin offering the 720,000 households of Orange County something very much like the Knight-Ridder service, using Sceptre terminals, next June or July. Times Mirror will include rental of the terminal in a hefty $30 monthly charge. There will be no special phone charges.

Chemical Bank, at the opposite end of the spectrum from the publishing companies, recently began selling a service that has nothing at all except at-home banking. Yet Chemical's Pronto costs the same $12 a month as Viewtron. It takes its instructions from the subscriber's home computer. At any hour of the day or night, the customer can check his balance, find out which checks have cleared, transfer money between accounts, and pay bills. So far bill paying is limited to 450 New York area merchants who have agreed to accept paperless electronic payments.

Chemical spent an estimated $20 million developing Pronto and has been selling franchises to banks around the country that would offer the service in their areas but use Chemical's computer. The price for a franchise: a $100,000 entry fee, plus annual charges. Only eight sales have been made so far, but Chemical hopes 35 banks will be offering Pronto before the end of 1984. John Farnsworth, 42, a senior vice-president in charge of Pronto, says he plans to "dimension-out" the service until it has all the bells and whistles that are so gaudily hung on Knight-Ridder's Viewtron. Chemical, in other words, wants to be a system operator.

So does Citibank. It began selling a service a year ago that, at least for a while, looked better than Chemical's. In Citi's HomeBase (all God's chillun got names), the subscriber's $10 a month buys a stripped-down version of Dow Jones News/Retrieval, as well as at-home banking, and permits him to pay bills to anybody (not just the 450 authorized recipients of Pronto payments). Citibank's service was announced with some fanfare and with demonstrations at computer stores. But the service really is just a test program, and Citibank has limited the number of subscribers to fewer than 1000.

The Citibank system has another oddity. It appears to offer customers the

Figure 3.5 How the systems perform their magic. Home transaction systems use computers that take orders from thousands of customer terminals and respond by sending information to the customers' home television or computer screens. News reports are stored in the central computer and delivered directly to subscribers. Other information, like the sorry state of your checking account, comes from remote sources, with the central-system computer serving as a "gateway" between the user and the service provider. Consumers can buy advertised goods and services with credit cards or bank cards.

ability to transfer funds electronically to anyone without writing a check. But in fact, the customer enters the amount, payee, and other information in his home computer, and in most cases Citibank then writes a check and mails it.

A group of banks gathered by Automatic Data Processing (ADP), a computing services company, will begin testing yet another transaction system to consumers next spring. About 20 banks have signed up, including Continental Illinois, Marine Midland, BankOhio, and Citytrust in Connecticut. In addition to at-home banking, the system will include news, weather, and stock market information packaged by Times Mirror. Consumers will have a choice between Sceptre and a similar Canadian terminal.

ADP starts with a considerable advantage in home banking. It is far and away the largest supplier of "pay by phone" services, by which customers instruct their banks to pay bills for them. ADP mans the phones and pays the bills for 85 banks, and has a memory file of 75,000 merchants and others to whom the people using pay-by-phone make payments. The ADP system will cost local banks about $400,000 for the first year's operation.

Other banks have decided that their proper place is at the tail of the publishers' kites. Banking services will be provided over the Knight-Ridder, Times Mirror, and Field Enterprises systems through VideoFinancial Services, a company owned by four large regional banks (Southeast in Florida, Banc One in Ohio, Wachovia in North Carolina, and Security Pacific in California). Four additional partners will be announced soon. Citibank asked to join, but the founding banks wouldn't have it. They feared that Citibank would use information gleaned from VideoFinancial to improve its competing service.

Wherever VideoFinancial contracts for the banking piece of a home transaction system, it invites local banks to offer the service to their depositors through its computer. The company running the local transaction system simply opens a switch—or gateway—that connects the consumer to VideoFinancial. The VideoFinancial computer verifies that the customer has enough money in his local bank account, executes the customer's instructions, and notifies his local bank of what happened. VideoFinancial collects a fee from the local bank. The local bank profits from fees charged to customers and by saving on clerical costs.

Other companies are waiting in the wings to start transaction systems of their own or offer unique services on somebody else's. E. F. Hutton has a gateway on the Knight-Ridder system. Hutton customers who subscribe to Viewtron in Miami will get a wide range of information on investments and the status of their accounts and will be able to write electronic checks against funds in their Asset Management Accounts. The service also will be available to any Hutton customer with a home computer for $17 a month.

Investors will soon be able to hear from Dean Witter via telephone line and the home computer. They'll be able to find out whether orders have been executed and how specific securities are doing—but not to place buy or sell orders. The same limitation will exist at Hutton and other full-price brokers: their registered reps insist on taking orders themselves to maintain customer relations.

A discount broker already accepts electronic buy and sell orders. C. D. Anderson & Co. in San Francisco began offering an at-home trading system last July and already has about 300 customers around the country. For a one-time charge of $300 and a minimum monthly usage fee of $10, an

Anderson client with an Apple- or IBM-compatible computer can call up the prices of stocks (with a 20-minute delay from the last trade on the New York Stock Exchange), track his own portfolio, and send buy and sell orders directly to Anderson's wire room for execution.

J. C. Penney, which has a field-tested though currently idle home banking system, also plans to go into home transactions in a big but still unformulated way. Penney got interested in the business as a means to bolster its catalogue sales. It has arranged a gateway on the Knight-Ridder system that will link shoppers directly to its own computer. Penney wants to sell merchandise over *every* other system as well. Penney also would like to become a system operator on its own, but may not be ready for that big gamble. Ralph Henderson, a Penney executive vice president, shudders when he reflects on the start-up costs of becoming an entrepreneur in home transactions. "We're looking," he says, "for some strategies to get into this business without betting the farm."

Just about everyone shares that uncertainty over how the economics of this new industry will work out. Some consultants are positively euphoric. Booz Allen & Hamilton predicts that the business will reach $30 billion in revenues by the mid-1990s, or nearly half again as much as the daily newspaper industry last year. Other consultants are more cautious, though still cheery. Michael Tyler, president of CSP International, a Manhattan firm that specializes in telecommunications, says the business may reach $4 billion in revenues—and 17 percent of US homes—by 1990.

Many executives who lived through field tests of the different systems also seem convinced that a lucrative opportunity awaits them. CBS participated in last year's test of the ADP system in Ridgewood, New Jersey. Vice President Harry Smith marvels at how consumers took to the home-shopping service included in the system.

"It's a very powerful tool," he says. "We were all surprised at the extent to which people went into the ordering process. They *liked* to do it." Henderson at Penney thinks service providers and system operators can collect substantial revenues by adding special services: "I'm a flyer. If I could file a flight plan with the FAA before leaving my home, I'd buy the service and gladly pay a premium." But as the teams move onto the field, CBS is not among the players and Penney is still in the locker room.

The tests showed that people do like the services, especially comparison shopping, paying bills via their banks, and up-to-the-minute news. However, many people involved with the systems doubt that consumers will pay enough to return a profit to the operators. Times Mirror found, for instance, that few people even in well-fixed Orange County would pay more than $30 a month for its system, including rental of a Sceptre terminal. "The problem is simple," says James Holly, head of the Times Mirror system. "The cost of providing this service to the consumer far exceeds what you can get from subscribers."

Times Mirror is counting on advertising revenues to put its system in the black. It plans to sell ads that will appear with news and other information that users request. However, the shopping services that consumers want most could get in the way of advertising sales. Many field tests showed that comparison shopping was a main attraction of the services. One complaint in the Ridgewood test was that the comparative price lists didn't cover enough products. Giving subscribers the price information they want could drive out advertising by all but the lowest-cost retailers.

Knight-Ridder puts little faith in advertising, and believes the significant profits will come from charging commissions on sales made through the service (at a rate of 1 per cent or $1 per sale, whichever is greater). But Knight's strategy also could run into problems. The discount and mail-order operations most likely to gain sales through home transactions may not be willing to pay commissions if they can reach home computer households without going through a system operator, as banks are doing.

The newspapers advantage in operating these systems is that much of the information to be provided is already on their computers and can be fed automatically to the home transaction customer. The banks' advantage, in theory, is that they will not only receive fee income from subscribers but also save on their own paperwork costs. But the banks aren't really ready to capitalize on the efficiencies of electronic transactions. Even banks that are geared up to provide instant transactions for a few hundred corporate treasurers couldn't handle the demands on their computers from thousands of home transaction customers. Says Albert Gillen, a stout and hearty broadcaster who runs Viewtron for Knight-Ridder, "The banks are crazy to get in this system. But you look in their back offices and they're in the dark ages."

Many executives in this infant industry are haunted by the fear that the inventors have once again trapped management into investing in something simply because it can be done. But Henderson, for one, is even more fearful that those who wait to get into the business will find themselves frozen out. "There is a window here," he says , "that will close in a couple of years." At Knight-Ridder, fully committed to a major investment here and now, executive vice president Norman Morrison says: "It's like the Normandy invasion. The boats are on the beaches." Chairman Alvah Chapman's attitude is one of calm confidence: "We're in a win-win situation. If this goes, we have a big moneymaker. If it doesn't, all the financial analysts will say, 'At least Knight-Ridder has stopped putting $12 million a year into that.'" Others in the business are as eager as Chapman to learn which way Knight will win.

Giants from different industries don't often contend for the same business. Banks, publishers, retailers, and data processors all feel theirs is the best perch from which to dominate the market for home transaction services. Some believe that system operators will grow rich by taking a tiny commission on each transaction; others are betting that the big profits will flow to service providers. What they're all after is that piece of the business where a single player has the most control over the product and the price.

Guide to Further Reading

General

Gerald W. Brock, *The Telecommunications Industry: The Dynamics of Market Structure* (Harvard University Press, Cambridge, MA, 1981). Definitive historical account of US telecommunications industry.

Leonard Lewin (ed.) *Telecommunications in the US: Trends and Policies* (Artech, Dedham, MA, 1981).

Joseph N. Pelton, *Global Talk: The Marriage of the Computer, World Communications and Man* (Sijthoff and Noordhoff, Holland, and Harvester Press, Brighton, England, 1981).

John Wicklein, *Electronic Nightmare: The New Communications and Freedom* (Viking, New York, 1981). On the threats and promises of the new telecommunications.

"The 'New Technology': Who Sells It? Who Needs It? Who Rules It?" A *Journal of Communication* special symposium, Autumn 1982.

Jean M. Ciano and E. Bryan Carne, "Telecommunications: The Next Generation, " and Shirley Fetherolf, "Telecommunications and the Future," in Howard F. Didsbury (ed.) *Communications and the Future* (World Future Society, Bethesda, MD, 1982).

Ken 'ichiro Hirota, "The Future for Telephony in Japan," *Telecommunications Policy*, December 1982.

Keiji Tachikawa, "Information Network System—New Telecommunications Converged With Computers," *Studies of Broadcasting*, March 1983 (Tokyo, Japan). Account of the Japanese "intelligent" network—well worth seeking out.

"Business Communications," International Data Corporation White Paper, *Fortune*, April 4, 1983 and March 19, 1984.

"World Telecommunications," *Financial Times* Special Survey, October 24 and 25, 1983.

John G. Posa, "Phone Not Going Digital," *High Technology*, May 1983.

US de-regulation/UK liberalization

Bro Uttal, "What's Ahead for AT&T's Competitors," *Fortune*, December 28, 1981.

"Telecommunications: The Battle for a Piece of the Action," *Business Week*, October 11, 1982.

Anne W. Branscomb, "Beyond Deregulation: Designing the Information Infrastructure," *The Information Society Journal*, No. 3, 1982.

Guy de Jonquieres and Paul Betts, "America's Communications Revolution," ninepart series in the *Financial Times*, January 10–24, 1983.

"Breaking Up AT&T," *Fortune* Special Report, June 27, 1983.

"Click! Ma is Ringing Off," *Time*, November 21, 1983.

"Deregulating America," *Business Week* Special Report, November 28, 1983.

Brian O'Reilly, "AT&T: What Was It We Were Trying to Fix?" *Fortune*, June 11, 1984.

Liberalisation of the Use of the British Telecommunications Network, Report to the Secretary of State for Industry by Prof. Michael E. Beesley (HMSO, London, January 1981).

Regulation of British Telecommunications Profitability, Report to the Secretary of State for Industry by Prof. Stephen Littlechild (HMSO, London, January 1983).

Private Line: The Future of British Telecom, an Anti-Report (Counter Information Services, London, October 1982).

Jason Crisp, "Even Freedom Has a Price" and "A Very Discreet Revolution," *Financial Times*, April 20 and 27, 1983.

Richard Brooks, "Can BT Reverse the Charges?" *Sunday Times*, October 16, 1983.

Guy de Jonquieres and David Freud, "BT: The 51 percent Solution," *Financial Times*, September 27, 1983.

Barry Fox, "Phone Users Will Depend on Their Referee," *New Scientist*, March 8, 1984.

Fibre optics

Jason Crisp, "It's All Done With Glass," *Financial Times*, August 3, 1981.

Peter Marsh, "Communications in the 1980s: Satellites or Fibre Optics?" *New Scientist*, July 23, 1981.

Les G. Gunderson and Donald B. Keck, "Optical Fibers: Where Light Outperforms Electrons," *Technology Review*, May–June 1983.

Jeff Hecht, "Fibre Optics Calls Up the Past," *New Scientist*, January 12, 1984. A history of fibre optics.

Roy Garner, "Light Links for Japan's Future," *Financial Times*, March 9, 1984.

"Fiber Optics: The Big Move in Communications—And Beyond," *Business Week*, May 21, 1984.

Cellular radio/Fancy phones

Aimée L. Morner, "The Looming Boom in Beepers and Car Phones," *Fortune*, December 14, 1981. Similar pieces appeared in *Time*, April 11, 1983, and *Business Week*, September 12, 1983.

John W. Dizard, "Gold Rush at the FCC," *Fortune*, July 12, 1982. Good account of cellular radio.

"The Race to Put a Phone in Every Car," *Business Week*, October 25, 1982.

Wilson P. Dizard, "Re-Inventing the Telephone," *InterMedia*, May 1983.

The *Financial Times* had features on cellular radio on October 1, 1982, December 6, 1982, February 8, 1983, and October 20, 1983, plus special Surveys on Mobile Communications on November 1, 1982 and May 15, 1984.
So-called fancy phones are described in: "Dial 'M' for Money," *Time*, January 31, 1983.
"Changing Phone Habits," *Business Week* cover story, September 5, 1983.
"Mickey Mouse Moves In," *Financial Times*, December 28, 1983.

Postal automation/Electronic mail/Teleconferencing/LANS

Jacob Rabinow, "Postal Automation: It Can Be Done," IEEE *Spectrum*, March 1981.
Implications of Electronic Mail and Message Systems for the US Postal Service (Office of Technology Assessment, US Congress, Washington DC, 1982).
New Technology, The Post Office and the Union of Post Office Workers, report prepared by the Science Policy Research Unit, University of Sussex, England. Published, with a very critical commentary (*Posts, Telecommunications and the New Technology*), by the Union of Communication Workers, London, September 1980.
Joel L. Fleishman, *The Future of the Postal Service* (Praeger, New York, 1983). An Aspen Institute study.
"Will Electronic Mail Deliver for Western Union?" *Business Week*, March 19, 1984.
William H. Dutton et al., "Utilization of Video Conferencing," *Telecommunications Policy*, September 1982.
Walter Kiechel III, "Teleconferencing to Stardom," *Fortune*, May 2, 1983. On psychological problems of electronic meetings. The *Financial Times* also had pieces on teleconferencing, August 24, 1983 and September 16, 1983.
John Lamb, "How Computers Can Talk to Each Other," *New Scientist*, September 9, 1982. On Local Area Networks (LANS). Also see *Time*, June 28, 1982 and D. N. Chorofas, *Designing and Implementing Local Area Networks* (McGraw-Hill, New York, 1984).
Herb Brody, "Reach Out and See Someone," *High Technology*, August 1983. On video conferencing.

Satellites/Satellite TV

Burton I. Edelson and Robert S. Cooper, "Business Use of Satellite Communications," *Science*, February 12, 1982.
Brian Wenham (ed.), *The Third Age of Broadcasting* (Faber, London, 1982).
John Cooney, "Lowering Skies for the Satellite Business," *Fortune*, December 13, 1982.
Kazuhiko Goto, "Japanese Project for Direct Broadcasting Satellite Service," *Studies of Broadcasting*, March 1983 (Tokyo, Japan).
Sarah Bartlett, "General Instrument Blind-Sides Comsat," *Fortune*, April 18, 1983.
L. Pollack and H. Weiss, "Communications Satellites: Countdown for INTELSAT VI," *Science*, February 10, 1984.
See also *New Scientist*, July 22, 1982, June 9, 1983, and September 22, 1983; *Business Week*, July 11, 1983, *Financial Times*, August 13, 1983; December 15, 1983, and January 4, 1984.

Cable/Cable TV

John Cooney, "Cable TV's Costly Trip to the Big Cities," *Fortune*, April 18, 1983.
"The Rush Into Cable Is Now Turning Into A Retreat," *Business Week*, October 17, 1983.
See also *Time*, September 27, 1982 and September 26, 1983.
Lawrence K. Grossman, "Cable Will Only Make TV Worse," *Washington Post*, July 25, 1982. Important, critical article by the president of the Public Broadcasting Service emphasizing that the fragmentation of audiences will make TV more like low-cost radio.
Home Entertainment in the 1980s, International Resource Development report (Norwalk, Connecticut, 1982).

UK developments are covered in *New Scientist*, April 1, 1982 and March 3, 1983; *New Statesman*, July 30, 1982 and April 8, 1983; *Sunday Times*, August 21, 1983; and the *Financial Times*, March 23, September 11, October 13, October 18, November 26 and December 4, 1982; April 28, April 30, September 3, and November 28, 1983, and June 13, 1984.

Videotex

"Window on the World: The Home Information Revolution," *Business Week*, June 29, 1981.
John Tydeman, "Videotex: Ushering in the Electronic Household," *The Futurist*, February 1982.
Roy A. Sutherland, "Home Banking: Electronic Money Invades The Living Room," *The Futurist*, April 1982.
Gerd D. Wallenstein, "Visions to Stay Home By: Visual Services for Home Delivery," *Telecommunications Policy*, June 1982.
John Tydeman et al., *Teletext and Videotex in the United States: Market Potential, Technology, Public Policy Issues* (McGraw-Hill, New York, 1982).
Stephen K. Badzik, "Videotex: Blessing or Bane for the 'Boob Tube'?" in Howard F. Didsbury (ed.), *Communications and the Future* (World Future Society, Bethesda, MD, 1982).
Michael Aldrich, *Videotex: Key to the Wired City* (Quiller Press, London, 1982).
James Martin, *Viewdata and the Information Society* (Prentice-Hall, Englewood Cliffs, NJ, 1982).
Efrem Sigel (ed.), *The Future of Videotext* (Knowledge Industry Publications, White Plains, NY, 1982).
Richard M. Neustadt, *The Birth of Electronic Publishing: Legal and Economic Issues in Telephone, Cable and Over-the-Air Teletext and Videotext* (Knowledge Industry Publications, White Plains, NY 1982).
Tim Robinson, "Dawn of the Electronic Newspaper," *New Scientist*, May 13, 1982.
"Videotex," a *Financial Times* special Survey, December 1, 1982.
Jeff Hecht, "Information Services Search for Identity," *High Technology*, May 1983.
"Electronic Publishing Moves Off the Drawing Boards," *Business Week*, August 8, 1983.

Business and financial information services are specifically dealt with in:

Raymond Snoddy and William Hall, "The Rise and Rise of the 'Information Factories'," *Financial Times*, August 13, 1983; and special FT Surveys on December 14, 1983 and December 6, 1984.

Shawn Tully, "The Big News at Reuters is its Money Machine," *Fortune*, August 8, 1983.

"The Selling of Reuters," *Time*, June 11, 1984.

"Publishers Go Electronic," *Business Week* cover story, June 11, 1984.

Part Two: The Human Interface

4 Computers in the Home

Personal Computers

Hoo-min D. Toong and Amar Gupta

A major surprise in the information technology revolution has been the remarkable rise of the personal computer. Enormous computing power can now be put on desks in homes, schools, and offices, at comparatively low cost. The personal computer has thus become the basic building block of the information society. The authors are both computer scientists with the Sloan School of Management, MIT, and this piece first appeared in Technology Review, *January 1983.*

Rapid advances in semiconductor technology since the 1960s have made possible tremendous increases in computing power at plummeting costs. Whereas 20 years ago a computer could be afforded only by large organizations, the minicomputers of the 1970s were purchased by departments and groups within such organizations. Now a personal computer is within the reach of all of us.

Today's personal computers offer significant computing capabilities, and they are easy to use, attractive, and increasingly popular for a broad range of day-to-day activities in both office and home. Indeed, they are dramatically changing many facets of our daily lives. Over 2 million personal computers are now in use in the US; sales were over $3 billion in 1982 and will exceed $6 billion in 1985. Personal computers are no longer a luxury but important tools for increasing productivity in the office, the most labor-intensive sector in the US economy.

The same capability for handling much of the paperwork of modern life—tax returns, mortgage calculations, and other financial transactions, for example—motivates people to use computers at home. However, a major segment of the market involves entertainment: personal computers offer a growing array of games, music, and art far more sophisticated, in terms of speed and capability, than video material based only on television. This advantage results from the ability of computers to store and quickly retrieve large amounts of information.

The future will bring many more uses for computers in both business and home. Indeed, the possibilities are virtually endless.

A great variety of computers, large and small, are now available. A personal computer is a machine or system meeting *all* the following major qualifications:

1 The price for the complete system is under $5000.
2 The system is designed to accept secondary memory devices to supplement the primary, built-in memory.
3 The user is expected to interact with the system continuously, not only at the beginning and end of a problem.
4 At least one general language (BASIC, FORTRAN, COBOL, PASCAL, ADA, or C) is available for this interaction.
5 The system is usable for a wide variety of problems and is not designed for any single application.
6 The computer is distributed through mass-marketing channels, with the marketing emphasis on the first-time computer user.

A typical modern personal computer consists of a circuit board with a silicon chip microprocessor and one or more memory chips attached. The microprocessor can perform hundreds of thousands of calculations every second, and the memory chips provide the primary storage for instructions and data. External storage devices, such as cassette tape units or small recording disks (floppy disks), augment the memory capacity and provide a storage medium that can be physically transferred from one personal computer to another. (Typical users begin with cassette units but soon change to disks to gain the advantages of greater speed and capacity.)

Input is through a typewriter-like keyboard unit. Output typically takes the form of words and numbers displayed either on a television screen or a similar specialized screen called a monitor. Most monitors are designed to display 24 lines of letters and figures, with each line containing a maximum of 80 characters. Adding a printer unit permits output in the form of a printed paper. A special device, called a "modem" (for modulator/demodulator), permits the computer to receive and transmit data over a conventional telephone line.

But the significant factor in defining a personal computer is not its physical features but the characteristics of the operating system. Designers of personal computers and software attempt to provide a friendly human–machine interface, even at the expense of brute computing power. Optional programs are also available so that the computer can be used for many different purposes. Although word processors and hobby computers have many characteristics of personal computers, they lack this flexibility.

Recent trends in microelectronics, memories, input–output mechanisms, and software suggest that the trend in microprocessors is toward larger

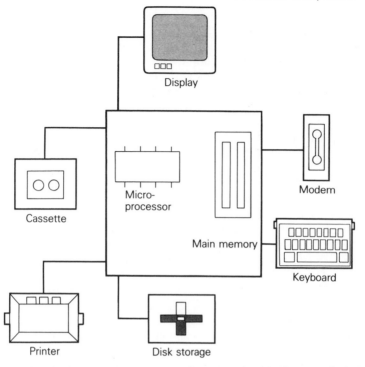

Figure 4.1 A typical personal computer configuration. A minimal system includes, in addition to microprocessor/memory, external cassette memory unit, monitor display, and keyboard. A printer permits output on paper, a modem permits communication between computers over a telephone line, and floppy disk storage can be added for increased memory.

"words" and higher circuit speeds. A computer capable of handling larger "words" is able to perform a complete operation in fewer machine cycles and to operate directly with larger memory. Both these assets enhance performance by increasing the speed of operations and the number that can be performed in a sequence without the operator's intervention. In addition to greater speed and memory access, these larger microprocessors have the advantage of greater accuracy.

The first wave of personal computers used 8-bit microprocessors—that is, microprocessors in which 8 binary digits (0 or 1) can be processed in parallel, giving the capacity to process in a single operation any number up to 256, or to address any of up to 256 points in the memory. Processing larger numbers with an 8-bit microprocessor requires multiple operations, which take more time. Newer systems use 16-bit microprocessors, and 32-bit microprocessors are now available.

Currently, 32-bit chips cost $250, compared with $50 and $8, respectively, for 16-bit and 8-bit chips. As technology improves and costs drop, 32-bit microprocessors will be standard in personal computers by 1990. Until then, 16-bit microprocessors will be the norm.

The primary memory in personal computers is of two different types: read-only memory (ROM) and random-access memory (RAM). In the former, information is fixed in the memory at the time of manufacture and is not lost when computer power is switched off. The role of such a ROM is to guide the computer through a fixed procedure, such as calculating square roots or translating a user's program into machine language. In a RAM, information such as special programs and data files can be "written in" or "read out" as frequently as desired, with any storage location directly accessible. RAMs are of two types: dynamic RAMs, which are cheaper but lose their stored information unless they are "refreshed" often, and static RAMs, which are costlier but do not need to be "refreshed." If power is lost, both types of RAMs lose their stored information.

Over the past decade, the number of memory circuits per unit of area on a chip has increased by a factor of 64, and cost on a unit basis has been reduced by a factor of 50. Both these trends will continue. So-called 64K dynamic RAMs (each contains 65,536 bits of information [$K = 2^{10} = 1024$; $64K = 64 \times 1024$]) are expected to remain popular until about 1984, when the much larger memory capacity of 256K dynamic RAMs is expected to become standard. Most manufacturers supply system programs on ROMs, and this practice will continue because programs are secure against power failures and users are less able to duplicate programs in this form.

Computers' main memories will be supplemented by secondary storage devices that offer larger and relatively inexpensive, though slower, capacity for long-term storage of program and data files. The most popular of these is a so-called "floppy disk"—a disk of mylar coated with magnetic material on one or both sides. Data are stored in a series of spots—either magnetized or demagnetized—along the concentric tracks. Heads for reading or writing data can be moved radially across the disk to reach a specified segment of circular track. Storage capacity depends on the format used for the stored data, the quality of magnetic surface, and the design of the reading–writing head. Floppy disks in current use typically have capacities of 1 to 4 million bits, sufficient to store 20,000 to 80,000 words of English text. During the next four years, higher-density floppy disks will be common. Indeed, disks offering capacities of 50 to 100 million bits are already becoming popular, but they are much costlier than floppy disks.

The primary display device used in all personal computer systems is a cathode-ray tube (CRT), either standing alone or as part of a television receiver. This system will continue through the foreseeable future. Output is typically presented in alphanumeric form—letters and numbers. Charts and game boards can be presented, but the memory and software required to display such graphical images is often complex and expensive. The letters and numbers in alphanumeric displays are patterns of dots programmed in special ROMs known as "character generators." The quality of the image depends on the number of points (or pixels) on the screen that can be addressed by the computer—that is, the number of points at which dots can be located. A typical low-resolution screen has a field of 6144 (128 times 48) pixels (*pic*ture *ele*ments—dots on the screen). High-resolution systems (100,000 pixels or more) allow sophisticated graphics for animation or detailed figures and may provide color as well.

The market and the players

The personal computer industry has grown as a direct result of the evolution of the microprocessor. This evolution began when Intel in 1971 packaged a complete, if somewhat limited, processor with a 4-bit word size in a single integrated circuit. The company followed with the first 8-bit processor in 1972 and an improved version in 1974. One year later Micro Instrumentation and Telemetry Systems, Inc. (MITS), an Albuquerque firm, developed the first personal computer around Intel's 8-bit processor. The basic system sold for $395 in kit form and $621 in assembled form, not including accessories (peripherals).

Though the MITS system is no longer manufactured, its method for connecting peripherals and the main computer has become an industry standard. Within three years after it was introduced, Radio Shack, Apple, and Commodore had entered the market. Now makers of large computers such as IBM and Honeywell, as well as minicomputer leaders such as Digital Equipment Corp. and Data General Corp., are also making personal computers, having observed their traditional markets being eroded by the new low-priced products. The results are rapid development of the market and growing popular interest and faith in personal computers.

The fact that the pioneering manufacturers did not survive beyond the initial phase of personal computer development results from the preoccupation with the needs of hobbyists. New entrants such as Radio Shack, Commodore, and Apple captured a major share of the market in 1978 by promoting fully assembled, ready-to-operate systems that were easier to use. Though the success of the companies now in the competition will depend at least partly on their financial and technical resources, their products' ease of use (or "user-friendliness," in trade jargon) is a significant competitive feature.

The personal computer market can be divided into four segments in terms of the computer's intended use: business, home, research–technical, and educational. The business segment, by far the largest at present, accounted for 750,000 sales (retail value of $2 billion) in 1982, or 54 percent of the units sold and 65 percent of the dollar sales. This segment will continue to be a major factor, with more sales than all other segments put together. No wonder, then, that the leading makers of personal computers are concentrating their sales efforts on the business sector.

There are 14 million businesses and 36 million white-collar workers in the

US, and many are present or prospective users of personal computers. Indeed, personal computers are especially suited to small businesses of ten people or less, including retailers, consultants, and professional offices such as those of lawyers, doctors, and dentists. Sales to larger organizations adopting the "computer-on-each-desk" concept will be a major factor by 1985.

The most visible segment of the personal computer "revolution" is in homes, where computers are used for entertainment and education, sending messages, and home finances. The development of appropriate software will soon enable sophisticated users to pay bills, manage bank accounts, compute taxes, and even buy household items without written transactions. The average cost of a home system was $1250 in 1982, when the home segment accounted for $465 million of sales (375,000 computers). The average cost will be less than $1000 in 1985 and $750 in 1990. Prices at the low end have already tumbled to under $300, and intense competition has made this market for small computers unprofitable for several producers.

The science segment accounted for 200,000 unit sales at a retail value of $470 million in 1982. Though the small computers available for technical work fulfill the definition of personal computers, they tend to be used for sophisticated calculations requiring more power as well as the capability to interface with other scientific instruments and higher-level languages. Thus, this market is likely to be served in the future by a series of highly specialized products.

The potential market for personal computers in education (60,000 units with a retail value of $120 million sold in 1982) has barely been tapped. In the US alone there are 3000 colleges, 29,000 secondary schools, 77,000 elementary schools, and 60 million students, all of whom could greatly benefit from ready access to computing power. For example, computer-assisted instruction would enable interactive learning in many different subjects at a pace tailored to each student's capability. Furthermore, the science and practice of computing—including programming skills—would be part of every student's education.

Computer manufacturers recognize that schools are the logical environment in which to develop computer skills, and they have reason to expect that students will purchase the machine on which they first learn computing. So manufacturers are offering many price incentives to the educational sector. Yet the educational sector will continue to account for only about 15 percent of total sales of personal computers for the rest of this decade, as school funding will be limited.

Although they want to serve all four segments, the industry leaders (Apple, Radio Shack, Commodore, IBM, and Xerox) are focusing on the business market. Vector Graphics is also emphasizing this market by tailoring personal computer systems to the needs of particular industries. Atari (Warner), Intellivision (Mattel), and Texas Instruments are focusing on the home market. Hewlett-Packard is offering a highly specialized compact product with built-in printer, tape memory, and video monitor especially for the scientific community.

In general, both prices and manufacturers' profit margins are falling, and mass marketing is becoming the rule. The major competitors will each spend over $10 million in 1983 on advertising aimed at expanding the market and establishing their place in it.

The dynamics of the entire personal computer market may be changed substantially by new entrants by 1990. Traditional barriers to entry into this field have crumbled. For example, manufacturing capability is not essential: the IBM personal computer is assembled almost entirely from premanufactured components not of IBM origin. By contrast, strength in marketing and distribution is a significant advantage, and many organizations with this capability may seek to replicate IBM's strategy for rapid entry into the field. Organizations such as General Electric, Procter and Gamble, Phillip Morris, and Du Pont all have the ability to enter the personal computer market in this way.

Selling personal computers

Strategies for marketing mainframe computers are not appropriate for marketing personal computers—profit margins are not large enough to justify hiring internal sales forces to sell directly to end users. As a result, producers are now experimenting with a wide variety of distribution strategies.

Franchised retail chains constitute a major distribution channel. For example, the Computerland chain sold $200 million worth of computers and related accessories in 1981. These stores distribute the products of many vendors, and their volume is large enough to support a technical and maintenance staff.

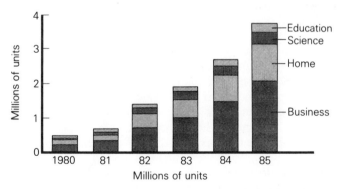

Figure 4.2 Estimated numbers of units to be sold in the US, by market segments. Personal computer sales have tripled since 1980, and this rapid growth will continue. Sales to business are more than half of the total, with education—constrained by shortages of funds—the market's poorest customer.

Manufacturer-owned retailed stores have been used successfully by Radio Shack. IBM, Digital, and Xerox have also opened such stores but only to supplement existing distribution channels. Except for Xerox, each manufacturer's stores sell only that manufacturer's products, and a prospective buyer is thus obliged to visit several stores to compare equipment of different makes.

Department store outlets have generally been unsuccessful. Mass merchandisers depend for profits on fast-selling commodities. According to computer industry data, personal computer buyers make four shopping trips totaling as much as seven hours when selecting their machines. These buyers also expect sustained support and maintenance services that department stores are unaccustomed to providing.

Office equipment stores specializing in copiers, typewriters, word processors, and other office equipment are well positioned to reach the most promising future market for personal computers and auxiliary equipment. If these stores can provide adequate servicing, they will become very popular.

Consumer electronics stores such as Tech Hi-Fi have been marketing personal computers with some success, but the lack of expertise at the store level has been a constraint. Japanese manufacturers have also established ties with such stores as distributors of other Japanese products. These channels will become a major factor as the Japanese increase their share of the personal computer market.

Independent retailers often lack the capital required to compete vigorously and are therefore not gaining in numbers and importance as quickly as other retail channels.

Catalog showrooms have been used by Texas Instruments, but personal computers require follow-up support that such showrooms have been unable to provide.

Mail order firms offering discounts ranging up to 30 percent appeal to price-sensitive customers, but their total lack of continuing support disenchants users. Price-cutting penalizes the full-service dealers on which manufacturers want to rely as major sales outlets. Accordingly, major companies are trying to discourage sales through mail order firms.

Direct sales staffs are used for large-volume sales to government, educational institutions, and major corporations who prefer dealing directly with the manufacturer. But such direct sales tend to antagonize dealers by depriving them of some of their most profitable opportunities, and profit margins are inadequate to support direct sales to small-volume and individual buyers.

Value-added houses serve specialized users with coordinated hardware and software, such as printing companies that need word-processing and typesetting capabilities but have little or no internal computing expertise.

Though franchised retail outlets are the largest sellers of personal computers in the US, the one compelling characteristic of personal computer distribution today is diversity. No one form of vendor or market approach is dominant.

Selecting a personal computer

No one should select a personal computer system without taking into account the tasks for which it will be used and the environment in which it will function. Even within a single business, personal computer users come from different areas—personnel, accounting, management, manufacturing, research, sales—and have different computing needs. Even within departments there are many possible uses. The personnel department will run personnel files on a regular basis but also may want to study policy questions—the likely benefits in terms of increased revenue from a larger sales force, for example. Furthermore, different users have different strategies for using a computer for *ad hoc* applications. Thus, the person who will use a personal computer should—almost by definition—choose it, partly on the basis of the services it will provide and partly on the basis of its apparent user-friendliness. Indeed, the machine's responsiveness to its user's needs and style is the critical technological breakthrough.

But it is not necessarily so simple. In many organizations, personal computers create and fill completely new needs that were not originally anticipated. Consider the example of a personal computer acquired to improve one lay worker's access to a main computer facility. While retrieving data from the main computer to run programs locally, this user finds that the personal computer can help evaluate alternative business strategies, answering, "what—if . . .?" questions instantaneously. The original application for which the personal computer was purchased becomes secondary.

Given that the uses and benefits of personal computers are hard to predict, and that a wide range of systems and software are available, how should a potential buyer decide what and when to buy? And should such a buyer postpone action, expecting costs to decline in the future as they have in the past?

There are two arguments against postponing purchase in anticipation of a lower future price. One is the high cost of waiting, based on the computer's great (and often unexpected) uses. The other is that prices are expected to decline less dramatically in the future than in the past: future systems will offer technological improvements that increase performance at unchanged cost. Prices seem likely to stabilize at around $1500 for a full personal computer system.

Future personal computers are likely to offer increased memory size, increased processor power (16-bit and even 32-bit instead of 8-bit microprocessors), improved printers, more powerful programming languages that give the user more compact and natural communication with the computer, improved user access (better keyboards supplemented or replaced by touch screen and voice systems), and greater flexibility (the computer will be able to intermix and manipulate text, numbers, and graphics more easily).

Yet today's personal computers may rapidly become obsolete, reflecting the rapid pace of change in semiconductor technology. Changes in the microprocessor, the heart of any personal computer system, are most likely—and most serious. A computer with a 16-bit microprocessor cannot operate on hardware and software designed for an 8-bit microprocessor. Though many vendors provide add-ons that enable a new microprocessor to support old software, there is never total compatibility.

Software for a personal computer is expensive, and for most users new software development is impractical. So the amount and utility of the software available with a particular system is the key question in the buyer's choice of a personal computer. The variation can be large: 11,000 programs are available for the Apple II and Apple II Plus computers, whereas fewer than 100 widely distributed programs exist for the IBM personal computer, a more recent introduction. Of course, the IBM programs may be just the ones the buyer needs, and more software will be developed for the IBM.

Price is also an important variable, including not only the manufacturer's suggested price but the range of discounts offered by different retail computer sources. But discount alone does not tell the story. The critical factor is the ability of the distributor to provide both routine hardware maintenance, usually at the retail outlet, and software support and maintenance. A company store is likely to offer very good service on that company's product line but not necessarily on any other line, while a retail computer store is likely to offer moderately qualified service and support for many different manufacturers' systems. Mail order sales offer very little service but usually have the highest discounts.

The ultimate question for the purchaser is that of system configuration: what combination of hardware and software is appropriate for a buyer's current and projected needs?

Perhaps the most important options are off-the-shelf software. To be sure they buy what they need, buyers should gain some experience with available programs before making the purchase; otherwise they risk having to modify their needs to fit the software they've purchased or embarking on the costly and time-consuming route of software development. The programming tools on most personal computers are primitive compared with the equivalents available for mainframes, and the widespread dearth of good programmers is a complicating factor. Indeed, trying to develop special programming for a personal computer is inadvisable except under extraordinary circumstances.

Memory capacity is another important variable. Most computers based on 8-bit microprocessors have no more than 48 or 64 kilobytes of built-in accessible memory (RAM), while the newer 16-bit processors can have 1, 4, or even 16 megabytes. (One byte is a string of 8 binary digits that represents one alphabetic character.) Additional memory can be inserted into most personal computers by plugging cards into slots or inserting extra chips directly onto the microprocessor board.

Storage capacity of the main memory for programs and instructions can also be supplemented by secondary devices, of which floppy disks are the most popular form. Disks with capacities ranging from 100,000 bytes to several megabytes now come in two standard sizes: 8 inches and 5¼ inches in diameter. But personal computer buyers should be aware that disks designed for one system will not generally operate in another maker's system because of different data-formatting conventions.

The critical performance characteristic of a floppy disk is the response time to a read–write request, which may range from 200 to 500 milliseconds. Though the response time for such a single file request is very short, a single compilation or word-processing command may involve 20 or more accesses to a floppy disk, and access time may account for 80 percent of the total command execution time. Where better performance and/or capacity are required, "hard" disks in the same format as the 5¼-inch and 8-inch floppy disks are useful. They offer storage capacities ranging from 5 to 50 megabytes and response time averaging 20 to 40 milliseconds, but their high cost (around $2000, including interfaces) can double a buyer's total investment in hardware.

All personal computer systems use cathode-ray tubes as output devices. If "hard copy" is required, buyers have a choice of several printing devices. Thermal printers, costing under $500, create images by applying points of heat to special paper. A dot-matrix printer, costing from $500 to $1500, has a vertical array of 7 to 10 dot-printing elements. These are activated by the computer as the printing head passes across the paper, forming alphanumeric characters and also graphs and drawings. The limited number of elements results in a somewhat stylized character, but the process is as fast as thermal printing—50 characters per second. If letter-quality printing is required, a printing mechanism with precisely machined character matrices that strike the paper through a ribbon is required. Such printers cost $750 or more and print between 30 and 90 characters per second.

Various accessories such as floppy disks and printers—or "peripherals"— are attached to personal computers through special "slot" receptacles. The IBM personal computer offers five such slots for peripherals such as printer, color display, dual floppy-disk controller, communications channel, and memory module.

All business-oriented personal computers offer text editing for word processing. Some permit use of a light pen to directly indicate on the screen the word to be edited, and a few check spelling against a mini-dictionary stored in the computer memory. Business data manipulation is facilitated by "spreadsheet" packages such as VISICALC—a very popular package that enables users to see on the CRT the impact of altering one figure (say, retail price) on all other figures (profit margin, corporate profit, return of investment). Software is now becoming available for generating displays such as those accompanying this article with only a few minutes' effort. For example, ExecuVision permits displays combining information in numbers, text, and pictures. It also provides animation that can add emphasis to presentations.

Though personal computers are today viewed primarily as stand-alone work stations for individual users, the ability to transmit and receive data and programs is an important attribute of most contemporary systems. Personal computers are already serving as intelligent terminals for retrieving and manipulating files from large computers. These networking and communication capabilities are important considerations in selecting a personal computer. Unfortunately, no industry standard for communication protocols and interfaces exists. Different vendors use different protocols, so it is difficult to link various personal computers in a single network.

Not just for playing games

Technological advances in communications are expanding the power of personal computers. For example, owners of personal computers can now receive current stock market quotations by telephone. Personal computers can be connected with nationwide electronic funds transfer systems, so users have access to full banking facilities. They may transfer funds from one account to another and buy and sell financial instruments.

In homes, personal computers are now primarily used for recreational activities such as computer games. But these applications should not be underrated. Computer games test and develop mental capabilities while generating familiarity with computer operations.

Home computers are also used for office work at home, and many other applications are—or soon will be—possible. For example, a tax return can be prepared with the aid of VISICALC. The user can analyze which forms and deductions to use for paying the minimum taxes. A personal computer can be used to balance checkbooks and plan investment strategies. And, of course, word-processing capabilities are useful in writing letters and professional papers. As commercial databases such as Dow Jones News/Retrieval and the Source are expanded and their subscription costs reduced, personal computers will be useful for locating stores and services, comparing prices, and placing orders. This would be especially appropriate for products or services whose prices fluctuate rapidly, such as those of the airline industry.

Computers are being used increasingly for security applications. Traditional home security systems are limited, often failing to distinguish between a natural event and an intrusion by an unwanted visitor. Computer-controlled home security systems can include a variety of routines to analyze the signals from sensors before generating an alarm. Such an analysis can distinguish, for example, between the entry of a cat and the entry of a human being, or between the noise of a telephone ringing and a door opening.

Another home-oriented application is a medical information system that performs some of the diagnostics of a medical doctor. Perhaps the most useful application of this is a database for poison or first-aid: the user reports what substance the patient has ingested or the nature of an accident, and the computer responds with a step-by-step first aid plan.

Children turn out to be heavy users of home computers. They enjoy computer games and seem able to learn computer languages faster than their elders. They find personal computers intriguing for recreation, school assignments, and self-paced courses that are now becoming available.

Even though today's personal computers are equivalent in basic computing power to the mainframes of the 1960s and the minicomputers of the 1970s, they should not be used as substitutes for the earlier machines. Rather, users should capitalize on the new technical developments, including the diminishing cost of computing, and the Englishlike languages that replace the cumbersome digital languages of early machines, to make personal computers supplement, not displace, mainframe computers. Personal computers should be used to analyze more issues in greater detail than ever before—not to do the more routine work typically assigned to mainframes.

In the 1960s and 1970s, computers were used principally for well-structured, periodic jobs such as payroll accounting. Today's personal computers are especially useful for more casual, *ad hoc* analyses and problem solving, such as studies of the impact of cost of living increases, new tax regulations, or alternative hiring and layoff policies.

Consider the stock analyst who uses data from company balance sheets to evaluate investment risks. The balance sheets contain detailed, current information on assets and liabilities. But interest rates fluctuate, markets rise and fall, taxes change, competition falters or gains. Each such change affects a company's current performance and future prospects. Using a central computing facility is an expensive way to analyze such problems, resulting in time delays and partial answers. But using a personal computer in the interactive mode linked to the central computer and a financial database makes it possible to understand the impact of changes quickly and accurately. When the power of our largest computers is placed at the service of people in their workplaces and homes, on such a direct, interactive basis—that is truly a revolution.

The Psychology of Personal Computers

Sherry Turkle

What people do with computers affects the way they see the world. Working with computers can also be a way of "working through" powerful feelings in a completely safe and controllable microworld, according to this study of first-generation computer hobbyists. The author is a sociologist in the Science, Technology and Society Program at MIT. This article is taken from her book, The Second Self: Computers and the Human Spirit *(Simon & Schuster, New York, 1984), and first appeared in* Social Studies of Science, *vol. 12, 1982.*

The material presence of the computer escapes no one; for a decade there has been talk about "computer revolutions" in business, banking, telecommunications, and education. Manufacturers, media, and computer scientists have presented the public to computers the way one might have presented Aladdin to his lamp. Here was the genie, the workhorse that we could mount today and ride onto the new millenium. By the end of the 1970s, with the mass manufacture of personal computers, the promises extended into the home. Home computers would teach us French, help us with financial planning, even do our taxes. For many Americans, the first thing that computers brought into their homes was not a more efficient medium for work but new worlds to conquer in play. "Space Invaders" became a household word for people with home computers or the machines' junior siblings, computerized video games. The experience of the decade has resulted in a widely shared public rhetoric. When people talk about the computer and their futures, they tend to fall back on the two images they know best: the computer as tool and the computer as toy; objective instrumentality and engrossing play. But there is another dimension. What people do with computers weaves itself into the way they see the world. People use a discourse about computers for thinking and talking about other things, about politics, religion, education, and about themselves and other people.

The subjective dimension of the computer presence is not merely a matter of discourse, of using the computer intellectually as a metaphor or model. There is another, and more emotionally charged, aspect which does not engage *ideas* about computers as much as the immediate quality of an individual's *experience* when working with them. Working with computers can be a way of "working through" powerful feelings. In this essay I develop this idea by using as a case study a group of computer users for whom issues related to control are particularly salient. For a first generation of computer hobbyists, controlling the computer is a way to deal with frustrations and desires, both personal and political, that have nothing to do with the computer *per se*.[1]

The subjects of my study are men and women who bought personal computer systems in the four years that followed the 1975 announcement of the "Altair"—the first computer small enough to sit on a desktop, powerful enough to support high-level language programming, and that you could build for only $420. My study began in 1978 with a questionnaire survey answered by 95 New England computer hobbyists (their names had been drawn from the roster of a home computer club and from the subscription list of a personal computer magazine), and continued during 1978 and 1979 with nearly 300 hours of conversation with 50 individuals who owned home computers.[2] What I found can be read historically: a study of the pioneer users of an increasingly ubiquitous technology. But most central to the intent of this essay is to use the story of the early hobbyists as a window into the highly personal ways in which individuals appropriate technologies. It is a case study of the "subjective computer," the computer as a material for thinking, for feeling, for "working through."

My emphasis on the subjective is at odds with a widespread ideology that quickly grew up around the emergent computer hobbyist culture. The Altair, aimed at a strictly hobby market, was followed by other small systems—the Pet, the Sol, and, most successfully, the Radio Shack TRS-80 and the Apple, marketed to less specialized audiences of small businessmen and curious householders. With this explosion of hardware came a lot of rhetoric about a personal computer revolution. Most of the talk, both from the companies that marketed the machines and from those who claimed to be the most visionary spokesmen for the people who bought them, was about all of the things that a home computer could do for you. The utilitarian, "genie in the bottle," ideology is expressed in the content of hobbyist conventions and magazines, filled with articles on how to make your home computer dim your lights, control your thermostat, run an inventory system for your kitchen or toolroom. And it is also found in writing on the personal computer from outside the hobbyist world. The view from the "outside" was well illustrated when, on May 14, 1979, the *Wall Street Journal* reported on its own little "evaluation experiment."

The paper had drafted one of its staff reporters, Mitchell Lynch, into the ranks of the "home computer revolution." It presented him with a TRS-80 and asked him to take it home. His assignment was to report back six months later on what it had been like. Lynch's story, a *Pilgrim's Progress* of little progress, ran under the head: "Computer Error: Trying to Use One in Your Own Home . . . Our Man Finds That He Can't Get It To Do Tax, Other Jobs . . ."

I got a spiffy $599 home computer for Christmas. And it stopped playing blackjack with me by New Year's Day, asked me "What?" about 1430 times by George Washington's Birthday, and mysteriously broke down and resurrected itself by Easter.

   I thought it would straighten out the family budget; it didn't. The ads said it would help educate the kids; it hasn't. Certainly it would do my tax returns; it can't. At least it would teach me to prepare those fancy programs for computers; it didn't. Indeed, if I hadn't been assigned to work with my computer, it would be gathering dust in my attic.

The terms of Lynch's evaluation are instrumental: what can the computer do? On these terms it fails, and when he goes to the experts for an explanation of what went wrong, he encounters an official "instrument" ideology from within: "Experts say that people like me have neither the technical training nor technical inclination to make a home computer strut its stuff." The point of having the computer, says the expert, is to make things happen. And, reassures the expert, they will happen with a better operator or a simpler computer.[3]

   This instrumental view is an important ingredient of the hobbyist ideology but it is not the whole story. In the course of my work I found a very different answer from within. Most hobbyists do make their computers "strut their stuff," but their sense of engagement and energy are found primarily in the non-instrumental uses of the technology. When asked in a questionnaire "What first attracted you to computers?" more than half the respondents gave reasons that were highly subjective. In response to an open-ended question, 26 percent said that they were first attracted to computers by an appeal that was intellectual, aesthetic, involved with the fun of what I would call "cognitive play." They wrote of "puzzle solving," of "the elegance of using computer techniques to handle problems," of the "beauty of under-standing a system at many levels of complexity." They described what they did with their home computers with metaphors like "mind stretching" and "using the computer's software to understand my wetware." Another 26 percent wrote of reasons for getting involved that seemed more emotional than intellectual. They wrote of the "ego boost" or "sense of power" that comes from knowing how to run a computer, of the "prestige of being a pioneer in a developing field," of the "feeling of control when I work in a safe environment of my own creation."

   The hobbyists who responded to my survey seemed familiar with Lynch's brand of scepticism, with people who ask them what they do with their computers and who won't take "cognitive play" for an answer. David, a 19-

year-old undergraduate at a small engineering school, put it this way: "People come over and see my computer and they look at it, then they look at me, then they ask me what useful thing I do with it, like does it wash floors, clean laundry or do my income tax—when I respond no, they lose interest." David said that when he started out, he was attracted to the computer because "I liked the idea of making a pile of hardware do something useful, like doing real time data processing . . . like picking up morse code with an amateur radio and transcribing it automatically into text," but in his list of things that he currently does with his computer, an instrumental discourse is most notable for its absence:

Conway's GAME OF LIFE in assembly code was a challenge, forced me to "think logically," and gave the pleasure of making something work the way I wanted it to . . . Having control from the bottom level of program for that game made me feel comfortable, safe, sort of at home.[4]

Thirteen percent of those who responded to my questionnaire told a similar story. Like David, they began their relationship with personal computation for instrumental reasons (they had an image of a job to do, a specific task), but they became absorbed by the "holding power" of something else. A full two-thirds of my survey sample either began with or ended up with a primary interest in what I have called the "subjective computer," the computer seen in its relationship to *personal meaning*. Clearly, to understand what people are doing with their home computers we must go beyond the "performance criteria" shared by the hobbyist magazines and the *Wall Street Journal*.

The simplest way of thinking about the subjective computer is through the metaphor of "the computer as Rorschach"—that is, seeing the computer as a projective screen for other concerns.[5] In the Rorschach test, one is presented with a set of ink blots and asked to make some sense of them. Some people see the blots as threatening; others see them as benign. Some focus on small, complex details; others on global form. So too, the computer presents us with an ambiguous stimulus. For example, although most people think of the computer as an object without real intentionality, they accord computers enough autonomy in action to make "blaming the computer" a commonplace of daily life. The very fact that the computer is a machine that touches on a sphere—intelligence—that man has always considered uniquely his is enough to make many people experience the computer as an object "betwixt and between," hard to classify, hard to pin down.

Sometimes people deny the irreducibility of computation by asserting that, no matter how complex the computation "product," a move in a chess game for example, "all the computer really does is add." Of course in a certain sense this is correct. But saying that a computer "decided to move the Queen by adding" is a little bit like saying that Picasso "created Guernica by making brushstrokes." Reducing things to this level of localness gives no satisfying way to grasp the whole. Just as in theoretical psychology there is a tension between the gestalt and the atomic, so too in computation there is a pervasive tension between the local simplicity of the individual acts that comprise a program or a computation, and what one might call the "global complexity" that can emerge when it is run.

The elusiveness of computational processes and of simple descriptions of the computer's essential nature, the tension between local simplicity and global complexity, all contribute to making the computer an object of projective processes, and exemplary "constructed object." Different people apprehend it with very different descriptions and invest it with very different attributes. In views of the computer's internal process, individuals project their models of mind. In descriptions of the computer's powers, people express feelings about their own intellectual, social, and political power—or their lack of it.

Looking at the computer as Rorschach, as projective, puts the emphasis on aspects of the individual—from cognitive style to personal fears—that are revealed through behavior with the machine. But of course the computer is more than a Rorschach. The Rorschach ink blots are evocative, revealing, but they stay on the page. They do not enter the life of the individual. The computer does. It is a constructive as well as a projective medium. For readers who have not had the experience of programming a computer, this idea may be sharpened by an analogy with another technology. In my own studies of people's emotional relationships with technologies, airplanes emerge as startlingly like computers in respect to the issues they raise for their hobbyists. Specifically, both are powerful media for working through the issue of control.

Local versus global and the question of control

There is always a compelling tension between local simplicity and global complexity in the working of a computer and in the appreciation of a computer program. Locally, each step in a program is easy to understand; its effects are well defined. But the evolution of the global pattern is often not graspable. You are dealing with a system that surprises. This play between simplicity and complexity allows programmers, as pilots do with flying, to make of computation very different experiences that provide a context for working through different needs in relation to issues of personal control.

Depending on how the programmer brings the computer's local simplicity and global complexity into focus, he or she will have a particular experience of the machine as completely understandable, under control, or as baffling, even as controlling. By focusing on the local, the line by line, you can feel in control. By focusing on the global, you can feel control slip away. In their style of programming a computer, people betray different levels of tolerance for temporary losses of control. Some will avoid it at all costs; but others will seek it out, and enjoy "playing" with sensations of risk and danger. And so different people end up with very different relationships to control and power in their programming work.

To illustrate this point, it is useful to begin with an example at one extreme. We see a first style in Howard, an ex-programmer, now a university professor, who describes himself as "having been a computer hacker."[6] Howard was not in my sample of hobbyists. He has a terminal at home which links to a large timesharing system, but when asked about home computers, he winced in distaste and said that he "wouldn't touch the stuff. It's too simple." His case, in which we see a love of programming for the feeling of "walking near the edge of a cliff," is meant as a contrast for what I found to be a prevalent and more conservative "hobbyist style."

Howard described his longtime fantasy that he would walk up to any program, however complex, and "fix it, bend it to my will." As he described his intervention, he imitated the kind of hand gestures that a stage musician makes toward the hat before he pulls out the rabbit. Wizards use spells, a powerful kind of local magic. Howard's magic was local too. He described his "hacker's approach" to any problem as a search for the "quick and dirty fix." For Howard, what was most thrilling about the experience of programming was "walking down a narrow line," using the program's flexibility (for him defined as the possibility of making a local fix) in a struggle to keep the whole under control. Weekends at the terminal with little to eat and little or no rest were frequent, as was the experience of not being able to leave the terminal while debugging a program, even when the obvious need was for sleep and looking at the whole in the morning, instead of trying to "fix it" by looking at it line by line all night. For Howard, the urgency of these encounters was tied to his sense that through them he was grappling with a computational essence—the struggle to exert control over global complexity by mastery of local simplicity.

A second programmer, Bob, is a computer professional, a microprocessor engineer who works all day on the development of hardware for a large industrial data system. He has recently built a small computer system for his home and devotes much of his leisure time to programming it. Whereas, for Howard, the excitement of programming is that of a high-risk venture, Bob likes it as a chance to be in complete control. Although Bob works all day with computers, his building and programming them at home is not more of the same. At work he sees himself as part of a process that he cannot see and over which he feels no mastery or ownership: "Like they say, I'm just a cog." At home Bob works on well defined projects of his own choosing, projects whose beginning, middle, and end are all under his control. He describes the home projects as a compensation for the alienation of his job. He works most intensively on his home system when he feels furthest away from any understanding of "how the whole thing fits together at work."

Howard and Bob have very different opinions about what is most satisfying about programming. These translate into different choices of projects, into different choices of programming language and level to program at, and ultimately into what we might call different computational aesthetics. Howard likes to work on large, "almost out of control" projects; Bob likes to work on very precisely defined ones. Howard finds documentation a burdensome and unwelcome constraint; Bob enjoys documentation, he likes to have a clear, unambiguous record of what he has mastered. Indeed, much of his sense of power over the program derives from its precise specifications and from his continual attempts to enlarge the sphere of the program's local simplicity.

Hobbyist programmers mean different things when they say that machine language programming "puts them in control." For some, the reference seems objective. Given the primitive higher-level languages available on the first and second generation of hobby computers, machine-level programming seems to them the best instrumental solution. For others, the issue is more subjective. Many hobbyists who said they felt uneasy using systems programs for which they didn't have the source code also admitted never looking at the source code listing which they felt they had to have. Having access to the code was symbolic. In other cases it was apparent that machine language programming was valued because the experience of doing it was pleasing in itself. It meant that the programmer was writing instructions that acted directly on the machine, no "building on top of" somebody else's interpreter. Most hobbyists have relationships with computation at work which involve sharing the machine with countless and nameless others. In personal computation they see a chance to be independent and alone. The machine comes to them virgin. They have full possession. Finally, there is the issue of asserting control over an inferior. Bob, like many of the other hobbyists I spoke with, is a middle-level worker in the computer industry. He does not feel very good about the importance of his job. Proving that he is "better than any dumb compiler" gives a sense of importance.

Using the computer to assert control was a central theme in my interviews with hobbyists. It was expressed directly, and also wove itself into four other issues that characterize the hobbyists' "subjective computer." These are using the computer to strengthen a sense of identity; to construct a completely intelligible piece of reality that is experienced as "transparent" and safe; to articulate a political ideology; and to experience a sense of wholeness that is absent in one's work life. It is these four issues to which I now turn.

Building identity

In achieving a sense of mastery over the computer, in learning about the computer's "innards," people are learning to see themselves differently. Among other things, they are learning to see themselves as "the kind of people who can do science and math." This was most striking among hobbyists who had no technical background. But it also came up among hobbyists who did see themselves as "technical people" but who for one reason or another had gotten "scared out of real science."

Barry is 28 years old, an electronics technician at a large research laboratory. He went to college for two years, hoping to be an engineer, then dropped out and went to technical school. He has always loved to tinker with machines and to build things. His current job is to calibrate and repair complex instruments, and he is very happy with it because he gets a chance "to work on a lot of different equipment." But he came to his job with a feeling of having failed, of not being "analytic," "theoretical," of not being capable of "what is really important in science."

Ever since I was a child I always had an interest in science, but I never had the opportunity or the passion to go back and finish college and get a real degree in science. I don't think I have a theoretical mind. I got all D's in mathematics. I have a more practical mind . . . I always had a great deal of difficulty with mathematics in college which is why I never became an engineer. I just could not seem to discipline my mind enough to break mathematics down to its component parts, and then put it all together and really do it.

Five years ago, Barry bought a programmable calculator and started "fooling around with it and with numbers the way I have never been able to fool around before," and says that "it seemed natural to start working with computers as soon as I could." To hear him tell it, numbers stopped being theoretical, they became concrete, practical and playful, something he could tinker with.

I'll pick up the calculator, and if I don't know how to do a problem I'll play with the calculator a few minutes, or a few hours and figure it out. It's not so much that the calculator does a particular calculation, but you do so many, have so much contact with the numbers and the results and how it all comes out that you start to see things differently . . . The numbers are in your fingers.

When the calculator and the computer made numbers seem concrete, the numbers became "like him," and Barry felt an access to a kind of thinking that he had always felt "constitutionally" shut out of: "When I write in assembler I feel that mathematics is in my hands . . . and I'm good with my hands."

Barry claims to have "grown out of" his aspiration to be an engineer. He says he doesn't keep engineering as a pipedream or think of his computer skills as something that could make it real. In terms of his career he says that "nothing has changed." But a lot has changed. Barry has always thought of himself as a bundle of aptitudes and ineptitudes that define him as the kind of person who can do certain things and cannot do others. Working with the computer has made him reconsider his categories.

I really couldn't tell you what sort of thing I'm going to be doing with my computer in six months. It used to be that I could tell you exactly what I would be thinking about in six months. But the thing with this, with the computer, is that the deeper you get into it, there's no way an individual can say what he'll be thinking in six months, what I'm going to be doing. But I honestly feel that it's going to be great. And that's one hell of a thing.

For Barry, the world has always been divided between the people who think they know what they'll be thinking in six months and those who don't. And in his mind, his home computer has gotten him across that line and "That's one hell of a thing." For Barry, part of what it means to have crossed the line is to start to call the line into question. When he was in school, his inability to do the kind of mathematics he had "respect" for made him lose respect for himself as a learner. The computer put mathematics in a form that he could participate in. Barry has three children, has bought them their own calculators, and encourages them to "mess around with the computer." He feels that they are going through the same problems with math and science that he had and he wants them to have "a better start." For Barry, the computer holds the promise of a better start, not because it might teach his children a particular subject, but because it "might change their image of themselves. They might think of themselves as learners."

Personal computers are certainly not the only hobby that people use to enhance their sense of identity. For my informants, "hobbies" have always been a way of life. Almost 90 percent of them had been involved in a hobby other than computation, most usually in another "technical" hobby, such as photography, ham radio, or model railroading. Fifteen percent of the hobbyists surveyed were using their computers to "augment" their participation in another hobby—for example, using the computer to keep an inventory of motorcycle parts, figure out ideal compression ratios for racing cars, interface with amateur radio equipment. For nearly a third of them, their home computer had completely replaced another hobby. People spoke of these abandoned hobbies as "fun" and as "good experiences," but their remarks about past hobbies underscored several ways in which in our day and time a computer hobby can be special. In particular, people spoke about their "switch to the computer" as making them part of something that was growing and that the society at large "really cared about."

Gregory is in his mid-forties, and has been in the electronics industry for all of his working life, as a technician, a programmer, and currently as a products designer. For two years, his computer shared space in his study with an elaborate model railroad system. A year and half before I met him he had bought a new hard copy printer and a graphics plotter. In the overcrowding that followed, the trains had finally found their way to storage in the basement.

Nobody ever really paid attention to my model railroad stuff, although a lot of the circuitry that I did for those trains was just as complex as what I'm now doing with my computer. But people would look at the train and they would say "that's cute." The computer is my own thing, but it's part of the real world too. When my kid got involved with my trains, he was just into my hobby. But if my kid becomes good at the computers, it will mean something.

I heard many echoes of Gregory's phrase, "it's part of the real world too." Hobbyists spoke about the computer offering them a connection with something beyond the hobby. For some, having a computer and "getting good at it" means crossing a frontier that separates "tinkering" from "real technology." They feel that the world sees their computer hobby as serious (several commented that friends and neighbours hardly even look at it as a hobby, as though the word were reserved for frivolities), and they start to see themselves that way too. Most first-generation hobbyists have technical educations, but many of them, like Barry, feel they have never been part of what is most exciting and important in the scientific and technical cultures. They see themselves as the low men on the totem pole. Working with computers, even small computers, feels technologically "avant garde." A smaller group of hobbyists (but a group whose numbers are growing as new generations of personal computers become more accessible to the nonspecialist) have always felt completely left out of the scientific and technical worlds. For them, owning a computer can mean crossing a "two Cultures" divide.

Alan, a 29-year-old high school French teacher who describes himself as "having a love affair with a TRS-80," has always felt he wasn't "smart enough to do science."

After Sputnik, when I was in grade school and then in Junior High, there was all that fuss, all the kids who were good in maths got to be in special classes. Rockets were going up . . . men trying to go to the moon. Decisions about things. Scientists seemed to be in charge of all that.

Alan majored in French ("It was easy for me . . . my mother is from Montreal") and took up carpentry as a hobby. And although he was good at it, it only reinforced his sense of not being able to do intellectual things, which in his mind meant not being able to "do anything technical." When Barry began to do mathematics with his calculator, he felt that he started to cross a line to become the kind of person who could expect change and excitement in his intellectual life. For Alan, his TRS-80 led him across a line to become a member of a different culture, a scientific culture, a culture of "powerful people."

For Barry, Alan, Gregory, relationships with computation enhanced self-image. There is another way in which working with a computer can influence an individual's sense of identity. Ideas about computers, about how they work and what they can and cannot do, can be used to assert ideas about people. They can provide metaphors for thinking about oneself. Many hobbyists, fascinated by the idea of someday being able to trace out the complex relationships of electronic events, machine and assembly language instructions, and higher-level language commands within the computer, used the image of these many levels of intelligence to think about how people might work, about how people might, or might not, be like machines. Some of this epistemology was implicit, for example when people made comments about "using the computer's software to think about my wetware." And sometimes, although less frequently, the issue became quite explicit. Conversations that began with descriptions of household "robotics" projects, such as a plan to build an energy monitoring system, led to reflections on how these projects would require programs that could represent the system's knowledge to itself, and from there into formal epistemological reveries: were these the kind of self-*representing* programs that ran inside people's heads? Do people have different kinds of self-representation programs for representing different kinds of knowledge, such as the knowledge of a dream and the knowledge of being awake? What kind of self-representation program might be running within people that allows them to remember and then forget their dreams?

Constructing reality

Hobbyist's descriptions of what it is like to work with their own computers frequently referred to the idea that the computer provides a safe corner of reality. Other hobbies can give a similar sense of security but often exact a price. For example, people can feel safe but limited. Alan, the French major, now "in love with his TRS-80," felt secure in his carpentry hobby, but he experienced it as a safety that came from refusing challenge. "It was an 'artsy' hobby. I couldn't see myself any other way." The computer is more likely than most other media to allow the experience of playing worlds (let us call them "microworlds") that are secure and also adventurous enough to allow for mind-stretching explorations. Almost all of the hobbyists I interviewed described some version of a limited, safe, and transparent microworld that was embodied in their personal computer. For Alan just the fact of working with a computer created such a world. For others, it was more specific: a morse code microworld; a text editor microworld; and, most generally, the assembly language microworld.

This use of the computer as a place to build a microworld is particularly salient for children when they are put in computational environments in which they have access to programming. Elsewhere I shall report on my study of elementary schoolchildren and adolescents who have gone through an experience with a LOGO computer system—a system designed to allow children to work in visually represented microworlds that are sufficiently constrained for a child to be able to understand and control them, yet sufficiently open-ended to give a child a real experience of intellectual power. Here I will mention only briefly one child and what she made of the computer as a medium for world building.

Deborah at 11 years old was the baby of her family, the youngest of three children. Her childhood had been spotted with illnesses which further exaggerated her "baby" position. The members of her family were always doing things for her, insisting that she was not old enough to do the things that she wanted to do most: take out the laundry, baby-sit, stay over at a friend's house, choose her own hair style and her own clothes. Dependent on others at home, very overweight, and with an image of herself as sick and weak, Deborah had little sense of her own boundaries, her ability to say no, to assert control. Even at 11, she had become involved with a crowd of older kids who were smoking, drinking, using drugs.

Towards the end of her eleventh year a LOGO computer came into Deborah's classroom as part of an educational experiment. At first, she found the computer frightening and threatening: until one day she hit upon the idea of confining the designs she made with the computer to ones in which the lines always came together in multiples of 30 degrees. She called it her "30 degrees world."

This restriction defined for her a space in which she felt safe but in which she was able to produce designs of great ingenuity and complexity. When I interviewed her two years later I found that she had used her experience with the "30 degrees world" as a kind of model, an experience-to-think-with. In her mind it represented how you could take control and make things happen by the judicious use of constraint. In Deborah's words, it was the first time that she ever "laid down her own laws." It was a turning point in her ability to take control of other situations. She lost 20 pounds, has given up smoking and drugs, and says that she "only sometimes" has a drink.

For Deborah and for many adult hobbyists the sense of safety with the computer derived from the feeling of working in a sphere of intelligibility and transparency, a sphere that is protected, much as the space of a psychotherapeutic or psychoanalytic relationship is set off, bracketed. People talked about feeling safe and secure in the world they had built with their home computers, a world where there were few surprises and "things didn't change unless you wanted them to." Of course, there was much talk of problems, of false starts, of frustrations. There are "bugs" in hardware and in programs. Things don't work; things go wrong. But bugs, with time, are either fixed or become "known" bugs. Joe is an insurance salesman in a small North California suburb who owns a second-hand Commodore Pet "with a lot of hardware problems." To Joe the bugs in his system "have become almost like friends": "I turn on the machine and I systematically check for my 'old friends,' and I swear, finding them there has a certain reassuring element."

Political ideology

The use of the computer as a medium for building a transparent and intelligible world brings us to how it can be used to think through questions of political ideology. Fred sells components for a large electronics supply house. He narrowly escaped starvation in a prisoner of war camp during World War II, and from that experience he says that he took "a sense of optimism." "I mean, if there is something out there and you want to do it— do it, understand it, act." Fred has tried to live that way. He is active in local politics; he keeps up with the news; he writes letters to the editor of his town newspaper. He bought his TRS80 on an impulse because "it seemed that you wouldn't be able to understand American society any more if you didn't know about computers." When it comes to working with his computer, Fred wants to know "exactly how things work": "There is a big gap in my own mind between the fact that an electrical circuit can be on or off and the binary number system . . . and again from there to the BASIC language. I've got to understand all of that."

When hobbyists like Fred spoke about "wanting to know exactly how things work" in their computers, they were usually talking about wanting to know how their systems were built up from level to level, from electrical circuit to high-level language command. Fred, for example, expressed sharp frustration at gaps in his ability to follow the system through: "I can't really follow the continuum. I'm going to a user's group meeting, I'm talking to people and reading books and some of it is helping, but I am really frustrated. I want to be able to follow the whole thing through."

Larry, 35, lives in a Boston suburb. His computer offered him the first way he ever found to challenge the school's judgments of his child's abilities. A year before I met him, Larry had bought an Apple computer for small business use and ended up bringing it home so that his children would be able to play with it. His 12-year-old son Todd had been judged "backward" by his teachers through six years of schooling. His math scores were low; he could barely read. But Todd picked up the Apple manual, taught himself how to use the game packages, and then taught himself how to program in BASIC. As far as anyone knows, the Apple manual was the first book that Todd had ever read. In three weeks Todd was writing his own games, games that demanded an understanding of variables and a knowledge of geometry that his teachers claimed he didn't have. Larry feels that, for the past six years, he has been intimidated by the school's evaluation of his son:

All these years they told me he was backward, so I believed he was backward. He certainly couldn't do math. And you couldn't get him to read. Now I think that he just wasn't interested. And then, maybe everybody treated him like he was stupid. That sort of thing can have an effect. I'm not going to sit still for it any more.

Larry is starting to demand more from his son's teachers. His experience with Todd has made him optimistic about what computers will mean for politics because "people will get used to understanding things, of being in control of things, and they will demand more." This optimism is widely shared among hobbyists, part of a distinct style of talking about computers and politics. Hobbyists, like Fred and Larry, take what is most characteristic about their relationships with the computer—using computers to build safe microworlds of transparent understanding—and turn it into a political metaphor. Hobbyists associate images of computational transparency and of "knowing how the machine works" with a kind of politics where relations of power will be transparent, where people will control their destinies, where work will facilitate a rich and balanced cognitive life, and where decentralized power will follow from decentralized information resources.

For many hobbyists a relationship with their home computer carries longings for a better and simpler life in a more transparent society. *Co-*

Evolution Quarterly, Mother Earth News, Runner's World, and *Byte Magazine* lie together on hobbyists' coffee tables. Small computers become the focus of hopes of building cottage industries that will allow people to work out of their homes, have more personal autonomy, not have to punch time cards, and be able to spend more time with their family and out of doors.

Some see personal computers as a next step in the ecology movement: decentralized technology will mean less waste. Some see personal computers as a way for individuals to assert greater control over their children's education, believing that computerized curricula will soon offer children better education at home than can be offered in today's schools. Some see personal computers as a path to a new populism: personal computer networks will allow citizens to band together to send mail, run decentralized schools, information resources, and local governments.

In sum, many of the computer hobbyists I have interviewed talk about the computers in their livingrooms as windows onto a future where relationships with technology will be more direct, where people will understand how things work, and where dependence on big government, big corporations, and big machines will end. They imagine the politics of this computer-rich future by generalizing from their special relationship to the technology, a relationship characterized by simplicity and a sense of control.

Alienation from work

Over 40 percent of those who responded to my survey worked, or had once worked, as computer programmers. The programmer is typically in a situation where he or she is in touch with only a very small part of the problem that is being worked on. Increasingly, programmers work in large teams where each individual has very little sense of the whole, of how it all fits together.[7] Programmers have watched their opportunities to exercise their skill as a whole activity being taken away (for those who are too young, the story of the process remains alive in the collective mythology of the shop). They have watched their work being routinized, being parcelled out into the well defined modules that make up the tasks of the structured programming team. They mythologize a golden age. This lived experience at work made programmers particularly sensitive to the parcellization of knowledge and to the alienation from a sense of wholeness in work. And they bring this sensitivity to their home computer hobbies.

Hannah worked as a programming consultant for a large business system for ten years before starting her own consulting company through which she offers her services to other computer hobbyists. To her, nothing is more depressing than working on a tiny piece of a problem. In her old job, "most of the time I didn't even know what the whole problem was." She likes working with computers at home because she has more control of her time and can spend more time with her family. But she says that what is most important about working with a personal computer is that "I can finally think about a whole problem." Hannah's feelings were widely shared among the hobbyists I interviewed, most notably among programmers, ex-programmers, and "team engineers." Images of lack of intellectual balance, of fragmentation, of not being connected, came up often, with the computer at home usually placed in the role of righting what had been wrong at work. As Hannah put it: "With my computer at home I do everything, I see my whole self, all my kinds of thinking."

For these people, having a computer at home meant "thinking experiences" where they could see what their real capacities were, where they had a chance to try things out. Karl, for example, who worked for a long time as a programmer and who is now an engineer in a microprocessor firm, had thought a lot about his mental "ecology."

In the first half of an engineering project there is literally nothing coming together and that's when I find that I need to go home and put something together. I used to make lamps like those in the living room out of glass bottles. But then towards the middle of an engineering project, things did start to come together and I would lose the need for cutting glass.

But if you never or rarely get to finish things at work, if your job is basically to make little pieces and it's somebody else's job to make them fit into a whole, then working with the computer at home can give you an experience of getting it all together. You do the whole thing—building up from machine code to finished project. It makes you feel in balance.

When I spoke with Karl he was at a point where everything at work seemed "pretty disconnected." Or, as Karl put it, if he hadn't had his home computer, it would have been "glass cutting time." Karl saw his current work with the computer as a corrective to fragmentation on the job. During our interview, he spoke to me about his current plans for revamping his computer system. Computer systems like his which have several components (keyboard, display screen, printer) have to be connected by using devices, usually simple circuits, called "interfaces." Karl's plan called for using a separate microprocessor as an interface for each component. He had conflicts about the "rationality" of his project. After all, he admitted, it was far from economical if one measures economy in terms of the cost of the hardware. Each interface circuit needed only a few specialized and inexpensive chips. There was no need for a general-purpose microprocessor at each node. Specialized chips could do the job more cheaply, but could not satisfy his desire to experience the system as maximally coherent. By replacing the special-purpose circuits by separate, but identical, general-purpose microprocessors, the whole system, at least in Karl's eyes, became uniform, intelligible, and systematic. Karl could not help concluding: "I guess you could say that my choice of projects is not always rational." But from the perspective of using a relationship with computation as a way of "working through" personal concerns, Karl's rationale was clear. For Karl the "inexpensive" solution, using a collection of opaque, *ad hoc* circuits, felt unintelligible. It felt to him like his work situation. His plans for his multiprocessor system were dictated by the logic of compensation rather than by the logic of material economy.

The mind and body of the machine

In studying the hobbyist experience I have found people, largely people with technical backgrounds, in intense involvement with machines. They describe their work (or rather their leisure) with the computer as different from what they have done before with other hobbies. They describe it as an involvement with greater personal consequence. Some of the sense of consequence comes from an historical moment: the computer hobby is seen as signifying a place in the "avant garde." Although in some circles "computer person" is a term of derision, the hobbyist experiences it with pride. Some of the sense of consequence comes from experiencing an individualistic and independent relationship with computation that can be mythologized as belonging to a now-past "golden age" of the programmer. But most of the sense of consequence comes from the holding power and intensity of the time spent with the computer. What is there about these people and these machines that makes possible relationships of such power and such intensity?

For me, the relationships that hobbyists form with their home computers can be partially captured with a metaphor of the "mind" and the "body" of the machine. The "mind" of the computer is that side of computation that involves thinking in terms of high-level programs. In this metaphor, relating to the "body" of the computer means not only working on hardware, but also, and indeed especially, working with programs in a way that is as close as possible to the machine code—that is to say, as close as possible to the core of the computer, its central processing unit (CPU). In terms of this metaphor I have found that the prototypical hobbyist is trying to get into a relationship with the body (rather than the mind) of the machine, to assert power and control in the relationship with the computer, and to create safe worlds of transparent understanding. In trying to find concepts for thinking more clearly about what draws the hobbyist to this kind of relationship with the CPU and about what its meaning might be, I find three issues particularly salient. I think, moreover, that, although I formulate them here in terms of computers, they are relevant to understanding relationships with other technologies as well.

The first issue goes back to control. The hobbyist complains of a work situation where everyone suffers from the constant presence of intermediaries. Bureaucracies stand between the programmer and the computer, a bureaucracy that schedules the computer, that decides its up and down time, that apportions the work for its software design and decides on priorities and procedures for access to it. At work, when something goes wrong with the system it is usually the fault of an intermediary person, one of the many "somebody elses" who deal with the machine. Or it may be the fault of a technical intermediary, one of the many elements in the computer system that mediate between the user and the bare machine: a compiler, an interpreter, an operating system, someone else's program. At home, the hobbyists feel themselves as working directly with the CPU, in complete and direct control of the machine's power. And when something does blow up, the situation has a special immediacy. It is between them and the bare machine.

When a FORTRAN program is run on a large IBM machine the events in the machine are far from being in one-to-one correspondence with the steps of code written by the programmer. Two factors contribute. First, it is in the nature of higher-level languages to work in a conceptual space different from that of the machine. FORTRAN works in a "formula" space, COBOL works in a "business" space, both very different from the space of bits and bytes. Second, the sense of indirect relationship is exacerbated when the compiled code is run by an operating system which allocates memory, mediates control of peripheral devices, and even interleaves the program with other programs. At home hobbyists can follow machine-language programs step by step as their instructions pass through the CPU. They can envision the changes in state of the whole system as being produced by specific actions of the CPU. And if they suspect that there is a bug in the hardware they can pull out an oscilloscope and see whether the CPU is doing what it should in response to a given instruction. They can figure out where the signals should be going, they can collect their own evidence for what is going wrong, trap and fix the bug themselves. Again and again in my interviewing I heard about the pleasures of debugging—of "going in with meters and scopes and tracking it down." The procedure exhilarates. With every successfully tracked bug comes an affirmation of direct control over the machine.

The issue of control was often explicitly recognized by the hobbyists I interviewed. But they lacked a language for naming a second issue which has to do with a notion referred to as "syntonicity" within the psychoanalytic tradition. Syntonicity implies that we should look for "body-to-body" identification in every powerful relationship with a technology—the body of the person and the body of the machine. It implies that we should understand the appeal of machine language in terms of people's ability to identify with what is happening inside the machine. The CPU of the hobbyist computer lends itself to personal identification with its primary action: moving something that is conceptually almost a physical object (a byte of information) in and out of some thing (a register) that is almost a physical place. The metaphor is concrete and spatial. One can imagine finding the bytes, feeling them, doing something very simple to them, and passing them on. For many of the people that I met in the hobbyist culture, getting into this kind of identification feels safe. It makes the machine feel real.

There is a third issue raised by the hobbyists' relationship to the CPU. It is an aesthetic one. The generation of hobby computers that was born in the 1970s are very primitive machines. The hobbyist thinks of much about them as "klugey," a computerist's way of saying that one is dealing with a compromise, a collection of patches whose structure has been dictated by arbitrary corporate decisions, by economic necessities. The corner of the hobbyist machine that seems to them to have the greatest "intellectual integrity," that distills what they feel to be a tradition of some of the best ideas in computer science, that comes closest to being "clean," is the CPU. And so it is natural for the hobbyist to seek the closest possible contact with it. For a culture in which there is a widely shared aesthetic of simplicity, intelligibility, control, and transparency, getting into the "un-klugey" part of the machine and working in machine code seems the most aesthetically satisfying way to use the personal computer as an artistic medium.

Concluding postscript

I would like to conclude with some personal reflections on personal computation and political metaphor. As an ethnographer, I use interviews and observations to enter sufficiently into people's lives to develop a sympathetic understanding of how they look at the world. And then I take this experience and try to distill from it those elements that will make the lives of the people I have been studying intelligible and meaningful to others. The hobbyists I interviewed are excited, enthusiastic, satisfied with what they are doing with their machines. It seems appropriate to report this enthusiasm and to try to capture a sense of the pleasures and satisfactions that these individuals are getting from developing "non-alienated" relationships with their computers, from "understanding" machine systems from the "bottom up," and from feeling satisfied that they have finally found models of transparency for thinking about the kind of political order they would like to live in.

But the picture also has a darker side. The "pleasing populism" of the hobbyists I interviewed is not in itself unproblematic. Will the individual satisfactions of personal computation (which seem to derive some of their power from the fact that they are at least in part responsive to political dissatisfactions) take the individual away from collective politics? People will not change unresponsive political systems or intellectually deadening work environments by building machines that are responsive, fun, and intellec-tually challenging. They will not change the world of human relations by retreating into a world of things. It would certainly be inappropriate to rejoice at the holistic and humanistic relationships that personal computers offer if it turns out that, when widespread, they replace religion as an opiate of the masses.

Notes

1 I use the word "hobbyist" in this essay to denote someone who owns a personal home computer, whether or not that computer was built from a kit.

2 The initial questionnaire was mailed to 215 hobbyists. Twenty-seven of the people interviewed were drawn from the original group of respondents and the rest were drawn from other sources—for geographical distribution (particularly to represent both the West and East Coast hobbyist "cultures") and to tap the personal computer "individualist," men and women who own personal computers but who don't like to attend meetings and who are not regular subscribers to personal computer magazines. In its first generation, the computer hobby world was a "technical" hobby culture, and like other such cultures—such as those around amateur radio and model railroading—it had a predominantly male population. This imbalance is reflected in my study. Four women responded to my survey. I met with two of them and with three other women hobbyists in the interview phase of my work.

3 My survey supports the idea that the satisfied personal computer consumer has technical training, and/or inclinations: 83 percent of those polled had majored in some scientific or mathematical field in college; 38 percent of them had pursued their scientific or technical studies to the graduate level; 28 percent had majored specifically in mathematics or computer science. Sixty percent of those polled actually made their living from some kind of work in the computer industry: building, selling, servicing, or programming the machines. For over a third of the hobbyists surveyed, when it comes to computers, they have the inclination rather than the training: 35 percent claim that everything they know about computers is self-taught.

4 The GAME OF LIFE refers to a popular class of computer demonstrations. In these demonstrations rules are applied repeatedly to patterns of objects on a checkerboard. Conway's ingenuity consisted of finding classes of simple transformations that would give rise to surprising, complex, and varied effects. The reference to "life" analogizes the dynamics of the game to the evolutionary image of the emergence of the complexity and variety of living organisms from very simple organisms. The popularity of the game springs from many sources: the biological referent fascinates, as does the interplay between simplicity and complexity. People describe a thrill from getting genuinely surprising results from the application of simple rules. Its popularity among hobbyists also reflects something about the kind of level programming required to run the game on a small computer. It is beyond the stage of routine programming, but not too far beyond. The main problem in programming the GAME OF LIFE is speed. The faster it can run, the more dramatic the effect. Having it run at an aesthetically satisfying speed on a small computer requires ingenuity. And this ingenuity has immediate, visible consequences.

5 See Sherry Turkle, "Computer as Rorschach," *Society/Transaction*, January-February 1980, 15–24.

6 For Howard, it was his style of programming that led him to identify with what for him was a computer "subculture," that of the hacker. His process of identification seemed analogous to that of a creative independent virtuoso who recognizes his peers not by the "job" they do, nor by their academic credentials, but because they share his sense of the personal importance, the urgency, of creating in the medium in which they work. Many hackers have dropped out of academic programmes in computer science in order to devote

themselves exclusively to computers. Based neither on a formal job nor on a specific research agenda, the coherency of the hacker subculture follows from a relationship with the "subjective computer"—that is, with a set of values, a computational aesthetic—and from a relationship with programming that may be characterized as devotion to it as a thing in itself. In university settings all over the country, where hackers are often the "master programmers" of large computer operating systems, academic computer scientists complain that the hackers are always "improving the system," making it more elegant according to their aesthetic, but also more difficult to use.

7 See Philip Kraft, *Programmers and Managers: The Routinization of Computer Programming in the United States* (Berlin, Heidelberg & New York: Springer-Verlag, 1977).

Teleworking from Home

Jack Nilles

Using computers to work from home could soon make daily long-distance commutes obsolete. Jack Nilles outlines the potential of telecommuting and discusses the pros and cons for employers and employees. He says the practice is likely to increase, with major consequences for businesses, unions, transportation systems, and even the landscape. The author is with the Centre for Futures Research, University of Southern California, and this first appeared in Technology Review, *April 1982.*

The Great American Dream is to live in a suburban setting at some little distance from the neighbors. The automobile provided this option, and cities changed as a consequence. Today microelectronics technology is beginning to do the same for information workers. The microcomputer is the analog of the automobile: the telephone line and the communications satellite serve as the highways. The major difference is that information is transported rather than information workers, who then have the potential to become "telecommuters." In principle, the telecommuter has access to anyone with a computer—with near-zero transit time.

While telework—the use of computers as a primary communications tool in the workplace—can be conducted from a centralized location just as easily as from a decentralized one, telecommuting implies decentralization. In effective telecommuting, if the information worker commutes at all, it is to the corporate office nearest his or her home. The organization essentially clones itself, locating regional work centers near the principal concentrations of its workers' residences. The regional centers are interconnected so that all accountants, marketers, salespeople, data entry clerks, and managers can communicate regardless of their individual locations.

The significant changes in work style made possible by telework, particularly telecommuting, are substantial. One of the major benefits is, of course, time saved by avoiding a daily commute that today averages about 18 miles per round trip. If, as is likely, the worker commutes to a place much closer to home, the savings are less—even zero, if the worker chooses to walk or bicycle to work instead of driving, but this alternative to hours tensed over a steering wheel may well be considered a benefit.

The result is direct as well as indirect monetary rewards for telecommuting employees. The most obvious is the possibility of eliminating the automobile from the household inventory, or of having the old one last a few years more because of reduced wear and tear, accompanied by smaller fuel and repair bills.

The information worker who works at home, or who personally provides the microcomputer for teleworking, may enjoy tax benefits as well. The cost of the equipment and computer programs, or software, are often deductible, as are house or apartment rental and maintenance expenses directly associated with the home office. The costs of an additional room and telephone for the telecommuter's office fit that criterion, as would additional insurance costs for any electronic gadgetry.

Housing itself is a growing problem, the central element of which is the high price of urban land. The location independence of telecommuting allows information workers seriously to consider relocating to small towns and cheaper real estate while retaining city jobs.

The primary concern of many employers considering the telecommuting option for their information workers is its effect on productivity. Although productivity is much more difficult to define when assessing information work than when accounting for the manufacture of products, it can be measured for support-level staff such as typists and data entry clerks. Our research indicates that teleworking is associated with significant (15 percent or more) increases in productivity in these areas. For management and professional workers the situation is more complex, since their work is less easily quantified, and, until very recently, many of the tools of telework were not tailored to higher-level work environments. Microcomputers and manager/professional-oriented software may eventually improve the office productivity of many of these workers, but anecdotal evidence indicates that telecommuting has already improved the productivity of programmers, scientists, and others whose jobs tend not to be highly dependent on face-to-face contact with colleagues.

Although information technology directly aids productivity, more subtle factors are also at play. For example, some telecommuting professionals say they feel a greater responsibility toward their jobs, and that they tend to be more diligent when given the privilege of working at home. These workers range from computer programmers to middle-level and upper-level executives. However, no extensive studies have yet been performed to test this and other conjectures. Teleworking is still too new, and because it triggers organizational change and deviations from traditional forms of management, few executives are eager to be the first to give it a try.

An even more urgent concern than productivity for some employers is that of worker availability. General unemployment figures notwithstanding, there are chronic shortages of some types of skilled information workers such as secretaries and computer programmers. Some workers, such as parents of school-age children and the physically handicapped, are unable to commute long distances or are available only part time. By moving the job to these workers via telecommuting the employer is freed from geographical restrictions in attracting competent and productive workers.

Operating costs in the central business district of a major metropolitan area are high. Rent or building ownership and other costs associated with attracting employees to these locations can constitute a substantial portion of corporate overhead, an expense telecommuting can reduce. Teleworking moves workers to lower-priced suburban properties—or even their own homes—at no additional cost to the employer, and can reduce office space per worker through office sharing by part-time telecommuters. In our 1973 study of a Los Angeles insurance company, we found that this factor alone (including the income from lease of the vacated downtown facilities) compensated for the costs of decentralization.

As with most technologies, there can be undesirable side effects to the benefits of teleworking and telecommuting. Most experiments in telecommuting now being conducted by US corporations involve secretarial and clerical employees working from their homes. These workers frequently complain of isolation from their fellows. This is another reason for stressing the local work center, as opposed to working at home, as the best mode of telecommuting—it is unreasonable to expect the average information worker to forgo all on-the-job social contact. An attractive compromise may be to encourage the worker to spend several days per week at home and a day or more at the local or central office to reaffirm collegial ties. Since the work product—information—is computer-based, the daily location of the worker—the home or office—may not be a major factor in efficiency.

Many people are intimidated by the apparent inscrutability of computers. This situation is changing as young information workers, who have grown up with TV and pocket calculators, enter the labor force and reach management and professional levels. By the end of this decade, a large percentage of the labor force will own personal computers, further diminishing the intimidation problem. And with increasing machine sophistication and greater attention to the design of computer software, computer-naive workers can deal more effectively and comfortably with computers at reasonable cost.

Nonetheless, information workers who are not computer-literate will be at a disadvantage if telework and other manifestations of the new information technologies become more widespread. The pace of displacement of these workers, though fairly slow, is accelerating, as is the rate at which these technologies are insinuating their way into the upper reaches of the corporate hierarchy. If there are no compensating expansions of the information market, telework-caused increases in productivity may result in unemployment.

Another potential problem of telework is that it can become compulsive and may exacerbate workaholism. The terminal or microcomputer in the den or living room may tempt the manager or professional to try "just one more" variation on that quarterly forecast. The convenience of the personal computer may also result in excessive polishing of the information product, polishing that previously was impossible or too inconvenient to bother with. In our surveys of personal computer owners, we found several who refused to take their machines home because they wanted clearly to separate their home and business lives. Telecommuting, in particular, can make that separation more difficult even while making work hours more flexible.

Foremost among issues that nag supervisors is that of control. A supervisor will ask, "How can I tell whether my employees are working if I can't even see them?" My usual rejoinder is, "How can you tell they're working when you *can* see them?" Except for routine secretarial and clerical work, where information results can be monitored almost continuously, supervision depends more on appearance than substance until the product appears. Here, too, the local work center provides an answer. Physical supervision of workers can be performed by anyone at the local center, and the quality of the information product can be checked by the functional supervisor, who may be located anywhere in the corporate telecommunications network.

Human organizations often have concerns that have nothing to do with efficiency or bottom lines. For example, concern with prestige can transcend other, more quantitative considerations. The corporate image, particularly as embodied in a building crowned with the corporate logo in the central business district, is not to be trifled with. While decentralization via teleworking has its attractions, the potential loss of the downtown corporate edifice may in some cases prove more important.

Another image-oriented issue is the replacement of the personal secretary by the typing pool and, more recently, the word-processing center. During this process the discovery was made that secretaries can do a lot more than take dictation and type letters—some erstwhile secretarial jobs have evolved into entry-level managerial positions. Meanwhile, managers and, to a lesser extent, professional workers are sometimes reluctant to be seen actually touching a computer keyboard, an act they consider damaging to their professional image. Nevertheless, managers are learning to do their own "typing" with the use of effective text-processing software in their personal computers. Office sharing may also threaten some workers' conception of status. Teleworking is likely to continue to erode many traditional worker distinctions.

A less obvious issue, but one that ultimately may be of greater importance, especially in Japan, is that of employee loyalty and the feeling of corporate identity. In an organization with a central location to which most employees report, workers may be more likely to identify their personal goals with those of the organization. Are telecommuting employees less likely to be loyal to the company than those in the central location? Are home telecommuters more likely to be, or to become, entrepreneurs selling their particular information services to the highest bidder? Will they be less enthusiastic or, more importantly, less productive in support of company goals?

The local work center may be a sufficient surrogate for the central office in developing company spirit. However, employees are also more likely to come in daily contact with employees from other organizations and to compare notes on employee benefits, working conditions, and the like. For organizations with deficiencies in these areas, telecommuting could have threatening effects.

Coupled with this could be the expansion of the movement by information workers to form a union. Directors of information-intensive companies in the US tend to be uncomfortable with the idea of unionization, and they also believe that decentralized organizations are more likely to become unionized than centralized ones. While there is little evidence for this, the feeling persists.

Historically, the influence of unions among information workers has been small. However, this influence is growing among clerical workers and professionals such as engineers and scientists. Unions may gain support and improve benefits for information workers through the advocacy of telecommuting, provided that, as the technologies are introduced in an organization, union members are adequately trained to use them. In the near term, however, unions are likely to be ambivalent about telecommuting, primarily because few have considered the process at all—they see neither great threats nor great opportunities in the process. Union officials we have talked with generally say that telecommuters could be covered through minor modifications to existing contracts. While telecommuting may be an attractive issue in unionizing centralized organizations—"Join the Union and Work Near Home"—the impediments to organizing already telecommuting-dispersed companies may be substantial.

Nevertheless, telecommuting may well provoke union-associated issues. Foremost among these is the apprehension about a return to the "sweat-shop" atmosphere through piecework payment systems, forcing some production costs onto workers, and the absence of enforced periodic rest breaks—people hunched over computer terminals for several hours need a break. Most of these issues are significant only for home telecommuters; it's unlikely that workers in dispersed centers would encounter them any more frequently than employees in traditional office settings.

A final impediment to the telework concept is tradition. People have been accustomed to leaving their homes to go to work ever since the early days of the Industrial Revolution. Cottage industries were supplanted by more efficient means of production that required workers to go to central locations to operate the machines. For information workers, that rationale is steadily becoming less compelling, but the pattern is set nonetheless.

Telecommuting makes possible a job migration without transportation. When the unavailability of low-cost transportation, the lack of local job opportunities, and the high levels of functional illiteracy make the escape from poverty or dependent living seem impossible, as in many urban areas, low-cost computers and software can provide entry-level information jobs. The critical point is that the inherent power of computer-based instructional techniques must be used to train these workers to at least a minimal level. Neither the software nor the private-sector or public-sector programs exist today to facilitate this transformation, yet the basic concepts and the technology do exist.

However, communication satellites make it possible to export jobs to other countries as well as to Appalachia. Thus, we as a nation are confronted with an interesting dilemma: do we use telecommuting to aid those who otherwise have no access to economic success, or do we use it to improve the economies of other nations? The technology has already escaped Pandora's box—if we don't use it, others will. The exported jobs may include not only those of entry-level information workers but those of higher-level employees as well. The net result could be an influx of job opportunities in the US, if we capitalize on our unique expertise in many aspects of the information industry to satisfy demand in other countries.

Widespread telecommuting can have major long-term effects on our national landscape. There are at least two possible types of change. The first consists of surrounding the local office complex with residential areas. This would result in the development of numerous city clusters, each with a population of from 100,000 to 200,000 (a good size for efficient provision of services such as law enforcement and street repair). The average commute to work in such a citylet would be three miles or less. Citylets would be interconnected by telecommunications networks and, of course, freeways for efficient, non-peak-hour automobile and truck travel.

Since the theoretical locational freedom made possible by telecommuting is limitless, a telecommuting society could scatter its dwellings more or less uniformly between its boundaries. But this would leave the remnants of city centers to blue-collar and other workers unable to telecommute, and would also require inefficient and energy-expensive systems for distributing materials and supplies to the "urban" population.

Both these extremes imply significant changes in existing central business districts. If many organizations were to flee the urban business district in favor of dispersed operations in the suburbs, blighted central-city areas could become still worse. One alternative might be to convert city buildings into multi-use facilities, bringing residences back to what are now office-city cores. In a reversal of the usual concept, workers might live in the city center and telecommute into the suburbs.

Information workers make good energy conservers. Niether the use nor the production of computers and telecommunications is energy-intensive. A typical computer terminal uses about 100 W of power, and a microcomputer with a relatively massive memory might use an additional 200 W. In most cases, this energy cost would be incurred anyway as individuals in an organization acquire their own computers; the added energy cost of telecommunications is on the order of a few watts.

Transportation is energy-intensive, accounting for about one-fourth of total national energy consumption. About one-ninth of this consumption is due to the urban automobile, which uses about 3 kWh of energy for each passenger-mile traveled. For every 1 percent replacement of urban commuting by telecommuting, the national gasoline bill would be reduced by 5.4 million barrels per year. In other words, if we replaced one-seventh of our urban commuting, we might no longer have to import oil. In addition, since pollution trends closely follow levels of energy consumption, air pollution should decrease as telecommuting increases.

Telework and telecommuting are not going to burst upon the scene overnight, but conditions are right for their influence to grow. Estimates indicate that telecommuters now number in the thousands. By 1990 there may be as many as 10 million telecommuters (mostly part-timers) if some additional conditions are met:

1 The information sector must continue to grow beyond its present point of slightly more than half the US labor force.
2 The telecommunications and microelectronics industries must continue to grow in power at their current rate of close to 30 percent per year, and telecommuting software must be made widely available and affordable.
3 Personal computers must also continue to grow in computational power per dollar. Personal computer software provides capabilities, including "friendliness," generally not available on mainframe (large industrial) computers.
4 Energy shortages may prove to be one of the greatest incentives, causing employers and employees alike to reconsider the high costs of commuting. An energy shortage sufficient to induce gasoline rationing would greatly increase telecommuting.

Telework and telecommuting are likely to increase at an accelerating pace over the next decade, spurred by many independent societal trends. This technology provides more options to both employers and employees, but the resulting changes, though sometimes subtle, can be ubiquitous, taking years before their effects are widely recognized. The costs and benefits of this rapidly disseminating technology must be defined and evaluated before any major and irreversible changes in the way society does business are implemented.

Second Thoughts on Moving the Office Home

William L. Renfro

The home office may free us from the daily commute to work, but we may be giving up a lot more than we gain—such as personal contact with colleagues and the boss. The author, who is president of a Washington, DC, firm of consultants, looks at some practical and psychological problems of working from home, while John Applegarth gives the other side of the story. Taken from The Futurist, *June 1982, and* Working Free: Practical Alternatives to the 9 to 5 Job *by John Applegarth (AMACOM, New York, 1982).*

You missed the 7:15 again. Forced to drive to work, you get stuck in heavy traffic. As you finally trudge into the office, past the glowering boss, you say to your officemate, "Won't it be nice when this place is automated and we can stay home and do our work in peace and quiet on computers?"

An appealing thought, perhaps, but an office in the home is a place where few people will be spending time in the next few years. *Los Angeles Times* columnist Jack Smith thought his life's dream had come true when he was given permission to write his column at home. Relieved of the burden of commuting to the office every day, he was free to work at his own pace.

Six weeks later, he had to return to the office temporarily because his granddaughter broke his typewriter. Once there, he discovered the rewards of office life that he had never appreciated: "the friendly faces, fresh in the morning; the clothes; the gossip; the flirtations; the benign conspiracies; lunch hour expeditions; the open forums on war and peace, Reaganomics, and the Rams quarterback controversy, none of which could be examined with such reckless spontaneity by anything canned for consumption on your home computer."

Smith felt rejuvenated when he came back to the office. His columns were richer. He had more ideas to work with. He recognized, as many home workers do, that he had been withering on the vine at home as though unwatered by contact with his colleagues. And he realized that he really came back to the office not just to write, but to renew his interaction with human life.

If sociologists are concerned today about the sense of isolation, aloneness, and anomie in our society, then they should be sending up warning flares about the sociological disaster the home office could bring. One of the principal functions of work in America, author Studs Terkel claims, is "schmoozing"—achieving a sense of companionship and togetherness among workers as they chat about their lives and gripe about common problems— and this function may die in the home office. If our primary links to the larger society are through our fellow workers, we will have to build institutions that will maintain human contact when the office is moved home.

Working in an office with other people shifts our focus away from ourselves and toward a larger group. This provides our first key link with society. The office offers a sense of team spirit and participation in something larger than the individual.

The feeling of "we're all in it together" at the office provides a necessary ingredient for the discipline and sacrifice so important in getting the job done. Your co-workers provide incentives and support. Sure, you feel a little slow this morning, but the others will be at the office on time. You want to show them that you can do your part and support the group, so you make the effort. At home, there is no incentive, no reward, no sense of personal accomplishment for making the extra effort.

The mere process of going to the office also has an important function: Your commitment to "be at work" for the day requires no further effort. At home, you have to make that commitment to work after each little distraction—the dog, the mailman, the kids, the dinner, the thousands of "honey-do" projects that can consume eight hours of work time before you know it.

Being "at work" provides a ready excuse to avoid those thousands of little errands or activities demanding your time. At the office, it is easy to say, "No, I can't do that—I'm working," but at home it is a difficult, seemingly arbitrary, but necessary task. Many women writers who work at home have long complained of the constant interruptions by friends and neighbors who simply don't understand a home worker's "office hours." The "imposed" routine of the office, of going to work, will have to be replaced by self-imposed discipline in the home office.

Even the notion that working at home saves a lot of money on the fuel it takes to commute back and forth is a delusion. You can spend much more on energy to heat or cool your home all day than you do on commuting a few miles.

And while we complain about the time lost in commuting, consider the many things we use commuting for: we plan our day, do some errands along the way, share thoughts with the car pool. The change of surroundings, the sense of going somewhere and of doing something, are part of the role that commuting plays in our daily lives, according to research by General Motors. We feel a sense of closure, of completeness when we leave work and head for home at day's end.

The sense of being "at home" will also change without this transition between home and office that commuting provides. It will be much harder to leave the headaches of work back at the office, and families may suffer. Home, that place where you feel absolute dominion over your environment, will for eight hours a day be dominated by some outside influence: your "castle" will be invaded by a boss calling to check up on your progress.

One of the major problems of computers has always been their requirement of formality: input has to enter through the keyboard in a specific way. In the language of information architects, this formality is a tremendous transaction barrier—something that makes computers hard to use. This kind of barrier means that the home worker will miss out on much of the informal communication that flows through an office.

Even the telephone, with its much lower barriers, does not permit the kind of accidental exchanges that happen in the hall, by the coffee pot, or at the copier. There is much more to communication between two people than the spoken word: there is eye contact, gesture, body language, inflection, tone, emotion. The computer, lacking these essential nonverbal elements of communication, is a barrier that the home worker must somehow overcome in order to sustain a relationship with the boss and with co-workers.

Studies conducted at IBM offices found that the contact between office workers decreases relative to the distance between their offices: double the distance between two workers and you cut the communication between them by a factor of four. Move them a floor apart and you cut contact by 90 percent. In other words, "out of sight, out of mind." The implications for the career of the home worker are obvious: if the boss can't see how well you're doing and how hard you're working, he may soon forget you and pass over you at promotion time.

But the biggest impediment to the home office may be the resistance of employers. Since the home worker loses access to the shared equipment and supplies of the office—the files, the copier, the bulky reference volumes, and so on—his productivity will suffer. Although the worker may argue that these things are not absolutely essential every day for every task, the employer has put them in the office for a reason: they make it possible for more people to be more effective and productive in their work.

The lack of contact with the worker presents another problem to the employer. More than 10 years ago, my department in a large corporation had an internal phone dictation service. Executives could call and control dictation recording equipment and get hard copy by inter-office mail within a few hours. But nobody trusted the privacy of the system for any important communication, and it was embarrassing to use such a fancy system for the minor memos. The result: the system didn't get used. This problem is compounded with the home worker. No responsible office manager will allow the free transmission to and from a worker's home of private, internal office communications, especially workers who are unknown, seldom (if ever) seen, and whose loyalty to the organization has not been established.

Certain economic and legal considerations that would have to be worked out with the home worker will make employers reluctant to give permission to employees who want to work at home. For example, the safety and security of expensive equipment kept in the home will be a difficult issue to settle. Jack Smith may pay the bill when his granddaughter breaks the typewriter, but not when a $10,000 communicating word processor goes on the blink. The home worker and the employer will have to settle questions of liability when, for example, thieves run off with the video display terminal, thinking they've stolen another television. Insurance and security systems will certainly add to the expense of the home office.

Many of the legal and economic issues could be resolved if the home worker were not an employee but an independent subcontractor who supplied his own equipment or leased it from the office or a third party. The government has been suspicious of such subcontractor relationships, however, and has made up a host of protective labor regulations. For example, the National Labor Relations Board has been arguing that women who knit wool hats in their homes for a Vermont company are employees and must be paid the minimum wage. The workers and the company involved have been arguing that the women are independent contractors who can set their own prices for their products and their time. Though this case has become a bit of a political football in the arguments over excessive government regulation, it could yet set an important precedent for the future of the home office.

Home office enthusiasts like to point out that office technology will open opportunities in the labor force for people often excluded in the past. Housewives who have a few hours to spare while the kids are at school could use their time to add needed dollars to the household income. The telephone cable can do more to reduce barriers for the disabled than all the barrier-free environments that technology has to offer.

But disabled people may discover that this is a dubious opportunity. When they try to overcome barriers to take part in society, other people must always keep them in mind. But if we cannot see the disabled, the elderly, and the infirm, and are not forced to deal with their special needs, then we as a society may very quickly forget about them.

People who are looking for a few hours' worth of work each day at home may find it difficult to obtain. The equipment necessary for the home office is only now becoming cost-effective for a 40-hour week. It would take four times as much investment in equipment to get the same productivity when the equipment is used only 10 hours a week. Not very practical when the prime interest rate is 17 percent.

The worker who must stay home because of children faces special problems. In the already cramped housing of today's young families, finding space for the home office will indeed be difficult. While the business may save a great deal in office space, and hence rental costs, it is not clear that this saving can or will be shared with the home worker. If an agreement between a business and its home employees could be worked out, "rental" income in the form of higher salaries could of course help with a family's mortgage; however, zoning laws against converting residential into commercial property are likely to prohibit any such agreement.

The "opportunity" for homebound workers has a very important social cost. Though these workers are undoubtedly a productive addition to the labor force, they will be taking jobs away from the lower levels of the office— jobs that have traditionally served as entry positions for lower-skilled workers. This could make the job market for the under privileged and for the unemployed teenager even smaller. Hardly a desirable social development.

How to Work at Home

1   Get dressed. Being "at work" is a role you assume daily, so you should wear a costume that suits the part. Wear whatever you would normally wear to the office.

2   Set up office hours. Post the schedule on your office door or where family and friends can see it, and by your phone so you can remember to tell others. Practice saying, "I'll be at work from 9 to 5." And *be* there.

3   Get to work on time. Punctuality is important to maintaining a routine, so even if getting to work only involves a 20-foot walk down the hallway, make sure you're never late.

4   Quit on time, too. Don't linger over office headaches through dinner time. When office hours are over, take a walk around the block. You must find a way to leave the office behind you—literally.

5   See people. It's easy to become isolated from the rest of society when you spend 24 hours a day in your house. Incorporate a game of squash into your morning routine, or a trip to the grocery store or newsstand each afternoon. Meet the office gang at a local tavern after work, and join professional associations so that you can maintain your business contacts as well as your personal ones.

Certainly there are specific kinds of work (such as computer programming and clerical tasks) that will easily be moved to the home. And perhaps the disabled, the infirm, and others who can now join the workforce without the burden of commuting will not be forgotten by society. Institutions can be built that will substitute for the traditional office's function of maintaining workers' links to society. But these adjustments require social change, which is traditionally much slower than technological change. Evolutionary social change has yet to catch up with the revolutionary technological changes happening right now.

Perhaps this social change is what we are seeing in the electronic arcades of shopping centres: the office workers of tomorrow are learning to love the technology. Rick Roelke, a friend of mine who is a computer game addict, has played with these games since they were first available. Now, from his condominium in the woods of New Hampshire, he writes computer programs. He could well be a pioneer of the electronic cottage industry of the future.

When the Rick Roelkes become more numerous and gain sufficient experience, they may build their own informal links to society and never know the office. They may never feel the loss of Smith's office adventures or Terkel's schmoozing. But they will never know a part of the human experience.

The same great claims made almost 20 years ago for the revolutionary impact of computers are being made again today for the computer in the office at home. America has shown a great tendency to whip itself up into these great frenzies of excitement over revolutionary developments. In 1964, RCA announced the commercialization of its picture phone with an advertisement showing the Manhattan skyline and the question, "Is this still necessary?" Nearly 20 years later, Manhattan still is—but the picture phone never was.

Those of us who are weary of commuting may have become too excited about the prospect of the home office. Until we do our homework—that is, until we have thought out all the consequences of working at home—let's just keep going to work *outside* the home.

The Other Side: What's Good About the Home Office
John Applegath

Do you wonder how some people manage to get so much done in a day?

One answer may be that they don't waste time getting back and forth to their place of work everyday, because they do their work at home. All those extra hours get used—for either work or pleasure.

Because my home serves as my office, I can be at work within 15 minutes of getting out of bed in the morning. Or I can wake up in the middle of the night with an idea and be at work in less than 5 minutes. Or, because I don't have to appear someplace at some fixed time, I can spend the early morning hours in bed reading or watching Phil Donahue or enjoying my wife's good company, and ease into working when I'm ready for it.

If I work until midnight to meet some important deadline, I am not faced with a long drive home—or a wait for public transportation alone in the dark. I can simply leave that neat, finally finished project on the desk and walk upstairs to bed.

With the availability of small computers that can be plugged into immense data banks, word-processing equipment, and other devices, more and more people are setting up what Alvin Toffler calls the "electronic cottage."

The new cottage industries may be found in a handsome brownstone in Chicago's Lincoln Park, beside a swimming pool in a rambling house in Palo Alto or Vancouver, or in a log house in a remote rural area. A surprising number of business firms and other organizations whose address is given as a post-office box number are actually individuals or couples working at home.

Of course, "working at home" shouldn't always be taken literally. Some people do at least part of their work in unusual settings. On occasions, "home" is Central Park for a consultant who finds that she does her work better in the fresh air and sunshine.

Some of my own best ideas got written down at the beach, or while I was sitting over a leisurely cup of tea in a pleasant cafe. Flexibility of time and place is the major advantage to working at home.

Jane Marriott, in the restaurant consulting business with her husband, likes the convenience: "It's not taking the time to get dressed, getting in the car, driving to your office, answering your calls. There's a lot of time that we both would waste. We've had a lot of discussion about whether or not to get an office, even close by. We both like to be able to work at any hour, dressed in any way we want. It really works out for me. I just walk into the studio in a bathrobe, or I can do it in the evening. I have a lot of materials accessible."

Pat Lee, a consultant and special projects manager, likes the flexibility: "It means if you want to work on a rainy day for 12 hours, and spend 12 hours in the sun the next day, you can. If you feel like sitting in your chenille bathrobe at midnight writing an employee handbook, and you really get into it and want to stay up till 4:00 in the morning doing it, that's terrific. You go ahead and do it. If you get really hot on a wage-and-salary study on a Saturday morning and you want to work through a whole weekend in order to take off Monday and Tuesday, you do it."

Author Jessica Lipnack, who also works at home with her husband, likes the opportunities for parenting: "The reason we bought this house was because of its size. We wanted to work at home. We use the two front-parlor rooms, and we each have individual studies on the third floor. I like having my work nearby. I find that very gratifying. We had an office for a long time, so I've also had the experience of going out of the house to work.

"We both really enjoy it in relation to our small children, because we've had a lot of access to the kids—they're seven months and three years. I'm nursing the baby, so it's made it possible for me to be one of the rare women who can work and nurse. That's a big issue for a lot of people with infants. So that's been wonderful. Being able to see them on and off all day and make choices about spending time with them has been great. You know, if something came up . . . if I wanted to spend some time, or if Jeff wanted to take Miranda off on a bicycle or something, we can do all of those things, and that is terrific."

Guide to Further Reading

Personal Computers

Ruth M. Davis, "Computers and Electronics for Individual Services," *Science*, February 12, 1982.

Hoo-min D. Toong and Amar Gupta, "Personal Computers," *Scientific American*, December 1982.

Jack M. Nilles, *Exploring the World of the Personal Computer* (Prentice-Hall, Englewood Cliffs, NJ, 1982).

Robin Bradbeer, *The Personal Computer Book*, 2nd edn (Gower Press, Aldershot, Hampshire, England, 1982).

Alfred Glossbrenner, *The Complete Handbook of Personal Computer Communications: Everything You Need to Go Online With the World* (St Martins Press, New York, 1983).

Alfred Glossbrenner, "Personal Computers: Passport to the Electronic Universe," *Technology Review*, May–June 1983.

Roy Mason with Lane Jennings and Robert Evans, "A Day at Xanadu: Family Life in Tomorrow's Computerized Home," *The Futurist*, February 1984.

Joel Dreyfuss, "What Will Send Computers Home," *Fortune*, April 2, 1984.

Gene Bylinsky, "High Tech Hits the TV Set," *Fortune*, April 16, 1984. On "Smart" TVs.

The Personal Computer Industry

The extraordinary progress of the personal computer industry itself is charted in: "Price War in Home Computers," *Time*, September 20, 1982; "The Year of the Mouse," *Time*, January 31, 1983; "Apple's Bid to Stay in the Big Time," *Fortune*, February 7, 1983; "Price War Blasts Open Home Computer Market," *Business Week*, June 13, 1983; "Shake-Out in the Hardware Wars," *Time*, June 27, 1983; "Sudden Shake-up in Home Computers," *Fortune*, July 11, 1983; "An Easy-Come, Easy-Go World," *Time*, September 5, 1983; "The Big Battalions Take Over", *Financial Times*, September 12, 1983; "Personal Computers: And the Winner is IBM," *Business Week*, October 3, 1983; "Mass-marketing the Computer," *Fortune*, October 31, 1983; "D-

Day for the Home Computer," *Time*, November 7, 1983; "Winners (and Losers) from IBM's PC Jr," *Fortune*, November 28, 1983; "Apple Launches Mac Attack," *Time*, January 30, 1984; "Apple Bites Back," *Fortune*, February 20, 1984.

Portable personal computers are described in Cary Lu, "Dawn of the Portable Computer," *High Technology*, September 1983, and "Here Come the Go-Anywhere Computers," *Fortune*, October 17, 1983. Business uses of personal computers (see also chapter 7 below, especially the piece from *Business Week*, August 8, 1983) feature in "Personal Computing: Productivity Tool for Business," an International Data Corporation white paper to management, *Fortune*, January 23, 1984; *Financial Times* surveys on "Personal Computers," January 18, 1983 and April 24, 1984; and "Is the Real Revolution in Personal Computers Just Beginning?" *Business Week*, October 31, 1983.

Teleworking

Malcolm Peltu, "New Life at Home for Office Workers," *New Scientist*, March 27, 1980.
Robert Cowen, "Cottage Computing: Glorifying the Trivial?" *Technology Review*, November–December 1981; and reactions to his argument (that a lot of work cannot be moved home) in the May–June 1982 issue.
Ove Svidén, "Automobile Usage in a Future Information Society," *FUTURES*, December 1983. Reports on a Swedish study which concludes that, overall, information technologies will not substitute for travel, only for the information carried on paper. Increasingly dispersed living will keep car usage high, but a lot depends on oil prices!
"It's Rush Hour for 'Telecommuting,'" *Business Week*, January 23, 1984.
"Telecommuting from a Flexiplace," *Time*, January 30, 1984.
Ursula Huws, "New Technology Homeworkers," *Employment Gazette* (London), January 1984).
Tammara H. Wolfgram, "Working at Home: The Growth of Cottage Industry," *The Futurist*, June 1984.

Video Games/VCRs

"Games That People Play," *Time* cover story, January 18, 1982.

Five articles in *Fortune* followed the ups and downs of the video games industry. Peter W. Bernstein, "Atari and the Video-game Explosion," July 27, 1981; Peter Nulty, "Why the Craze Won't Quit" and Andrew C. Brown, "Cashing In on the Cartridge Trade," November 15, 1982; Lisa Miller Mesdag, "How Coleco Plans to Keep Flying High," March 7, 1983; and Gary Hector, "Atari's New Game Plan," August 8, 1983; so also Herb Brody, "Video Games Enter Technology Time Warp," *High Technology*, June 1983, and "Video Games Go Crunch," *Time*, October 17, 1983.

Newspapers and magazines have carried lots on the booming market for VCRs, the Betamax/VHS battle, and the problem of video pirates. For an account of the history of video recorders and how they work, see Barry Fox, "The Home's Most Complex Box," *New Scientist*, December 1, 1983.

The prospects for video discs were discussed in *Time*, March 23, 1981; *New Scientist*, July 30, 1981 and October 21, 1982; and *Fortune*, November 2, 1981.

5 Computers in Schools

Here Come the Microkids

Frederic Golden

The information technology revolution has spawned a new generation of computer-literate kids who know more about computing than their parents—and even their teachers. But experts disagree as to whether this is a good thing or a bad thing. The author discusses why some kids take so readily to computers and reviews what is happening in us schools. This article first appeared in Time, *May 3, 1982.*

The day is officially over at Benjamin Franklin Junior High School in Ridgewood, NJ, but a handful of students are still hard at work. They are "Muller's disciples," followers of a popular math teacher named Bob Muller, 30, who heads Benjamin Franklin's computer program. Oblivious to the clang of the last class bell, the disciples are hunched over their desktop computers, while long reams of paper clatter out of printers and green phosphorescent TV screens dance with ciphers and letters.

George Mamunes, 14, a gangling ninth-grader dressed in flannel shirt, blue jeans and hiking boots, knits his thick, dark eyebrows while putting the finishing touches on a computer program, already nearly 300 lines long. For those uninitiated in the special languages of the computer age, it looks like a hopeless mess of numerical gibberish. But when completed, these arcane instructions should produce a computer image of the heart detailed enough to show every major artery and vein, as well as valves and chambers. The electronic heart is part of a teaching tool George is putting together for eighth-grade biology classes.

A few feet away sits Pam Miller, 14, a ninth-grader with long, brown hair draped far down her back. She is operating a computer program—or software—that simulates the workings of a nuclear reactor. Today she is fine-tuning the section that governs the control rods, those regulators of the reactor's nuclear fires. Tapping away at the keyboard, Pam explains: "You have to maximize the power output without destroying the reactor."

Suddenly, flashing numbers burst upon the screen. "There," says Pam, her face lighting up. "Reactor overheated. Power output low. Reactor core damaged. Melt-down!" A disaster that she has brought on intentionally, just to show how it could happen.

Other disciples, seated at terminals scattered around the room, are no less absorbed. Meilin Wong, 15, chic in blue velour blouse, jeans and Bass moccasins, is trying to figure out what went wrong with her business data management program. She is an old hand at such troubleshooting, having spent much of last semester "debugging" a program that, when printed out, stretches over 30 ft. Jim McGuire, 13, is creating a video game called Spaceship, which will let electronic star warriors zap a boxy-looking orbital intruder. A more mundane program is emerging from 15-year-old Dave McCann's terminal: a verb test for seventh- and eighth-grade Spanish classes. Off in a corner two youngsters are putting the impish face of *Mad* magazine's cartoon hero, Alfred E. Neuman, onto the computer screen.

Says Muller, as he presides proudly over these after-hours computer converts: "No one told them they have to be here. They're not usually doing assignments. They're experimenting. They're letting their imaginations run free."

Muller's disciples are not all math whizzes. Or straight-A students. Or particularly precocious. They are reasonably normal youngsters who have grown up with computers. For them, in ways that few people over 30 can understand, manipulating these complex machines is as natural as riding a bike, playing baseball or even solving Rubik's cube. Like thousands of others across the country, they are part of a revolutionary vanguard: the computer generation. Not only is this generation propelling traditional education down promising avenues, it is tugging at the entire social fabric, foreshadowing changes at least as startling and momentous as those ushered in by a new generation of automobile users more than a half-century ago.

In the classroom, where youngsters are being introduced to the machines as early as kindergarten, they astound—and often outpace—their teachers with their computer skills. After school they gather at the mushrooming number of computer stores (more than 1500 at last count) or join the computer clubs that are becoming a regular part of the youthful landscape. Huddling around any available machine, they argue over their programs, thresh out computer problems and swap software as intensely as kids once haggled over baseball cards. In the summer, they may even go off to computer camps, another growth industry, and if they are Boy Scouts, they may try for a computer merit badge.

During mischievous moments, they may tinker with one another's programs, writing in steps that will flash an unexpected insult or obscenity across a buddy's video screen. Some try to pick the encoded electronic locks on copyrighted software, taking glee in outwitting their elders, or spin fanciful plots to break into computer networks. A few turn their skills to profit by showing baffled businessmen how to get idle, new computers to run, or by establishing Delaware-based corporations to market their own software creations. To the bafflement of their parents, they talk in a jargon of their own ("Hey, Charlie, you should have POKEd instead of PEEKed").

As with so many other changes in contemporary life, the spark for this revolution is technological: a bit of silicon sophistication variously known as the personal, home or microcomputer. No larger than an attaché case, apart from its video screen, this mighty mite packs the computing power of machines that two decades ago occupied a full room. Yet the micros, as they are affectionately called, are a relative bargain to buy and are becoming steadily cheaper. Many models cost under $1000, bringing them within reach of schools, parents or the children themselves. Last week, in the sharpest price break yet, Timex announced it will begin selling a small home computer for a suggested retail price of $99.95.

But size and price cannot explain why computers have taken such a strong hold on so many youngsters. Certainly their interest has been stirred by a related rage, video games, whose computer-generated flashes, zaps, and pings have not only all the appeal pinball machines had for their elders but go a significant step further: they pique young minds to learn more about all that electronic prestidigitation. But many experts, and most of the young operatives, agree that the overwhelming attraction of the machines is the lure of control, the pleasure of being able to think out and then make something happen, a satisfaction all too often denied children.

Lewis Stewart, 14, a black ninth-grader at Manhattan's PS 118, reads at a fifth-grade level; yet mastering his school's computers was literally child's play for him. Recognized by students and teachers alike as his school's best computer programmer, Lewis works afternoons as an instructor for a computer consulting firm, introducing younger children to the machines. Last year his employers sent him to Chicago, where he displayed his special teaching gifts before a meeting of educators. As Lewis told *Time* correspondent Peter Stoler, "I love these machines. I've got all this power at my fingertips. Without computers, I don't know what I'd be. With them, I'm somebody."

Perhaps because of the faintly macho side of computers, the bug seems to strike many more boys than girls in the pre-adolescent years. Says Steve Siegelbaum, Lewis's teacher: "Maybe it's because boys are pushed more toward math and logic than girls are. Maybe it's because boys are just more aggressive."

Paradoxically, the computer passion is often stirred in youngsters who seem least likely to be interested in high tech. Jay Harstad, 12, of Minnetonka, Minn., litters his house with poems and sketches but will do almost anything to avoid doing his math homework. Yet Jay is one of the Gatewood Elementary School's premier computerniks and regularly helps teachers introduce fourth-graders to the machines. At West High School in Wausau, Wis., Chris Schumann, 16, a junior, has made a name for himself by translating musical notes into digital form and getting a computer to play

Bach and Vivaldi through its loudspeaker. Originally, Chris regarded computers as remote and forbidding, but that changed when he was introduced to his first micro. "It looked real friendly," he says. "It didn't overpower you. It wasn't this ominous thing but something you could get close to."

The closeness can be contagious. Explains Nick Newman, 15, Muller's chief disciple at Ridgewood: "The more you do on the machine, the more enjoyable it gets. It becomes habit-forming." In Alpena, Mich., youngsters who had learned computer skills in junior high were devastated when they got to senior high school and found too few machines to go around. Says Alpena Elementary School principal Burt Wright: "I've got high school kids begging to come in after school and use our machine." The truly addicted—known half scornfully, half admiringly as computer nerds—may drop out almost entirely from the everyday world. In Lexington, Mass., one legendary 16-year-old nerd got so deeply immersed in computers that he talked to no one, headed straight to his terminal after school and barely sat down for meals. The only way his father could get him away from the terminal was to go down to the cellar and throw the house's main power switch, cutting off all electricity.

Barry Porter, 14, of San Francisco, is a computer-age truant, so attached to the machine that he often skips school, rarely reads anything other than computer manuals and hangs out with his pals in the Market Street computer store, often plotting some new electronic scam. Barry (not his real name) currently boasts an illicit library of about 1000 pirated (i.e. illegally copied) programs worth about $50,000 at retail prices, including such software gems as VisiCalc, the popular business management and planning program. Before security was tightened up, he regularly plugged his computer into such distant databanks as the Source (which provides news bulletins, stock prices, etc.) via telephone without paying a cent.

No one can say exactly when the computer generation began—certainly not earlier than the 1960s, when computers began appearing in schools. But even computer whizzes in their twenties are acutely aware of how soon they are likely to be outstripped by today's grade schoolers. Says Steven Jobs, 27, the multimillionaire co-founder of Apple Computer Inc.: "These kids know more about the new software than I do." New York computer executive Charles Lecht goes further: "If you were born before 1965, boy, you're going to be out of it."

Where their parents fear to tread, the microkids plunge right in, no more worried about pushing incorrect buttons or making errors than adults are about dialling a wrong telephone number. Says mathematician Louis Robinson, IBM's resident computer sage: "They know what computers can and cannot do, while adults still regard them as omnipotent." Says Hughes Aircraft Chairman Allen Puckett, who shares an Apple with son Jim, 12: "A lot of adults grew up in a slide-rule world and still reject computers. But computers are as natural to kids as milk and cookies."

More and more members of the computer generation are tasting the heady pleasure of teaching their own teachers how to use the machines and, if they are lucky enough to have computers at home, instructing their parents as well. Says Ridgewood's Newman, a regular teacher of teachers: "It's a sort of mutual doorway. The barriers between adult and child, between teacher and student, are broken, and it's person to person. Nobody's looking down on anyone; they're looking each other right in the eye."

Often adults find it easier to ask a child how to do something than to ask another adult. Says University of Kansas education professor Mary Kay Corbitt: "One adult student of mine brought her son to computer class, and I discovered that he was doing her assignment while she watched. Two weeks later she overcame her anxieties and was participating fully." Confronted with the strange and unsettling world of the computer, teachers can get a useful perspective on what it is like to be a student again. After taking part in an elementary course in programming, Lois Brown, 54, a Wausau grade-school teacher, is thoroughly chastened. "Now I realize how little kids feel when there's a concept they don't understand. I sat in that course not wanting anyone to know all the things I didn't understand."

Despite their obvious wariness of computers, parents are taking the lead in getting them into the schools. In Florida, communities have staged cake and candy sales, carnivals and tree plantings, weekend car washes, even a bike-a-thon to raise funds to buy computers. Says Marilyn Neff of Miami: "We feel computers will be the new paper and pencil." Of the 250 computers in the schools of Utica, Mich., more than two-thirds have been purchased by parent-sponsored fund drives. Says Utica Principal Paul Yelinsky: "Moms and dads are coming in and telling the counsellors they have to get their kids in computer classes because it's the wave of the future." So important is computer literacy that the Alfred P. Sloan Foundation is beginning a major program to get even such traditional liberal arts schools as St John's College in Maryland to begin giving courses in it.

Though many schools began purchasing computers with federal aid, budget cutbacks are drying up that well. Apple's Jobs points out that other nations, especially Britain, France and the Soviet Union—though surprisingly not the electronics-minded Japanese—are paying far more attention to computer education than is the US. Earlier this year, Jobs persuaded California Congressman Pete Stark and Missouri Senator John Danforth to introduce bills in Congress that would allow computer manufacturers to take a hefty tax write-off for any machines they donate to elementary and high schools. Under the present law, full deductions for such scientific equipment are allowed only if it is given to colleges and universities.

Jobs originally spoke of giving an Apple to every public elementary and secondary school in the country, more than 80,000 computers worth as much as $200 million retail. Now he thinks private schools should be included and is encouraging other manufacturers to join in the program as well. Meanwhile, Apple's archrival, the Tandy Corp., maker of the Radio Shack computer line, is taking a different tack: it has pledged $500,000 in equipment to spur development of educational programming, or courseware, for the classroom.

Many of the approximately 100,000 computers now in US schools—roughly one for every 400 students—are in affluent suburbs like Ridgewood, a national leader in computer education. But the machines are also found in the unlikeliest of places. On a Chippewa Indian reservation in Wisconsin, computers are being used by young members of the tribe to learn their ancient and nearly forgotten language. Alaska's small rural schools have been ordering computers to meet a special need: they allow students of different ages and abilities in the same small classrooms to learn at their own pace. Dubuque, Iowa, where the *New Yorker* founding editor Harold Ross disdainfully located his provincial old lady, has 13 machines and another 20 on order. Bill Holloway, a professor of computer education at the University of Kansas, calls the spread of small computers in the classroom nothing less than an avalanche. According to various industry studies, there may be from 300,000 to 650,000 computers in the schools by 1985.

So far, the most common, and least interesting, way to use school computers is in direct drill and review. The machine simply quizzes, prods and grades the student, very much like a robot teacher. Hundreds of programs of this type are available for popular computers like the Apple II Plus, Radio Shack's TRS-80 and the Commodore PET. But many of these programs are little more than computerized rehashes of the old classroom flash cards that go back to the days of McGuffey's readers. One notable difference: today when the student answers correctly, the screen will light up with WOWS, HOORAYS or smiling animals. Wrong answers may produce sad or scowling faces, perhaps accompanied by a falling tear.

Partly because of teachers' fears—of the machines and for their jobs—and partly because of the poor quality of software, the frequently heralded electronic revolution in the classroom has been slow to occur. Now, however, it is being pushed along by steady improvements in teaching programs, thanks to imaginative enterprises like the Minnesota Educational Computing Consortium. One of its more refreshing drills: a program called Wrong Note, which helps teach sight reading of musical scores. As a simple tune emanates from the computer's loudspeaker, matching notes appear on the screen, but sometimes the quiz intentionally errs and obliges students to find the false note. In order to do so, they can order up a repetition of the tune as often as Bogie and company did in *Casablanca*. Says Kenneth Brumbaugh, director of the consortium's instructional services: "Imagine asking a teacher to play it again and again!"

Even very young children can profit from such exercises. At the Nordstom Elementary School in Morgan Hill, Calif., a suburb of San Jose, Colin Devenish, 7, is working with a classmate on the arithmetic drill, honing his skills in addition and subtraction. Unlike youngsters doing such drilling in the past, Colin seems to be enjoying himself enormously. Why? "Because," he replies mischievously within earshot of his teacher, "the computer doesn't yell."

Computers, operated only by touching a few buttons, are also remarkably effective devices for educating the handicapped. At the California School for the Deaf in Fremont, Rhonda Revera, 16, has worked with computers for five years, studying every subject from fractions to spelling. Rhonda offers a paean to the machine in sign language: "Computer makes me remember. It is fast, easy and better than writing on paper."

Still another important use of computers is as a remedial tool. At Manhattan's P.S. 118, Lewis Stewart has not only improved his command of the language with his work on computers, but has also prepared practice exercises for classmates with even more serious reading problems. One is a spelling drill with a special incentive built into it: if all the answers are correct, a video game pops onto the screen as a reward. When one youngster worked his way through the drill, even classroom hecklers were impressed. Said one: "Hey, Old Wentworth's getting better."

More entertaining and demanding are think tank-type strategy games like Geography Search, which launches competing teams on a Columbus-like voyage of exploration. They must make their way across the Atlantic, taking into account currents and winds, finding their longitude and latitude by means of star patterns and the length of a shadow thrown by a stick at high noon (methods that worked for Columbus, after all), and coping with such unforeseen perils as an outbreak of scurvy, an attack by pirates and a tropical storm. Only shrewd planning, wise choices and cooperative action ensure survival. The simulated voyage becomes uncannily real to the participants. Says the game's creator, Thomas Snyder, 31, who heads Computer Learning Connection, Inc., of Cambridge, Mass.: "When they get near the end and the computer finally shows them another ship near by, they act as if they had actually spotted a ship at sea."

Until a few years ago, the few computers available in secondary schools were essentially "dumb" terminals linked by telephone lines to a large, centrally located machine that served a variety of users through an arrangement called time-sharing. All the courseware was stored in the big computer's powerful memory, which could be tapped at will by students and teachers. The most successful example of such a system—and the one still used by Wisconsin's Chippewa Indians—is PLATO (for Programmed Logic for Automatic Teaching Operations). Developed in the 1960s by the University of Illinois and Control Data Corp., PLATO is an exemplary teacher containing more than 8000 hours of courseware, much of it in a continuous curriculum. Thus, if a youngster forgets a point from an earlier lesson, PLATO will search its prodigious memory and patiently recapitulate.

But such time-sharing schemes are extremely expensive, since they require open lines to the central computer. They also can become backed up at peak hours, and do not always lend themselves readily to what is the most intellectually demanding use of the computer: learning how to program it.

For this, the inexpensive, easy-to-operate personal computer, entirely self-contained and relying on equipment immediately at the student's side, is an ideal instrument—much more "user-friendly," as manufacturers like to say, than big machines. Yet even with a handy micro, programming can overwhelm the uninitiated. The programmer and computer must "speak" a common language.

In the early days of the digital computer, this was extremely difficult. The machine reduces all the information it receives, whether it arrives as letters, numbers or graphic symbols, into the simplest possible electronic statements: either a yes or a no, represented by pulses of high or low voltage. To command the machine in its own internal language meant writing out endless strings of one or zeros, called bits and bytes, symbolizing those yes or no statements. But scientists soon began creating alternate languages for communicating with the machines that vaguely resemble everyday speech.

The most popular of these computer tongues is BASIC (for Beginner's All-purpose Symbolic Instruction Code). Developed at Dartmouth by mathematician John Kemeny and his colleague Thomas Kurtz to let even the least mathematically gifted student converse with the university's computers, it is "understood" by virtually all of today's personal computers. To show just how easy the language is, Kemeny offers this extremely simple lesson in programming: tell the computer to find the square roots (*i.e.*, the numbers that, when multiplied by themselves, yield the original numbers) of eleven successive values, say 20 through 30.

The entire operation can be accomplished for a program of just four steps:

```
1 FOR N = 20 TO 30
2 PRINT N, SQR(N)
3 NEXT N
4 END
```

Translated into everyday language, the first line tells the computer to let N stand successively for 20 through 30. The second instructs the machine to print the first value of N (that is, the number 20), compute its square root (SQR) and print out the result. The third tells the computer to go on to each of the succeeding values all the way through 30. Finally, the program tells the computer to call it a day, its job having been done. Even the smallest machine can do such calculations in a flash, compared with the hours of work they might require of human computers.

To preserve their creativity, the students can readily store their programs on magnetic tape or on a small, 45 rpm-size plastic record called a floppy disc—which is not, as some parents believe, a new form of back injury. Then when the occasion arises for using the program again, the computer operator merely loads the instructions back into the machine and punches in some new values for N. The same broad principles apply to the creation of all software, even complex simulations like Geography Search.

Literal-minded brutes that they are, computers do exactly what they are told. No more and no less. But youngsters of even the most tender age are surprising educators by showing they can master the beasts with startling ease. Computer software expert Leona Schauble of the Children's Television Workshop (producers of *Sesame Street*) recalls getting an eight-year-old boy at Manhattan's Little Red School House started on a simple computer game. The game generated an image of a frog that would leap up and catch a butterfly, provided the right buttons were hit. After a few minutes, she checked back and found the frog jumping in slow motion. When she asked the youngster what happened, he replied, "Well, I wanted to make the frog catch more butterflies. So I got a listing of the variables and slowed him down." In other words, the youngster had broken into the game's program and changed it to suit himself.

To instruct very young children, even Kemeny's BASIC is much too mathematical. Instead, more and more schools are turning to an innovative computer language called LOGO (from the Greek word for reason), developed by Seymour Papert and his colleagues at MIT. A mathematician who studied with the Swiss psychologist Jean Piaget, Papert has become something of a guru of the computer generation, predicting that the machines will revolutionize learning by taking much of the mystery out of mathematics, science and technology. Says he: "The computer can make the most abstract things concrete."

With a deceptively simple set of commands, LOGO enables youngsters who know nothing of geometry and algebra, and barely know how to read, to manipulate a triangular figure, dubbed the Turtle, on a computer screen and trace all manner of shapes with it. At the Lamplighter School in Dallas, teachers using LOGO get youngsters of three or four to write simple computer instructions. In one game, they maneuver "cars" and "garages" on the computer screen in such a way that the cars are parked inside the garages. While playing with LOGO, the youngsters learn simple words, the difference between left and right, and geometric concepts that they would not ordinarily encounter until junior high.

The machines crop up in the lives of youngsters even before they enter school—and sometimes before they learn to walk or talk—in the guise of such siliconized gadgetry as Little Professor and Speak & Spell. With a few presses of the button, these computerized games produce flashing lights, squealing sounds and disembodied voices that inculcate the rudiments of spelling and calculating. A record of sorts may have been set by Corey Schou, a computer scientist at the University of Central Florida in Orlando: he rigged up a home computer so his five-month-old daughter could operate it by pressing buttons in her crib and changing the designs on a nearby screen. Says the proud papa: "Basically, it's an electronic kaleidoscope, another diversion, another learning device."

Whatever it is, it prepares youngsters for all those buttons they will encounter soon enough in and out of school. Parents and teachers may shudder at the thought, but it is only a short hop from skilful operation of a video game to learning fundamentals of programming. Says MIT sociologist Sherry Turkle, 33, who has been studying the youthful computer culture for five years: "The line between game playing and programming is very thin. Programming takes what is powerful about games—this articulation of knowledge, this learning about strategy—and carries it to a higher level of power."

By the time the youthful programmers reach the eighth or ninth grade, their skills may reach a marketable level. In Chicago, Jonathan Dubman, 14, and Kay Borzsony, 13, have formed a company called Aristotle Software to sell their own computer games and graphics programs. Says Kay: "The nice thing about the computer business is that there is no real bias against children. In the computer magazines, you read articles by 12- and 13-year-olds." Laura Hyatt, 15, of Ridgewood, helps a stymied local insurance office figure out how to use its software. Says she: "It's better than babysitting." And, at $3.50 an hour, somewhat more profitable.

The prodigy of prodigies may be Eugene Volokh, 14, of Los Angeles. A Russian émigré, he earns $480 each week by doing 24 hours of programming for 20th Century-Fox, while carrying a full load of courses as a junior at UCLA. This year Greg Christensen, 18, of Anaheim, Calif., could make $100,000 in royalties from a video game he developed that was bought by Atari. Other youngsters are waiting at the sidelines in hopes of catching up with these young entrepreneurs. Every Tuesday night, Scott Whitfield, 13, and his brother Shawn, 11, appear at the Menlo Park, Calif., public library to get computer instruction. Says Scott: "We'll probably never get a job if we don't learn how to use a computer."

Not all youngsters take equally to the machines. In a typical computer class, only about one in five students becomes seriously involved. Says Steven Scott, 16, of Wausau's West High: "Either you get the hang of it or you don't." Even so dedicated a computernik as Ridgewood's Nick Newman finds programming interesting only for a purpose. His own goal is to apply his computer knowledge to a career in science or medicine.

Whatever these youngsters make of their computer experiences, they will surely confront the world differently from their parents. The precise, orderly steps of logic required to use and program the machines promise to shape—and sharpen—the thought processes of the computer generation. Indeed, the youngsters playing all those strategy games are doing precisely what corporations do when they plan to launch a new product, or what military leaders do when they devise strategies to confront a potential foe.

Whether such abilities will change the world for the better is another matter. Princeton psychologist George Miller, for one, has doubts that "a few years of thinking like a computer can change patterns of irrational thought that have persisted throughout recorded history." Other social critics ask if clear thinking is enough—if, in fact, there might not be a danger in raising a generation to believe that it has the analytical tools to contemplate any problem. Says MIT computer science professor Joseph Weizenbaum: "There's a whole world of real problems, of human problems, which is essentially being ignored." It is still impossible, after all, to reduce a human relationship to a printout or to solve a moral question by bits and bytes.

Some critics predict a future not unlike that portrayed in Isaac Asimov's *I, Robot*, a science fiction novel set in a society so thoroughly computer-dominated that the people cannot do arithmetic. Humanist critic George Steiner acerbically calls the computer generation the advance guard of a breed of "computer-mutants." Says Steiner: "They will be out of touch with certain springs of human identity and creativity, which belong to the full use of language rather than mathematical and symbolical codes."

Many others are much more sanguine. University of Chicago philosopher of science Stephen Toulmin predicts that computers will "re-intellectualize" the television generation. "TV relieved people of the necessity to do anything," says Toulmin. "Computers depend on what you do yourself." Catholic theologian David Tracy argues that "using computers sharpens the mind's ability to deal with our world: the world of technology."

The final word may be simpler, and not pronounced by elders who find a devilish soul in the new machine. More so than adults, the young know the computer for what it is. Says a 10-year-old at Manhattan's Bank Street School: "It's dumb. I have to tell it everything." They also know something important is afoot. Says Shawn Whitfield: "When I grow up it's going to be the Computer Age. It won't affect parents. They're out of the Computer Age. They had their own age."

Computers for Children

Seymour Papert

Inventor of the computer language, LOGO, and follower of Piaget, Seymour Papert is Professor of Mathematics and of Education at MIT. In this clear statement of his philosophy, he argues that it is possible to design computers so that learning to communicate with them becomes a natural process. And learning to use computers can change the way children learn everything else. Taken from the introduction to Mindstorms: Children, Computers and Powerful Ideas *(Basic Books, New York, 1980 and Harvester Press, Brighton, England, 1980).*

Just a few years ago people thought of computers as expensive and exotic devices. Their commercial and industrial uses affected ordinary people, but hardly anyone expected computers to become part of day-to-day life. This view has changed dramatically and rapidly as the public has come to accept the reality of the personal computer, small and inexpensive enough to take its place in every living room or even in every breast pocket. The appearance of the first rather primitive machines in this class was enough to catch the imagination of journalists and produce a rash of speculative articles about life in the computer-rich world to come. The main subject of these articles was what people will be able to do with their computers. Most writers emphasized using computers for games, entertainment, income tax, electronic mail, shopping, and banking. A few talked about the computer as a teaching machine.

This book too poses the question of what will be done with personal computers, but in a very different way. I shall be talking about how computers may affect the way people think and learn. I begin to characterize my perspective by noting a distinction between two ways computers might enhance thinking and change patterns of access to knowledge.

Instrumental uses of the computer to help people think have been dramatized in science fiction. For example, as millions of *Star Trek* fans know, the starship *Enterprise* has a computer that gives rapid and accurate answers to complex questions posed to it. But no attempt is made in *Star Trek* to suggest that the human characters aboard think in ways very different from the manner in which people in the twentieth century think. Contact with the computer has not, as far as we are allowed to see in these episodes, changed how these people think about themselves or how they approach problems. In this book I discuss ways in which the computer presence could contribute to mental processes not only instrumentally but in more essential, conceptual ways, influencing how people think even when they are far removed from physical contact with a computer (just as the gears shaped my understanding of algebra although they were not physically present in the math class). It is about an end to the culture that makes science and technology alien to the vast majority of people.

Many cultural barriers impede children from making scientific knowledge their own. Among these barriers the most visible are the physically brutal effects of deprivation and isolation. Other barriers are more political. Many children who grow up in our cities are surrounded by the artifacts of science but have good reason to see them as belonging to "the others"; in many cases they are perceived as belonging to the social enemy. Still other obstacles are more abstract, though ultimately of the same nature. Most branches of the most sophisticated modern culture of Europe and the US are so deeply "mathophobic" that many privileged children are as effectively (if more gently) kept from appropriating science as their own. In my vision, space-age objects, in the form of small computers, will cross these cultural barriers to enter the private worlds of children everywhere. They will do so not as mere physical objects. This book is about how computers can be carriers of powerful ideas and of the seeds of cultural change, how they can help people form new relationships with knowledge that cut across the traditional lines separating humanities from sciences and knowledge of the self from both of these. It is about using computers to challenge current beliefs about who can understand what and at what age. It is about using computers to question standard assumptions in developmental psychology and in the psychology of aptitudes and attitudes. It is about whether personal computers and the cultures in which they are used will continue to be the creatures of "engineers" alone or whether we can construct intellectual environments in which people who today think of themselves as "humanists" will feel part of, not alienated from, the process of constructing computational cultures.

But there is a world of difference between what computers can do and what society will choose to do with them. Society has many ways to resist fundamental and threatening change. Thus, this book is about facing choices that are ultimately political. It looks at some of the forces of change and of reaction to those forces that are called into play as the computer presence begins to enter the politically charged world of education.

Much of the book is devoted to building up images of the role of the computer very different from current stereotypes. All of us, professionals as well as laymen, must consciously break the habits we bring to thinking about the computer. Computation is in its infancy. It is hard to think about computers of the future without projecting onto them the properties and the limitations of those we think we know today. And nowhere is this more true than in imagining how computers can enter the world of education. It is not true to say that the image of a child's relationship with a computer that I shall develop here goes far beyond what is common in today's schools. My image does not go beyond: it goes in the opposite direction.

In many schools today, the phrase "computer-aided instruction" means making the computer teach the child. One might say the *computer is being used to program* the child. In my vision, *the child programs the computer* and, in doing so, both acquires a sense of mastery over a piece of the most modern and powerful technology and establishes an intimate contact with some of the deepest ideas from science, from mathematics, and from the art of intellectual model-building.

I shall describe learning paths that have led hundreds of children to becoming quite sophisticated programmers. Once programming is seen in the proper perspective, there is nothing very surprising about the fact that this should happen. Programming a computer means nothing more or less than communicating to it in a language that it and the human user can both "understand." And learning languages is one of the things children do best. Every normal child learns to talk. Why then should a child not learn to "talk" to a computer?

There are many reasons why someone might expect it to be difficult. For example, although babies learn to speak their native language with spectacular ease, most children have great difficulty learning foreign languages in schools and, indeed, often learn the written version of their own language none too successfully. Isn't learning a computer language more like the difficult process of learning a foreign written language than the easy one of learning to speak one's own language? And isn't the problem further compounded by all the difficulties most people encounter learning mathematics?

Two fundamental ideas run through this book. The first is that it is possible to design computers so that learning to communicate with them can be a natural process, more like learning French by living in France than like trying to learn it through the unnatural process of American foreign-language instruction in classrooms. Second, learning to communicate with a computer may change the way other learning takes place. The computer can be a mathematics-speaking and an alphabetic-speaking entity. We are learning how to make computers with which children love to communicate. When this communication occurs, children learn mathematics as a living language. Moreover, mathematical communication and alphabetic communication are thereby both transformed from the alien and therefore difficult things they are for most children into natural and therefore easy ones. The idea of "talking mathematics" to a computer can be generalized to a view of learning mathematics in "Mathland"; that is to say, in a context which is to learning mathematics what living in France is to learning French.

In this book the Mathland metaphor will be used to question deeply engrained assumptions about human abilities. It is generally assumed that children cannot learn formal geometry until well into their school years and that most cannot learn it too well even then. But we can quickly see that these assumptions are based on extremely weak evidence by asking analogous questions about the ability of children to learn French. If we had to base our opinions on observation of how poorly children learned French in American schools, we would have to conclude that most people were incapable of mastering it. But we know that all normal children would learn it very easily if they lived in France. My conjecture is that much of what we now see as too "formal" or "too mathematical" will be learned just as easily when children grow up in the computer-rich world of the very near future.

I use the examination of our relationship with mathematics as a thematic example of how technological and social processes interact in the construction of ideas about human capacities. And mathematical examples will also help to describe a theory of how learning works and of how it goes wrong.

I take from Jean Piaget[1] a model of children as builders of their own intellectual structures. Children seem to be innately gifted learners, acquiring long before they go to school a vast quantity of knowledge by a process I call "Piagetian learning," or "learning without being taught." For example, children learn to speak, learn the intuitive geometry needed to get around in space, and learn enough of logic and rhetorics to get around parents—all this without being "taught". We must ask why some learning takes place so early and spontaneously while some is delayed many years or does not happen at all without deliberately imposed formal instruction.

If we really look at the "child as builder" we are on our way to an answer. All builders need materials to build with. Where I am at variance with Piaget is in the role I attribute to the surrounding cultures as a source of these materials. In some cases the culture supplies them in abundance, thus facilitating constructive Piagetian learning. For example, the fact that so many important things (knives and forks, mothers and fathers, shoes and socks) come in pairs is a "material" for the construction of an intuitive sense of number. But in many cases where Piaget would explain the slower development of a particular concept by its greater complexity or formality, I see the critical factor as the relative poverty of the culture in those materials that would make the concept simple and concrete. In yet other cases the culture may provide materials but block their use. In the case of formal mathematics, there is both a shortage of formal materials and a cultural block. The mathophobia endemic in contemporary culture blocks many people from learning anything they recognize as "math," although they may have no trouble with mathematical knowledge they do not perceive as such.

We shall see again and again that the consequences of mathophobia go far beyond obstructing the learning of mathematics and science. They interact with other endemic "cultural toxins," for example, with popular theories of aptitudes, to contaminate people's images of themselves as learners. Difficulty with school math is often the first step of an invasive intellectual process that leads us all to define ourselves as bundles of aptitudes and ineptitudes, as being "mathematical" or "not mathematical," "artistic" or "not artistic," "musical" or "not musical," "profound" or "superficial," "intelligent" or "dumb." Thus, deficiency becomes identity and learning is transformed from the early child's free exploration of the world to a chore beset by insecurities and self-imposed restrictions.

Two major themes—that children can learn to use computers in a masterful way, and that learning to use computers can change the way they learn everything else—have shaped my research agenda on computers and education. Over the past ten years I have had the good fortune to work with a group of colleagues and students at MIT (the LOGO[2] group in the Artificial Intelligence Laboratory) to create environments in which children can learn to communicate with computers. The metaphor of imitating the way the child learns to talk has been constantly with us in this work and has led to a vision of education and of education research very different from the traditional ones. For people in the teaching professions, the word "education" tends to evoke "teaching," particularly classroom teaching. The goal of education research tends therefore to be focused on how to improve classroom teaching. But if, as I have stressed here, the model of successful learning is the way a child learns to talk, a process that takes place without deliberate and organized teaching, the goal set is very different. I see the classroom as an artificial and inefficient learning environment that society has been forced to invent because its informal environments fail in certain essential learning domains, such as writing or grammar or school math. I believe that the computer presence will enable us to so modify the learning environment outside the classrooms that much if not all the knowledge that schools presently try to teach with such pain and expense and such limited success will be learned, as the child learns to talk, painlessly, successfully, and without organized instruction. This obviously implies that schools as we know them today will have no place in the future. But it is an open question whether they will adapt by transforming themselves into something new or will wither away and be replaced.

Although technology will play an essential role in the realization of my vision of the future of education, my central focus is not on the machine but on the mind, and particularly on the way in which intellectual movements and cultures define themselves and grow. Indeed, the role I give to the computer is that of a *carrier* of cultural "germs" or "seeds" whose intellectual products will not need technological support once they take root in an actively growing mind. Many if not all the children who grow up with a love and aptitude for mathematics owe this feeling, at least in part, to the fact that they happened to acquire "germs" of the "math culture" from adults, who, one might say, knew how to speak mathematics, even if only in the way that

Moliere had M. Jourdain speak prose without knowing it. These "math-speaking" adults do not necessarily know how to solve equations; rather, they are marked by a turn of mind that shows up in the logic of their arguments and in the fact that for them to play is often to play with such things as puzzles, puns, and paradoxes. Those children who prove recalcitrant to math and science education include many whose environments happened to be relatively poor in math-speaking adults. Such children come to school lacking elements necessary for the easy learning of school math. School has been unable to supply these missing elements, and, by forcing the children into learning situations doomed in advance, it generates powerful negative feelings about mathematics and perhaps about learning in general. Thus is set up a vicious self-perpetuating cycle. For these same children will one day be parents and will not only fail to pass on mathematical germs but will almost certainly infect their children with the opposing and intellectually destructive germs of mathophobia.

Fortunately, it is sufficient to break the self-perpetuating cycle at one point for it to remain broken forever. I shall show how computers might enable us to do this, thereby breaking the vicious cycle without creating a dependence on machines. My discussion differs from most arguments about "nature versus nurture" in two ways. I shall be much more specific both about what kinds of nurturance are needed for intellectual growth and about what can be done to create such nurturance in the home as well as in the wider social context.

Thus this book is really about how a culture, a way of thinking, an idea comes to inhabit a young mind. I am suspicious of thinking about such problems too abstractly, and I shall write here with particular restricted focus. I shall in fact concentrate on those ways of thinking that I know best. I begin by looking at what I know about my own development. I do this in all humility, without any implication that what I have become is what everyone should become. But I think that the best way to understand learning is first to understand specific, well-chosen cases and then to worry afterward about how to generalize from this understanding. You can't think seriously about thinking without thinking about thinking about something. And the something I know best how to think about is mathematics. When in this book I write of mathematics, I do not think of myself as writing for an audience of mathematicians interested in mathematical thinking for its own sake. My interest is in universal issues of how people think and how they learn to think.

When I trace how I came to be a mathematician, I see much that was idiosyncratic, much that could not be duplicated as part of a generalized vision of education reform. And I certainly don't think that we would want everyone to become a mathematician. But I think that the kind of pleasure I take in mathematics should be part of a general vision of what education should be about. If we can grasp the essence of one person's experiences, we may be able to replicate its consequences in other ways, and in particular this consequence of finding beauty in abstract things. And so I shall be writing quite a bit about mathematics. I give my apologies to readers who hate mathematics, but I couple that apology with an offer to help them learn to like it a little better—or at least to change their image of what "speaking mathematics" can be all about.

In the Foreword of this book I described how gears helped mathematical ideas to enter my life. Several qualities contributed to their effectiveness. First, they were part of my natural "landscape," embedded in the culture around me. This made it possible for me to find them myself and relate to them in my own fashion. Second, gears were part of the world of adults around me and through them I could relate to these people. Third, I could use my body to think about the gears. I could feel how gears turn by imagining my body turning. This made it possible for me to draw on my "body knowledge" to think about gear systems. And finally, because, in a very real sense, the relationship between gears contains a great deal of mathematical information, I could use the gears to think about formal systems. I have described the way in which the gears served as an "object-to-think-with." I made them that for myself in my own development as a mathematician. The gears have also served me as an object-to-think-with in my work as an educational researcher. My goal has been the design of other objects that children can make theirs for themselves and in their own ways. Much of this book will describe my path through this kind of research. I begin by describing one example of a constructed computational "object-to-think-with." This is the "Turtle."[3]

The central role of the Turtle in this book should not be taken to mean that I propose it as a panacea for all educational problems. I see it as a valuable educational object, but its principal role here is to serve as a model for other objects, yet to be invented. My interest is in the process of invention of "objects-to-think-with," objects in which there is an intersection of cultural presence, embedded knowledge, and the possibility for personal identification.

The Turtle is a computer-controlled cybernetic animal. It exists within the cognitive minicultures of the "LOGO environment," LOGO being the computer language in which communication with the Turtle takes place. The Turtle serves no other purpose than of being good to program and good to think with. Some Turtles are abstract objects that live on computer screens. Others, like the floor Turtles shown in the frontispiece, are physical objects that can be picked up like any mechanical toy. A first encounter often begins by showing the child how a Turtle can be made to move by typing commands at a keyboard. FORWARD 100 makes the Turtle move in a straight line a distance of 100 Turtle steps of about a millimeter each. Typing RIGHT 90 causes the Turtle to pivot in place through 90 degrees. Typing PENDOWN causes the Turtle to lower a pen so as to leave a visible trace of its path, while PENUP instructs it to raise the pen. Of course, the child needs to explore a great deal before gaining mastery of what the numbers mean. But the task is engaging enough to carry most children through this learning process.

The idea of programming is introduced through the metaphor of teaching the Turtle a new word. This is simply done, and children often begin their programming experience by programming the Turtle to respond to new commands invented by the child such as SQUARE or TRIANGLE or SQ or TRI or whatever the child wishes, by drawing the appropriate shapes. New commands, once defined, can be used to define others. For example, just as the house in figure 5.1 is built out of a triangle and a square, the program for drawing it is built out of the commands for drawing a square and a triangle. The figure shows four steps in the evolution of this program. From these simple drawings the young programmer can go on in many different directions. Some work on more complex drawings, either figural or abstract. Some abandon the use of the Turtle as a drawing instrument and learn to use its touch sensors to program it to seek out or avoid objects.[4] Later children learn that the computer can be programmed to make music as well as move Turtles and combine the two activities by programming Turtles to dance. Or they can move on from floor Turtles to "screen Turtles," which they program to draw moving pictures in bright colors. The examples are infinitely varied, but in each the child is learning how to exercise control over an exceptionally rich and sophisticated "micro-world."

Readers who have never seen an interactive computer display might find it hard to imagine where this can lead. As a mental exercise they might like to imagine an electronic sketchpad, a computer graphics display of the not-too-distant future. This is a television screen that can display moving pictures in color. You can also "draw" on it, giving it instructions, perhaps by typing, perhaps by speaking, or perhaps by pointing with a wand. On request, a palette of colors could appear on the screen. You can choose a color by pointing at it with the wand. Until you change your choice, the wand draws in that color. Up to this point the distinction from traditional art materials may seem slight, but the distinction becomes very real when you begin to think about editing the drawing. You can "talk to your drawing" in computer

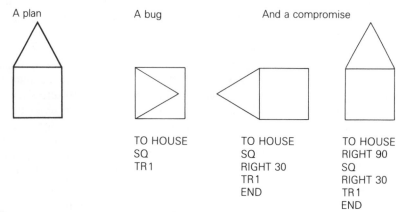

Figure 5.1 Drawing a house with the Turtle.

language. You can "tell" it to replace this color with that. Or set a drawing in motion. Or make two copies and set them in counter-rotating motion. Or replace the color palette with a sound palette and "draw" a piece of music. You can file your work in computer memory and retrieve it at your pleasure, or have it delivered into the memory of any of the many millions of other computers linked to the central communication network for the pleasure of your friends.

That all this would be fun needs no argument. But it is more than fun. Very powerful kinds of learning are taking place. Children working with an electronic sketchpad are learning a language for talking about shapes and fluxes of shapes, about velocities and rates of change, about processes and procedures. They are learning to speak mathematics, and are acquiring a new image of themselves as mathematicians.

In my description of children working with Turtles, I implied that children can learn to program. For some readers this might be tantamount to the suspension of disbelief required when we enter a theater to watch a play. For them, programming is a complex and marketable skill acquired by some mathematically gifted adults. But my experience is very different. I have seen hundreds of elementary schoolchildren learn very easily to program, and evidence is accumulating to indicate that much younger children could do so as well. The children in these studies are not exceptional; or, rather, they are exceptional in every conceivable way. Some of the children were highly successful in school, some were diagnosed as emotionally or cognitively disabled. Some of the children were so severely afflicted by cerebral palsy that they had never purposefully manipulated physical objects. Some of them had expressed their talents in "mathematical" forms, some in "verbal" forms, and some in artistically "visual" or in "musical" forms.

Of course, these children did not achieve a fluency in programming that came close to matching their use of spoken language. If we take the Mathland metaphor seriously, their computer experience was more like learning French by spending a week or two on vacation in France than like living there. But like children who have spent a vacation with foreign-speaking cousins, they were clearly on their way to "speaking computer."

When I have thought about what these studies mean I am left with two clear impressions. First, that all children will, under the right conditions, acquire a proficiency with programming that will make it one of their more advanced intellectual accomplishments. Second, that the "right conditions" are very different from the kind of access to computers that is now becoming established as the norm in schools. The conditions necessary for the kind of relationships with a computer that I will be writing about in this book require more and freer access to the computer than educational planners currently anticipate. And they require a kind of computer language and a learning environment around that language very different from those the schools are now providing. They even require a kind of computer rather different from those that the schools are currently buying.

It will take most of this book for me to convey some sense of the choices among computers, computer languages, and, more generally, among computer cultures, that influence how well children will learn from working with computation and what benefits they will get from doing so. But the question of the *economic* feasibility of free access to computers for every child can be dealt with immediately. In doing so I hope to remove any doubts readers may have about the "economic realism" of the "vision of education" I have been talking about.

My vision of a new kind of learning environment demands free contact between children and computers. This could happen because the child's family buys one or a child's friends have one. For purposes of discussion here (and to extend our discussion to all social groups) let us assume that it happens because schools give every one of their students his or her own powerful personal computer. Most "practical" people (including parents, teachers, school principals, and foundation administrators) react to this idea in much the same way: "Even if computers could have all the effects you talk about, it would still be impossible to put your ideas into action. Where would the money come from?"

What these people are saying needs to be faced squarely. They are wrong. Let's consider the cohort of children who will enter kindergarten in the year 1987, the "Class of 2000," and let's do some arithmetic. The direct public cost of schooling a child for thirteen years, from kindergarten through twelfth grade, is over $20,000 today (and for the class of 2000, it may be closer to $30,000). A conservatively high estimate of the cost of supplying each of these children with a personal computer with enough power for it to serve the kinds of educational ends described in this book, and of upgrading, repairing, and replacing it when necessary would be about $1000 per student, distributed over 13 years in school. Thus, "computer costs" for the class of 2000 would represent only about 5 percent of the total public expenditure on education, and this would be the case even if nothing else in the structure of educational costs changed because of the computer presence. But in fact computers in education stand a good chance of making other aspects of education cheaper. Schools might be able to reduce their cycle from 13 years to 12 years; they might be able to take advantage of the greater autonomy the computer gives students and increase the size of classes by one or two students without decreasing the personal attention each student is given. Either of these two moves would "recuperate" the computer cost.

My goal is not educational economies: it is not to use computation to shave a year off the time a child spends in an otherwise unchanged school or to push an extra child into an elementary school classroom. The point of this little exercise in educational "budget balancing" is to do something to the state of mind of my readers as they turn to the first chapter of this book. I have described myself as an educational utopian—not because I have projected a future of education in which children are surrounded by high technology, but because I believe that certain uses of very powerful computational technology and computational ideas can provide children with new possibilities for learning, thinking, and growing emotionally as well as cognitively. In the chapters that follow I shall try to give you some idea of these possibilities, many of which are dependent on a computer-rich future, a future where a computer will be a significant part of every child's life. But I want my readers to be very clear that what is "utopian" in my vision and in this book is a particular way of using computers, of forging new relationships between computers and people—that the computer will be there to be used is simply a conservative premise.

Notes

1 Piaget is at the center of the concerns of this book. I make a slightly unorthodox interpretation of his theoretical position and a very unorthodox interpretation of the implications of his theory for education. The reader who would like to return to the source needs some guidance because Piaget has written a large number of books, most of which discuss particular aspects of children's development, assuming that the others have been read as a theoretical preface. The best short book about Piaget is M. Boden's *Piaget* (London: Harvester Press, 1979). A good starting place for reading Piaget's own texts is with H. E. Gruber and J. J. Voneche (eds), *The Essential Piaget: An Interpretive Reference and Guide* (New York: Basic Books, 1977). My own "short list" of books by Piaget that are most readable and provide the best philosophical overview of his ideas are: *The Child's Conception of the World* (New York: Harcourt, Brace, 1929); *The Child's Conception of Physical Causality* (New York: Harcourt, Brace, 1932); *The Psychology of Intelligence*, trans. Malcolm Piercy and D. E. Berlyn (New York: Harcourt, Brace, 1950); *The Origins of Intelligence in Children,* trans. Margaret Cook (London: Routledge and Kegan Paul); *Introduction à l'epistémologie génétique* (Paris: Presses Universitaires de France, 1950); *Insights and Illusions in Philosophy,* trans. Wolfe Mays (New York: World Publishing Co., 1971); *The Grasp of Consciousness,* trans. Susan Wedgwood (Cambridge: Harvard University Press, 1976). For a critique of the "Piaget Curriculum Developers," of whom I have said that they are "standing Piaget on his head," see G. Groen, "The Theoretical Ideas of Piaget and Educational Practice," *The Impact of Research on Education,* ed. P. Suppes (Washington DC: The National Academy of Education, 1978).

2 LOGO is the name of a philosophy of education in a growing family of computer languages that goes with it. Characteristic features of the LOGO family of languages include procedural definitions with local variables to permit recursion. Thus, in LOGO it is possible to define new commands and functions which then can be used exactly like primitive ones. LOGO is an interpretive language. This means that it can be used interactively. The modern LOGO systems have full list structure; that is to say, the language can operate on lists whose members can themselves be lists, lists of lists, and so forth.

Some versions have elements of parallel processing and of message passing in order to facilitate graphics programming. An example of a powerful use of list structure is the representation of LOGO procedures themselves as lists of lists so that LOGO procedures can construct, modify, and run other LOGO procedures. Thus LOGO is not a "toy," a language only for children. The examples of simple uses of LOGO in this book do however illustrate some ways in which LOGO is special in that it is designed to provide very early and easy entry routes into programming for beginners with no prior mathematical knowledge. The subset of LOGO containing Turtle commands, the most used "entry route" for beginners, is referred to in this book as "TURTLE TALK" to take account of the fact that other computer languages, for example SMALLTALK and PASCAL, have implemented Turtles on their systems using commands originally developed in the LOGO language. The TURTLE TALK subset of LOGO is easily transportable to other languages.

It should be carefully remembered that LOGO is never conceived as a final product or offered as "the definitive language." Here I present it as a sample to show that something better is possible.

Precisely because LOGO is not a toy, but a powerful computer language, it requires considerably larger memory than less powerful languages such as BASIC. This has meant that until recently LOGO was only to be implemented on relatively large computers. With the lowering cost of memory this situation is rapidly changing. As this book goes to press, prototypes of LOGO systems are running on a 48K Apple II system and on a TI 99/4 with extended memory. Readers who would like to be kept informed of the status of LOGO implementations can write to me at LOGO project, MIT Artificial Intelligence Laboratory, 545 Technology Square, Cambridge, Mass. 02139. See S. Papert et al., *LOGO: A Language For Learning* (Morristown, NJ: Creative Computing Press, 1981).

3 The history of the Turtle in the LOGO project is as follows. In 1968–9, the first class of twelve "average" seventh-grade students at the Muzzy Junior High School in Lexington, Massachusetts, worked with LOGO through the whole school year in place of their normal mathematics curriculum. At that time the LOGO system had no graphics. The students wrote programs that could translate English to "Pig Latin," programs that could play games of strategy, and programs to generate concrete poetry. This was the first confirmation that LOGO was a learnable language for computer "novices." However, I wanted to see the demonstration extended to fifth-graders, third-graders, and ultimately to preschool children. It seemed obvious that, even if the LOGO language was learnable at these ages, the programming topics would not be. I proposed the Turtle as a programming domain that could be interesting to people at all ages. This expectation has subsequently been borne out by experience, and the Turtle as a learning device has been widely adopted. Pioneer work in using the Turtle to teach very young children was done by Radia Perlman who demonstrated, while she was a student at MIT, that four-year-old children could learn to control mechanical Turtles. Cynthia Solomon used screen Turtles in the first demonstration that first graders could learn to program. At the other end of the age spectrum, it is encouraging to see that Turtle programming is being used at a college level to teach PASCAL. See Kenneth L. Bowles, *Problem Solving Using PASCAL* (New York: Springer-Verlag, 1977). Controlling Turtles has proven to be an engaging activity for retarded children, for autistic children, and for children with a variety of "learning disorders." See for example, Paul Goldenberg, *Special Technology for Special Children* (Baltimore: University Park Press, 1979). Turtles have been incorporated into the SMALLTALK computer system at the Xerox Palo Alto Research Center. See Alan Kay and Adele Goldberg, "Personal Dynamic Media" (Palo Alto, Calif.: Xerox, Palo Alto Research Center, 1976).

4 *Touch Sensor Turtle.* The simplest touch sensor program in LOGO is as follows:

```
TO BOUNCE                Comments
REPEAT                   This means repeat all the individual steps
  FORWARD 1              The turtle keeps moving
  TEST FRONT TOUCH       It checks whether it has run into something
  IF TRUE RIGHT 180      If so, it does an about-turn
END
```

This will make the Turtle turn about when it encounters an object. A more subtle and more instructive program using the Touch Sensor Turtle is as follows:

```
TO FOLLOW
REPEAT                   Comments
  FORWARD 1
  TEST LEFT TOUCH        Check: Is it touching?
  IF TRUE RIGHT 1        It thinks it's too
                         close and turns away
  IF FALSE LEFT 1        It thinks it might
                         lose the object so it turns toward

END
```

This program will cause the Turtle to circumnavigate an object of any shape, provided that it starts with its left side in contact with the object (and provided that the object and any irregularities in its contour are large compared to the Turtle).

It is a very instructive project for a group of students to develop this (or an equivalent) program from first principles by acting out how they would use touch to get around an object and by translating their strategies into Turtle commands.

Educational and Social Implications

Christopher Dede

Education technologists have cried wolf many times in the past. But now developments in computer technology really do mean that big changes in the education system are underway. The author, who is Professor of Education and Futures Research at the University of Houston, presents a cogent analysis of the many consequences for society, both intended and unintended. This article first appeared in Programmed Learning and Educational Technology, *November 1981.*

In this issue of *Programmed Learning & Educational Technology*, a variety of new technologies is discussed. If applied to education, these devices may reshape both the delivery systems used to convey instruction and the subject matter of the traditional curriculum. Potential consumers of new educational technologies include very young children, students at every level of formal education, recipients of industrial or professional training, the aged, adults engaged in nonformal learning activities—in short, virtually everyone in the society. Thus, the size, method of operation and content of education may alter dramatically.

Such a major shift would create numerous educational, social and ethical consequences: some intended, some not. Research in the field of technology assessment has indicated that the unintended, second-order effects of a technological innovation on society are frequently more influential, long-term, than its direct and deliberate effects. For example, in many crowded metropolitan areas one can travel by car no faster than by horse—the greater speed of the automobile has been lost through congestion—but automotive pollution and petroleum availability remain as major societal concerns. Thus, attempting to anticipate all the likely implications of a technological advance may be well worth doing, as seemingly minor changes in the method of implementing a technology may have major long-term consequences, and some technologies may cause such high eventual costs to society that the short-term benefits they offer are not worth acquiring.

An overview of some illustrative potential implications of the new educational technologies for society is given below (a complete discussion would require far more space). These effects of technological innovation can be divided into two categories: outcomes likely regardless of method of implementation, and outcomes highly contingent on the particular implementation strategy. The former set of effects can be used to calculate the overall costs/benefits of using a given technology; the latter analysis can help to determine the best approach for introducing the technology into educational practice.

Effects likely regardless of implementation strategy

Futures researchers tend to use "short term" and "long-term" on a more sweeping scale than generally practised. For the purposes of this discussion, short-term will refer to effects within a three-year to ten-year timescale; long-term will indicate consequences that may take several decades fully to appear.

Short-term effects
One important outcome of using the new educational technologies will be that a larger proportion of the society will have access to instruction. This expanded clientele for education will be interested primarily in informal activities; the learning experience will not necessarily involve credentialing, nor will formal schooling be necessary. Instead, a variety of people, in their homes or places of employment, will be able to interact with some device capable of improving the quality of their work, personal interactions or leisure activities. Barriers of inconvenience (in schedule or distance), cost and total time commitment will be reduced; the diversity of possible offerings will be expanded.

The number of people who ultimately opt to participate in some form of schooling will probably slowly rise, as nonformal instruction whets the interest of some learners toward greater depth and sophistication. Even when equivalent learning experiences are available through technological means, many who now attend schools will choose to remain in the classroom, as access to human interaction is one of the predominant factors underlying their decision to attend. Thus, these technological innovations will not, on balance, compete with existing educational institutions, but rather will open up new markets and increase the base of interest on which the schools draw. In turn, society will benefit from the greater general knowledge of its citizenry.

A second short-term consequence of implementing the new information technologies in education will be high initial capital investments in development and delivery systems, followed by an overall reduction in the daily operating costs of formal schooling. These savings will ensue from the substitution of machines for human effort, moving schools from their present labour-intensive position to a more capital-intensive stance. In the classroom, this will be accomplished partly by higher student–teacher ratios, with machines assuming some of the burden of instruction, and partly by extra-school use of instructional technology with a reduction in length of the school day.

Over the past two centuries, and especially during periods of inflation, technological innovation has proven economically profitable for other sectors of society because:

1 the rate of increase productivity is higher for machines than for people (in computer-assisted instruction; for example, productivity is increasing about 10 percent per year);

2 the cost of people rises more rapidly than the cost of machines—especially devices that are not energy-intensive (e.g., costs of micro-computers are dropping rapidly, while teacher salaries are rising);

3 the availability of a technological delivery system can create new markets (as discussed above); and

4 centralized and standardized production of materials offers certain economics of scale (as described below).

The magnitude of these financial savings will be dependent on how large a total clientele is attracted and what mix of machine to human interaction is selected, but substantial cost reductions are likely in the schools with almost any implementation strategy.

The usual objection to this scenario of fiscal savings is that students must have the constant supervision of an adult to learn. It is argued that students will run riot if left largely unsupervised with machines as their major learning resource. The school is seen as a detention center and teachers as babysitters; instructional technology is perceived as fundamentally uninteresting, requiring the "stick and carrot" as motivational supplements.

In fact, properly done and applied to appropriate content, teaching devices can be more interesting and motivating than the typical teacher (who is overworked, underpaid and responsible for far too many students simultaneously). If learning is set within a context of "play," students will need no more supervision than they do for recreational activities: young children will require some amount of adult attention, older students relatively little. Some learners—either because of emotional problems or idiosyncratic learning styles—will require a traditional classroom context, but most students could readily and profitably adapt to a mixture of machine and human instruction.

To raise the capital necessary for initiating the large-scale usage of technology in schools and delivering instruction to extra-school markets, a variety of innovative funding strategies will be needed. "School bonds" may take on a whole new purpose—not to construct new buildings, but to purchase computers or satellite-receiving antennas. Information technology corporations may become a source of long-term, low-interest loans. Industries may engage in "partnerships for productivity" with educational institutions, providing funding in exchange for help with work-related training.

A third short-term implication of the new educational technologies will be that, to realize these cost reductions, large numbers of instructional devices must be sold and curricula will have to be centrally produced. The "learning curve" for integrated circuit production is such that costs drop very rapidly as the number of units produced increases. Specialized devices requiring only a limited number of units are thus inordinately expensive; general-purpose equipment which can be sold nationwide is remarkably cheap.

The economics of scale for curricular and instructional materials are even more extreme. Software is very costly to produce (especially sophisticated television material or artificial intelligence-based computer programs), yet incredibly inexpensive to use and reuse. Rapidly outdated or highly specialized curricula are thus relatively impractical, unless the projected market base is huge; however, compared with individual human teaching, preparation of high-quality "perennial" broad-based subject matter is quite cost-efficient.

These economic characteristics have a number of implications for education and for society. With centralized production of standard software, inequalities from variations in regional curricula will be reduced and instruction will be, on balance, of higher quality. However, local control over educational content will be eroded, and the diversity of approaches an individual teacher can take towards a subject will be circumscribed. Only the most advanced—and expensive—technological instruction styles (artificial intelligence approaches, videodisk with microcomputer-controlled branching) have intrinsic user flexibility and adaptability. Thus, while developing a variety of instructional strategies to match the diversity of human learning styles will be technically quite feasible, from an economic standpoint the temptation will be to have only one software package.

Moreover, the need for rigorous quality control to eliminate error and propaganda will be much greater than at present. When a significant percentage of the nation's students obtain their information and thinking skills from the same, rigidly programmed instructional package, any bias introduced—deliberately or accidentally—will require a great deal of effort collectively to unlearn. Further, revision of content will be relatively expensive for filmed material (although not for computer software) and altering the instructional approach used will require redoing the unit at enormous cost. Therefore, updating materials or incorporating new ideas will be difficult (as it is now, with human teachers, but for different reasons).

Similar problems exist with instructional hardware. Given the need for economics of scale, the range of economically feasible designs is limited. Adding innovations to whatever devices have been distributed will be difficult. The spectrum of teacher options will be reduced (although in practice, with large class sizes, legislated requirements and regulations, and prescribed textbooks, teachers have few options anyway).

In brief, the educational quality of the device (or instructional unit) and the profit margin of the manufacturer will be inversely related. Great care will need to be taken in reaching some moderate policy so that, on the one hand, instructional technology vendors do not abandon the market and, on the other hand, quality of instruction is not sacrificed. Some public-interest agency may be required to establish standards that can accomplish this outcome.

A fourth short-term consequence of the new instructional technologies will be the necessity for massive changes in both pre-service and in-service teacher training. Educators must be prepared so that they can do low-level maintenance and programming of these new devices (extensive high-level programming by individual teachers, while somewhat desirable from an educational point of view, vitiates any economic benefits of the technology). The diagnosis, evaluation, and remediation of student performance will be very different for these new types of instruction and will require extensive retraining of current teachers. Educators will need to have an accurate sense of the strengths and weaknesses of these technologies and must know when to refuse to substitute machine instruction if human interaction is more appropriate.

The teacher role will become less boring, as computers take over many "busy work" chores of record-keeping and information management, but will demand more creativity and initiative (in addition to some degree of competence and comfort with machines). As a result, the personality types attracted to teaching, the financial remuneration needed to keep quality personnel, and the intellectual demands of the profession will all change. Existing regulations, union resistance, university conservatism, bureaucratic inertia, and the general community unrest when any shift is proposed in educational policy will all make such necessary retraining difficult.

Personnel shifts may cause considerable tension in the first few years of technological implementation. When the dust settles, as many (or more) educational jobs will exist as at present; but a considerable number of these positions will be supported by an expanded clientele for educational services and will be located outside the school system in industries, communities and the media. The difficulties of pre-service teacher training for such a wide variety of potential professional roles will become immense, and the minimum time necessary for at least some forms of certification is likely to increase.

Overall, the impact of instructional technology on educational employment will be not so much to reduce jobs as to alter roles and to shift employment to educational agents other than schools. Different implementation strategies may meet with varying degrees of resistance (as discussed later), but any approach will encounter a high degree of inflexibility and conservatism from many educators. As a result, despite the temptingly large budgets of schools, the predominant initial usage of instructional technologies may well come in homes and industries acting independently of the formal education system.

These generalizations as to inevitable short-term effects of technological innovation will vary somewhat for each particular type of technology. For computers:

1 Hand-held microprocessors (such as Texas Instruments' "Speak and Spell") will be inexpensive enough to produce and program so that relatively specialized devices will be economically feasible and updating of design and software fairly easy. Little teacher training will be required to add such machines to the classroom repertoire.

2 Stand-alone microcomputers (such as the Apple II), to achieve economics of scale, will have to be general-purpose in both hardware and software capabilities. The difficulties in standardization, updating, and teacher training discussed above will apply to these devices and will be increasingly evident as the sophistication and power of these machines grow.

3 Large computers (linked to homes and industries via telephone terminal or cable/television) are intrinsically capable of multiple special-purpose applications, but require enormously complex software whose depreciation time, to be cost-effective, must span a number of years. The most advanced programming for these machines will probably strive to be usable without teacher retraining (although attempts at "teacherproof" curricula have been notably unsuccessful in the US). Compared with instructional usage, computer conferencing and record-keeping and applications will be relatively easy to alter and update.

All forms of television and videodisk/microcomputer programming will be very expensive to revise and improve (high-quality educational television programming in the US currently costs over $200,000 per half-hour). Teacher training will be essential if these media are to be used well in classroom settings.

Long-term effects
Once these short-term impacts of instructional technology have emerged, "second-order" consequences will begin to appear in society. One of these consequences, likely to be independent of the implementation strategy used, is a differentiation of "education" and "training," with education being done by people and training by machines. Any type of instructional programming will be most effective for subject matter with a limited range of right answers (training), because defining an incorrect learner response and channeling student effort in an appropriate direction are much easier. While creating good "multiple right-answer" instructional units (education) is technically possible, the difficulties and costs of doing so are prohibitive compared with using human teachers. Thus, training in subjects such as reading, basic maths, accounting, descriptive botany, and so on will be done by technology;

and education in creative writing, clinical psychology, ethnography, etc. will be done by teachers. Most curricula will be a carefully differentiated mixture, with training initially predominant, then ever-increasing amounts of education being added. Training will take place primarily in extra-school settings; education mostly in schools (or their future equivalents).

This shift is likely to make possible much greater efficiency in learning. Training is best done on an individual basis in spontaneous, relatively small chunks of time, rather than in classrooms with large numbers of students and a prescheduled, extended block of time to be filled. Moreover, the process of creating an instructional programme for subject matter involves enough thought and detailed specification that the resulting product is likely to be a quantum jump ahead of existing materials. Education, too, will become easier when each of a group of learners has been individually certified to have the requisite skills and knowledge to tackle more advanced subject matter. As a result of all these effects, the present curriculum may be taught in a third of the time it takes currently (especially where training is predominant, as in the early primary grades). In turn, this will allow both financial savings from students graduating earlier and the inclusion of additional material in the school curriculum.

A second long-term effect of the new instructional technologies will be the gradual emergence of a new definition of "intelligence." Historically, our concept of intelligence, the cognitive skills we most value, has altered with each new innovation and information management technology. For example, at the end of the nineteenth century, having a near-photographic memory was an important component of intelligence because the rapid data-retrieval systems we have now were not available. Ten thousand years ago, before the development of writing, total recall was essential, since all knowledge was accessible only through the oral tradition. Several decades from now, memory skills may become a trivial part of intelligence through advances in computer data storage.

In general, the intellectual skills most important to possess will become increasingly centered on analysis, synthesis, and evaluation. As tremendous amounts of data become easily accessible through technology, and as options in the presentation of that information multiply, being able to select the best option from the menu of choices, analyze the data's implications, synthesize it with other knowledge, and evaluate the outcome will be crucial to successful intellectual endeavor. (For example, a writer expert in using a word processor exercises a set of skills completely different from one using a typewriter.) Those portions of education that stress fluency in lower-order cognitive skills will thus gradually be de-emphasized in favor of building proficiency in higher-order abilities.

A third long-term consequence of the new educational technologies will be a higher overall rate of societal change through both technological innovation and social invention (i.e. both new techniques for manipulating the physical environment and new collective patterns of interaction among people). Most futures researchers believe that the developed countries are in the midst of a technological revolution comparable in scope to the industrial and agricultural revolutions. Information is seen as the incipient key economic resource (just as capital has been in the industrial period and land in the agricultural). The values of society—barring a normative transformation—are projected as still acquisitive, competitive, sensate, and materialistic. Hence, the information technologies are viewed solely as a means to intensify the speed with which the present ethical framework shapes reality to its desired goals, rather than as developments that will intrinsically alter these goals themselves.

Some of this faster rate of societal change will come from a better understanding by the citizenry of the strengths and weaknesses of technology. Almost everyone will be interacting with instructional devices on a regular basis, learning from and programming machines. As people constantly see what these devices are and are not capable of accomplishing—and that they are not magical in nature—a greater sophistication in developing, assessing, and controlling technologies will emerge.

Another factor in increasing the rate of societal change will be the continual resocialization possibilities that instructional technology opens for adults. No longer will a high school diploma signify the effective end of an adult's responsibilities to update his/her knowledge, skills, and awareness of societal issues. Society will benefit as some previously undereducated adults become more capable of coping with laws, bureaucracies, banks, and voting.

"Citizen's literacy" will emerge as a powerful tool for handling societal crises. (For example, a two-hour comedy programme on constructive coping with petrol shortages could be quickly distributed if a petroleum availability crisis developed.) Some degree of social inertia and conservatism will be eroded as new ideas and social experiments are transmitted via informal instructional units. The decision-making capabilities of citizens will be enhanced, and a moderate degree of decentralization of power may ensue (so long as the information transmitted is knowledge rather than propaganda).

Coupled with this faster rate of societal change will be an increasing homogenization of different nations and cultures. As the centralization of production grows, instructional units will become ever more standardized; whatever cultural differentiation takes place in programming to adapt to different audiences may be as much "window dressing" as actual. This growing uniformity will speed the rate of global change and facilitate intercultural communication, but will also reduce the pool of diversity and pluralism on which the human race can draw. If economic factors are too major a determinant of the quality of instructional content, a "lower common denominator" effect on programming may occur, with a resultant loss in worldwide human potential.

A fourth long-term implication of the new instructional technologies will be the evolution of a fundamentally different model of teaching/learning. Along with the differentiation of training and education discussed earlier will come (at minimum):

1 a centralization of curriculum development and financing approaches;
2 a decentralization of the learning environment into homes, communities, and industries;
3 a decredentialization and informalization of the educational experience;
4 new types of government regulations to allow educators to interface with public utilities as communication channels;
5 a privatization of the educational enterprise, as information technology vendors get involved;
6 new types of diagnostic, assessment, and evaluation strategies in response to larger grading pools and altered definitions of learner effectiveness;
7 new "machine-coupled" teaching strategies;
8 new administrative networks, with the erosion of many middle management positions as increased information transfer becomes possible without intermediary functionaries;
9 new types of people attracted to the various educational professions, with different skills and salary requirements;
10 a revolution in the process and content of teacher training/certification.

In brief, sweeping changes in roles and relationships will occur for parents, teachers, administrators, industries, publishers, media, and government.

Such a new educational model is long overdue, as rapid social change coupled with fierce resistance to educational innovation has made existing approaches to schooling obsolescent. Once the economic benefits of redoing educational practice become apparent, ideological objections to change will crumble under pressure from society, which has seen costs spiral over the past several decades and has little confidence in many present instructional methods. As a result, the shift depicted above will take time to initiate, but once begun will develop fairly rapidly. A key to successful evolution will be a balanced policy of trade-offs between economic considerations and educational quality (as discussed earlier).

All of these "inevitable" long-term effects will be intensified or reduced by the particular implementation strategy chosen, but none are truly avoidable. As such, they represent illustrative "bottom-line" costs and benefits of a commitment to technological innovation in education.

Effects dependent on implementation strategy

A variety of plausible alternative scenarios can be constructed for the emergence of information technology as a major educational tool. One metaphor used by future researchers is to view the future as a "tree." We stand on the trunk of the tree (the present) looking upward towards the branches (the major likely alternative futures). Each step we take up the trunk towards the branches (each decision we make in the present) chops off a branch (greatly reduces the probability of a cluster of alternative futures). By the time we reach the branches—when the future becomes the present— all the branches are gone but one (the new trunk), and a new set of alternative futures stretches upward.

Who the major actors are in implementing information technologies in education and how these actors interrelate will be crucial determining factors in which alternative future emerges into reality. In the US, some major potential "players" are the equipment vendors, the media, the textbook publishers, the federal government, the formal educational establishment (teachers, administrators, state school officers), and the individuals know-ledgeable in software production (i.e. artificial intelligence experts, television directors). Space does not permit an explication of all plausible alternative implementation scenarios, but an illustrative sample can be given for the US.

Scenario one: no federal role   In the absence of any major federal encouragement for the development of education technology in schools, the full potential of the microcomputer to aid education is not realized. Information technology corporations concentrate on the home market; since the affluent are the major consumers of these products, recent gains in educational equity are reversed. State and local education agencies show some interest in these technologies, but significant innovation takes place only sporadically, in isolated pockets of interest (owing, in part, to the lack of equipment standardization). The textbook publishers do not play a major role because of lack of capital, so their well-developed distribution systems are not utilized. The media continue to concentrate on entertainment rather than instruction.

Scenario two: block grant programs   The federal government uses large grants to state and local education agencies in an attempt to encourage usage of educational technology. Limited controls (some orientation to equity, the handicapped, and the gifted) are placed on these funds. Information technology corporation work shifts from the home to the school market in response, but centers on promotion rather than development. Materials produced are of low sophistication; no standardization takes place. Hardware is sold first, with software packages promised later. A few franchised teacher-training operations spring up, but are of low quality; in general, the approach is to make things "teacherproof." A few publishers try to compete on their own; most sign on with a particular information technology corporation and are used primarily to aid in promotion. Penetration into the schools is excellent, but materials end up gathering dust fairly quickly (as with the post-Sputnik curriculum reforms).

Scenario three: creation of a distribution system   The federal government guarantees a market and a distribution system for industry-developed hardware and courseware, so long as it meets certain capability specifications (standardization, some equity orientation, a "conservatively progressive" approach to sophisticated design). The home market is put on the back burner, as the information technology corporations put most of their effort into school-orientated hardware design (their strength), with some effort going into meeting software specifications. After a fast start, especially on equipment, real progress slows to a crawl owing to lack of industry knowledge of the existing curriculum, student capabilities, and learning theory. The dearth of people adept in both education and computing becomes a real stumbling block. The textbook publishers are anything but helpful, fighting for their economic lives with charges of a "national curriculum." The eventual products emerge behind original time projections, heavily hardware-orientated and quasi-teacherproof, reasonably progressive in their sophistication but ill-tailored to inclusion in the existing curriculum, and with little supplementary teacher training. Penetration into schools is difficult; much ends up on the home market instead, amid recriminations.

Scenario four: long-term loans   The federal government supplies extended low-interest loans for software development to textbook publishers, with a few strings attached (mild equity orientation, some required standardization). Some publishers make deals with information technology corporations for equipment; most "go it alone." The quality of courseware developed is very uneven, but overall is low-grade and unsophisticated. This is due in part to a rush to be first into the market, in part to a commitment to current low-quality curriculum materials, in part to lack of information technology expertise. Some standardization takes place, but is far from complete; teacher training efforts are reasonable in scope, but very rudimentary. Penetration into the schools is moderate, but the overall impact on the curriculum and on student learning is minimal. Information technology corporations concentrate their efforts on the entertainment market.

As can be seen from the above brief sample, the plausible combinations and permutations of alternative implementation scenarios are too complex to explicate further here. Moreover, any such discussion would probably be too idiosyncratic to the us educational system to be useful in a broader context. A few generalizations can be made, however, about potential effects highly contingent on the implementation strategy that emerges.

Short-term effects
One short-term consequence of any implementation strategy will be that the primary initial clientele served will depend on who the major actors are in pioneering large-scale instructional technology usage. Whatever population the innovating group is accustomed to serving will be the initial recipient of the new technologies. As can be partially seen from the sample scenarios above, an innovation effort led by educators would serve primarily the schools; textbook publishers the schools; equipment vendors the family, military, and industry; the media the family; and the federal government

some mixture of these consumer groups (depending on the particular government agencies chosen to spearhead the effort).

This initial choice of clientele will be very influential in determining the later evolution of instructional technology. For example, if resistance from the schools were high enough and marketing not targeted to educators, a substantial number of families in the US might be the first major users of educational technology. Their purchases would be made without any particular expertise in curriculum development or instruction; the materials bought would probably be advertised as supplementary aids for diagnosis, remediation and enrichment of the student. In turn, some schooling systems (primarily those serving an affluent clientele who could afford these devices) would be pressured by the community not to use instructional technology in the schools, but to modify the existing curriculum to adapt to the existence of these extra-school learning aids. This type of indirect educational reform would take place haphazardly across the nation, without any particular knowledge on the part of the decision-makers, and in a manner likely to polarize educators against instructional technology. Alternatively, were the initial market to be the schools, the later evolution of educational technology might be quite different—better in some ways, worse in others.

Because these early stages of implementation are so crucial in determining who the eventual clients will be and what services they will be offered, present-day decisions will have a profound future impact. Given that each of the "players" described earlier has some special interest of its own, a good societal implementation strategy might be to have some nonprofit public interest group coordinate information exchange among the major actors and evolve some recommendations on standards for marketing, government regulation, product characteristics, and so on. In this manner, an optimal implementation approach might be achieved through coordinated action with a market framework.

A second short-term effect contingent on the style of implementation that evolves will be the extent to which human interaction is reduced in the learning process. If machines are simply substituted for people without compensatory shifts in the human teaching that remains, personal contact and the affective skills learned through modelling others' behaviour will be partially lost. In a world daily growing more impersonal, the retention of large amounts of human interchange in learning seems important, both for socialization purposes and to enhance the quality of life.

Some degree of person-to-person interchange can be incorporated into instructional programming if new types of communication skills are developed that allow being person-orientated and affective even when interacting via the computer or television. After all, this is what an adept media personality is able to accomplish; studying these figures may reveal how sociability, social presence, and affect can be incorporated into machine-mediated communication. Certainly, people have adapted to using letters and telephones without completely losing the human touch; training via these new technologies could also become more personal if some thought is given to development.

Even with affectively orientated programming, however, important amounts of human contact will be lost unless time spent with human instructors is more intensely person-centred in compensation. At present, as the number of students per classroom increases, the human element in teaching is being eroded even without the intrusion of machines. Some of the financial savings derived from technological innovation could be used to reduce class sizes for human instructors. Revision of current teaching practice to take advantage of recent findings in social and humanistic psychology would also allow intensification of personal contact. Such a thrust should be a major component of any technological implementation policy.

Long-term effects

One long-term consequence heavily dependent on the type of implementation that evolves will be the equity (or inequity) with which the advantages of the instructional technologies are distributed to learners across the nation. In the US, several decades of work to equalize educational opportunity and (to some extent) educational achievement may be lost if a *laissez-faire* approach is adopted to dissemination of these new devices. The aged, the handicapped, the poor, and minority groups could all benefit greatly from access to instructional technology, but do not have the financial resources to compete with the affluent as potential clientele.

Market forces, if the sole criterion for implementation strategy, will dictate that the educational hardware and software produced be designed for the needs of the largest and richest body of consumers: the middle and upper-class majority culture. Not only will less fortunate groups lack the capital to invest in purchasing equipment and programmes, but also the materials developed may well be directed toward a different cultural background and different educational needs. Two levels of schooling could easily emerge in a society: one geared to students who have instructional technologies available in the home and school, another relying solely on traditional methods of teaching. Given the potential these technologies have for improving the educational process, the latter group of students would be placed in an intrinsically inferior position.

Moreover, early childhood experience with instructional technology may create an advantage that a deprived individual can never overcome as an adult. Learning computer programming, for example, is somewhat similar to learning a second language: easy when young, far more difficult when older. The disadvantages to society of having a two-tier system of education—with its loss of available human potential—would be profound.

On the other hand, if equal access to high-quality instructional technologies designed to meet the needs of diverse groups were guaranteed, educational discrimination and inequality in society might be reduced more quickly than at present. As the economic situation of education worsens, poor and minority students are the first to suffer: these technologies could alleviate that problem. Some of the causes of unequal achievement may be related to a small range of teaching strategies trying to service a broad spectrum of learning styles; through individualization, technology could reverse that situation. By extending communication networks across cultures, classes, and generations, instructional technologies could improve the integration of the elderly, minorities, the poor, and the handicapped into society.

A subsidized and organized process of technological implementation will be necessary if equity is to be helped, rather than hindered, by these new devices. Incentives must be provided to manufacturers to produce software for different needs and cultural backgrounds; many schooling systems will require additional funds to purchase equipment and programmes; parents will need training to help them maximize their children's benefit from exposure to these machines. For both ethical and economic reasons, society would be wise to invest some resources in improving the training of all its citizens, but the choice to proceed with this type of implementation must be made quickly before *laissez-faire* marketing begins.

A second long-term implication contingent on the implementation approach chosen will be whether a knowledge coordination sector is deliberately created as a method for national strategic planning. At present, knowledge is produced and disseminated in most societies in a relatively haphazard manner, with no overall perspective on what types of ideas and skills are needed or how these can best be created in the population. If information is to be the key economic resource of the future, research and development its major means of production, and human capital its ultimate source, then a society would be well advised to attempt to maximize all these factors through some form of central coordination. The information technologies, properly implemented, offer a powerful means for achieving this end.

Such a knowledge coordination sector would have several functions:

1 anticipating societal needs for knowledge;
2 developing in educational institutions the capacity for training appropriate levels of human resources;
3 assessing the capability of current institutional mechanisms to generate needed knowledge and augmenting this ability where necessary;
4 organizing the dissemination to citizens of vital knowledge so that it is fully utilized.

This type of systemic approach would represent a commitment to education in its broadest sense as a fundamental reconstructive force for society.

The information technologies, once established in a nationwide instructional network, would make such a knowledge coordination sector feasible. The economic advantages—and the general improvement in quality of life made possible by a more literate citizenry—seem to justify the effort involved in implementing this type of national strategic planning. Should this be a goal of society, the initial implementation of instructional technology would have to be organized so as to facilitate the formation of this knowledge production and dissemination system.

The effects of instructional technology described in this section are all largely dependent on which alternative implementation strategy is used. None of the positive consequences are likely to emerge without some active intervention taking place; many of the negative outcomes will happen unless steps are taken to block their occurrence. Decisions made today, often without thinking about long-range implications, will be very powerful in shaping our future options in these areas.

Conclusions

A number of implications for educational policy can be drawn from this assessment of the new instructional technologies. First, given the importance to society of the effects this article describes, a thorough study of the consequences of implementing the information technologies in education needs to be made. One such research project is currently being sponsored by the Office of Technology Assessment, US Congress; similar studies in other national settings would be of great value. Given how crucial early implementation decisions will be in shaping the ultimate outcome, the moment-to-moment approach characteristic of most educational decision-making seems particularly unsuited to this situation.

Second, while maintaining an adequate profit margin to encourage investment will be important to the growth of instructional technology, too much emphasis on purely financial factors may vitiate some of the most valuable potential contributions of these technologies to society. Ensuring all groups equal access to technological services, orientating machines and curricula to diverse needs and backgrounds, updating and improving the initial instructional products, keeping a human presence part of the educational experience, and creating high quality of hardware and software are important goals that will be expensive for developers to achieve. Policy decisions on issues of implementation will need to strike a balance between these goals and ensuring reasonable profits.

Third, as with all educational innovations, considerable resistance to instructional technology can be expected from the formal schooling system. If this inertia and opposition becomes severe, developers may bypass the school and direct their marketing efforts primarily to homes and industries. In turn, extra-school use of these devices will mount a different, indirect set of pressures on formal education. The history of educational change suggests that the eventual outcome of an innovation will be of most social benefit if educators are involved from the beginning stages. Hence, educational decision-makers need to recognize that instructional technology will be influential with or without their support and should work to make its influence on schooling as positive as possible.

Fourth, given that educational technology is likely to increase existing rates of societal change and cultural homogenization, all members of society need to discuss how much stronger these two trends should become. "Progress" brings mixed benefits at best, and some degree of strategic planning is necessary to ensure that technological change results in a desirable future. Educators have a special responsibility to alert other sectors of society to the probable consequences of implementing the instructional technologies.

Finally, none of the major actors in the development of educational technology—equipment vendors, educators, textbook publishers, the media, the federal government—are neutral. All have their individual special interests to consider, and the public good is unlikely to emerge from a *laissez-faire* free-for-all among them. Creation of some nonprofit, public interest agency to mediate among these actors may be the optimal long-term strategy for the national benefit.

The world is facing a period in education similar to the introduction of the printing press five hundred years ago. In converting to the use of books, people confronted problems similar to those discussed here: a potential loss of the human factor, the necessity for a new educational model, career shifts, massive needs for capital investment, equity issues, etc. Although decades passed before books were used to their full instructional potential, the shift to the printed word for information dissemination ultimately did result in progress, increased learning, and exciting new frontiers for education. Given a comparable opportunity, will we rise successfully to the challenge?

Guide to Further Reading

Articles

"School Computers Score at Last," *Business Week*, July 27, 1981.
Ludwig Braun, "Computer-aided Learning and the Microcomputer Revolution," *Programmed Learning and Educational Technology*, vol. 18, No. 4, 1981.
S. Reed, "Schools Enter the Computer Age: The Impact," *New York Times* Spring Survey of Education, April 25, 1982.
"Peering Into the Poverty Gap," *Time*, November 15, 1982.
Information Technology and Its Impact on American Education (Office of Technology Assessment, US Congress, Washington DC, 1982).
William Work, "Communication Education for the Twenty-first Century," *Communication Quarterly*, vol. 30, No. 4, 1982.
Christopher Roper, "French Flock to Computer Centre," *New Scientist*, February 10, 1983.
Ian Anderson, "Californian Schools Reap Bumper Harvest of Apples," *New Scientist*, March 3, 1983.
Greg Kearsley and Beverly Hunter, "Electronic Education," in *High Technology*, April 1983.
Peter Marsh, "A Brighter Future for the Computer Kids," *New Scientist*, August 18, 1983.
J. Megarry, "Educational Technology, Promise, and Performance," *Programmed Learning and Educational Technology*, vol. 20, No. 2, 1983.
"Slugging It Out in the Schoolyard," *Time*, March 12, 1984.

Computer camps for children and adults are described in *Time*, August 3, 1981 and August 22, 1983.

Books

J. Howe and P. Ross, *Microcomputers in Secondary Education* (Kogan Page, London, 1981).
Kenneth Owen (ed.), *Videotex in Education: A New Technology Briefing* (Council for Educational Technology, London, 1982).

Peter Coburn et al., *A Practical Guide to Computers in Education* (Addison–Wesley, Reading, MA, 1982).

Christopher Smith (ed.), *Microcomputers in Education* (Ellis Horwood, Chichester, England, 1982).

Roy Garland, *Microcomputers and Children in the Primary School* (Falmer Press, London, 1982).

Alan Maddison, *Microcomputers in the Classroom* (Hodder & Stoughton, Sevenoaks, England, 1983).

David Hawkridge, *New Information Technology in Education* (Croom Helm, London, 1983).

Ray Hammond, *Computers and Your Child* (Century, London, 1983).

Tim O'Shea and John Self, *Learning and Teaching with Computers: Artificial Intelligence in Education* (Harvester Press, Brighton, England, 1983).

6 Factory Automation

Computers in Manufacturing

C. A. Hudson

Computers are now widely used in product design and on the factory floor. During the next decade, they will become more and more involved at every stage of the manufacturing process. In telling the story so far, C. A. Hudson provides a good account of CNC, CAD, and CAM and points up some of the problems encountered when introducing automation. The author is Manager of Technology, Industry Automation Division, Westinghouse Electric. This article first appeared in Science *magazine, February 12, 1982.*

While the US remains among the most productive nations in the world, other industrialized countries are quickly closing the gap. In Japan manufacturing productivity is currently growing at the rate of 4.1 percent a year. France and Germany have manufacturing productivity growth rates of 4.9 and 5.0 percent a year, respectively. Meanwhile, in the US the rate of growth in manufacturing productivity has fallen sharply. From 1969 to 1973, output per man-hour increased at a compound rate of 2.9 percent. From 1973 to 1979, the gains dwindled to 1.6 percent a year.[1]

In the US the program to reverse this productivity pattern significantly must rely on the continued development of advanced technology and its application. Perhaps the most important element in this reliance on innovation is increased factory automation and a growing use of computers and microprocessor technology in manufacturing. Today, we are on the technological and sociological edge of a dramatic increase in the use of computers in our factories. This will have a profound impact on the nation's productivity growth in the next decade.

Within the next 10 to 15 years, four evolutionary trends will meet on the factory floor: (1) the increasing power and simplification of computers, (2) a widespread appreciation of the practicality of computerized manufacturing and robotic applications, (3) a new realization of the impact of computers on people and of people on computers, and (4) a growing awareness of the urgent need for manufacturing innovation in our society.

Many computerized factory systems exist today as islands of automation. The immediate task of the scientific and technical communities is to use the increased power and simplicity of computers to link these elements into an integrated system. Making use of low-cost computer hardware to perform more and more jobs will make such an integrated factory system economically viable.

Some difficult technical problems remain. We must develop generally accepted, standardized interfaces between computerized design engineering and computerized manufacturing, between individual machines and machining centers, and between computers and the people using them. We must also refine the present state of application technology and reduce the cost of the factory automated system through the increased use of computers. Advances in factory automation are dependent on advances in computer technology.

Computer evolution

The technological changes in the computer field during the past several years clearly equal any technological change that has occurred in our society over the past 100 years. The introduction of high-performance, low-cost microprocessor and storage technology has dramatically improved and enhanced the functions and capabilities of computer software and hardware. With today's increased power, manufacturing computer systems can be made more adaptable to the manufacturing environment, thus cutting systems engineering costs and time per installation.

The cost of computer hardware itself has been steadily declining. This trend will continue with the introduction of new technologies such as very high-performance microprocessor chips based on very large scale integration (VLSI). On the other hand, the cost of the human and software resources for systems engineering and programming has gone up. In 1955, 85 percent of the total cost for processing information was hardware; it is estimated that by 1985 hardware will account for only 15 percent of this total cost. To improve manufacturing productivity we must reverse this trend by optimizing the use of our human resources and taking advantage of the increased computer power and reduced hardware costs.[2]

The need to make increased use of available computer power is heightened by the decline in available technical manpower. The National Science Foundation found the annual growth rate of scientific and R&D personnel between 1954 and 1969 to be 5.9 percent. Some 556,000 employees were involved in technical work in 1969, but the number fell to 517,000 in 1973 and then grew to only 610,000 by last year—a rate of only 2.8 percent annually.[3]

The increased capability of computer technology and broad availability of application software packages, including the expanded use of problem-oriented languages and database software, can reduce the cost of program development and maintenance by a factor of 10. The graphics capabilities of engineering computers can link integrated design and drafting systems to manufacturing systems. With these "user-friendly" approaches to provide a bridge between man and computer, the user can interact directly with the manufacturing system without traditional interfaces and jargon-heavy manuals.

Computers and the applications they support can now be considered for functions or activities that only recently were impractical. The opportunity for the application of this technology, for all practical purposes, is unlimited.

Manufacturing automation

Advances in the two primary elements of factory computerization—computer-aided design (CAD) and computer-aided manufacturing (CAM)—will create a new industrial revolution. By integrating design with manufacturing, we can not only turn out new product designs much faster, but also program the computer to make sure the designs provide quality and reliability as well as the lowest possible manufacturing costs. CAD/CAM is the integrated use of advanced computer technology in engineering and manufacturing. It is a common database of part and product geometry and related information which makes it easier to translate a creative idea into a final product at a reduced cost.

With CAD, a user can define a part shape, analyze stresses and other factors, check mechanical actions, and automatically produce engineering drawings from a graphics terminal. When CAD is combined with the CAM system, the user can also manipulate nongraphic data such as bills of material, shop information, and cost factors. The end result is greater design flexibility and what is referred to as designing to cost.

The CAD functions can be grouped in four categories: design and geometric modeling, engineering analysis, kinematics, and drafting.

1 In design and geometric modeling, the designer describes the shape of a structure with a geometric model constructed graphically on a cathode-ray tube. The computer converts this picture into a mathematical model, which is stored in the computer database for later use. Many other design functions depend heavily on the model. It can, for example, be used to create a finite-element model for stress analysis, serve as input for automated drafting to make a drawing, or be used to create numerical control tapes for the factory.

2 After the geometric model has been created, the engineer can easily calculate such things as weight, volume, surface area, moment of inertia, or center of gravity of a part. But the most powerful method of analyzing a structure is probably finite-element analysis. In this technique, the structure is broken down into a network of simple elements that the computer uses to determine stresses, deflections, and other structural characteristics. The

designer can see how the structure will behave before it is built and can modify it without building costly physical models and prototypes. This procedure can be expanded to a complete systems model, and the operation of a product can be simulated.

3 With computer kinematics, the user can examine the effects of moving parts on other parts of the structure or design and can analyze more complex mechanisms.

4 Finally, the CAD system automatically drafts drawings for use in manufacturing.

Computer-aided design is a good example of the transition of expensive, state-of-the-art computer technology to a commercial, economically justifiable system. Recent advances in CAD technology have increased the productivity and effectiveness of design engineering groups. Such systems will be even more common in the next five to ten years.

Manufacturing groups can draw on the geometric and numerically coded description produced by CAD to create numerical control tapes, which allow direct computer control of shop machines, determine process plans and scheduling, instruct robots, computerize testing, and in general improve the management of plant operations.

Computer-aided manufacturing has five main functions: tool design, machine control, process and materials planning, robotics, and factory management.

1 Manufacturing engineering and tool design deals with the machines and fixtures needed to make a new product. In effect, the set of machines, tooling, and fixtures is a new product, and all the techniques of CAD are used in fashioning it. The CAD techniques are then used to simulate plant operation and the integration of machines and materials handling.

2 Machine automation consists of a chain of increasingly sophisticated control techniques. At the lower end of the spectrum are fixed automation with relays or cams and programmable controllers, where relays have been replaced by electronics. Moving up the spectrum, numerical control (NC) refers to controlling a machine with prerecorded, numerically coded information to fabricate a part. In this case, the machine is hardwired and not readily reprogrammed. In computer numerical control (CNC) the machine is directly controlled by a mini-computer, which stores the machining instructions as software that is relatively easy to reprogram. Because of the computer control, CNC has the advantages of much higher storage capability and increased flexibility. Virtually all numerical control is computer-based, yet only ten years ago CNC was an expensive exception.

3 Process planning considers the detailed sequence of production steps from start to finish. The process plan describes the state of the workpiece at each work station. An important element in process planning is group technology, in which similar parts are organized into families to allow standardized fabrication steps; this permits significant savings by avoiding duplicate tooling and systems engineering. Most automated process-planning

systems use a retrieval technique based on part families and existing databases for standard tooling and fabrication processes. Materials planning or manufacturing resource planning is concerned with the precise flow and timing of manpower, materials, and processes; it is a detailed look at how everything comes together. The ultimate goal is to have continuous use of all production equipment, no bottlenecks, and a minimum inventory.

4 Because they are widely applicable, robots have a distinct advantage over specialized, highly engineered manufacturing systems. The economic advantage of a mass-produced, readily adaptable robot over a one-of-a-kind system with a great deal of engineering content is obvious. Robots are now being used to perform materials-handling functions in CAM systems. They can select and position tools and workpieces for NC or CNC tools, operate tools such as drills and welders, or perform test and inspection functions. Through visual or tactile sensors, the robot can manipulate objects. Through its computer intelligence, it can inspect the object and provide the machine with corrective feedback or actually reprogram the machine or change the tooling.

5 Factory management coordinates the operations of an entire plant. Factory management systems tie together individual machine tools, test stations, robots, and materials-handling systems into manufacturing cells and the cells into an integrated whole. An integrated CAM system of this sort is usually hierarchical, with microprocessors handling specific machining functions or robot operation, middle-level computers controlling the operation and work scheduling of one or more manufacturing cells, and a large central computer controlling the overall system.[4]

Reliability is greatly improved by structuring the control system correctly. Local, distributed control (with defined responsibilities) reports up to a supervisory control that, in turn, is linked to a managerial computer. This parallels the structure of the typical industrial organization.

Ultimately, the digital output from the CAD computer will be simply plugged into the CAM system to reprogram the plant's manufacturing computers. In such an integrated system, the databases will be organized in a way that avoids redundancy and reformatting of information. And any change in one part of the system will automatically revise dependent or related information in other parts of the system.

Bridging the CAM and CAD systems will be one of our major jobs in the future. A fundamental difference that has to be reconciled is that CAD makes use of a pictorial, graphics-oriented computer database while CAM involves a great deal of text-oriented information. In other words, we need to find a way for the computer doing the drawing to speak the same language as the computer directing the manufacturing plant.

Layering is one way to link these systems. Layering is a particular technique for structuring the CAD and CAM databases. It enables various people to input data without losing control of the overall design and manufacturing process. Equally important, it enables shop people to see information that is meaningful to them without having to sort through and understand the rest of the information that is normally included in a drawing.

To do this, all information is organized in an arrangement resembling layers, or slices, inside the database. The engineer or users in other departments of an organization can provide pertinent information or examine any or all layers of information according to their particular needs. As an example, a printed circuit board may have 250 to 300 layers of information. A manufacturing engineer inputs layers of information that deal with fabrication and assembly. In turn, machine operators concerned with the details of the drilling and cutting configuration may access layers dealing with this part of the drawing. Other layers provide information pertinent to the needs of the purchasing department or component assemblers.

Another major effort to integrate computer systems is an Air Force program called ICAM (integrated computer-aided manufacturing). This is a practical attempt to greatly shorten the time span for the implementation of compatible and standardized computer-manufacturing techniques and to provide a unified direction for industry. The ICAM program provides seed money for the establishment, within private industry, of modular subsystems designed to computerize and tie together various phases of design, fabrication, and distribution processes and their associated management hierarchy. As appropriate, these mutually compatible modules will be combined to demonstrate a comprehensive control and management package capable of continual adjustment as production needs and the state of the art change.

The ICAM program is divided into five major parts.

1 Defining the manufacturing architecture. This permits a concentration on problems of generic scope and wide applicability as the basis for later projects in integration, support, and application systems and demonstrations.

2 Developing integration methodology. This activity provides a bridge between industry and ICAM for the transfer of ICAM technology for the integrated factory of tomorrow. The projects addressed include establishing factory simulation techniques, ICAM implementation techniques, configuration management, modeling tools, software integration simulation, automated systems engineering methodology, and various system analysis and design capabilities.

3 Establishing support systems. This is concerned with the portion of the ICAM system involving computer operations, including both software and hardware and both operational and managerial aspects of computerized manufacturing.

4 Establishing application systems. This includes such items as manufacturing cost and design guides, the design-manufacturing interface, manufacturing standards, group technology concepts, and scheduling and process planning. Under an ICAM contract, the National Bureau of Standards considered standards in computer communications, languages, and networks to identify potential conflicts within an integrated manufacturing environ-

ment. Other areas of concern include robotics, prototype integrated production cells, integrated materials-handling and storage systems, and integrated manufacturing control and material management.

5 Demonstrating the ICAM program. The ultimate goal in ICAM is the use of totally integrated manufacturing systems by industry in the completely automated factory.[5]

Robotics

The problems encountered in trying to integrate advanced computer concepts, new manufacturing technologies, and the shop floor are clearly evident in the evolution of robotics. Robots are classified according to the way we provide them with information and the amount of self-adaptability they possess. The most comprehensive categorization of robots is provided by Japan's Industrial Robot Association:[6]

1 Manual manipulators are worked by an operator.
2 A fixed-sequence robot has a manipulator that repetitively performs successive steps of an operation according to a predetermined sequence which cannot be easily changed.
3 A variable-sequence robot is similar to the fixed-sequence robot except that the set information can be easily changed.
4 A playback robot reproduces, from its computer memory, operations that were originally executed under human control.
5 An NC robot is a manipulator whose tasks are programmed by using numerical control tapes or cards.
6 An intelligent robot, using sensory perception, detects changes in the work environment and proceeds accordingly, using its decision-making capability.

Industry today is focusing on the development of NC and intelligent robots. There has been a growth in the number of firms that manufacture and sell such robots which is reminiscent of the proliferation of minicomputer companies in the 1960s.

Basically, robots are microprocessor-controlled mechanical devices that perform a function or provide an intelligent interface between machines and processes. They can be intelligent enough to make on-the-spot manufacturing "decisions." But for robots to become practical, we must reduce their size, mechanical complexity, and installed cost—primarily through the expanded use of computer and control technology.

Robots can duplicate human manipulative skills with accuracy and precision. Their flexibility and versatility, as opposed to hard automation, make robots ideally suited to the kinds of small batch jobs that constitute the bulk of industry's manufacturing activity. Today, robots, are freeing people from jobs that present serious health hazards, are mundane, or are highly repetitive. In most cases their use is justified for non-economic reasons.

In the US industry has been slow to adopt robotics. This reluctance appears to be due primarily to the large initial investment and the general availability of relatively inexpensive manual labor. Why install a $100,000 or $150,000 robot to perform a $25,000-a-year job? Also, the majority of today's robots are monsters: bulky, unwieldy mixtures of hydraulic and mechanial contraptions with a machine tool heritage.

This situation is changing. Robots are becoming more streamlined, and, when they are manufactured in large quantities, will rapidly decline in cost. Many technologically innovative firms are entering the business. Equally important, system engineering, which represents as much as two-thirds of the cost of a robotic application, is being greatly reduced. It is not difficult to imagine that in a short time the cost of a typical robotic system will be paid back in one or two years. In the next decade the cost of a robot is likely to be down to $10,000 to $20,000, while skilled labor costs might easily be $25 or $30 an hour. When this economic threshold is reached, there will be a virtual flood of robotic applications.

When this happens, robots will play an important part in the totally integrated factory of the future. Most of our plants will have a direct numerical control supervisory computer that coordinates the activities of several NC and CNC machines or hardwired machining centers and robots and connects all the machines into a system. The robot interface will handle the transfer of material and, with newly developed sensory capabilities, will also perform the in-line inspection of parts. For the near future, however, our factories will be some particular mix of machines, robots, and people that makes the most economic sense.

Eventually—in a decade or so—robots will fill a void in the supply of skilled labor. There has been a shift in the labor force from blue-collar to white-collar workers and from production jobs to service jobs. Currently, about two-thirds of our work force and 85 percent of all college graduates are employed in a service-related activity. The total service-oriented labor force is expected to increase by 20 percent in the 1980s—to about 85 million people. This shift to a service economy, coupled with the slowdown in the growth of the US population, suggests that many businesses will find factory labor in short supply: this is already the case in Sweden today.

At a leading university in Japan, there is a robot with human-type hands and legs, TV-camera eyes, artificial ears and mouth, and touch and joint sensing. These technologies are combined to provide the robot with some of the capabilities of a two- or three-year-old child. For example, when ordered to fetch an item in the room, the robot looks around the room and finds the article, walks to it, picks it up, and brings it back. If the robot does not understand a command, it speaks up.

Such robots are essentially showcase examples: they are not appropriate for the majority of industrial applications. In fact, a universal person-like robot would make little sense except for very limited, specialized applications. At present, the major applications for robotics are in arc welding and material transfer. In the future, the major application will be in assembly. Artificial intelligence is a worthwhile goal, but for the moment industry has more than enough applications for "dumb" robots. Employment of robots in these applications is actually limited by the extensive engineering required to put them to use.

It is estimated that by 1990 two-thirds of the robots sold to industry will be off-the-shelf, modular units rather than specially designed systems. Looking at automated systems generally, by the end of the decade, 30 percent of our systems will consist of hard automation, about 20 percent will be adaptive control, and the remaining 50 percent will be systems of a universal programmable nature.

Respondents to a study conducted by the Society of Manufacturing Engineers ranked the technical and performance barriers constraining the rapid utilization of robots in US industry.[7] The leading technical barriers were mechanical manipulation, vision systems, tactile systems, sensory systems, programming, and control systems. The primary performance barriers included accuracy, speed, and the ratio of capacity to size.

Robots will have a much greater impact in manufacturing when their total installed cost is reduced and they are more easily programmed. As with CAD and CAM systems, the most difficult thing about putting in a robot is interfacing it with the factory—both the machines and the people.

Research today is focused on the development of (1) equipment that will make greater use of computer technology to cut the cost of systems engineering and power electronics, improve servo-motor technology, and rapidly move from hydraulics to electrics; and (2) sensors that will enable robots to perform more reliably and with greater precision and adaptability.

A good overview of sensor types has been provided by Bejczy:

The nonvisual sensor information is used in controlling the physical contact or near-contact of the mechanical arm/hand with objects in the environment. It is obtained from proximity, force-torque, and touch-slip sensors integrated with the mechanical hand. These sensors provide the information needed to perform terminal orientation and dynamic compliance control with fine manipulator motions . . .

Terminal orientation and dynamic compliance control are essential and intricate elements of manipulation. Soft and adaptive grasp of objects, gentle load transfer in emplacing objects, assembling or disassembling parts with narrow tolerances, and performing geometrically and dynamically constrained motions (like opening or closing a latch or fitting two parts together) are typical examples of manipulator control problems that challenge both sensor and control engineering . . .[8]

Vision systems close the control loop and allow the robot to interact in a dynamic, changing environment. A second use of vision will be for the critical inspection of the batch-produced parts. The position and orientation of the part can be used in advanced automation systems to direct a robot manipulator to pick up the part for an assembly or transfer operation.

The Robotics Laboratory at Westinghouse is working on state-of-the-art applications in many of these areas. Systems and development engineers are working on the integration of controls, tooling, processes, computer, and other elements of the automated factory. Specialists in robotics are concentrating on developing and applying high-speed vision systems, tactile and force feedback sensors, high-performance electric servo systems, adaptable programmable assembly techniques and computer control, and artificial intelligence.

One of these projects is called APAS (automated programmable assembly system). Funded in part by the National Science Foundation, APAS is a pilot program in which robots are used to assemble components into the end bells of the Westinghouse line of fractional-horsepower motors. It is a development project intended to transfer newly developed technology to the factory. The fractional-horse power motors are currently assembled in batches averaging 600 units at a time. There is a 20-second assembly time per motor to put together 30 different parts, and there are 13 changeovers a day to handle 450 different motor styles.

The first section of the line puts parts on the motor end bells. To start the 15-second subassembly operation, a vision system in conjunction with a five-axis PUMA (programmable universal machine for assembly) robot inspects the end bells to make sure they are the style currently being assembled. The end bell is then oriented and placed on a pallet. The next step in the assembly is the insertion of the uppermost components: a thrust washer, a bearing cap, and a felt washer. An auto-place robot picks up the parts and loads them onto an anvil. The end bell is moved into the station, and the parts are passed on. At the same time a semisolid lubricant is injected.

At the next station, four screws, a plastic plug, and a contact point are inserted. Following this there is a complicated assembly procedure. In order to assemble all the different styles of end bells, several styles of certain parts are required. More precisely, six styles of mounting rings and three styles of dust caps are needed. Programmable feeders are used to accomplish the feeding and orienting of all these parts. At this station, a PUMA robot picks up a mounting ring, dust cap, and felt washer on an oil finger from the programmable feeder and places them on an insertion device, where they are fitted onto the end bell.

At the final station, a vision system and a PUMA robot are used to perform the final inspection of the end bell, pick it up from the pallet, and remove it from the system.

The computer control and sensory parts of APAS are the most revolutionary elements of the application. A distributed microprocessor system is needed to handle many simultaneous tasks in the short 15-second assembly time, and a visual sensory system is required to provide orientation and feedback. One master microprocessor controls the entire system. Under its control are three types of smaller microprocessors for vision control, local process control, and robot path control. All these controllers work in conjunction with the master computer to coordinate the inspection and assembly procedures. The vision system on this project recognizes randomly oriented parts on an assembly line after a multipass learning cycle controlled by an operator. It can also be used to rotate the part to any given angle.[9]

Fostering innovation

The technology exists in many parts of the world to achieve significant advances in many areas of factory automation. Many European countries are working on numerical control, process planning, and group technology approaches rather than the graphics approach emphasized in the US. They are well advanced in integrating CAD and CAM systems.

In Japan, major advances are achieved through the efforts of the Ministry of International Trade and Industry (MITI). For example, hundreds of millions of dollars and some of the finest minds in business and universities are being applied to the task of developing a fifth-generation computer. Another example is a $60 million government-funded project to develop a flexible machine system. This system will use high-energy lasers to manufacture small batches of machined parts with assembly line efficiency. The project involves more than 500 engineers from 20 Japanese companies, and it could revolutionize much of manufacturing.

MITI makes use of Japan's homogeneity and organizational milieu. It would be inappropriate for the US to adopt the same methods to foster manufacturing innovation. However, we must recognize the urgency of the problem. The US is one of the few major industrialized nations in the world without a significant coordinated industry–government–university program directed at improving manufacturing technology. We need a national strategy for productivity improvement that brings together government, business, labor, and academia in a cooperative, rather than adversary, relationship. We will have to remove many of the disincentives to innovation and find new ways to capitalize on our diversity and our proven creative and inventive abilities. Just as MITI capitalizes on Japan's homogeneity, we must find new ways to foster, encourage, and channel our innovative diversity.

At present, Westinghouse is working with the National Science Foundation and the universities of Rhode Island, Florida, and Wisconsin on technology development programs. Along with the Robotics Institute of Carnegie-Mellon University, we are developing "seeing," "feeling," and "thinking" robotic systems for several of our factories. We are also very interested in the Air Force's ICAM program to coordinate sophisticated design and manufacturing techniques now used by industry on a piecemeal basis. This program attempts to integrate design, analysis, fabrication, materials handling, and inspection and to develop hardware and software demonstration manufacturing cells in selected aerospace plants.

Sociological impact of automation

The easier it is for people to use computers, the broader will be their applications in manufacturing. We are moving away from the airplane cockpit approach, with rows of complex devices, to create simple computer tools. Compare the ease with which we use personal computers today with the way we approached computers in the early 1970s. The same changes will occur in manufacturing.

The computer has three language levels: machine language, programming language such as COBOL or FORTRAN, and user interface or problem-oriented language. Designing the computer so that it can be quickly used by someone familiar with a problem—but not with computers—is the most difficult of all programming tasks. Today, computers "converse" with users in pictures, in ladder diagrams, or in the secretarial language of word processing. Voice recognition systems will free workers' hands to perform other tasks and make it even easier for them to use computers. In the future, when research efforts begin to pay off in systems with some understanding of natural language and with "common sense," communications with computers may become as simple as talking to a three-year-old.

An important area of concern in manufacturing is worker safety. Automated systems and robots must be able to work side by side with humans. Major developments are taking place in the sensor area, particularly with proximity detectors and ultrasonic sensors, to make robots more suitable for inclusion in existing factories.

The application of automated systems in manufacturing will have several major affects on the people involved in production. It will make our jobs more interesting and challenging; it will enhance job security; and it will multiply the productivity increases.

Workers today are looking for greater job satisfaction through greater involvement and increased sophistication. New technologies provide this added dimension to the workplace. For instance, draftsmen use CAD today to perform work that was normally performed by engineers just five or ten years ago. Engineers, in turn, are freed to delve into even more technically sophisticated areas. As a peripheral advantage, the critical need for technical manpower is partially satisfied.

To manage technological change, we must manage our human resources better. For example, we must commit ourselves to ensuring that none of our workers is laid off because of technological changes, as long as they are willing to be retrained and accept new job assignments. Our experience at Westinghouse has been that employees displaced by robots normally move up to better, more challenging work. We should also rethink who can do what job. For instance, there is a tremendous potential for productivity improvement if the person who knows the machine better than anyone else also has the skills to program the machine while it is working on another job and the skills to debug the programs on the spot.

In Westinghouse, by putting the programs for people first, we expect to multiply the productivity improvements that are gained through technology and capital investments. With participative management, for instance, employees welcome advanced technology because they feel in charge of it. Only when these programs are in place will we emerge from the showcase and token automation phase that manufacturing is presently in.

Today, scientific work in the application of computers to factory automation is in the embryonic stage. We are on the verge of seeing the cost of NC, CNC, and robotics become low enough for these systems to be economically justifiable for many more applications. The cost will continue to decline as application problems are resolved and the computer becomes an understood and respected partner in the manufacturing environment. When this happens, our nation's productivity will be greatly enhanced.

Notes

1 "The Reindustrialization of America," *Business Week*, June 30, 1980, p. 9.
2 "Information Systems Planning Guide," unpublished internal manuscript, Westinghouse Electric Corporation.
3 "The Reindustrialization of America," p. 9.
4 J. K. Krouse, "CAD/CAM"—Bridging the Gap from Design to Production," *IEEE Trans. Proc. Commun* **PC-33**, 191 (1980).
5 "ICAM Program Prospectus, Manufacturing Technology," US Air Force, September 1979.
6 "A Report on Robotics in Japan," *Robotics Today*, **3**, 26 (Fall 1981).
7 Delphi Study, Society of Manufacturing Engineers, Dearborn, Mich., 1981.
8 A. K. Bejczy, *Science*, **208**, 1327 (1980).
9 R. Sugarman, *IEEE Spectrum*, **17**, 53 (September 1980).

Industrial Robots on the Line

Robert Ayres and Steve Miller

Robots have been the subject of much speculation in the past, but now they really are proliferating. Authors Ayres and Miller, respectively professor and lecturer in the Department of Engineering and Public Policy at Carnegie-Mellon University, explain the current functions and future potential of robots. They argue that government, employers, and labour unions must cooperate to ease the massive transition problems. This article comes from Technology Review, *May–June 1982.*

Industrial robots are not humanlike androids that stroll around and converse as R2D2 did in *Star Wars*. They are machine tools—programmable manipulators that can move parts or tools through pre-specified sequences of motions. Like more familiar machine tools, a robot can repeat the same task for prolonged periods with great precision, but a robot's tasks can be extremely complex. Moreover, a robot can be "taught" new tasks and can use accessory tools to extend its range of abilities. State-of-the-art robots—mostly in research labs—even have crude senses of "sight" and "touch" and limited capability to coordinate their manipulators with sensory information.

As yet, robots at work in industry cannot react to unforeseen circumstances and changing environments, nor can they improve performance based on prior experience. Because of these limitations, today's robots are mostly used in repetitive, "preprogrammable" tasks such as spot welding, grinding, spray painting, stacking, and the loading and unloading of machines.

The next generation of robots will have greatly improved vision and sensory feedback, enabling them to perform many more of the tasks still performed by production workers on the factory floor. Thus, robotics has the potential greatly to increase productivity and wealth, but also promises significant social impacts, such as displacement of workers and shifts in the structure of the world economy.

Throughout history, societies have functioned by forcing large numbers of people to perform dull, dirty, dangerous, and demeaning but necessary tasks. Factories, in particular, have traditionally used humans to handle materials, load and unload machines, operate tools, and assemble parts. Such tasks make use of human motor skills and eye–hand coordination but often require little judgement or creativity.

Robots can fill such jobs, providing an immediate overall societal benefit. In a particularly exotic example, robots may someday be used to handle dangerous radioactive materials, including wastes in disposal facilities. Such a robot was built in 1958 by Hughes Aircraft to handle radioactive materials at the Atomic Energy Commission facilities in Albuquerque, NM. Mobile robots would also be extremely useful in exploration, mining, construction, and maintenance and repair work. For example, the US Department of Energy is currently evaluating the potential of robotics in nuclear reactor maintenance, and the Navy and others are actively developing unmanned submersibles—a kind of robot—for both military and civilian purposes. If we succeed in creating orbiting space colonies or industrializing the moon, asteroids, and other planets, it will be with major assistance from robots. The *Viking 2* lander that touched down on Mars in September 1976 was the first of such exploration robots, and similar extended missions are now being planned.

Future robots may serve disabled people such as paraplegics, and especially quadriplegics, full time. For example, voice-activated robots now being developed in the US and Japan might eventually perform a variety of tasks from feeding to page-turning.

In 1978, Quasar Industries of Rutherford, NJ, announced its intention to mass-produce a household android "within two years." The project was a hoax but stimulated press interest, if only because of the vast potential market. And Niemen-Marcus Department Stores introduced a household "robot"—actually a remote-controlled device—in their 1981 catalog. Meanwhile, Joseph Engelberger, president of Unimation, the largest US manufacturer of robots, has promised he will soon have a robot (to be named Isaac, after Asimov) that will serve coffee in his office.

Robotic roots

The term *robot*, from the Czech word *robota*, or serf, was popularized a half-century ago by Czech playwright Karel Capek in his play *R.U.R.* But the concept of programmable machinery dates back to eighteenth-century France, when Bouchon, Vacaunson, Basile, Falcon, and Jacquard developed mechanical looms controlled by punched cards. In the US in the mid-nineteenth century, Christopher Spencer invented the Automat, a programmable lathe that made screws, nuts, and gears. Its cutting patterns could be varied to the end of a rotating drum. Mechanical controls were standard in the machine-tool industry until the 1950s.

Mechanical manipulators also have a long history. In 1892, Seward Babbitt of Pittsburgh patented a rotary crane with a motorized gripper for removing hot ingots from furnaces. The first jointed mechanical arm (specialized for spray painting) that could repeat a series of preset motions was developed by Willard Pollard in 1938.

But the robotic age really began in 1946 when George Devol developed a general-purpose playback device for controlling machines. And in 1954, Devol patented the first manipulator with a playback memory that controlled movements from one point to subsequent points. Devol's early patents were sold to Consolidated Diesel Corp. (Condec) and led to the formation of Unimation, Inc., Condec's robot division. Between 1954 and 1963, Devol and several others patented the major features of the first generation of robots.

These early robots had computerlike functions such as electronic memory, but the pertinent components were permanently wired to perform specific sets of tasks. Software-controlled robots were not commercialized until the early 1970s. The first—controlled by a minicomputer—was offered in 1974 by Cincinnati Milacron, and microprocessor-controlled robots followed several years later. Such "softwired" robots—really specialized "peripherals" of general-purpose computers—enjoy the benefits of huge amounts of "memory" or data storage, and are consequently far more flexible than machines controlled only by specialized electronic logic circuits. The best of today's robots can work in several coordinate systems, are easily reprogrammed, utilize sensors, and to some extent respond to variations in "real time"—that is, as they occur.

How many are there, and where?

Industrial robots in the US are undergoing a virtual population explosion; their numbers have increased from 200 in 1970 to about 4500 today. At the end of 1980, almost 30 percent of the US robot population was owned by only six firms, three in the auto industry. However, robot use has spread to a larger number of firms over the past several years.

Most potential robot users fall within a group of five industries known as "the metalworking sector." As specified by the Standard Industrial Classification Code, the five are primary metals, fabricated metal products, machinery (except electrical), electrical and electronic equipment, and transportation equipment. As of 1980, there was only 1 robot for every 1300 production workers in these industries, and less than 1 robot for every 3000 production workers throughout all manufacturing.

The Japanese have been more aggressive in applying robots to industry: as of the end of 1981, there were nearly 14,000 programmable robots in Japan.

Larger estimates of the Japanese robot population stem from a difference in definition: the Japanese include some 65,000 nonprogrammable manual manipulators and fixed-sequence machines in their own total, but the US definition excludes these.

What robots can and cannot do

Because they are slow and have limited accuracy, force, and versatility, present robotic manipulators are unsatisfactory for many applications. There are also significant trade-offs among the various measures of performance. For example, extremely accurate robots with very small payloads allow only limited kinds of movements and a relatively tiny working volume. Such accurate but motion-limited robots may be appropriate for some operations with very small parts, such as assembling watches and cameras. On the other hand, robots that can move large payloads through greater distances are not usually very precise, but they can be useful in heavy manufacturing industries.

Robots' accuracy is limited by the precision and durability of their power transmission mechanisms, such as hydraulic lines and valves, gear trains, belts, chains, and linkages. All such devices have inherent slop, and they wear with use, making things even more difficult for the control system. New robot designs, such as a direct-drive electric manipulator developed at Carnegie-Mellon University, eliminate transmission mechanisms and pave the way for a new generation of lightweight, high performance robot arms.

Most robots in use today must be "taught" or "walked through" a task in complete detail. However, many industrial tasks such as welding, grinding irregular surfaces, cutting logs, and fitting pieces of cloth together require adjustment as the task proceeds. Such tasks are easy for humans but exceedingly difficult for robots, which must be able to sense key attributes of workpieces to make the necessary adjustments. Robots must also be able to know when a workpiece is damaged or of inferior quality, perhaps removing it from the line, and to recognize when a job is finished.

To do these tasks, robots need sensors that can measure workpieces and their orientations and compare such information with predetermined standards. Such sensors can provide a stream of raw data to the robot's control system, where data processors filter, enhance, and interpret the information, make appropriate decisions, and implement revised instructions.

Computer vision systems that distinguish silhouettes in black and white have been in use since 1973. These systems depend on special lighting and other techniques to produce a high-contrast image. Several commercially available systems that use this technique can inspect, count, locate, and orient parts as well as guide a manipulator to an object, making corrections as needed. More advanced vision systems that can recognize shapes, deduce details in shadowed areas, determine distances stereoscopically, and see in three dimensions are under development.

Researchers are now developing robotic software that can "learn from experience" and do high-level planning. Given a goal and the necessary sensory input, such machines could modify or even create an instructional program, essentially learning on the job. In comparison, today's robots require each step of an instruction to be laid out in complete detail.

Improvements in programming languages can greatly ease this transition to more "intelligent" robots. *Explicitly programmed* languages require the human operator to specify manipulator positions and trajectories. In comparison, new high-level *world-modeling* languages, which contain simple but powerful instructions, automatically generate manipulator positions and trajectories. But such languages can be used only with robots controlled by a general-purpose computer, and a very few currently installed robots have the capability. As a result, today's robots, with their limited data-processing capacity, cannot recognize and pick desired parts from bins, nor can they perform complex assembly and disassembly work routinely.

Contrary to popular belief, American manufactured goods are not primarily mass-produced: between 50 and 75 percent of the dollar value of durable goods manufactured in the US are "batch-produced." In batch production, a relatively small number of products are made over a period of a few weeks or months at most. Despite improvements in computer-controlled machine tools and robots over the past 20 years, most batch production is still quite labor-intensive.

Shifts in consumer preferences and a growing demand for customized products are forcing manufacturers simultaneously to increase both product variety and quality. Thus, a more flexible manufacturing technology is needed because production runs will be shorter and changeovers more frequent. Most important, the need for extensive retooling to accommodate production redesign must be reduced or eliminated. These simultaneous requirements have pushed existing production technologies and management techniques to their limits.

But a robotized factory need have no such constraints. Indeed, "flexible" automation—based on clusters of multipurpose, easily reprogrammed, computer-controlled machines—is ideally suited to batch production. But robots are not yet cost-effective in custom applications in which a great deal of time may be needed for skilled machinists to set up the machines to produce very few copies of the product. Also, the time needed to write the necessary computer programs could exceed the machine operation time. It is usually easier for a machinist to make such custom pieces manually than for engineers to "translate" the procedures into a program of instructions suitable for use by a robot.

Robots are not yet cost-effective in most mass-production applications either, because specialized mass-production machinery can operate at higher speeds and perform more efficiently. But mass-production machinery, such as high-speed transfer lines used to manufacture automobile engines and transmission, has other limitations. Such machinery is custom-built for a single product and can accommodate few, if any, variations. Indeed, an assembly line is virtually a huge specialized machine that can produce only a single product. As a result, substantive design changes in mass-produced goods are costly to implement. It is generally cheaper to scrap the specialized machinery and rebuild the system from scratch.

Thus, mass production would be cheaper if the capital equipment were itself mass-produced and more flexible. The virtue of programmable, general-purpose robots in that a standardized unit may be utilized in many different configurations. Specialization is achieved through changes in software rather than hardware.

The factory of the future

There could be no more dedicated and untiring factory worker than a robot. Robots can repeat tasks such as spot welding and spray painting flawlessly on a variety of workpieces, and they can be quickly reprogrammed to perform entirely new tasks. However, once installed, a particular robot is likely to be a specialist in a particular application because of mobility constraints. Thus, the capabilities of today's programmable machines are not yet fully exploited.

In the next few years, we can expect to see many industrial robots installed in medium-batch manufacturing plants. Robots will feed workpieces to clusters of automatic machines in "work cells." Such work cells may be serialized to form a "closed-loop" manufacturing system controlled by microprocessors. However, stand-alone robots will still be crucial in carrying out preprocessing functions such as cutting raw bar stock, and also for supplementary functions such as heat treating, surface plating, and assembly. Today, human operators still have to measure workpieces manually to ensure that such "closed-loop" machining operations meet specifications. Eventually, when work cells are fully automated, robots will perform such routine measurement and inspection.

The factory of the future will consist of such closed loops linked together in a flexible, computerized manufacturing system. Precursors of such systems have already been built in the US, East Germany, and Japan. These flexible, computerized manufacturing systems could be quickly and cheaply modified to make changes to existing products and even to produce entirely new products inexpensively. In the ideal flexible manufacturing system, the average unit cost of producing one thousand (or million) copies of a product could well approach the average unit cost of producing each of a thousand different products!

Machine utilization would be higher in robotic production systems. In comparison, manually operated machine tools in today's metalworking industries are idle far more than they are in use—from 70 to 95 percent of the time in small job shops and batch production, and between 60 and 80 percent of the time in typical mass-production plants. This downtime stems from scheduled maintenance time, incomplete use of the second and third shifts, plant shutdowns, scheduling inefficiencies, and setup time. And because of the limitations of manual materials-handling systems, there is typically a large work-in-process inventory on the shop floor.

Significantly, manufacturers' durable goods, including machine tools and other capital equipment, are almost entirely batch-produced. The use of robots and computer control could greatly boost the efficiency with which such equipment is produced. Thus, the price of capital goods in relation to other production factors can be expected to decline fairly sharply over the next half-century.

It is difficult to overstate the potential significance of increased productivity on the nation's economy. Manufactured goods might become cheaper as the capital equipment used in production declines in cost. Also, consumer demand might be stimulated by decreasing prices. But because old or worn-out items could be more cheaply replaced, that demand could take new directions. For example, people might increase their demand for educational opportunities or leisure time as their real buying power increases. Finally, a lower real cost for manufactured goods might even decrease the rate of inflation. If inflation is caused by too much money chasing too few goods, a boost in productivity could be an effective way to break out of the vicious cycle.

The robot revolution

As part of the recent Carnegie-Mellon University study called *Impacts of Robotics on the Workforce and Workplace*, members of the Robot Institute of America were asked to rank the factors influencing their decision to install robots. Of the respondents, 19 were robot-users and 19 were considering adoption.

Survey respondents overwhelmingly ranked efforts to reduce labor cost as their main reason for purchasing robots. A 1980 survey conducted by the Charles Stark Draper Laboratories of Cambridge, Mass., also determined that direct labor cost was the primary motivation for using assembly robots. Users frequently pointed out that the return on their investment in robotics would not be favorable without a resulting dramatic decrease in direct labor costs.

One executive speculated that inexperienced users weigh only direct labor costs because they do not know what other categories of cost will be affected. He said that his firm had learned how to quantify indirect benefits such as improved quality and reduction in materials requirements. But other experienced users did not report this kind of learning, and the benefits of improved product quality and increased flexibility were generally considered "nebulous."

Some respondents indicated that they also take into account broader strategic concerns such as long-term competitiveness, but only one firm said outright that it had invested heavily in robotics to improve the quality and competitive standing of its product. That firm was also the only one to emphasize strongly other "intangibles" such as improved production flexibility, and it was alone in not evaluating robot applications, primarily on the basis of conventional return-on-investment calculations.

The firms were asked to estimate what percentage of jobs within a given occupational title could be done by a non-sensor-based robot similar to most of those on the market today ("Level 1" robots), and by the next generation of robots with rudimentary sensing capabilities ("Level 2"). Based on these results, we estimate that Level 1 robots could theoretically replace about 1 million operators, and Level 2 robots could theoretically replace 3 million of a current total of 8 million operators. However, this displacement will take at least 20 years. By 2025, it is conceivable that more sophisticated robots will replace almost all operators in manufacturing (about 8 percent of today's workforce), as well as a number of routine nonmanufacturing jobs.

The private and public sectors should make a concerted effort to plan for these changes. The transition will not be catastrophic if workers are properly trained and directed toward growth areas. In fact, this transition will probably be less dramatic than the impact of office automation. By 2025, most current operators will have retired or left their jobs, and robot manufacturing, programming, and maintenance itself will provide some new jobs, although most will probably not be in manufacturing. Growth sectors in the economy, including undersea and space exploration, may also provide many new jobs. Therefore, young people entering the labor force in the near future will have to learn marketable skills other than welding, machining, and other tasks that are being robotized.

Even though the adjustment problems seems manageable, the potential for social unrest in specific locations cannot be dismissed quite so lightly. Over half of all the unskilled and semiskilled workers in jobs that could be replaced by robots are concentrated in the five major metalworking sectors. Almost one-half of all production workers in these five industries are geographically concentrated in five Great Lakes states—Indiana, Illinois, Michigan, Ohio, and Wisconsin—plus New York and California. The metalworking sector accounts for a large percentage of total manufacturing employment in these states. Therefore, the impact of not improving the productivity and competitive standing of these industries will be concentrated in the same few states.

There may also be a disproportionate impact on racial minorities and women. Nonwhites account for only 11 percent of the national workforce but comprise between 15 and 20 percent of manufacturing operators and laborers. And women employed in semiskilled and unskilled manufacturing jobs are less likely to be represented by labor organizations than their male counterparts, leaving them more vulnerable to displacement by technological innovation.

Minimizing this displacement through attrition does not seem feasible.

According to the Bureau of Labor Statistics data, only 1 to 3 percent of metal working employees leave their place of work as a result of quitting, discharges, permanent disability, death, retirement, and transfers to other companies. (These figures include only people who actually leave their establishment, not those who change jobs within one company.) The vast majority of manufacturing workers still have 20 or more years of active work life left. As of 1980, between two-thirds and three-fourths of these workers were less than 45 years old, which means that barely a third of the workforce would have retired normally by the year 2000. Although skilled workers are usually older, they are not as likely to be replaced by robots in the near future.

Union response

Labor unions will be heavily involved in the move toward robotics. Over one-third of all wage and salary workers and a significantly higher proportion of production workers (85 percent of motor-vehicle equipment operators, 52 percent of laborers, 47 percent of other durable-goods operators, and 41 percent of non-durable-goods operators) are represented by labor organizatons. Over 90 percent of those represented are actual union members.

Unions and employers have devised contract provisions to soften the impact of technological change, as well as to allow workers to share in the benefits from improved productivity. For example, the United Autoworkers Union's "wage-improvement factor" explicitly calls for an annual benefit based on a percentage of increased productivity, independent of cost-of-living considerations.

Companies can spread fewer available jobs among a greater number of employees by giving workers paid time off in addition to legal and religious holidays. The UAW negotiates such paid personal holidays when productivity is increasing within a plant and unit labor requirements are decreasing. Other unions have implemented a similar strategy by increasing standard vacation time. The UAW is also the principal national advocate of providing supplemental benefits in addition to unemployment compensation for workers who are laid off.

"Transitional allowances" can ease the impact on workers transferred from one plant to another. In four union contracts we reviewed, individual allowances range from $500 to $1760. In some cases other benefits such as seniority also follow transferred employees. Severance pay may also provide the firm with a quick but costly means of reducing the size of its workforce—workers may be paid lump sums to leave their jobs and may also receive a percentage of their pension benefits.

Advance notice to workers of technological change is often a stipulation in negotiated contracts. The UAW and International Association of Machinists require committees composed of both union and management representatives to study technological developments. (However, the current contract of the International Brotherhood of Electrical Workers gives management the sole right to administer the introduction of new technology.)

Retraining provisions have also been negotiated as the responsibility of the employer by three of the four unions we studied. As a result of the recently ratified Ford–UAW Agreement, a joint Union–Management Employee Development and Training Program will be established. This center will arrange for or provide "training, retraining, and development assistance for employees displaced by new techologies, new production techniques, and shifts in customer-product preference. Similar efforts will be undertaken for employees displaced as a result of facility closings or discontinuances of operations." In addition, the UAW has run its own training and retraining programs.

The integrity of the bargaining unit is another important bargaining issue. For example, the UAW has made several agreements stating that all jobs previously in a bargaining unit will stay in the unit. Thus, if an operator in a bargaining unit is replaced by a robot, then the robot's operator will also be in the unit.

Investing in human capital

On the whole, private industry is doing very little to prepare workers whose jobs may be eliminated or substantially changed by the use of robots. To ease the transition and ensure optimal productivity from our "human capital," the following goals warrant serious consideration.

1 Affected industries should identify vulnerable categories of workers well before their jobs are actually eliminated.
2 Industry, government, and labor unions should cooperate in planning for long-range employment needs and publicizing new job-skill requirements.
3 Education and training facilities should be established to retrain workers caught in shrinking skill categories in more marketable skills. Otherwise, unemployment is likely to become pervasive, especially among the least-skilled workers.
4 Industry and government separately or jointly should create facilities to locate suitable jobs for displaced workers and to help pay the costs of relocation.
5 Unemployment compensation should be legislatively restructured into a job security fund financed jointly by workers and employers (similar to Social Security). The fund would be used to pay for transportation, maintenance, and retraining of displaced workers on a sliding scale, with benefits proportional to seniority.

The transition to the factory of the future is occurring now. Manufacturers, government, and labor unions can ill afford to ignore the challenge of cooperatively upgrading the skills of the nation's workers when an unprecendented need for people with new skills is imminent. If appropriate measures are not taken, the nation will experience unnecessary economic distress and lost opportunities.

Flexible Manufacturing Systems

Gene Bylinsky with Alicia Hills Moore

The next stage in the race to the completely automatic factory is the introduction of so-called flexible manufacturing systems or FMS. As experienced technology-watcher Gene Bylinsky shows in this piece, which appeared in Fortune *on February 21, 1983, FMS brings us very close to the long-heralded workerless factory. The Japanese, it seems, are in the lead.*

Flexible manufacturing systems—FMS is the bristly acronym most commonly used—complete a process of factory automation that began back in the 1950s. First came numerically controlled machine tools, which performed their operations automatically according to coded instructions on paper or Mylar tape. Then came computer-aided manufacturing, or CAD/CAM, which replaced the drafting board with the CRT screen and the numerical control tape with the computer.

The new systems integrate all these elements. They consist of computer-controlled machining centers that sculpt complicated metal parts at high speed and with great reliability, robots that handle the parts, and remotely guided carts that deliver materials. The components are linked by electronic controls that dictate what will happen at each stage of the manufacturing sequence, even automatically replacing worn-out or broken drill bits and other implements.

Measured against some of the machinery they replace, flexible manufacturing systems seem expensive. A full-scale system, encompassing computer controls, five or more machining centers, and the accompanying transfer robots, can cost $25 million. Even a rudimentary system built around a single machine tool—say, a computer-controlled turning center—might cost about $325,000, while a conventional numerically-controlled turning tool would cost only about $175,000.

But the direct comparison is a poor guide to the economies flexible automation offers, even taking into account the phenomenal productivity gains and asset utilization rates that come with virtually unmanned round-the-clock operation. Because an FMS can be instantly reprogrammed to make new parts or products, a single system can replace several different conventional machining lines, yielding huge savings in capital investment and plant size.

Flexible automation's greatest potential for radical change lies in its capacity to manufacture goods cheaply in small volumes. Since the era of Henry Ford, the unchallenged low-cost production system has been Detroit-style "hard" automation that stamps out look-alike parts in huge volume. There is little flexibility in hard automation's transfer lines, which get their name from the transfer of the product being worked on via a conveyor from one metalworking machine to another. But such mass production is shrinking in importance compared with "batch production" in lots of anywhere from several thousand to one.

Seventy-five percent of all machined parts today are produced in batches of 50 or fewer. Many assembled products too, ranging from airplanes and tractors to office desks and large computers, are made in batches. Even such stalwarts of inflexible mass production as the automakers are developing systems to produce more low-volume models for small market segments.

In the past, batch manufacturing required machines dedicated to a single task. These machines had to be either rebuilt or replaced at the time of product change. Flexible manufacturing brings a degree of diversity to manufacturing never before available. Different products can be made on the same line at will (see figure 6.1). General Electric, for instance, uses flexible automation to make 2000 different versions of its basic electric meter at its Somersworth, New Hampshire, plant with total output of more than 1 million meters a year.

The strategic implications for the manufacturer are truly staggering. Under hard automation the greatest economies were realized only at the most massive scales. But flexible automation makes similar economies available at a wide range of scales. A flexible automation system can turn out a small batch or even a single copy of a product as efficiently as a production line designed to turn out a million identical items. Enthusiasts of flexible automation refer to this capability as "economy of scope."

Economy of scope shatters the tenets of conventional manufacturing. There is no long trip down the learning curve on the factory floor, thanks to the unprecedented precision the system brings to each step of the manufacturing process, from machining to inspection. (There will, of course, be a learning curve in product design, but much of the learning will take place more quickly and cheaply on a computer.) The manufacturer will be able to meet a far greater array of market needs, including quick-changing ones—even the needs of markets the company is not in now. He can keep up with changing fashions in the marketplace—or set them himself by updating his product or launching a new one. He has many more options for building a

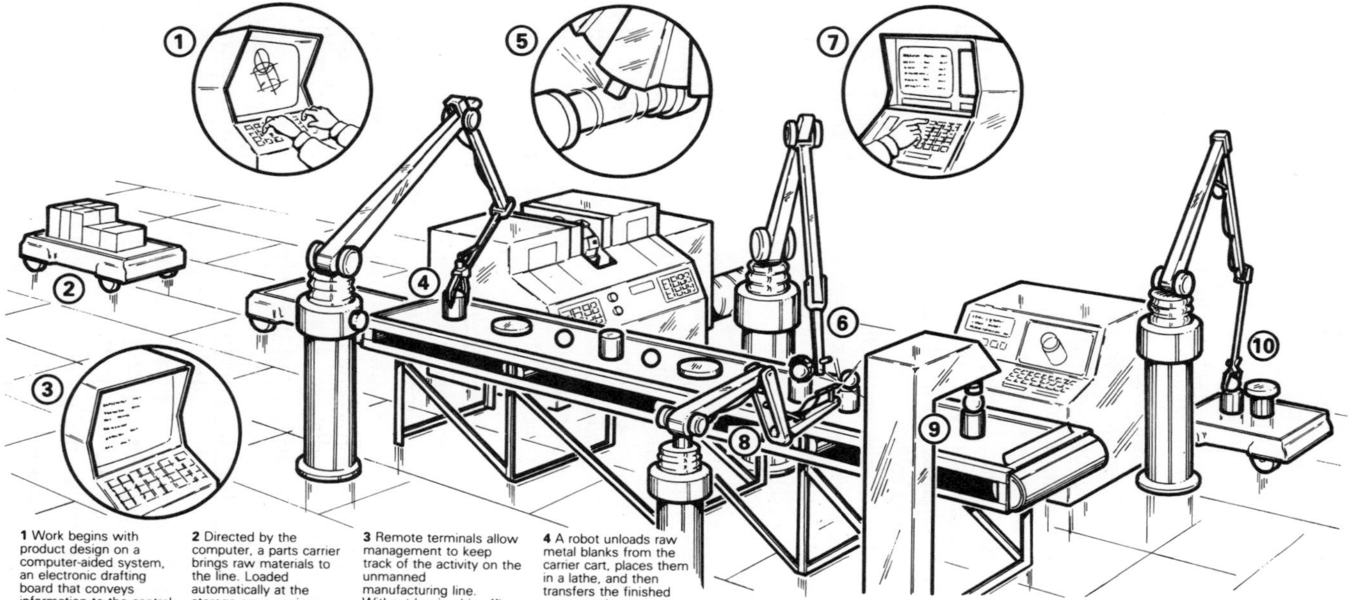

1 Work begins with product design on a computer-aided system, an electronic drafting board that conveys information to the central system.

2 Directed by the computer, a parts carrier brings raw materials to the line. Loaded automatically at the storage area, carriers usually are guided by low-frequency radio signals transmitted through a wire buried in the floor.

3 Remote terminals allow management to keep track of the activity on the unmanned manufacturing line. Without leaving his office a manager can ask the robots what they have done for him today—say, how many products they have turned out.

4 A robot unloads raw metal blanks from the carrier cart, places them in a lathe, and then transfers the finished part onto the conveyor. The "pick and place" robots are programmed for their tasks, but new ones under development will be guided by vision or touch.

5 A revolving holder supplies the appropriate tools for each part to be machined. Directed by the central controller, the lathe automatically picks the right tool and performs the prescribed cutting operations.

6 An assembly robot puts the parts together. The joining of complex parts is more difficult to automate than machining.

7 An electronic foreman —a computer terminal known as a programmable controller —directs the work. Reprogrammed at its keyboard by a human supervisor the controller can change the number and type of products being made.

8 A welding robot joins the parts, making as many welds as necessary on all sides. Until the advent of flexible automation, such robots usually were stand-alone, single-purpose units installed as replacements for human welders.

9 The newly made product is scrutinized by a camera containing a semiconductor chip that can "see" and instantly measure deviations from standards. Automatic inspection is an area where US manufacturers excel.

10 A robot places each product in precisely the right spot on an automatic cart that will carry it off to a shipping area. Should something go wrong, red lights will alert human "tenders," who walk through the plant monitoring production.

Figure 6.1

new plant: FMS frees manufacturers from the tyranny of large-scale investments in hard automation, allowing construction of smaller plants closer to markets.

Flexible manufacturing is the ultimate entrepreneurial system: it will allow fast-thinking manufacturers to move swiftly into brand-new fields and to leave them just as swiftly if need be—at the expense of less agile older producers. As the new tools come increasingly into use, "some companies will find themselves blind-sided by competitors they never imagined existed," says Joseph D. Romano, a vice-president at A. T. Kearney Inc., management consultants.

Flexible manufacturing systems were developed in the US more than ten years ago by Cincinnati Milacron, Kearney & Trecker, and White Consolidated. The US remains a world leader in the technology: the major machine tool builders are being joined by new suppliers with great financial resources and technical abilities, such as GE, Westinghouse, and Bendix. The most unusual new venture is GM's linkup with Fanuc Ltd, Japan's leading robot maker, to form a new company, GMF Robotics. The joint venture will bring together GM's considerable capabilities in design and software and Fanuc's expertise in building and applying robot systems. GMF plans to start building products next year.

However, most of the action in flexible automation is now in Japan, and both American and European manufacturers will soon start feeling the pressure. Like many other manufacturing technologies conceived in the US—among them numerically-controlled machine tools and industrial robots—the FMS was greeted with a yawn by US manufacturers. The Japanese have become the implementers par excellence of this new type of factory automation, not because they are great technical innovators, which they admit they are not, but because they have moved fast in putting the new systems into their factories. Once again, the path to success in a new manufacturing method leads through those Japanese factories set up as spotless little towns with flower beds and tree-lined streets.

A visitor to Japan these days finds the new manufacturing system turning out parts for machine tools in Nagoya, electric motors near Mount Fuji, diesel cylinder blocks in Niigata, and many other products elsewhere. In most cases these plants run on three shifts. During the day skeleton crews work with the machines. At night the robots and the machines work alone.

In Fanuc Ltd's cavernous, bumblebee-yellow buildings in a pine forest near Mount Fuji, automatic machining centers and robots typically toil unattended through the night, with only subdued blue warning lights flashing as unmanned delivery carts move like ghostly messengers through the eerie semidarkness. This plant, one of two in the Fuji complex, makes parts for robots and machine tools (which are assembled manually, however). The machining operation, occupying 54,000 square feet, is supervised at night by a single controller, who watches the machines on close-circuit TV. If something goes wrong, he can shut down that particular part of the operation and reroute the work around it.

Some Americans think that Fanuc's Fuji complex is just a showcase. Some showcase. The total cost of the plant was about $32 million, including the cost of 30 machining cells, which consist of computer-controlled machine tools loaded and unloaded by robots, along with materials-handling robots, monitors, and a programmable controller to orchestrate the operation. Fanuc estimates that it probably would have needed ten times the capital investment for the same output with conventional manufacturing. It also would have needed ten times its labor force of about 100. In this plant one employee supervised ten machining cells; the others act as maintenance men and perform assembly. All in all, the plant is about five times as productive as its conventional counterpart would be.

Across the street, 60 machining cells and 101 robots toil in a big two-storey facility automatically machining parts and assembling them into 10,000 electric motors a month. There is nothing else like it in the world. Men perform maintenance functions here in the daytime. The robots work through the night, in silence marred only by hydraulic sighs and the sibilance of those automatic carts. The first floor of the plant contains the machining cells and 52 robots. Machining is carried out on about 900 types and sizes of motor parts, in lots ranging from 20 to 1000 units. Machined parts are temporarily stored in an automatic warehouse; they are automatically retrieved when they are scheduled for assembly on the second floor.

Yamazaki Machinery Works Ltd operates a flexible automation plant near Nagoya that makes parts of computerized numerically-controlled lathes and machining centers; the latter combine several metalworking machines and incorporate automatic tool changers. In the daytime 12 workers man the $20 million plant. At night only a lone watchman with a flashlight is on duty while the machines keep on working.

A conventional machining system with similar production volume, according to Yamazaki, would require 215 workers and nearly four times as many machines, and would take three months to turn out the parts the new plant makes in three days. The company estimates that over five years of operation its plant will produce after-tax profits of $12 million, compared with $800,000 for a conventional plant that size. Yamazaki is now transferring this technology to its machine-tool-making plant in Florence, Kentucky—bad news for Yamazaki's American competitors.

But the most astonishing Japanese automated factory will be started up next month by Yamazaki about 20 miles from its headquarters near Nagoya. This will be what Tsunehiko "Tony" Yamazaki, the personable senior executive managing director, describes as his company's twenty-first-century factory. The new plant's 65 computer-controlled machine tools and 34 robots will be linked via a fiber-optic cable with the computerized design center back in headquarters. From there the flexible factory can be directed to manufacture the required types of parts—as well as to make the tools and fixtures to produce the parts—by entering into the computer's memory names of various machine tool models scheduled to be produced and pressing a few buttons to get production going. The Yamazaki plant will be the world's first automated factory to be run by telephone from corporate headquarters.

The plant will have workmen, to be sure: 215 men helping produce what would take 2500 in a conventional factory. At maximum capacity the plant will be able to turn out about $230 million of machine tools a year. But production is so organized that sales can be reduced to $80 million a year, if need be, without laying off workers. The Yamazaki plant illustrates yet another aspect of economy of scope: with flexible automation, a manufacturer can economically shrink production capacity to match lower market demand.

Though Japanese machine tool makers are the most ambitious installers of flexible automation, they are by no means alone. FMS is spreading throughout Japanese manufacturing, with Panasonic, Mitsubishi, and other consumer and industrial goods producers installing the new systems.

So far, nothing even remotely comparable is happening in manufacturing in the US or anywhere else in the world. Disturbingly, all of US industry can boast only about 30 flexible manufacturing systems in place; in Japan one large industrial company, Toyoda Machine Tool Co., has more than 30. Frets David Nitzan, director of industrial robotics at the research and consulting firm SRI International, "We are facing another sputnik—a Japanese sputnik."

The growing Japanese lead underscores frequently heard charges that US managers are too remote from technical disciplines to appreciate the potential of such new technologies, and too engrossed with short-run financial results to invest in them. More often than not, machine tool makers report, executives of US manufacturing companies look at something like a flexible machining system only in relation to the narrowly defined functions of the conventional tools it might replace—not for its potential to provide a different, and far more efficient, organization of the manufacturing process. "Cost accounting is a very poor language to communicate new ideas in," observes Paul R. Haas, vice president of Kearney & Trecker's special-products division.

One consequence of this myopia is apparent in the ageing of the US machine tool stock. "Many American factories are barrier reefs—one old, tired technology piled on top of another," says James A. Baker, the executive vice president in charge of GE's drive to develop automated systems. "Even when they build new factories, Americans tend to use the same old machines, shipping them from the old plant to the new plant."

Uncle Sam is the Methuselah of machine tools: more than 34 percent of US tools are 20 or more years old, the highest proportion in any major industrialized nation. Even England is better off; only 24 percent of its machine tools are similarly ancient. In Japan, only 18 percent of machine tools are 20 or more years old; 61 percent are less than 10 years old, against

31 percent in the US. In fact, fewer than 4 percent of machine tools installed in the US are numerically controlled—though the concept has been commercially available for a quarter of a century.

The art of managing the factory is no less antiquated. Fixated by the short run, managers have pursued all sorts of piecemeal efforts to hold down costs without stopping to map out systematic ways of organizing the factory floor for more efficiency. Too often production procedures in US factories appear to be little more than accretions of *ad hoc* solutions to problems ranging from space shortages to union-dictated work rules. Managers focus obsessively on chipping away at direct labor costs rather than exploring better ways to organize the work force—or investigating the extent to which new technologies are making direct labor costs less important.

Japanese corporate leaders, by contrast, tend to be sympathetic with technical disciplines—a far greater proportion of them are engineers—and they have more freedom to incur short-run costs in pursuit of long-run strategic objectives. Perhaps most important of all, a technology like flexible automation is a logical extension of a manufacturing philosophy that views the production of goods as a seamless activity that starts with product design and ends with support in the field—a philosophy, as the Japanese put it, of "making the goods flow like water."

"Japanese management takes a holistic view of manufacturing," says James F. Lardner, a Deere & Co. vice-president who supervised a major restructuring of the farm equipment company's manufacturing operation. "They apply logic and common sense to their problems rather than laboratory investigations and discounted cash-flow calculations."

Even before they began to adopt flexible automation, Japanese plants typically employed far fewer people for a given output than American and European plants. The Japanese were able to do that by reorganizing production—including the placement of machine tools on factory floors. In the US and Europe machine tools are usually grouped by type, and parts are directed to them as required. The Japanese instead place different types of machines together so that each given part can be processed in one place. Much quicker than anyone else, too, the Japanese have taken to such important concepts as "group technology"—the grouping of similar parts into families for easier manufacture and better inventory control. Like so many other Japanese manufacturing methods, group technology isn't new; the idea evolved in the 1920s in Germany.

A central element of the Japanese manufacturing philosophy is the famous just-in-time concept, the system in which materials and components are delivered as required on the shop floor, not accumulated and stored for future use. Since one-third of factory space is usually employed for storage, the savings are substantial. But there is much more. By reducing inventories to the lowest level at which operations can be sustained, the Japanese force their manufacturing organizations to deal with problems previously hidden. For example, the Japanese already practised preventive maintenance on their machine tools to a degree unknown in the US. Just-in-time has forced them to do even better, because the flow-through of products requires every single machine to function perfectly all the time.

Yet the Japanese do nothing that Westerners can't—if they only decide to do it. Marvelously efficient factories using the latest automated equipment exist in the US and Europe, towering like islands of excellence in a sea of stagnation and yielding remarkable benefits to their owners. Take Deere & Co.'s giant new tractor assembly plant in Waterloo, Iowa, which need not take a back seat to any plant in any industry anywhere. In the past few years, Deere has restructured the Waterloo complex from a gigantic, somewhat chaotic job shop into a world-class producer. Deere has poured $500 million into the complex, $150 million of it into the 2.1 million square foot assembly plant. Chassis and engines received from sister plants are joined with tractor cabs and bodies made at Waterloo into gleaming mechanical behemoths. Almost all the materials handling at Waterloo is under computer control. Each part or subassembly—engine, transmission, wheels, and so on—is automatically assigned to a specific customized tractor ordered by a dealer; it is retrieved from storage and delivered automatically to the assembly line just when it is needed. Putting just-in-time to work, Deere has cut inventory in some areas by as much as 50 percent, saving millions of dollars.

Flexible automation allows Deere to build a tractor at least twice as fast as before. And it has given the company a new agility: Deere can now successfully compete not only against other big manufacturers but also against "short-liners" that make only one farm implement in higher volumes. What's more, the company is right now is bidding on a defense contract worth hundreds of millions of dollars. If it wins the contract, it will start making bulldozers, graders, and other heavy construction equipment on its new flexible automation lines—putting new pressure on such established manufacturers as Caterpillar and Case.

But if manufacturers think that they have to pour hundreds of millions of dollars into flexible automation to reap its rewards, they are mistaken. They can begin by acquiring smaller machining centers to modernize portions of their operations. Clifford R. Meyer, president and chief operating officer of Cincinnati Milacron, recalls his delight at recently visiting one of his client companies, which occupies a garage at the end of an alley in a Los Angeles suburb. Inside, a father–son entrepreneurial team mans $450,000 worth of the latest computerized machine tools—successfully competing as a parts supplier to aerospace companies against much larger firms with older production equipment.

Furthermore, not all plants that install flexible automation have to be started from scratch. GE, Ford, and GM are among the manufacturers that have successfully revitalized old plants by installing new machinery. Like many other American companies, GE had for years lagged in automating its own factories. Some old GE plants, James Baker notes, "make Santa's workshop look like the factory of the future." Today, however, GE can boast

Automation creeps into Europe
Ann M. Morrison

Though Europe is still the world's leading maker and exporter of machine tools, flexible manufacturing systems (FMS) are limited to a few showplaces. Companies like Italy's Comau, France's Renault, Britain's Kearney & Trecker Marwin (in which the US firm of Kearney & Trecker has a minority interest), and West Germany's Burkhardt & Weber, among others, offer world-class factory systems. But few Europeans are buying.

Mired in a trough of world recession and high interest rates, manufacturers are reluctant to scrap traditional machines. Says Reinhold Helm, a Düsseldorf-based vice president for A. T. Kearney Inc., a US management consulting firm, "With the average age of all European machinery at 17 years, industry has problems coming to terms with today, let alone tomorrow."

The common European approach is to automate gradually, but a few companies have installed complete systems. A new three-man plant in Colchester, England, turns out the same number of gears in three days that 30 men used to produce in a month. Renault's Bouthéon plant near Saint-Etienne fabricates and assembles truck gearboxes, but it could make any kind of cast-iron or aluminium part that can be contained in a 1600 millimeter cube. At Messerschmitt-Bölkow-Blohm's plant in Augsburg, West Germany, parts for the Tornado fighter are produced on a line so flexible that an American observer remarked, "It could just as well have been making parts for lawn mowers."

With companies slow to jump in, much of the leadership in FMS comes from governments. Companies in Britain buying a flexible manufacturing system can get an outright grant for up to one-third of the cost. Most of the money to equip the Colchester plant came from the government. The West German Ministry for Research and Technology is putting together a system composed of 13 machine tools built by seven separate West German companies, with robots supplied by an eighth. It will be able to produce a wide range of gears. France is putting $350 million into FMS and robot research over the next three years in a joint project involving universities, the National Center for Scientific Research, and industry. It also gives some small grants directly to manufacturers. These programs may not necessarily be equitable ("What happens to the company that applies for a grant after the money runs out?" asks Lucien Rama, secretary of the European Committee for Cooperation of the Machine Tool Industries), but they might at least keep the Japanese in sight.

notable successes in converting some of its old plants into what it calls "factories *with* a future"—a marketing slogan the company uses to impress upon potential customers that they need not build entirely new factories *of* the future. Its meter plant in New Hampshire, modernized at a cost of $25 million, is the epitome of an antiquated multistory mill building. Another ancient GE plant, the Erie, Pennsylvania, locomotive facility, is being transformed with a $300 million investment into an ultramodern automated factory—inside if not on the outside. Building a batch of locomotive frames formerly took about 70 skilled machine operators 16 days; the newly automated factory will turn out these frames in a day—untouched by human hands. The displaced workers are being retrained for other, more sophisticated jobs. As a general matter, in fact, flexible automation threatens employment less than might be supposed. The US faces a shortage of skilled machinists for the rest of the decade, and automation of assembly, where semiskilled jobs predominate, will proceed much more slowly than automation of machining.

GM is also advancing. Last October it installed its first flexible automation system, an Italian-built Comau system with three machining centers, at the Chevrolet Gear and Axle Division in Detroit. Almost immediately GM discovered just how valuable the new manufacturing flexibility can be. When an outside supplier failed to deliver a front-axle component up to quality standards, GM brought the job in-house. It designed and built the tooling for the component on the FMS in ten weeks—a job that normally would have taken up to a year.

As yet, such examples are rare exceptions. The majority of US manufacturers either do not yet grasp the significance of the new technologies, or do not want to invest while money is still expensive and future markets uncertain. Moreover, there are still no turnkey flexible manufacturing systems available. Installation of big systems requires skilled people not all companies may have on hand.

But as GE's Baker argues: "We're running out of time and excuses." The price of delay may be disaster, as the experience of some American machine tool manufacturers show. During the 1970s the Japanese became the world's first mass producers of computerized machine tools. While the majority of US machine tool makers—small companies, in the main—plodded along using old technologies and methods, the Japanese redesigned their tools for easier manufacturing with flexible automation and equipped these tools with advanced yet simple electronic controls that the humblest job shop could understand. Then, in the late 1970s, the Japanese caught the US toolmakers napping in the midst of the capacity crunch, with delivery times stretched up to two years.

The Japanese, to be sure, played an additional trump. Their machine tool industry had been mobilized by Japan's Ministry of International Trade and Industry (MITI) into a cartel and bolstered with millions of dollars in government funds, as shown in documents obtained in Japan by lawyers for Houdaille Industries, the US machine tool maker that has asked President Reagan to deny investment tax credits for some Japanese tools sold in the US.

This double-barreled assault left US makers of numerically controlled machining centers and lathes in shambles. By the end of last year the agile Japanese had captured more than 50 percent of the US market for those machines. The US machine tool industry has lost its erstwhile position as the world's leading producer to competitors not only in Japan but also in Europe and even Taiwan.

The National Academy of Engineering, usually not given to alarmist statements, in a report soon to be issued calls the Japanese and other foreign inroads "a very threatening development that could seriously endanger the future economic security of the American industry." The reason for concern is simple: the machine tool industry is central to the growth of all manufacturing.

There will be more Japanese surprises in the years to come. This October, for instance, the Japanese government's mechanical engineering laboratory near Tokyo will unveil a small prototype factory of the future where novel, laser-equipped machine tools will take production automation a big step beyond where it is now. The new machines will perform metalworking processes now done separately—such as turning, drilling, and milling—all at once, cutting batch production time of metal parts in half and reducing the number of production processes by 60 percent. Lab director Minori Kanai, with 400 researchers on the $60 million project, says he doesn't know of any similar research into factory automation on this scale anywhere else in the world, and he predicts that the brand-new tools will be on the Japanese market in three to five years.

Guide to Further Reading

General

Articles
Peter Marsh, "Britain Advances in Computerized Factories," *New Scientist*, March 19, 1981.
"The Speed-up is Automation," *Business Week* Special Report, August 3, 1981.
Gene Bylinsky, "A New Industrial Revolution is On the Way," *Fortune*, October 5, 1981.
Donald Gerwin, "Do's and Don'ts of Computerized Manufacturing," *Harvard Business Review*, March–April 1982.
Robert Ayres and Steven Miller, "Robotics, CAM and Industrial Productivity," *National Productivity Review*, Winter 1981–2.
Paul Kinnucan, "Flexible Systems Invade the Factory," *High Technology*, July 1983.
John Bell, "Designers Get the Picture," *New Scientist*, March 24, 1983.
Bela Gold, "CAM Sets New Rules for Production," *Harvard Business Review*, November—December 1983.
Geoffrey Charlish, "FMS—a Way of Thinking," *Financial Times*, November 3, 1983.
"Manufacturing Automation," *Financial Times*, Surveys, July 16, 1982 and January 12, 1984.
John K. Krouse, "Automation Revolutionizes Mechanical Design," *High Technology*, March 1984.

Books
Peter Marsh, *The Robot Age* (Abacus, London, 1982).
G. Halevi, *The Role of Computers in Manufacturing Processes* (John Wiley, New York, 1982).
Graham Winch (ed.), *Information Technology in Manufacturing Processes* (Rossendale, London, 1983).
A. Sorge, G. Hartmann, M. Warner and I. Nicholas, *Microelectronics and Manpower in Manufacturing* (Gower Press, Aldershot, England, 1983).

Robots

Articles

"The Robot Revolution," *Time* magazine cover story, December 8, 1980.
Hazel Duffy, "The Robot Comes of Age," *Financial Times*, May 19, 1981.
Michael Cusumano, "Robots Step Out of the Factory," *New Scientist*, January 6, 1983.
James S. Albus, "Robots in the Workplace: The Key to a Prosperous Future," and Vary T. Coates, "The Potential Impacts of Robotics," in *The Futurist*, February, 1983.
John Thackray, "America's Robotics Rising," *Management Today*, February 1983.
Keith Rathmill, "The Great Robotics Explosion," *Management Today*, July 1983.
Peter Bruce, "Robots: Why Tomorrow Has Been Delayed," *Financial Times*, September 8, 1983.
Christopher Clarke and Jeremy Cecil-Wright, "The Road to Robotics," *Management Today*, January 1984.

Books

Joe Engelberger, *Robotics in Practice* (AMACOM, New York, and Kogan Page, London, 1981).
G. L. Simons, *Robots in Industry* (NCC Publications, Manchester, England, 1981).
Exploratory Workshop on the Societal Impacts of Robotics (Office of Technology Assessment, US Congress, Washington DC 1982).
Robert Ayres and Steven Miller, *Robotics: Applications and Social Implications* (Ballinger Press, Cambridge, MA, 1983).

US competitiveness fears

Articles

"The Reindustrialization of America," *Business Week* Special Issue, June 30, 1980; later published by McGraw-Hill, New York, 1981.
Robert A. Leone and Stephen P. Bradley, "Toward an Effective Industrial Policy," *Harvard Business Review*, November–December 1981.
Robert B. Reich, "Industrial Policy: Ten Concrete, Practical Steps to Building a Dynamic, Growing and Fair American Economy," *The New Republic*, March 31, 1982.
Ronald E. Müeller and David H. Moore, "America's Blind Spot: Industrial Policy," *Challenge*, January–February 1982.
William Serrin, "Collapse of Our Industrial Heartland," *New York Times Magazine*, June 6, 1982.
"Manufacturing Is a Flower," *Time*, March 26, 1984.

Books

US Industrial Competitiveness: A Comparison of Steel, Electronics and Automobiles (Office of Technology Assessment, US Congress, Washington DC, July, 1981).
Ira C. Magaziner and Robert B. Reich, *Minding America's Business: The Decline and Rise of the American Economy* (Harcourt, Brace, Jovanovich, New York, 1982).

Barry Bluestone and Bennett Harrison, *The Deindustrialization of America: Plant Closures, Community Abandonment, and the Dismantling of Basic Industry* (Basic Books, New York, 1982).

Michael le Boeuf, *The Productivity Challenge: How to Make it Work for America and You* (McGraw-Hill, New York, 1982).

Terry F. Buss and F. Stevens Redburn, *Shutdown in Youngstown: Public Policy for Mass Unemployment* (State University of New York Press, Albany, NY, 1982).

William J. Abernathy, Kim B. Clark and Alan M. Kantrow, *Industrial Renaissance: Producing a Competitive Future for America* (Basic Books, New York, 1983).

Robert B. Reich, *The Next American Frontier* (Times Books, New York, 1983 and Penguin, 1984).

Japan

"Japan's Strategy for the '80s," *Business Week* Special Issue, December 14, 1981.

William Ouchi, *Theory Z: How American Business Can Meet the Japanese Challenge* (Addison-Wesley, Reading, MA, 1981).

Richard Tanner Pascale and Anthony G. Athos, *The Art of Japanese Management: Implications for American Executives* (Simon & Schuster, New York, 1981, and Penguin, Harmondsworth, England, 1982).

Richard D. Robinson, "Can the Japanese Keep it Up?" *Technology Review*, August–September 1982. A very interesting piece on the Japanese success story, highlighting low Japanese military spending and their avoidance of the high costs of adversarial relationships (management–labour, business–government) pertaining in the West.

Gene Gregory, "Japan's Industrial Revolution: The Next Phase," *The Futurist*, August 1982.

Richard J. Schonberger, *Japanese Manufacturing Techniques* (Free Press, New York, 1982).

Britain

Peter Marsh, "A Fresh Start for British Machine Tools," *New Scientist*, February 26, 1981.

George Avlonitis and Stephen Parkinson, "The Automatic Factory Dawns," *Management Today*, May, 1981. On why the UK lags.

Geoffrey Charlish, "Land of Hope But Not Much Automated Glory," *Financial Times*, February 26, 1982.

Jim Northcott with Petra Rogers, *Microelectronics In Industry: What's Happening in Britain*, final, summary report of a project by the Policy Studies Institute (PSI, London, March 1982).

Jim Northcott and Petra Rogers, *Microelectronics in British Industry: the Pattern of Change*, follow-up report (PSI, London, March 1984).

7 Office of the Future

The Mechanization of Office Work

Vincent E. Giuliano

The office is the primary locus of information-handling, the activity that is coming to dominate the US economy. Vincent E. Giuliano, a senior researcher with consultants Arthur D. Little, Inc, describes the evolution of the information-age office through the pre-industrial and industrial stages. As the new information technology replaces paperwork, productivity, customer service, and job satisfaction can be improved. Taken from Scientific American, *September 1982.*

Mechanization was applied first to the processing of tangible goods: crops in agriculture, raw materials in mining, industrial products in manufacturing. The kind of work that is benefiting most from new technology today, however, is above all the processing of an intangible commodity: information. As machines based mainly on the digital computer and other microelectronic devices become less expensive and more powerful, they are being introduced for gathering, storing, manipulating, and communicating information. At the same time information-related activities are becoming ever more important in American society and the American economy; the majority of workers are already engaged in such activities, and the proportion of them is increasing. The changes can be expected to alter profoundly the nature of the primary locus of information work: the office.

An office is a place where people read, think, write, and communicate; where proposals are considered and plans are made; where money is collected and spent; where businesses and other organizations are managed. The technology for doing all these things is changing with the accelerating introduction of new information-processing machines, programs for operating them, and communications systems for interconnecting them. The transformation entails not only a shift from paper to electronics but also a fundamental change in the nature and organization of office work, in uses of information and communications, and even in the meaning of the office as a particular place occupied during certain hours.

Office mechanization started in the second half of the nineteenth century. In 1850 the quill pen had not yet been fully replaced by the steel nib, and taking pen to paper was still the main technology of office work. By 1900 a number of mechanical devices had established a place in the office, notably Morse's telegraph, Bell's telephone, Edison's dictating machine, and the typewriter.

In 1850 there were at most a few dozen "writing machines" in existence, and each of them was a unique, handmade creation. Typewriters were among the high-technology items of the era; they could be made in large numbers and at a reasonable cost only with the adoption and further development of the techniques of precision manufacturing with interchangeable parts developed by Colt and Remington for the production of pistols and rifles during the Civil War. By the late 1890s dozens of companies were manufacturing typewriters of diverse designs, with a variety of layouts for the keyboard and with ingenious mechanical arrangements. (Some even had the type arrayed on a moving, cylindrical element and thus were 70 years ahead of their time.) By 1900 more than 100,000 typewriters had been sold and more than 20,000 new machines were being built each year. As precision in the casting, machining, and assembly of metal parts improved and the cost of these processes was lowered, typewriters became generally affordable in offices and homes. The evolution of typewriter usage was comparable to what is now taking place—in only about a decade—in the usage of office computers and small personal computers.

With the typewriter came an increase in the size of offices and in their number, in the number of people employed in them and in the variety of their jobs. There were also changes in the social structure of the office. For example, office work had remained a male occupation even after some women had been recruited into factories. (Consider the staffing of Scrooge's office in Charles Dickens' *A Christmas Carol.*) Office mechanization was a force powerful enough to overcome a longstanding reluctance to have women work in a male environment. Large numbers of women were employed in offices as a direct result of the introduction of the typewriter.

The first half of the twentieth century saw a further refinement of existing office technologies and the introduction of a number of new ones. Among the developments were the teletypewriter, automatic telephone switching, ticker tape, the electric typewriter, duplicating machines and copiers, adding machines and calculators, tape recorders for dictation, offset printing presses small enough for office use, and data-processing equipment operated with punched paper cards. The new devices were accompanied by a rapid expansion in the volume of office communications and in the number of people engaged in white-collar work.

The first computers in offices were crude and very expensive by today's standards. By the mid-1960s most large businesses had turned to computers to facilitate such routine "back office" tasks as storing payroll data and issuing checks, controlling inventory and monitoring the payment of bills. With advances in solid-state circuit components and then with microelectronics the computer became much smaller and cheaper. Remote terminals, consisting of either a teletypewriter or a keyboard and a video display, began to appear, generally tapping the central processing and storage facilities of a mainframe computer. There was steady improvement in the cost effectiveness of data-processing equipment. All of this was reflected in a remarkable expansion of the computer industry. The late 1960s and the 1970s also saw the advent of inexpensive copiers, minicomputers, small and affordable private automated branch exchanges (electronic switchboards), the word processor (the typewriter's successor) and then, toward the end of the 1970s, the microcomputer.

An anthropologist visiting an office today would see much that he would have seen 25 years ago. He would see people reading, writing on paper, handling mail, talking with one another face to face and on the telephone, typing, operating calculators, dictating, filing, and retrieving files from metal cabinets. He would observe some new behavior too. He would see a surprising number of people working with devices that have a typewriterlike keyboard but also have a video screen or an automatic printing element. In 1955 the odds were overwhelming that someone working at an alphabetic keyboard device was female and either a typist or a key-punch operator. No longer. The keyboard workers are both female and male, and the typewriterlike devices now accomplish an astonishing variety of tasks.

Some of the keyboard workers are indeed secretaries preparing or correcting conventional correspondence on word processors. Other workers are at similar keyboards that serve as computer terminals. In one office they are managers checking the latest information on production performance, which is stored in a corporate data base in the company's mainframe computer. Economists are doing econometric modeling, perhaps calling on programs and data in a commercial service bureau across the continent. Librarians are working at terminals connected to a national network that merges the catalogues of thousands of participating libraries. Attorneys and law clerks are at terminals linked to a company whose files can be searched to retrieve the full text of court decisions made anywhere in the country. Airline personnel and travel agents make reservations at terminals of a nationwide network. Some of the devices are self-contained personal computers that engineers and scientists, business executives, and many other people depend on for computation, data analysis, scheduling, and other tasks.

Many of the users of terminals and small computers can communicate with one another and with their home offices through one of the half-dozen "electronic mail" networks now in existence in the us. A surprising number of people are doing these things not only in the office but also at home, on the factory floor, and while traveling. This article was written with a portable personal computer at home, in a hotel in Puerto Rico, and at a cottage in New Hampshire. I have drawn on information from personal files in my company's mainframe computer and have also checked parts of the text with colleagues by electronic mail.

What all of this adds up to is a shift from traditional ways of doing office work based mainly on paper to reliance on a variety of keyboard-and-display devices, or personal work stations. A work station may or may not have its own internal computer, but it is ultimately linked to a computer (or to several of them) and to data bases, communications systems, and any of thousands of support services. Today the work stations in widest service handle written and numerical information. In less than a decade machines will be generally available that also handle color graphics and store and transmit voice messages, as the most advanced work stations do today.

My colleagues and I at Arthur D. Little, Inc., expect that by 1990 between 40 and 50 percent of all American workers will be making daily use of electronic-terminal equipment. Some 38 million terminal-based work stations of various kinds are by then likely to be installed in offices, factories, and schools. There may be 34 million home terminals (although most of them may not function as full work stations). In addition, we expect there will be at least 7 million portable terminals resembling today's hand-held calculators, most of them quite inexpensive.

Until recently most work stations and their supporting devices and data-base resources were designed to serve a single purpose: to prepare text, access stock market data, or make air travel reservations, for example. The stockbroker's terminal started out as a replacement for the ticker tape, the word processor as a replacement for the typewriter. The first terminals therefore served as complete work stations only for people who were engaged in a more or less repetitive task.

Now the capabilities of the work station have been extended by developments in the technology of information processing, in communications, and in enhancements of the "software," or programs, essential to the operation of any computer system. A variety of resources and functions have become accessible from a single work station. The stockbroker can not only check current prices with his terminal but also retrieve from his company's data base a customer's portfolio and retrieve from a distant data base information on stock-price trends over many years. Millions of current and historical news items can also be called up on the screen. He can issue orders to buy or sell stock, send messages to other brokers, and generate charts and tables, which can then be incorporated into a newsletter addressed to customers.

It is not only in large corporations that such tools are found. Low-cost personal computers and telecommunications-based services available to individuals make it possible for them to enjoy a highly mechanized work environment; indeed, many professionals and many office workers in small businesses have work station resources superior to those in large corporations where the pace of office mechanization has been slow.

By the year 2000 there will surely be new technology for information handling, some of which cannot now be foreseen. What can be predicted is that more capable machinery will be available at lower cost. Already a personal computer the size of a briefcase has the power and information storage capacity of a mainframe computer of 1955. For a small computer an approximate measure of performance is the "width" of the data path, that is, the number of bits, or binary digits, processed at a time. Computational speed can be represented roughly by the frequency in megahertz of the electronic clock that synchronizes all operations in the central processor. Memory capacity is expressed in bytes; a byte is a group of eight bits. The customary unit is the kilobyte, which is not 1000 bytes but rather 2^{10}, or 1024. Only three years ago a powerful personal computer had 48 kilobytes of working memory and an 8-bit processor running at a rate of one megahertz.

Today about the same amount of money buys a machine with 256 kilobytes of working memory and a 16-bit processor chip that runs at 4 megahertz or more. Storage capacity and processing power will continue to increase—and their costs will continue to decrease—geometrically. By the year 2000 memory and processing power should be so cheap that they will no longer be limiting factors in the cost of information handling; they will be available as needed anywhere in an organization. The next 20 years will also see the continuing extension of high-capacity communications, of networks for the exchange of information between work stations and other computers, and of centralized data banks. Together these developments will provide access to information, to processing capacity, and to communications facilities no matter where the worker is or what time it is.

New technology inevitably affects the organization of work. One can define three evolutionary stages of office organization, which I shall designate pre-industrial, industrial, and information-age. Each stage is characterized not only by its technology but also by its style of management, personnel policies, hierarchy of supervisory and managerial staff, standards of performance, and human relations among office workers and between the workers and their clients or customers.

The first two stages correspond to the well-understood artisan and industrial models of production; the nature of the third stage is only now becoming clear. The operation of a pre-industrial office depends largely on the performance of individuals, without much benefit from either systematic work organization or machines. The industrial office organizes people to serve the needs of a rigid production system and its machines. The information-age office has the potential of combining systems and machines to the benefit of both individual workers and their clients.

Most small business, professional, general management, and executive

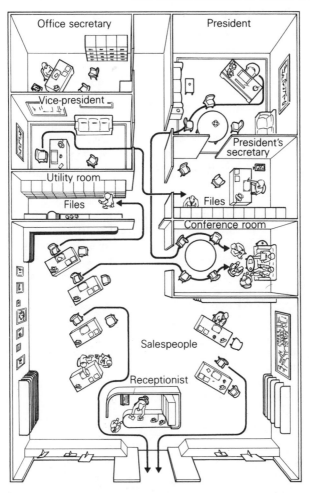

Figure 7.1 Pre-industrial office organization dates back to the mid-nineteenth century but is still typical of most professional, small-business, and even corporate management offices today. It is represented here by a hypothetical real estate brokerage. There is little systematic organization. Each person does his job more or less independently, moving about as necessary to retrieve a file, take a client to see a property, or attend a meeting where the sale of a house is made final. Individuals can have different styles of work, and human relations are important. The pre-industrial model of office organization can still be effective for some small operations. Conversion to information-age methods is fairly easy.

offices are still at the pre-industrial stage. In a pre-industrial office little conscious attention if any is paid to such things as a systematic flow of work, the efficiency or productivity of work methods, or modern information technologies. What information-handling devices are present (telephones, copiers, and even word processors) may be central to the operation, but there is no deliberate effort to get the maximum advantage from them. Good human relations often develop among the employees; loyalty, understanding, and mutual respect have major roles in holding the organization together. An employee is expected to learn his job, to do what is wanted and needed, and to ask for help when it is necessary. Varied personal styles of work shape the style of the operation and contribute to its success.

Pre-industrial office organization generally works well only as long as the operation remains small in scale and fairly simple. It is inefficient for handling either a large volume of transactions or complex procedures requiring the coordination of a variety of data sources. If the work load increases in such an office, or if business conditions get more complex, the typical response is to ask people to work harder and then to hire more employees. Such steps are likely to be of only temporary benefit, however. Without the help of additional systems or technology, effectiveness and morale may soon begin to break down.

One response to the limitations of preindustrial office organization has been to bring to bear in the office the principles of work simplification, specialization, and time-and-motion efficiency articulated for factory work some 70 years ago by Frederick W. Taylor. The result is the industrial-stage office, which is essentially a production line. Work (in the form of paper documents or a folder of papers related to one customer) moves from desk to desk just as parts move from station to station along an assembly line. Each worker gets a sheaf of papers in an "in" box; his job is to perform one or two incremental steps in their processing and then to pass the paper through an "out" box to the next person, who performs the next steps. Jobs are simple, repetitive, and unsatisfying. A worker may do no more than staple or file or copy, or perhaps check and confirm or correct one element of data. And of course everyone has to work together during the same hours in the same office to sustain the flow of paper.

The production-line approach has been considered particularly suitable for office activities in which the main job is handling a large volume of customer transactions, as in sending out bills or processing insurance claims. Many large production-line offices were instituted in the early days of computerization, when information had to be gathered into large batches before it could be processed by the computer; input to the machine then took the form of punched cards and output consisted of large books of printouts. Because early computers could do only a few steps of a complex process, the industrial office had to shape people's tasks to fit the needs of the machine. Computers and means of communicating with them have now been improved, but many large transaction-handling offices are still stuck at the industrial stage.

The industrial model of office organization is based on a deliberate endeavor to maximize efficiency and output. To create an assembly line the flow of work must be analyzed, discrete tasks must be isolated, and work must be measured in some way. There is a need for standardization of jobs, transactions, technologies, and even personal interactions. A fragmentation of responsibility goes hand in hand with bureaucratic organization and the

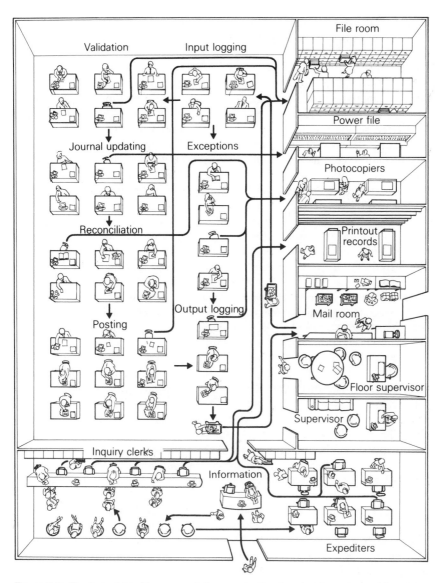

Figure 7.2 The industrial office, essentially a production line, has been favored for operations handling a large number of transactions, as in this, the claims adjustment department of an insurance company. Tasks are fragmented and standardized. Documents are carried from the mail room to the beginning of the production line and eventually emerge at the other end; the flow is indicated by the arrows. Successive groups of clerks carry out incremental steps in the processing of a claim; in general, they leave their desks only to retrieve files or examine computer printouts. If clients make inquiries, they are dealt with by clerks who may be able in time to answer a specific question but can seldom follow through to solve a problem. The work is usually dull. The flow of information is slow, and service is poor.

proliferation of paperwork. Most of the workers have little sense of the overall task to which they are contributing their work, or of how the system functions as a whole.

The industrial office has serious disadvantages. Many errors tend to arise in a production-line process. Because of the subdivision of tasks, efforts to correct errors must often be made without access to all pertinent information, with the result that the errors are sometimes not corrected but compounded. Moreover, production-line operations can be surprisingly labor-intensive and costly. As more people are hired to cope with an error rate that increases faster than the volume of transactions, the cost per transaction increases and efficiency declines.

Effective people do not want to stay in boring jobs; people who do stay often lack interest in their work, which becomes apparent to the customer. Even if workers do their best, the system may defeat them, and customer service is likely to be poor. Because a given item can take weeks to flow through the pipeline it is often difficult to answer customer inquiries about the current status of an account, and even harder to take corrective action quickly. For example, a clerk may be able to check a sales slip and agree that a customer's bill is incorrect; in many instances, however, the clerk is able to change the account only by feeding a new input into the production line, with little assurance that it will have the desired effect. As a result, the billing error can be adjusted incorrectly or can be repeated for several months.

In the mid-1970's the recognition of these limitations, combined with the availability of new work-station information systems, motivated a few progressive banks and other services organizations with a heavy load of transactions to take the next step: they converted certain departments to a mode of operation more appropriate to the information age. The information-age office exploits new technology to preserve the best aspects of the earlier stages and avoid their failings. At its best it combines terminal-based work stations, a continuously updated data base, and communications to attain high efficiency along with a return to people-centered work rather than machine-centered work. In the information-age office the machine is paced to the needs and abilities of the person who works with it. Instead of executing a small number of steps repetitively for a large number of accounts, one individual handles all customer-related activities for a much smaller number of accounts. Each worker has a terminal linked to a computer that maintains a data base of all customer-related records, which are updated as information is entered into the system. The worker becomes an account manager, works directly with the customer, and is fully accountable to the customer.

Information is added incrementally to the master data base. The stored data are under the control of the worker, who can therefore be made responsible for correcting any errors that arise as well as for handling all transactions. Since information is updated as it becomes available, there is no such thing as "work in process," with its attendant uncertainties. An inquiry or a change in status can be handled immediately over the telephone: the sales

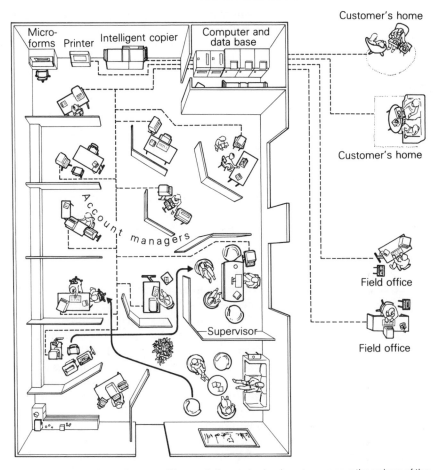

Figure 7.3   The information age office exploits new technology to preserve the values of the pre-industrial office while handling a large volume of complex information. The drawing shows an information age claims adjustment department. Each adjuster mans a work station, which is linked to a compuer that maintains and continually updates all client records. Each adjuster can therefore operate as an account manager, handling all operations for a few clients rather than one repetitive operation for a large number of clients. Necessary action can be taken immediately. Forms are updated and letters are written at the same work station that gives access to stored data, and the forms and letters can be printed automatically. The same facilities are available to adjusters visiting a client's home or working in one of the company's field offices (right). The work is more interesting, service to clients is improved, and costs are reduced.

slip can be inspected, the customer's account can be adjusted and the bill that is about to be mailed can be corrected accordingly.

The design of effective systems and the measurement of productivity are still important in the information-age office with a large volume of transactions, but the context is different from that of the industrial office.

Productivity is no longer measured by hours of work or number of items processed; it is judged by how well customers are served. Are they satisfied? Are they willing to bring their business back? Are they willing to pay a premium for a high level of service?

To the extent that the answers are "yes," the company gains an important competitive advantage. Even if cost-cutting is not the only objective, the company can expect dramatic savings in personnel costs. Staff reductions of as much as 50 percent have been common in departments making the changeover to a work-station system. Those employees who remain benefit from a marked improvement in the quality of their working life.

The benefits of the information-age office are not limited to the transaction-intensive office. A similar transformation can enhance productivity, effectiveness, and job satisfaction in offices concerned with management, general administration, and research. Most such offices are still in the pre-industrial stage. They can be transformed to the information-age stage by the introduction of such person-centered technologies as the work station and electronic mail.

Once most of the activities of a job are centered on the work station the nature of the office can be transformed in still another way: there is no longer any need to assemble all workers at the same place and time. Portable terminals and computers, equipped with appropriate software and facilities for communication (including the telephone), create a "virtual" office, which is essentially anywhere the worker happens to be: at home, visiting a client or customer, in a hotel or even in an airplane. The remote work station can communicate electronically with the central office, and so it extends the range of places where written and numerical material can be generated, stored, retrieved, manipulated, or communicated.

The effects of small-computer technology on the locale of work are analogous to those of the telephone. Because of the almost universal distribution of telephones it is not necessary to go to the office to call a customer or a co-worker, but until now it has been necessary to go there to write or dictate a letter, to read mail or to find something in a file. Now the work stations and ancillary electronic devices of an automated office can be linked to external terminals and personal computers. The job is no longer tied to the flow of paper across a designated desk; it is tied to the worker himself. The individual can therefore organize his own time and decide where and when he wants to do his work. Individuals who work best early in the morning or late at night can do so. A project team I have been working with for about a year has members in several East Coast and West Coast cities and rural areas, and we communicate regularly by electronic mail. The cost of the correspondence is about a tenth of the cost of regular mail per item, and it turns out that about half of the messages are generated outside of offices and outside of conventional working hours.

What will happen to the physical office? It has its virtues, after all. The office provides a home for organizations, a place for people to come together face to face, and a work-oriented environment away from home. Many people need the structure of an office schedule; they like (or at least they are accustomed to) compartmentalization of the day and the week into time for work and time for other activities. Another role for the office is to house centralized forms of communications technology, such as facilities for video conferences, that are too expensive for the home. For these reasons and others I think the physical office will remain a part of working life, at least for as long as I am working. There will be continuing change, however, in how often some workers go to the office and in why they go there.

Many powerful factors are operating together to propel the transformation of office work. A complex set of feedback loops links economic and social change, new developments in information technology, the widespread adoption of the technology, and the introduction of the new office organization the technology makes possible. The large number of information workers, for example, stimulates interest in enhancing their productivity. The concern for productivity serves to increase demand for technologies that can reduce the cost of handling information. Thus several trends reinforce one another to generate an ever stronger market for information products and services. The infiltration of the new devices into the workplace in turn creates an environment in which working electronically is the normal expectation of the worker.

Economics is a major factor. It is becoming far cheaper to communicate electronically than it is to communicate on paper. The transition to word processing from multidraft secretarial typing can reduce secretarial costs from more than $7 per letter to less than $2. Even more dramatic savings are associated with electronic mail, which can bring the cost of sending a message down to 30 cents or less. Electronic filing, in which a "document" is stored and indexed in a computer memory, brings further savings. (The highest-cost activities in manual correspondence are making multiple copies, filing them, and retrieving them.) Such obvious reductions in cost are overshadowed by the savings in the time of managers and executives, the largest element by far in the cost of running an office.

The savings are becoming more significant each year as the cost of the electronic technology is reduced. For example, fast semiconductor memory is a tenth as expensive now as it was in 1975; the cost will drop by another factor of 10 by 1995. The result has been to bring into the individual consumer's price range information-handling capabilities that only a few years ago called for very expensive equipment.

As the market for mechanized work stations expands, more money is invested in research and development for communications, electronics, software, office mechanization systems, and the like. The time span between the development, introduction, and obsolescence of a product becomes shorter. Each year brings a new generation of semiconductor devices; each generation makes possible a new set of applications. The dramatic improvement in products in turn builds demand for them and strengthens the trend toward office mechanization.

Whether a company's business is in farming, mining, manufacturing, transportation, or retailing, its management, marketing, distribution, and other operating controls are basically office-centered, information-handling activities. As the number of blue-collar workers decreases, the proportion of white-collar workers even in manufacturing organizations continues to increase. In virtually all commercial enterprises one finds executives, managers, clerks, and secretaries; in most organizations there are also more specialized information workers, such as engineers and scientists, attorneys, salesmen, librarians, computer programmers, and word processors. These people constitute the human capital resources that can make an information-intensive economy viable.

Yet a tendency to think of white-collar workers in offices as support personnel, outside the economic mainstream, has tended to inhibit the transformation of office work. Physical activities that produce food, minerals, and manufactured goods have been regarded as the only truly productive ones, whereas the handling of information has been considered necessary but essentially nonproductive. This way of looking at things (which may have been appropriate in an industrial society) persists today, even in the minds of economists who call for the "reindustrialization of America." It deeply affects the thinking of corporate management.

Even though most work in American society is information work and most such work is done in offices, the benefits of an increase in the productivity of office workers are not always within the field of view of managers. For those who retain a pre-industrial view of office organization, the very concept of productivity seems irrelevant or inappropriate in the context of offices or information work. Those who have an industrial office orientation tend to focus on labor saving measurements; the installation of new technology and a system for exploiting it is evaluated only in the context of cutting visible office costs.

It is in offices that the basic decisions are made that determine the cost effectiveness of an entire organization. The office is the place where the timeliness of a decision or of a response can have immense consequences. If the office is ineffective, the organization must be ineffective. As it happens, moreover, a high degree of mechanization of the kind described in this article is much less expensive in the office than analogous mechanization is in the factory or on the farm.

The mechanization of office work is an essential element of the transformation of American society to one in which information work is the chief economic activity. If new information technology is properly employed, it can enable organizations to attain the following objectives:

1 a reduction of information "float," that is, a decrease in the delay and uncertainty occasioned by the inaccessibility of information that is being typed, is in the mail, has been misfiled or is simply in an office that is closed for the weekend;
2 the elimination of redundant work and unnecessary tasks such as retyping and laborious manual filing and retrieval;
3 better utilization of human resources for tasks that require judgment, initiative, and rapid communication;
4 faster, better decision-making that takes into account multiple, complex factors; and
5 full exploitation of the virtual office through expansion of the work-place in space and time.

The Automated Office: Myth Versus Reality

A. Richard Immel

Progress toward the office of the future has been slower than many expected. Word processing has gained acceptance, but electronic mail and video conferencing have not. A. Richard Immel also says that the whole notion of office "productivity" needs examination. This article first appeared in the May 1983 issue of Popular Computing, *a magazine to which the author is a contributing editor.*

> *"Tell me, then, what you think of the . . ." There Dr Breeze paused, took a long breath, and puffed himself up to say in a stentorian voice, "the Office of the Future?"*
>
> *Perhaps it was the gathering blue dusk, the sense of disorientation created by the cluttered room, or the jet lag, the tiredness, the mental strain. Anyway I hear myself say:*
>
> *"There isn't going to be an Office of the Future."*
>
> *"Oh!" said Dr Breeze in mock astonishment. "You must not let anyone hear such words." Blushing, he went to the door and closed it carefully, after looking to the left and to the right to make sure no one had heard my statement. When he came back, he spoke low, in conspiratorial tones.*
>
> *"Who are you, some kind of anarchist? Everybody is working on the Office of the Future, everybody is investing in it, and now you're saying it isn't going to exist. They'll kill you if they find out."*
>
> —imaginary dialogue between Jacques Vallee and Dr Breeze, the Wise Gnome of Washington [from Jacques Vallee's book, *The Network Revolution: Confessions of a Computer Scientist*, And/Or Press Inc., Berkeley, CA, 1982].

It all started back in 1964, when IBM stuffed a magnetic-card recording device into a Selectric typewriter. The company called it a word processor, and offices haven't been the same since. Nowadays, of course, word processing is humdrum. The new catchwords are *multifunction workstation, knowledge workers, type I* and *type II offices*, and, the ultimate, *the automated office*: a paperless, push-button, electronic world where high finance meets high technology in a triumph of white-collar productivity. In the office of the future, company execs will send personal memos to employees 10,000 miles away with the touch of a button and instantly extract vital data for a board meeting from the company's electronic files. They will browse electronically through video calendars and tickler files to schedule meetings with subordinates or check on overdue reports. And they will chat simultaneously with division chiefs in Pocatello, Auckland, and Buenos Aires via a televised conference at company headquarters.

This is all possible using today's technology, yet how many companies are doing all of this?

None. Zero. Zip.

That's not to say nothing is happening. Electronic technology is creeeping into the business office as phone systems become computerized, word processors replace Selectric typewriters, and middle managers smuggle in personal computers to run spreadsheet programs like Visicalc. Electronic mail is being used here and there, and a few chief executive officers have even gone online by installing terminals in their executive suites.

Offices are gradually being automated, but there hasn't been the wild rush that many observers predicted a few years ago. Although the companies that sell equipment are still trying to convince corporate America that it needs fancy electronics to remain competitive, few businesses are buying that argument anymore, at least not in its entirety.

Offices are changing, of course, and much of this change has to do with silicon technology. "Office automation will play itself out in the micro revolution," asserts Patricia Seybold, a consultant and editor of *The Seybold Report on Office Systems*. She's talking about the microcomputer chip that will eventually change the way the business office works. It's just that the changeover is now coming more slowly than many thought it would several years ago.

Seybold now says we're in "year one of a ten-year evolution," adding that she's "surprised by how long it takes people to absorb technology." She sees three main reasons for the present sluggishness: the economic recession, uncertainty about what office automation really is, and potential buyers' doubts about whether the new technology will pay for itself.

Jack Adams, a partner and office automation specialist with Price Waterhouse and Company, agrees that "progress is being made, but at a much slower pace than we had assumed. I think people were dazzled by the possibilities of technology, but they underestimated the amount of time needed to introduce change." One result of that, Adam notes, can be seen in equipment sales patterns. "What's happened to many manufacturers is that sales of word processors and associated devices are going very well, but some of the sales expected of other technological advances in the 'office of the future' are not going very well."

This doesn't mean the market for the new office equipment is dead—far from it. Market researchers are still pumping out forecasts of enormous growth in sales of office technology over the next five to ten years. In some cases, the numbers are astronomical: International Data Corporation, for example, projects cumulative expenditures of $180 billion for high-technology office equipment during the five years ending in 1986. Another marketing firm, International Resource Development Inc., sees expenditures for office automation rising to an annual level of $36 billion by 1990, three times the present yearly sales.

The basis for these forecasts is the expectation that companies are desperate to do something to improve white-collar productivity. As the American economy has become more service-oriented, salaries of white-collar workers have grown to become the dominant business expense. In some industries such as insurance, the white-collar payroll can account for 70 percent of a company's expenses. Given that statistic, however, it's not altogether clear whether this is a "problem" that office automation can remedy, or just a condition of certain kinds of businesses.

Indeed, many feel that perhaps the amount of change that's really needed has been overestimated and that manufacturers and vendors of office automation equipment have pushed what they have to offer too hard. For example, when IBM first brought out its magnetic-card typewriters in the 1960s, the company persuaded customers to cluster equipment in word processing centers to take advantage of what were thought to be economies of scale.

"It was a miserable failure," says Jay Kunin, a management consultant with Boston's Hammer & Company. "Managers said, 'You damned well won't take away *my* secretary,' and the secretaries said, 'I'm damned well not going to give up my job and sit in a typing pool all day.'" Most companies that tried IBM's suggestion have abandoned the idea of word processing centers because the internal disruptions they caused actually lowered productivity rather than increased it.

Centralized word processing is on the way out and no one is lamenting its demise, but the thinking that created it in the first place still exists. Many people fear that the office of the future could become a replay of the industrial assembly line and create a new battleground for labor and management. Organizations such as 9 to 5 (the National Association of Working Women) actively support improvements in equipment design and legislation that would guarantee clerical workers adequate health protection. "The equipment is neutral; it's how it's used that can be a problem," says a spokesperson for the 12,000-member union.

Despite the availability of a relatively wide range of new office technology, no company has yet put it all together. That's because it won't all *work* together. In the absence of a dominant supplier, each major manufacturer wants its equipment or its software to be the industry standard. The result is a technological Tower of Babel. In many cases even a company's own products are incompatible with each other. For example, Xerox, which has made a major commitment to wedge itself into the office automation market, introduced three major products in the past year that are not fully compatible with each other (a personal computer, an office word processor, and a state-of-the-art work station called the Xerox Star).

Companies are lagging behind their customers' needs in other respects as well. They've been slow to design devices that are truly easy to use and consistent in function. All telephones and typewriters are essentially the same, but every word processor is different. Difficult situations arise when different brands or models of word processors are thrown together in an organization. While two keystrokes on one word processing system will delete a word, on another system the exact same keystrokes will erase the entire document. This lack of consistency is analogous to different auto companies putting the brake pedal and the accelerator in different places.

Electronic mail and video conferencing

"We've also seen a lot of failures in electronic mail," says Kunin. An electronic mail system uses desktop terminals attached to a communications network (usually telephone lines) to send messages directly from terminal to terminal. The problem is that the cost of each terminal—anywhere from $1000 to $3000—makes it difficult to justify the purchase of such a system if handling messages is all it does.

"It's a chicken-and-egg situation," claims Kunin. "The system has no value until you have enough people on it, but you can't justify putting enough people on the system until it has value to them."

Electronic mail presents conceptual problems too. To the extent that it removes communications barriers by channeling messages directly to an executive, it short-circuits more typical—and perhaps more desirable—office communication patterns. For the most part, senior executives neither read their mail nor write their own letters and memos (their secretaries do most of both); if there's no way to screen it, users of electronic mail can be deluged with unwanted messages. Moreover, electronic filing systems create the problem of controlling unauthorized access to confidential data.

Voice mail, on the other hand, *is* a practical technology because everyone already has the terminal—a telephone. A voice mail system converts messages into digital form and stores them in a computer's memory; the receiving party can recall a message by telephone at any time and have it reconverted into its voice form. Also called voice store and forward (VSF), this technology—rather like an elaborate answering machine—puts an end to "telephone tag," the all-too-common problem of two people failing to reach each other because one is out when the other calls back. While it's not good for all occasions, voice mail is a tremendous improvement in most cases, and, unlike electronic mail, it's relatively cheap: VSF services are offered for a monthly fee; all you need is a Touch-Tone telephone system.

Video teleconferencing is another intriguing experimental technology, but again it's hard to justify because it's so expensive. Such a system requires a full TV channel to send and receive the signals, which generally means a trip to the phone company's facilities. Besides the expense of several thousand dollars an hour, video teleconferencing raises confidentiality questions because the entire session has to be monitored by phone company technicians (unless your company happens to have a two-way satellite-antenna link on the roof). The benefits of teleconferencing are yet to be proven. Avon Products is one example of an early user that has cut back because the video meetings lacked the creativity that comes from personal contact.

Foremost McKesson, a $4.5 billion drug, food, beverage, and chemical concern, has also experimented with teleconferencing. "It has a place as a communications tool," says Charles Hoerner, manager of office systems, adding that it could replace a significant amount of corporate travel. But Hoerner points out that it's not suited at all to situations in which the conferees don't know each other, or where persuasion or selling needs to be done.

Another office technology that's created more problems than solutions is local networking. The idea is to provide a way for all of the various pieces of electronic equipment in an office to communicate with each other and share resources. It's a great idea that will someday be absolutely essential, but right now there are dozens of competing systems in use (Ethernet, Wangnet, ARCnet, Omninet, and SNA, to name a few) and virtually no standards or compatibility. Each manufacturer hopes to dominate the industry, but no one company has the marketing muscle arbitrarily to impose its system.

Given such confusing choices, it's a wonder to some that office automation has gotten as far as it has. Large-scale automation is generally making its biggest gains in organizations with high percentages of clerical workers—banks, insurance companies, oil corporations, and a few US government agencies.

For example, the San Francisco-based Bank of America, whose 85,000 employees and 3000 offices in 100 countries make it the nation's largest commercial bank, has more than ample motivation to control its office expenses. Frank Gatewood Jr, vice-president in charge of integrated office systems, says that after a slow start the bank is now making a major commitment to automation: already it has more than 1000 work stations in place (mostly word processors) and 6000 users of an expanding, worldwide electronic mail system. About 1500 personal computers (Apple and IBM) are also used in the bank. Gatewood reports that the budget for automation doubled last year and will probably double again this year. By the end of the 1980s, Gatewood says the bank will have invested $100 million in office automation, mostly in basic areas such as word processing, electronic mail, and information processing (i.e., using desktop terminals to work with data stored in the bank's larger computers).

Bank of America has avoided the network dilemma by deciding not to decide. "We've found that for most of our telecommunications the telephone satisfies our needs," Gatewood says. "We don't see the need for a local network. Moreover, we hesitate to jump in because there are no standards. As long as we don't have to make a decision we aren't going to make one." Despite the fact that he used to work for IBM, Gatewood doesn't have a terminal in his own office yet. His main source of data is the telephone and a set of telephone directories stored on microfiche.

Another of the true pioneers in advanced office automation is Edward Scott Jr. As Assistant Secretary of the US Department of Transportation, Scott gained the reputation of being the government's most efficient bureaucrat. Responsible for 20,000 people and a budget of $17 billion, Scott became frustrated in the late 1970s with the mountains of enervating governmental paperwork. He figured that computers might help, but he was equally frustrated by consultants who talked about terrific office systems not yet available commercially. Scott finally got a special appropriation from Congress and built his own.

The system Scott developed became known as TAOS (Transportation Automated Office System). It was run by a minicomputer linked to terminals on the desks of his managers. Developing TAOS changed Scott's life. "I became a convert, a zealot," he says. Convinced that office automation will be "an important influence in our society," Scott left government service and set up his own company (now a part of Computer Consoles Inc.). Desk Power, his microcomputer-based executive office system, supports functions such as word processing, electronic filing, database access, and calendars on up to 16 terminals.

Productivity in the office: a fuzzy concept

Scott feels it will take another 10 or 15 years for technology like the system he developed to be accepted in most offices, but he has no doubt it will happen. "People simply can't afford not to have office automation," he says.

How does he know that? He doesn't, of course, and neither does anybody else. "Anybody who's used one of these systems will tell you the same thing—it's not susceptible to productivity analysis," Scott says. "The best minds have tried and they always get back to mushiness. This is very subjective stuff."

Because it is so difficult to demonstrate the value of office automation, Scott seeks out potential customers who don't need to be convinced. "I don't think you can sell this technology to nonbelievers," he says. "Our strategy is to go after the people who think it's real and meaningful and convince them that we have a better product."

In the expanding world of the office of the future few subjects create more heat than the question of productivity. How does a company justify spending hundreds of thousands of dollars on equipment that may not pay its way?

The figure for total white-collar salaries in the nation is put at more than $1 trillion annually, and it is rising. Using that number as a touchstone, Booz, Allen & Hamilton, the New York management consulting firm, has been pushing a detailed 15-company productivity study it completed two years ago. Booz, Allen researchers tabulated how "knowledge workers" (professionals, managers, salesmen, executives, etc.) spend their time each working day. The report concluded that installing appropriate electronic automation would eliminate anywhere from 10 to 30 percent of these workers' nonproductive time and that this would increase net income by more than enough to pay for the new equipment.

Critics, on the other hand, say that, while office automation may yield benefits, it's very difficult to pinpoint them. Hammer & Company's Jay Kunin, for example, maintains that measuring managerial and professional productivity is fundamentally impossible. "At best that's the wrong approach; at worst it's getting you into questions that shouldn't be asked— unless for some reason you want to try to quantify the productivity of your managers by how many minutes they spend on the telephone," Kunin says.

Kunin's own approach is to look at the overall performance of the entire organization. Instead of using technology to do the same old things faster or more efficiently, Kunin recommends re-examining the basic premises behind what a company is doing and how it gets it done. Each organization is different, he claims, and electronic technology is far from being a panacea.

Price Waterhouse, a firm that also does a considerable amount of consulting on office automation, takes a similar approach. Improving the productivity of the organization may involve office automation, but the initial step is the old-fashioned one of determining what functions might be eliminated. As automation specialist Jack Adams says, "You'd be surprised how many there are. A well-conceived program [for increasing productivity] can pay for itself in cost savings in four to six months without any significant use of office automation."

Adams says the company that decides to automate should do it a little bit at a time, starting with word processing, electronic mail, and voice mail. "Those are the things that are percolating to the surface as bread-and-butter applications," he says. "You can talk about the office of the future, but the body of the work system is word processing and its variations."

One would think that productivity in word processing could certainly be measured. Alas, even that has its soft edges. Keystrokes and error rates can be and are counted by office managers, making it possible to know if a secretary or clerk is putting out more or less paper per hour, but this is a valid measure only if the document being prepared is a final product. In cases where it is not, word-processing equipment may actually hinder productivity. Letters and documents tend to be rewritten more often, sometimes *much* more often, when the word processor replaces the typewriter. Whether the improvement in quality justifies the added cost is a difficult question.

Kunin cites the case of a big law firm that, after shifting to word processors, found some documents were being revised up to 40 times. In such a situation, is the fortieth revision significantly better than the thirty-ninth, or the thirty-third? "Who knows?" Kunin asks.

Bank of America, on the other hand, thinks automation is worth the expense. Frank Gatewood reports that one of the bank's loan units can now produce the paperwork involved in a commercial loan in one third the time it took with a typewriter. In one case an entire loan package was turned around in three days instead of two or three weeks. "That represented $65,000 in income to the bank in short-term loans," he says. "Word-processing technology lets you be very nimble; you're able to go after business you couldn't go after before."

The executive computer

The joker in the office automation deck is the personal computer. Just how many are now being used in businesses is hard to say. Some estimates put the figure at close to 1 million (equivalent to the number of installed word processors, by the way) and predict a rise to 10 million over the next 10 years.

Many corporations don't even know how many personal computers are being used in their offices. Even when the number is known, just how these machines fit into the office automation puzzle is unclear. Jack Adams of Price Waterhouse thinks personal computers are a healthy development: "The microcomputer encroaches on everything—data processing, word processing, distributed processing, and time-sharing."

Dan Ferris, a San Francisco software consultant, believes that the uncontrolled proliferation of personal computers could cause problems for many companies. Based on a study of 20 companies in the San Francisco area, Ferris predicts that, as managers try to do more and more with their little computers, the companies' data-processing departments will be forced to take over software support of the machines. This, he says, will result in "painful solutions" because of the extra burden placed on already over-worked data processing departments.

Thomas Drohan, president and chief executive officer of Foremost McKesson, has other concerns. His company brought in about 20 IBM personal computers last year as an experiment to see what they could do for managers. Even though Drohan recently got one in his own office, he has no thoughts of using it to run his company. But he does hope to replace "tons of marketing plans, projections, and interpretations of business data" with it.

Foremost's foray into microcomputers is still very much in the experimental stage, Drohan emphasizes. One major question that still has to be addressed concerns access to stored data. Drohan sees it partly as an internal security problem, but he thinks it's also important that information access priorities don't become indicators of status in the organization, like keys to the executive washroom. "That would be counterproductive," he says. "After all, the whole point of the personal computer is to break down that sort of thing."

Another awkward situation the personal computer is helping to break down is that old executive suite syndrome known as keyboard phobia. Fear or dislike of keyboards isn't limited to top management, of course, but it's a significant factor in keeping terminals out of the reach of executives. The affliction seems to be the result of two things: the feeling that typing (*keyboarding* is the euphemism) is secretarial work, and the fact that most executives never learned how to type.

Keyboard phobia is curable, and it will probably die out by itself as younger executives who are more comfortable with such things climb the corporate ladder. Beyond that, it's probably only a superficial reason why the vast majority of top managers aren't going to be using computers or terminals of any kind in their offices for a long time to come. No matter what is said about executive information systems and decision support systems and whatever else they're called, chief executives simply don't need terminals to run their companies.

At Bank of America, Frank Gatewood is spending millions on office automation for senior administrators, including three electronic calendars just to keep up with bank president Sam Armacost's hectic schedule. However, Gatewood doesn't see any of this moving through the door into the boss's office for a long time. "Top executives do not need up-to-the-minute, real-time information," he explains. Their world, he says, involves "policy direction, historical performance, and prospects for the future." For such concerns they don't need a computer on the desk. "The executive terminal is an idea ahead of its time," Gatewood asserts. "I think it's 10 to 15 years away. It's going to happen, but for now our goal is to improve staff performance in support of that executive. That we know how to do."

But there was a twinkle in [Dr Breeze's] eye that encouraged me to go on.

"If what you call an office is a square room with telephones and typewriters in it, then I agree with all the companies like Xerox and IBM that are building word processors and duplicators and voice recognizers and other electronic whatnots to expand their capabilities. That particular office of the future is going to happen of course. Whether we like it or not."

"So what are you worried about?"

He goaded me on, dizzy as I was, and I felt the tenseness in his old voice.

"I just don't think it's that simple. An office has people in it, you see. They have lives and ambitions and needs and emotions. They interact. Most of the 'productivity' everybody talks about comes from the interaction, not from the ability to write and process pieces of paper. And those machines will do nothing for the people interaction. So the benefits from having all that hardware may not happen for 10 years, or 20 years. . . ."

"And when they do happen?"

"By then other things will have changed drastically under pressure of the technology. So the real problem is much larger than making offices more productive. The real problem is to decide right now what human quantities and freedoms are

worth fighting for, and which ones are not. Because when we are through with this digital transformation there won't be a stone left standing on another stone of our social edifice. There may still be a future, but there won't be an office to put in it."
—from *The Network Revolution: Confessions of a Computer Scientist,*
by Jacques Vallee

Personal Computers Invade Offices

Business Week

The so-called office revolution is happening, but in ways that weren't envisaged. Company executives, besieged by equipment suppliers and bewildered by the choice, have postponed the purchase of complete systems—or forsaken them altogether. Many have simply taken their personal computers into the office instead. This Business Week *cover story appeared on August 8, 1983.*

The personal computer is sweeping into corporate offices on the wings of a popular revolt. Managers, fed up with corporate computer centers that need months to do a job, are taking matters into their own hands and buying their own machines. The results already are astonishing. Even those managers who type only by the "hunt and peck" method are now using their desktop computers to do everything from analyzing financial data to sending their mail electronically and tapping into large corporate databases. "With only a cheap personal computer and a communications link," says Gus V. Morck, manager of telecommunications at Atlantic Richfield Co., "you now have what people have only been talking about—the office of the future."

But such declarations of computing independence are creating what can best be called "computer shock." Corporations are discovering personal computers popping up everywhere, and it is nearly impossible for them to get the desktop units to work together or to link up with the large corporate information systems. And as personal computers proliferate, companies fear they will lose the security and accuracy of valuable corporate data files. Executives are confused, too. Besieged by hundreds of equipment makers—each with its own office systems approach—managers are finding it increasingly difficult to pick the right computer for their own use.

The variety of office automation products now on the market is so overwhelming, says Joseph Ferreira, a vice-president at Diebold Group Inc., that executives "can't handle all the [choices] coming to them—they are suffering option shock." Pounding on corporate doors are more than 150 makers of personal computers, several dozen established manufacturers of computers and office equipment, and some 100 companies selling the communications gear needed to link the machines together.

Personal computer shock is forcing many corporate customers to rethink their plans to install office automation and their ideas on how to handle the growing profusion of desktop computers throughout their organizations. Already, a three-way battle among the office automation, telecommunications, and data processing departments has erupted at many companies over control of the personal computer influx. The only solution for many corporations, says Randy J. Goldfield, president of Omni Group Ltd, a New York office automation consultancy, is "to build new organizations to support the new [equipment]."

The corporate user is not the only one hit by personal computer shock. Some equipment makers are seeing their office markets rapidly dry up as the desktop computer replaces specialized word processors, office automation systems, and data processing terminals. This is forcing an increasing number of manufacturers to come up with new product and marketing strategies to take the personal computer into account. "The personal computer has been accepted as *the* workstation [in the office]," says Amy D. Wohl, president of Advanced Office Concepts Inc., a Pennsylvania office automation consulting company. As a result, she says, the personal computer will "permanently alter how office automation will work."

The office of the future had long been talked about as an environment where everyone, from the clerks to the chief executive, would be linked in a centralized electronic network with all the intelligence and memory located in the hub of the system, a large mainframe computer. These networks would quickly move corporate data to a display screen on an executive's desk, whiz correspondence throughout the company, and even keep track of such individual data as appointment calendars and telephone messages. In most cases, the corporate data processing department was expected to put this network together and buy the equipment. But despite all the hoopla, few office systems of any size have been put into operation. Most corporations have automated only the clerical worker, not the manager.

Without a doubt, the personal computer is rapidly changing the picture. "The personal computer has taken office automation out of the hands of secretaries and brought it to the manager," says Margaret H. Levine, manager of office systems for Columbia Pictures Industries Inc. This year, US business executives and professionals will buy more than 1.85 million desktop machines, more than double the number purchased last year, predicts Framingham, Mass., market researcher International Data Corp.

In this massive buying spree, the technicians running the central data processing departments have lost control to the laymen users. Last year,

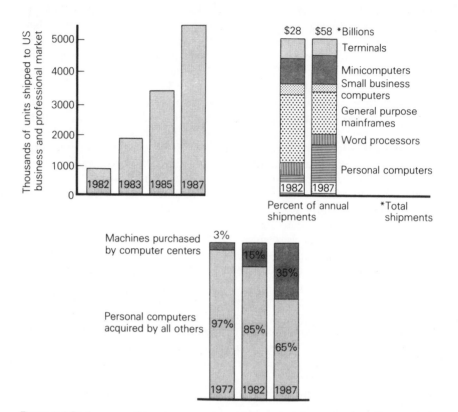

Figure 7.4 Businesses will buy far more personal computers than any other office automation gear—and individuals will purchase most of them.

85 percent of the desktop computers were bought by users, either individuals or departments, outside the data processing establishment, IDC estimates. Says consultant Wohl: "With the personal computer [now available], managers weren't willing to wait for the big system implementation anymore."

Once a manager buys a desktop computer, he quickly finds that it can serve as an electronic window on the entire business world. From that single unit, a busy professional can make financial calculations, send electronic mail, put together reports, and retrieve information from the corporate database and from outside computer files—commodity or stock prices, for example. But to accomplish this impressive set of tasks, users must be able to hook up with the corporate database and with other computers throughout the country—indeed, around the world. And there lies the rub.

Communications increasingly is becoming a major new issue. The computer industry would have corporate users of personal computers believe that their stand-alone units can easily be linked together into big networks. The reality is that such computer networks are still rudimentary. "There's no clear-cut discipline that lets [a broad range of personal computers] work easily together," maintains Charles A. Jortberg, an office automation consultant in Peabody, Mass. "People are working like hell to get to that." Seymour Merrin, a Westport, Conn., computer dealer, tells his corporate clients "not to buy networking." Says Merrin: "We are not truly satisfied with any network. We believe the answers are just beginning to come out."

The lack of a single, effective solution to the communications problem is causing widespread confusion. Unlike highfidelity stereo products, which are built to standards so that any brand turntable can play any record on the market, computers have no such standards. As a result, a customer generally has to buy just one brand of equipment, or else he cannot connect the multitude of units spreading throughout the office.

In order to provide customers with a more complete solution to their office automation problems, most equipment suppliers are beginning to offer a broader line of products or are making it easier to hook their products into an office system. This includes:

1 *Linkups*. Personal computer makers such as Apple Computer Inc. are adding features so that their desktop machines can talk to large mainframes.

2 *Office systems*. Computer and office automation companies are beginning to pitch comprehensive office systems that offer everything from personal computers to large central computers as well as the communications to connect all the equipment. This list of companies includes Burroughs, Data General, Digital Equipment, International Business Machines, Sperry, and Wang.

3 *Adding "intelligence"*. Makers of computer terminals and word processors are all adding personal computer features to their products. Terminal makers such as Televideo, Lear Siegler, and Applied Digital Data Systems offer souped-up "intelligent" terminals. Makers of word processors, such as Lanier Business Products, NBI, and CPT, are adding data processing and communications functions to their products.

4 *Telephone terminals*. Telecommunications companies, such as American Telephone & Telegraph, Northern Telecom, Rolm, and Mitel, are marketing computerized private branch exchanges (PBXs) and combination telephone–terminals that can move both voice and data signals between telephones and computers.

This glut of options is causing most companies to take longer than ever to decide what to buy. Companies such as Cleveland's AmeriTrust Co. are "moving cautiously," says Andrew P. Strada, senior vice-president for bank operations administration. "One of our fears is that [our equipment] will become obsolete," he says. "The buying cycle [for office automation equipment] has been spread out from 3 months to 18 to 24 months," asserts Goldfield of Omni Group.

If customers do not limit their search, warns consultant Ferreira of Diebold Group, they will suffer "paralysis by analysis." Already many new companies are having a tough time cracking the office systems market because corporate customers are "arbitrarily eliminating 80 percent of their choices," says Ferreira.

Many large corporations have already organized office automation committees to figure a way to cope with option shock. But these committees are discovering that "no single vendor, regardless of claims, offers the full robust product line needed to automate the office," declares Ronald R. Hollis, senior vice-president and head of systems and management information services at First Chicago Corp. So, rather than make a commitment now to one big integrated system, many companies are giving their managers personal computers instead.

First Chicago is "letting the wildflowers grow," Hollis says, by permitting each department to experiment with its own personal computers and other office equipment. Like many other office automation committees, Hollis' group specified only that a department limit its hardware selections to just one or two specific brands. "By selecting a primary vendor, we minimize the eventual problem of [system] integration," he says.

As neat an interim solution as the personal computer might be, it does open up a Pandora's box as far as data security is concerned. "A major management worry is [preserving] the integrity of information in the company, and the personal computer threatens that integrity with a proliferation of little files," says Robert G. Costain, a researcher in the office automation market at Dataquest Inc. Some companies have already begun to make new policies on what data can be kept in a personal computer user's electronic files and how it can be protected from misuse.

The potential problems are so alarming to some corporations that they see the need for significant corporate restructuring. "In the 1970s, three autonomous organizations were responsible for data processing, word processing, and telecommunications," notes Peter A. Wright, an industry analyst with the Gartner Group. But now, the personal computer is tearing down those corporate distinctions. "Microcomputers may be the vehicle that ultimately merges the decision-making of the management information system, office automation, and telecommunications centers into a centralized organization," says Wright.

This new corporate organization not only would pull together the three existing departments' managers but also would require a new group of information managers to supervise the use and protection of valuable, and often proprietary, corporate data. It would be dramatically different from today's data processing department. Rather than being responsible only for processing information, the new group would become the custodians of corporate information. "There is probably going to have to be some kind of information resource management group prepared to deal with such issues as the organization, implementation, technology, training, and development—a group that will take a holistic approach," says Omni's Goldfield.

One company starting to move in this direction is Sun Co. In January, the Radnor, PA, energy company formed an interdepartmental group to determine the best way to integrate its office systems, telecommunications, and data processing. "The main reason [for setting up the group] is the recognition on the part of the corporation that you have to put [those technologies] together," says Edward B. Parish, Sun's manager of electronic communications and office systems.

With more and more companies trying to get a handle on the personal computer revolution, many industry watchers believe that corporations will start moving again toward large centralized data processing systems. "The personal computer [fad] has 18 months to run, then it will slow down," asserts Aaron C. Goldberg, an industry analyst at IDC.

One reason why companies might move toward the more expensive and more centralized systems is the hidden costs of maintaining a network of personal computers. "The personal computer is perceived as a $2000 to $5000 expense," notes Gartner Group's Wright. But users also need peripherals such as memories and printers, and software. By the time a user totes up those additional purchases and buys communications links, each desktop computer could end up costing more than $20,000, Wright estimates. "Many large companies may not realize the ultimate costs that the original $3500 investment may bear."

Even if the centralized systems reappear, they will bear the stamp of the personal computer. For one thing, managers weaned on personal computers may object to going back to a central system that would recreate the same kind of frustrating delays that caused them to buy the personal computer in the first place. Sooner or later, the personal computers will plug into the integrated systems.

After much initial resistance, some data processing managers are coming around to the desktop models. "When the data processing guy finds that small systems give him a way to take care of individuals with less impact on his organization, he'll take them as a reasonable route," says Wohl of Advanced Office Concepts. Some of them even acknowledge that personal computers are more cost-effective. "You can take a million dollars and spend it on one large facility," says consultant Jortberg, "or spread it out and put in five small installations with four to five times the data processing power."

Whatever approach corporations choose, they will need to move quickly to get their strategies in place for installing office automation systems and personal computers. Some companies already realize that if they do not set policies and equipment standards now they could find themselves left with serious communications headaches and equipment that does not work together. "Establishing policies before personal computers become wide-spread could prevent major liabilities," says Gartner Group's Wright. "Corporations do not have the luxury of time."

Suppliers battle it out

Not too many years ago, the information processing market was neatly segmented. Mainframe computer maker competed against mainframe computer maker. And the same was true of minicomputer manufacturers, terminal makers, and word processor companies. But the personal computer has radically changed all that: now every manufacturer competes against every other manufacturer in the office automation business.

The major battleground in this new war is the office desktop. As managers began to purchase their own inexpensive personal computers and to slip them into their offices, the small machines staked out that valuable desktop turf. Now all of the equipment makers are fighting for that space. But there can be only one winner, because managers do not want to clutter their desks with several black boxes—they want one machine that does everything. And the player who gets his product on the desk will have the brightest prospect of selling the remainder of the office system. "The big prize is to capture the new user's loyalty, so that other, more expensive products can be sold," says Peter A. Wright, an analyst with Gartner Group, a Stamford, Conn., market research company.

The personal computer is winning the battle. "The personal computer has become the *de facto* work station [in the corporation] due to its low price, wide availability, and ease of use," says Aaron C. Goldberg of International Data Corp. While desktop computers represented only 14 percent of the value of all processors sold to businesses last year, they will account for one-third of the market by 1987, predicts IDC.

This radical shift is forcing computer and office equipment makers of every stripe to make fundamental changes in their product lines to survive. They are all scrambling to introduce a personal computer or to add personal computer functions to their current products, for only by doing so can they hope to compete in the office. "Personal computing functions are the key to capturing the desktop in the office," says office automation consultant Patricia B. Seybold, editor of *The Seybold Report on Office Systems*. "The personal computer is becoming part of pop culture."

The revolution is also forcing manufacturers to reshape their marketing strategies. Once, selling a computer was a matter of persuading the corporate data processing manager and a few of his aides. But now manufacturers are spending millions of dollars on splashy advertising and promotional campaigns intended to reach executives directly. "I'm amazed at how much vendors are calling on managers all over the company," says Alan Gross, an information systems manager at International Paper Co. "It used to be that the [computer manufacturers] couldn't waste time on anyone else [besides the data processing director]. Now they're mass marketing and advertising on TV," he notes.

That blitz of ads, on television and in the business press, is confounding and confusing nontechnical executives. "I read one airline magazine where there were 38 different computer systems advertised," notes office automation consultant Charles A. Jortberg, founder of Jortberg Associates in Peabody, Mass. "The ads all read the same; they all promised the same things to the user."

Largely because of that confusion, the biggest winner so far in the battle for desktop space is undoubtedly International Business Machines Corp., with its ubiquitous Personal Computer. The premier computer marketer's sheer clout in the information processing industry gave the PC—a late entry into the fray—instant credibility. It is fast grabbing market share from competitors and setting hardware and software standards for the rest of the industry, just as IBM's mainframe computers did in the past. IBM has already taken nearly 25 percent of the personal computer market, having shipped an estimated 350,000 units since the PC was introduced in August 1981. For many users, "the low-risk decision is to standardize procurement on IBM," says Wright of Gartner Group.

IBM itself is unexpectedly feeling the effects of the personal computer revolution. The PC is not only replacing some of its Model 3270 computer terminals, but also cannibalizing sales of its Datamaster small business computer and the Displaywriter, its word processor. The computer giant has already converted at least one of its Datamaster production lines to make the PC, which is still in short supply. "The Personal Computer is a fundamental part of our office systems strategy. It will be IBM's premier work station," says Francis G. Rodgers, IBM's marketing vice-president.

Spurred on by IBM's blessing of the personal computer phenomenon, smaller makers are moving quickly to add personal computing functions to their own product lines. For example, NBI Inc., the Colorado manufacturer of word processors, has just introduced a new machine that runs the same programs as IBM's PC, with only minor modifications, and offers its own brand of sophisticated word processing. At the same time, Applied Digital Systems, Esprit, and Lear Siegler, all once-booming terminal makers, are adding intelligence to their terminals to make them perform more like personal computers.

The makers of office systems designed around a central computer also recognize that their products will no longer sell without personal computing features. Some companies—Data General, Digital Equipment, and NCR, for example—are marketing their own desktop machines. Others, such as Sperry Corp., have introduced devices that can turn their terminals into personal computers.

Robert C. Miller, senior vice president at Data General, which introduced several new personal computers on July 20, admits that his company "could have done a lot more business in the office" if it had brought out a successful personal computer earlier. "There's no question the personal computer helps complete a vendor's story for the office," he says.

Most of the more established manufacturers are betting that personal computers will end up as just one part of the coming comprehensive corporate office automation systems. "What we're really seeing is not personal computers vs. other systems," says Thomas S. Kavanagh, president of NBI, "but the idea of personal computing—something that happens up front and not in the back room of traditional data processing."

"The key is to link [personal computers] to other offerings," says IBM's Rodgers. "The company that can provide the best [networking] will maintain the leadership role." In fact, personal computers will speed up the installation of integrated office systems, predicts Henry Ancona, manager for office automation and information systems at Digital Equipment Corp. in Maynard, Mass. "Customers buy personal computers, then they immediately want to exchange data with one another and with other departments. So then," he says, "they will move to integrated systems, using the personal computer as a terminal."

As confused customers try to choose the office automation vendors they will deal with, the manufacturers with the broadest product lines will probably be the big winners. "Most customers are adopting a two- or three-vendor strategy. They look for vendors with the widest selection," asserts Samuel F. Galgaino, vice-president for product marketing at Wang Laboratories Inc. Among the companies that will succeed, predicts the Gartner Group's Wright, are IBM, Digital Equipment, Hewlett-Packard, and Wang. The reason, he says, is that these companies have done the best job of assembling a broad array of products and developing brand recognition among customers.

To ensure survival in this environment, the less dominant manufacturers are entering into joint ventures to beef up their product lines. For instance, to get expertise in communications and develop new products to link its office equipment, NCR Corp. has invested in Ztel Inc., the new Andover, Mass., maker of computerized private branch exchanges that can carry both voice and computer data over telephone lines. NCR also has purchased Applied Digital Data Systems Inc., the terminal maker. Sperry has bought 15 percent of Trilogy Systems Corp., the supercomputer maker, and 13 percent of Magnetic Peripherals Inc., a joint venture of Control Data Corp. and Honeywell Inc. that manufactures data storage devices.

Less diversified manufacturers "will have to carve out specialized niches among smaller users," asserts Wang's Galgaino. But smaller companies insist that, once standard methods of connecting different brands of office equipment are developed, the market niches under the umbrella of those standards will be very profitable.

The leading example of this is Apple Computer Inc. The California company, which all but launched the personal computer revolution, intends to participate in the office automation boom, even though it does not offer a broad product line. It plans to do so by offering personal computers that are far more advanced than its competitor's machines and then adding features that will allow the machines to communicate with larger systems using IBM's Systems Network Architecture. To do just that, Apple has just made its Lisa computer compatible with the IBM PC. Explains E. Floyd Kvamme, executive vice-president for marketing and sales at Apple: "The key is to provide systems capable of being connected to networks."

In the long run, industry experts predict, the move to large, integrated office automation systems will undoubtedly bring a full-scale shakeout among the smaller entrants. "There's going to be a fallout of a lot of companies in the next three to four years," asserts Robert L. Puette, general manager of Hewlett-Packard's personal office computer division. "To survive, small companies have to provide a unique contribution, since they can't provide a comprehensive package of goods," he says. "Those that don't will be gone in time."

"The problem was, we had no plan"

Moving to a new building in Columbus, Ohio, seemed to give American Electric Power Co. the perfect opportunity to automate its headquarters and put in a full-fledged office system. So more than a year ago, Chairman W. S. White Jr assigned his executive assistant, Bruce A. Renz, to find out if "office automation is real," and if so, how the company should approach it. Renz formed a four-man task force, which decided that it was indeed real. But, Renz adds, "we did not conclude that putting a personal computer on [every] desk was the right answer."

Like many companies, AEP "was in fact [already] deeply immersed in office automation by some definitions," Renz says. "The problem was, we had no plan. We felt strongly about the need to have control over the process, something that is lacking with personal computers [by themselves]. Our concern was 10 guys might sit in their offices solving the same problem."

As a result, AEP is now installing a centralized system from Data General Corp. that will tie into AEP's mainframe computers and give the company control over the corporate database. Up to 1000 engineers, lawyers, finance specialists, and secretaries will use terminals to tap into the new system, which could cost as much as $2 million. The company calculates it will save money by cutting close to half of the $500,000 to $1 million it spends annually on outside computer services.

"We cannot afford to wait until Utopia comes"

Despite the growing difficulty of choosing among the many brands of office equipment now flooding the market, "we see the payout as so great that we cannot afford to wait until Utopia comes" to automate offices, says Mayford L. Roark, executive director for systems at Ford Motor Co. As a result, the auto maker is putting heavy emphasis on the distribution of data processing to departments throughout the company.

Early this year, Ford set up an Office Productivity Center with a technical staff of seven to evaluate equipment and decide what would best serve the company's needs. That group is still laying out Ford's office automation strategy, but the auto maker is now adding about 100 personal computers a month, bought under a volume purchase agreement with Victor Technologies Inc. Ford has already installed 2000 or more word processors and microcomputers throughout its North American operations. Roark believes that this equipment—which is being used mostly by financial and materials analysts and others who need to make projections and follow trends—will pay for itself in only a year.

But once Ford managers get used to working with desktop computers, they begin to find out what the bigger machines can do for them as well. So, while smaller models have proliferated, so have large mainframe machines. In fact, Roark says, the installation and use of big computers at Ford is undergoing "the most rapid growth we've ever had."

"We're in the caveman era now"

As the "one-man show" coordinating office automation at Diamond Shamrock Corp., Robert J. Rudkin has his hands full. "We're in the caveman era now" when it comes to linking 150 word processors and 60 personal computers into a network at the Dallas chemical and oil company, he concedes.

Computers from International Business Machines, Wang, Apple, Tandy, and Datapoint proliferated in the past year at Diamond Shamrock when austerity programs prevented departments from adding people but allowed them to buy the productivity-boosting machines to get the job done. The computers and word processors, which cannot all communicate with one another, were bought with no master plan, and Rudkin is "trying to connect it all together and make it work." His first goal is to get the machines to share the same information via a computerized private branch exchange from Northern Telecom Inc. that can transmit both voice and computer data over telephone lines.

Rudkin was not able to specify what brands of computers the departments could buy, because each group wanted the freedom to automate in its own way. That generated some ill will among the company's data processing professionals. Now, however, department heads and data processing experts "are starting to come together," Rudkin says. After departments using different software came up with conflicting results from the same data, they realized, "that they can't operate in a vacuum," he says. "We tend to take the personal computer as gospel. We know now that's not always true."

Guide to Further Reading

General

Articles
R. J. Spinrad, "Office Automation," *Science*, February 12, 1982.
Bro Uttal, "What's Detaining the Office of the Future?" *Fortune*, May 3, 1982.

International Data Corporation "White Papers to Management" on office automation systems appeared in *Fortune*, October 5, 1981, October 18, 1982, and October 3, 1983.

The *Financial Times* (London) ran special surveys on the use of computers in offices and business on April 13, 1982 and April 11, 1983; on personal computers in business, April 24, 1984; and on office equipment, November 19, 1984. Alan Cane tracked progress with "Electronic Office May Be a Little Delayed," October 15, 1980; "The Electronic Office: Who Needs It?" June 25, 1982; "Office of the Future Bonanza is Delayed," November 22, 1982; and "The Promised Bonanza—At Last," January 23, 1984.

Dwight B. Davis, "Personal Computer Networks Go On-Line," *High Technology*, March 1984.

"Linking Office Computers: The Market Comes of Age," *Business Week*, May 14, 1984.

Joseph D. Becker, "Multilingual Word Processing," *Scientific American*, July 1984.

Books
Wilbert O. Galitz, *Human Factors in Office Automation* (LOMA, Atlanta, 1980).
David Birchall and Valerie Hammond, *Tomorrow's Office Today* (Business Books, London, 1981).
Denis Jarrett, *The Electronic Office* (Gower Press, Aldershot, England, 1982).
Andrew Doswell, *Office Automation* (John Wiley, New York, 1983).
John J. Stallard, E. Ray Smith and Donald Reese, *The Electronic Office: A Guide for Managers* (Dow-Jones Irwin, New York, 1983).

Case Studies (British)

Introducing New Office Technology, IDS Study no. 252 (Incomes Data Services, London, 1981).

John Steffens, *The Electronic Office: Progress and Problems* (Policy Studies Institute, London, 1983).

Word Processing Applications in Local Government (Local Authorities Management Services and Computer Committee, London, 1982).

Dudley Gibson (ed.), *Wordprocessing and the Electronic Office* (Council for Educational Technology—Microelectronics Education Programme, London and Newcastle, 1983).

The Impact of Advanced Information Systems and *A User Perspective on Information Technology* (both from National Economic Development Office, London, 1983).

Debate

Barbara Garson, "The Electronic Sweatshop," *Mother Jones*, July 1981. Office worker lays bare the realities of some modern office work. Concludes that the office of the future is really no better than the factory of the past.

Marvin Kornbluh, "The Electronic Office: How It Will Change the Way You Work," and Don Mankin, Tora K. Bikson and Barbara Gutek, "The Office of the Future: Prison or Paradise?" *The Futurist*, June 1982.

Office Automation: Jekyll or Hyde? proceedings of the International Conference on Office Work and New Technology, Boston, November 1982 (National Association of Working Women, Cleveland, Ohio, 1983).

John Thackray, "White Collar Blues," *Management Today*, March 1980.

David A. Buchanan and David Boddy, "Advanced Technology and the Quality of Working Life: The Effects of Word Processing on Video Typists," *Journal of Occupational Psychology*, Vol. 55, 1982.

Reports on the possible health hazards from VDUs appeared in *New Scientist*, January 27, 1983; *The Times* (London), June 28, 1983; and both *Business Week* and *Time*, July 25, 1983. See also a very interesting discussion of office ergonomics:

Walter McQuade, "Easing Tensions Between Man and Machine," *Fortune*, March 19, 1984.

So-called "terminal phobia" is discussed in *Business Week*, March 29, 1982; *Time*, July 19, 1982; and in Bro Uttal, "Linking Computers to Help Managers Manage," *Fortune*, December 26, 1983.

8 Banks, Shops, Hospitals

Electronics in Commerce

Martin L. Ernst

The world of commerce is far more automated than the actual production of goods. Banking, transportation, and distribution are being transformed by electronics. The author, a senior researcher with consultants Arthur D. Little, Inc, also shows how entirely new services—like credit cards and electronic airline reservations— have thus been brought into being. But the revolution is not without its problems. Reprinted from Scientific American, *September 1982.*

Commerce encompasses all the economic interactions among the members of a society. It is information-intensive: it requires that information be made available on goods, their prices, and their utility, and the modern means of paying for goods requires that financial information be transmitted between the parties involved. Moreover, the storage and transportation of goods call for detailed records such as schedules and manifests. Commerce has therefore been quick to adopt new forms of technology for processing information. Clay tablets found in the Middle East furnish strong evidence that writing originated with commercial records. More recently, the sales history of computers shows that the large-scale industrial application of computers began with commercial functions.

Today, when commerce is responsible for more than 35 percent of all employment in the US, the new electronic technologies dominate the mechanization of financial institutions. In other commercial sectors nonelectronic machines play a larger role. Beyond this some central themes emerge that are useful for understanding the mechanization of commerce: how it arises, how it is implemented and what its impacts are likely to be. Here I shall give a series of examples to develop these themes. The examples will come from four major areas: finance, transportation, the distribution of goods and, lastly, communications.

One theme concerns the role of government in commerce. Traditionally, governments have performed certain commercial functions or have intervened in them. Good coinage, honest weights and balances, well-maintained roads and port facilities, and equitable (or at least consistent) judicial systems played a major part in establishing the success of city-states and later of nations. Hence most aspects of finance, transportation, and communications in the US have been subject to considerable regulation, and the first-class mail service is operated by the government as a monopoly. In many countries the same activities are conducted almost entirely by government organizations, and in virtually all countries a major part of the legal structure is devoted to codes concerned with commercial transactions. Inevitably, then, government intervention has influenced the adoption of new technology in commerce, sometimes favorably, sometimes not. The basis on which an industry is subsidized or regulated establishes incentives that favor the rapid adoption of some technical applications and the slow adoption of others.

A second theme is the almost universal requirement in commerce for interactions among independent parties. Banks must cooperate with one another, and moreover they must have a considerable degree of uniformity in their basic procedures if checks are to be a useful way to pay bills. Transportation companies, wholesalers, and warehouse operators must work together to ensure the flow of goods from factories to stores. Even the US Postal Service and the Bell System, which hold a dominant position in their field of communications, must rely on others for significant parts of their total function. For example, the Postal Service relies on airlines and on the manufacturers of sorting machines. The need for cooperation often limits individual organizations in their choice among available technologies. Often a move toward mechanization can be made only after broad agreement within an industry or across several industries, and this requirement can dictate the pace and the nature of change.

In the light of this second theme, the third is ironic. Many of the most effective mechanizations in commerce have been based on quite modest devices. Many others have been based on new configurations of standard machines. Often they are machines that have long been used in other economic sectors.

Banking

Among financial industries banking offers the widest scope for examining the nature and impact of mechanization. A good place to begin is with a familiar item: the checks people and businesses employ to receive and make payments. Offered by some 14,000 commercial banks (and now by some 5,000 savings banks through Negotiable Orders of Withdrawal, which are the equivalent of checks), checks and their processing have always been labor-intensive and form an obvious target for mechanization.

Regardless of where it is deposited, a check, or at least the information on it, must be returned to the person who wrote the check by way of his bank.

The journey is sometimes a long one. The mechanization of the process began years ago with the introduction of electromechanical check-sorting equipment and bookkeeping machines. Then the federal government began producing almost all its checks on punch-card stock. These early steps, however, were fairly limited. The start of modern check processing was the introduction of magnetic-ink character recognition (MICR) encoding on all checks. These machine-readable numbers identify the account on which the check is drawn and the bank in which the account is held. Complete machine processing becomes possible when numbers that identify the receiving account and the bank are entered into computerized records by the bank at which the check is first deposited and the amount of the check is MICR-encoded on the check itself.

The introduction of MICR took almost 15 years. It began in the early 1950s with a series of studies sponsored by several bank associations and the Federal Reserve System. In 1958 standards for coding, the position of codes on checks, an acceptable printing ink, and similar requirements were established. Another ten years were to pass before essentially all banks employed MICR, but with the larger banks taking the lead some 85 percent of all checks were being encoded with MICR numbers by 1963. The process illustrates the long time it takes when a large number of organizations must be brought together to make standard changes, even if the changes appear to be simple.

MICR's stream of machine-readable data speeded the development of computer-based record-keeping for virtually all checking accounts. And the benefits of having mechanized account records spread rapidly to other bank activities. Computer-based equipment for tellers was introduced early in the 1960s to speed the entry of data and the processing of transactions that originate at a teller window. More recently, the existence of computer account records has facilitated the growth of automatic teller machines (ATMs). These have become increasingly popular; cash withdrawals and certain other operations, such as transfers between accounts at a single bank, can now be done at any time rather than only during banking hours.

The ATMs illustrate a characteristic of much of the mechanization of commerce. In an ATM transaction the customer himself punches a set of codes into a terminal; thus he provides his bank with machine-readable data. In effect, he is doing work for which the bank previously had to use its own staff. This type of "labor sharing" is becoming increasingly common. It is part of a trend that has been under way for decades; in the 1930s, for example, self-service in supermarkets began to replace what used to be the function of store clerks.

A logical next step in the evolution of check-processing will be for payments to be made by means of an electronic terminal in the home. The terminals themseves are technically quite straightforward. Their installation nonetheless awaits several developments. First, the population of home terminals must be large enough to make a bill-paying service economically feasible. Second, enough banks and merchants must agree on the terms of the service to give it a reasonable degree of universality. Finally, consumers must have reasons to want to use the terminals.

Potentially there are incentives for all the participants in such systems. Viewed in their totality, the costs of checks are not small. When a check is used to pay a bill, for example, the cost is about $1. Somewhat more than half of that cost is incurred by the banks, and the remainder is incurred by the biller and the payer for postage, paper, and printing. The current costs, however, are almost invisible to the consumer: they are "bundled" into the total cost of bank and merchant services. Since most people are comfortable writing checks, many will be reluctant to change unless they get some benefit in the form of lower costs, greater convenience, or extra service. In effect, people may insist that they be rewarded for the labor they provide when they handle transactions electronically.

Service to individuals is only part of a bank's operations; relations with business organizations are at least as important. In this regard a major factor in mechanization has been the drive by businesses to improve their cash management. As interest rates have risen, the number of ways in which a business can earn (or save) interest on short-term loans has grown. To earn or save the interest, however, the business must know its cash position accurately and be able to move its cash around quickly and economically. To meet these needs a variety of electronic terminal systems have evolved that enable businesses and other organizations to communicate directly with bank computers. By means of these terminals, a business can keep track of its liquid assets almost minute by minute and can issue instructions for transferring them to where they can be best employed.

In order to move the funds rapidly, banks in turn must be interconnected with one another by computer-to-computer telecommunications networks. The oldest and most important network is FedWire. FedWire, which began in 1918 when the Federal Reserve System leased a set of telegraph lines, serves to make settlements of the payments between banks that result from the totality of checks and other transactions individual banks have processed. The settlements are final, in that the interbank payments transacted by way of FedWire are guaranteed by the Federal Reserve System. Other major wire systems are CHIPS (Clearing House Interbank Payment System), which is operated by the New York Clearing House Association; BankWire, which is managed by an open consortium of US banks, and SWIFT (Society of Worldwide Interbank Financial Telecommunication), which originated as a European international system but now has many US members. Among these, CHIPS provides the most dramatic example of growth and utilization.

CHIPS was established in 1970. It replaced an earlier system in which paper checks had been carried by messengers from major banks to the New York Clearing House, which provided a facility for settling local interbank accounts among its members. As the traffic in checks grew, it became increasingly difficult for clerks in the individual banks to process the day's transactions with any assurance that outgoing payments (for which their bank would be responsible) were adequately covered by deposits and payments flowing in. The danger was either that the local checking system would lose its timeliness or that clerks would be forced to rely on their own judgment to decide whether or not to forward specific outgoing payments. In forwarding some payments the clerks rather than senior bank officers could be authorizing large amounts of credit.

The number of CHIPS transactions has increased by a factor of 20 in the first ten years of the system's full operation. In terms of the dollar value handled its growth has been even greater, amounting to an average annual increase of some 40 percent. So rapidly has the flow of money into and out of business accounts increased that a typical member of CHIPS will process dollar values each day that can be tens of times the total worth of the bank itself. Meanwhile, consumer-oriented automatic clearinghouses have been created to handle electronically operations such as the direct deposit of paychecks in an employee's bank account and the payment of periodic consumer bills for mortgages, rent, and utilities. In the case of direct deposit, payroll data are transmitted to a clearinghouse, which rearranges the information so that each bank gets the individual payroll payments for the accounts it maintains. Direct deposit, encouraged strongly by the federal government for its own employees and for Social Security payments, is growing fairly rapidly; so far the consumer-payment side is clearly less popular.

Overshadowing all these changes has been the growth of bank-sponsored credit cards. Banks were relatively slow to offer credit cards in force. Individual banks had offered cards in the 1950s, and some regional systems evolved late in that decade, but the development of the two national bank systems, called Bank Americard and Master Charge at first, did not take place until the late 1960s. Growth after that was rapid, and the card systems have long since replaced many merchants' credit operations as well as cash and checks.

What has all this meant for banks, their customers and society in general? For one thing, it is clear that without mechanization many of the paper-based systems simply could not have coped with the growth in the use of bank services over the past two decades. Not only would labor costs have risen greatly, but also, given the need for the careful checking and balancing of individual entries the banking industry requires, the sheer handling of paper documents would have made it impossible to keep the conduct of transactions timely. The result would no doubt have been some combination of higher prices, poorer service, and limited growth. Instead, productivity in the processing of checks has probably more than doubled between 1960 and 1980. The growth in productivity attributable to the wire networks has almost certainly been even larger.

In a subtle way mechanization has contributed to a change in the nature of banking. Becauses of government regulations that restrict the activities of banks, the banks have lost ground to other financial organizations as intermediaries between those who have money available and those who want to borrow money. For example, large corporations now transact short-term loans with each other rather than conducting the equivalent transactions with banks. The loans are mediated by underwriting and securities companies. Moreover, money-market funds have replaced savings accounts for many individuals. Meanwhile, however, and partly as a result of high capacity, efficiency, and low costs, the banks have increased their activity as movers of funds. In the process the operations of the financial intermediaries that now compete with banks have been supported and even facilitated.

Employment in banking has not suffered from mechanization. Because increases in productivity have been accompanied by even larger increases in the demand for bank services, employment grew by 50 percent between 1970 and 1980, from slightly over a million to nearly 1.6 million. Much of the current flow of money through banks is a response to high inflation and volatile interest rates, and the flow may drop if a stabler economic environment returns. Regulatory reforms may remove the restrictions banks now face. The industry has nonetheless been restructured, and a complete return of banks to their traditional practices is unlikely.

Transportation

The distribution of goods has always required financial services; in fact, many bank and insurance practices arose in response to the needs of those who transport goods. The mechanization of transport, however, clearly calls for more nonelectronic equipment than banking does, although transport too makes intensive use of electronic systems. The combination of nonelectronic and electronic equipment can be seen in the recent growth of containerized shipping. Containers themselves are an ancient form of technology, but their current level of service owes much to the availability of telecommunications and computers.

The modern container-ship era started in the late 1950s, when Malcolm McLean acquired the Pan Atlantic Steamship Company, a shipping organization that connected ports in the US Northeast with ports on the Gulf of Mexico and in Puerto Rico. Having previously operated a trucking line, McLean decided to offer a "ferry service" for trucks, but one that left the truck wheels ashore. By having flatbed trailers carry containers overland and by lifting the loaded containers between ship and shore, he speeded loading and discharge and could move goods in containers from their origin to their destination. A similar service between the US West Coast and Hawaii was inaugurated by the Matson Navigation Company at about the same time. Both services flourished, and others began to emulate them.

The subsequent progress of containerization was difficult. The standardization of container sizes and of the fittings with which cranes lift the containers was an early requirement. Regulation also influenced the pace: Puerto Rican and Hawaiian routes were domestic and relatively unregulated, but US international shipping services were both highly regulated and highly subsidized by the government. The subsidy mechanism tended to discourage the adoption of container technology; for example, it subsidized US operators for crew costs higher than those of foreign operators but not for the costs of developing more fully mechanized ships and services.

Institutional barriers were also numerous. Shipping has many participants: shippers, freight forwarders (consolidators of cargo who make money from the differential in transportation rates between small quantities and larger ones), overland carriers, insurance companies, longshoremen, pier operators, agents and customs brokers (who are hired to expedite the formalities at the dockside), and the shipping lines themselves. There are likewise many regulators: the Federal Maritime Commission, the Maritime Administration, the Interstate Commerce Commission, and the US Customs Service. Each of these parties was affected differently by the introduction of containers, with some gaining advantages and others feeling threatened. Although the net financial effect of containerization was a considerable savings in costs, the distribution of the savings was worked out only slowly.

Containerization offers many benefits. Cargo is better protected from damage and pilferage and so insurance rates can be lower. Handling times and costs are greatly decreased; indeed, in many instances containers can move from their origin to their destination without being opened en route. From the ship operator's point of view, the major gain is the far greater utilization of his primary asset: his ships. With cargo-handling rates in port increased by a factor of 10 or more (from about 500 tons per day for conventional ships to 5,000 for container ships), container ships can spend far more time at sea earning money and less time in port, where all they accrue are costs.

To support container services a variety of telecommunications and computer systems are crucial. Since the ships are large, fast, and spend little time in port, their cargo manifest is seldom complete when they are ready to leave. Thus manifests are best transmitted by telecommunications from computer to computer rather than being sent by air mail to destination ports. Even more important, it is necessary to keep track of the multitude of containers. Who owns each one? What type of container is it? Where is it? What does it carry, and from where to where? What is its condition? Does it need repairs? Who owes whom how much money for its use or maintenance? Where should it go next when it is empty? Is the supply of empties in a region adequate or inadequate? Where can empties be obtained? Small container-ized shipping services can be operated without computers, but the large pools of containers necessary for efficient major operations would become hopelessly confused without computerized control.

The container ships themselves are increasingly mechanized. Typically, the engine room in the newer ones goes unmanned on the night shift, because the instrumentation and control systems on the bridge are adequate for most operations. Crews have dwindled in size (although they remain above the minimum levels set by the US Coast Guard), and because of the small amount of time the ships spend in port crews get long vacations and are given quarters more reminiscent of cruise ships than of traditional freighters. Matching the decrease in port time has been a decrease in the number of berths needed at a port. To be sure, considerable space must be available to marshal and store containers, but this space often is provided at new terminals on the outskirts of port cities. The inner-city waterfront previously devoted to shipping thereby becomes available for redevelopment as commercial, residential, and recreational areas.

Since all ocean transportation requires some associated overland movement, containerization has also had an impact on trucks and railroads. Railroads in particular can take advantage of the large flows of cargo arriving on container ships by introducing efficient unit trains that move the containers to major inland destinations. Such services now operate coast to coast in the US, where they also are being utilized intensively to deliver bulk cargoes such as coal and ores. The cars making up bulk-cargo unit trains are coupled together for an extended period. The couplings allow each car to rotate about the long axis of the train, so that they can be unloaded by rotation without the labor-intensive operations of decoupling and recoupling, an example of a simple technology having quite impressive results.

New electronic equipment is being applied in all forms of transportation. Computer-based systems are beginning to provide brokerage services that bring together empty trucks and cargo awaiting movement. Microprocessors installed in vehicles now analyze data they get from sensors to improve the performance of all types of engines. They will come to have a role in other forms of vehicle control. In one potential application microprocessors would be given information on the "consist" of a train (the order and weight of each car in the train) and use it to guide the train's braking. The results would likely be lower fuel consumption and less wear and damage to both the train and its cargo. Up to now, however, the most widespread application of new electronic systems undoubtedly has been in air transportation.

Consider electronic reservation systems. The first was introduced in 1963 by American Airlines, Inc. It was devised in a joint effort with IBM. Before then the airlines employed massive paper-based systems to keep detailed passenger records and had electromechanical display boards in their larger reservation offices to show available seats on planes. The displays were hard to read from a distance; in fact, agents sometimes had to rely on binoculars. The correlation between paper records and electromechanical ones often was quite poor. Between 1960 and 1980 the number of passengers on US airlines grew by a factor of five. The speed, accuracy, ease of use, and cost efficiency of electronic reservation systems clearly facilitated this growth. Since air transportation was an expanding industry, employment in it also increased (by 15 percent between 1970 and 1980).

More than most activities, transportation must be viewed as a system, since it often shunts goods from one mode of conveyance to another as it takes them from origin to destination. This suggests that the deregulation of transportation now under way in the US will yield fewer and larger companies, with some of them operating in several modes. It seems likely that mechanization will accelerate the process, since the organizations best able to exploit the benefits of mechanization will usually be those with a degree of control over the entire system of moving goods rather than just one mode of it.

Distribution

In each aspect of the final distribution of goods (that is, in wholesaling, warehousing and retailing) a variety of types of mechanization are in progress. An example in wholesaling is intercompany electronic data interchange. Here, the intent is to mechanize all aspects of order processing, including not only the transmission of orders to sellers but also the presentation to buyers, in electronic form, of current information on prices, discounts, special offers, and the like. The technology is based on now-standard telecommunications networks connecting the computer terminals of the participants, but once again the critical requirement is for standardization.

The requirement arises because individual businesses usually have their own format for preparing price lists, orders, and invoices. This creates no problems for people but presents major difficulties for computers. Usually an agreement on standards and formats is reached by trade associations. For example, the Food Marketing Institute, the Grocery Manufacturers of America, Inc., and four other trade organizations have cooperated in developing standards for transactions among distributors, brokers and manufacturers of food. These standards make up the Uniform Communications System (UCS) for the grocery industry. A slightly different approach is being taken for transportation documentation, an aspect of commerce that includes tariffs, manifests, control documents, and arrangements for billing and payment. In this case the effort is being spearheaded by the Transportation Data Coordinating Committee, a nonprofit organization supported by dues from shippers, carriers, and other members. The approach most often taken in all these efforts has been to standardize communication formats and protocols so that each participant is then able to develop computer programs that will translate between his private format and the communication standard.

In warehousing, computers serve a wide variety of record-keeping and scheduling activities. A particularly valuable application is maintaining records of the location of items in storage. This makes possible random storage location, which can be almost twice as efficient as dedicated storage in utilizing the space in warehouses. Another major application is the mechanization of handling operations. The possibilities include aids to essentially manual operations by means of fork lifts and similar equipment that incorporate sensors and microprocessors; operator-controlled but nonetheless semiautomatic order-picking systems, and completely mechanized order-picking systems that can include automatic palletizers and depalletizers, devices that move cartons onto and off portable platforms. Fully automatic specialized warehouse systems were designed as early as 1958, but most of the efforts since then have been devoted to semiautomatic systems because such systems retain greater flexibility for responding to changes in the sizes and shapes of the objects being handled in the warehouse.

In retailing, the focus of mechanization has been the cash register and other point-of-sale systems. The retailing of food offers the best example, largely because of the introduction of the Universal Product Code (UPC) bars. Like the machine-readable numbers on checks, they are an instance of standardization for the sake of mechanizing the input of data for electronic processing. UPC bars are the set of thick and thin lines now printed on essentially all prepared food items. At the checkout counter they enable the clerk to identify each purchase to an electronic terminal by passing the item over a photoelectric laser scanner built into the counter. The terminal then retrieves the price of the item from the store's central computer and prints it on a sales slip. At the same time the purchase record can be entered in an automatic inventory control and reordering system. That system in turn can produce data for market research, cost control in the store, and shelf-space allocation. The least popular aspect of UPC is the display of prices. The merchant wants to post prices only on the shelves where items are stocked and save the labor cost of having them stamped on the items themselves. In some states, however, the merchant is being required by law to continue putting the price on each item. Nevertheless, the product code systems are cost-effective, and after a slow start they are spreading fairly rapidly.

Other forms of mechanization in stores are quite common. Terminals have been developed to validate checks that a customer wants to cash by searching a data base to see if any of his previous checks have ever "bounced." Still other forms of mechanization are envisioned. Technically feasible, but not yet in significant service, is a terminal at the checkout counter that is connected to a local bank network. By inserting a plastic card known as a debit card, the customer can directly debit his checking account by the amount of a purchase rather than paying for that purchase in cash or with a check.

Telecommunications

The last aspect of commerce I shall touch on is telecommunications. It is an aspect of high importance. For one thing, telecommunications is one of the fastest growing industries in the world. It has a range of recognized technical alternatives and potential applications that will take several decades to fully exploit. (The telephone network is so extensive that changes cannot come quickly. Imagine the effort that would be required to extend fiber-optic lines into every household and connect them to produce a visual-communication system.) At the same time, the availability of adequate telecommunications is a prerequisite for many of the forms of mechanization described above.

Without a long history of incremental mechanization, the telephone service we accept rather casually today would long since have become uneconomic or impossible to maintain. The most obvious episode in this history was the introduction of the dial systems, which took the labor of establishing a telephonic connection between two parties and transferred it from the telephone operator to the caller. Over the years, the dialing system has been expanded in geographic scope to both national and international operations. Moreover, the efficiency of long-distance operations has been enhanced by the installation of computer-based systems for recording calls and mechanizing the preparation of telephone bills. Within the central telephone offices a series of technical advances has marked the transition from electromechanical switches to solid-state ones. The newer switches have lower maintenance requirements, are faster, have better transmission characteristics, and more efficiently collect and maintain data for usage analysis and billing.

Many of these developments are based on sophisticated technology. The telephone system also illustrates, however, that major advances in productivity can be made with quite simple devices. Originally all telephone installations were "hard-wired," that is, the telephone lines were more or less permanently connected to the instruments. Then in the 1960s telephone companies introduced plug-in extension telephones. They nonetheless required that at least one hard-wired telephone be installed at each service number. More recently practice has changed further: the telephone wiring now put into houses has universal sockets that enable users to install their own telephones. Although this last change reduced pressure from manufacturers selling equipment designed to interconnect with telephone company equipment, a major motive was the growing resistance of government regulators to granting rate increases to cover escalating telephone installation charges. Many of the telephone companies have now developed computer data bases that record the location of all the wiring and jack sockets in a home, so that people changing their residence can determine their needs in advance and arrange to pick up telephones at a telephone store and install the instruments themselves. A relatively simple set of changes in industry practice has thus allowed labor-sharing and a reduction in installation personnel.

Some consequences

Viewing the mechanization under way in commerce, one is reminded of some of the paintings of the elder Brueghel, with their almost frenetic energy and their display of the diversity and detail of human activities. Commerce is of course only one aspect of life and not the totality Brueghel viewed, and this narrower focus is largely responsible for the similarities I have noted that run through much of its mechanization.

First, although commercial activities provide most of our economic infrastructure, they also must rely on it. A technical innovation cannot be exploited widely if the infrastructure to support it is not available. To be sure, some institutions can take advantage of mechanization better than others, but the others cannot be left out or the process will fail to achieve the universality needed to make it effective. This is the reason the development of formal or informal standards is critical to widespread mechanization whether the standards are formats for data input and telecommunications or specify the size, shape, or other characteristics of physical objects. The adoption of standards can trigger an entire set of steps toward mechanization. Still, the introduction of standards is not without risks. Standards introduce a rigidity that may be regretted later if technical advances open up better possibilities. Difficulties also can arise when different standards conflict. Banks have chosen magnetic technology for the input of data; retail stores have chosen optical technology. This may turn out to raise a barrier wherever a combined system is sought.

Second, mechanization calls for new institutional relations. Since commerce is a system and its benefits are distributed among many participants, agreement much be reached (or forced) on how the benefits of an innovation are to be shared. Otherwise there may be disincentives for any single participant to invest in the innovation; too few of the benefits may be gained by those who take the risks. The role of government in speeding, delaying or biasing mechanization, and therefore in the new institutional relations, is pronounced. In telecommunications and the financial industries much of the current deregulatory trend in the US can be traced to technical advances that either could not be exploited well in the regulated environment or could be used to bypass the intent of existing regulation. In the latter case the nation was being left with the rules but without the benefits the rules were supposed to encourage.

The benefits of the mechanization of commerce take many forms. Although labor productivity is an obvious example, the results of mechanization have almost always included advances in qualities such as capacity, speed, responsiveness, reliability, and economy. These advances have sometimes helped to increase demand, so that displacement of labor has been avoided and employment has actually increased. In any case, the level of activity and quality we now take for granted in many industries simply would not have been possible without mechanization.

So much for the benefits; what about the drawbacks? Some of them are related specifically to commerce; others belong to broader concerns about the impact of mechanization on society. In the first category the most publicized concern is the vulnerability of individuals that may result from mechanization in the various financial industries. This concern has already led to consumer protection legislation covering the loss of credit cards, the actions required when errors in billing are found, and similar matters. There remain, however, fears about fraud, theft, and invasion of privacy.

There are a variety of measures to alleviate these fears. Some measures are legislative, some are technical. They will of course have a price: higher costs for financial institutions and their customers, and more complex demands on consumers for identification during the entry of transactions. So far, competition among financial institutions has discouraged individual efforts to implement the technical measures. The greatest current problem is the lack of a formal system for reporting and aggregating data on the frequency and scale of undesirable incidents. They will never be eliminated, but they can be controlled. One can anticipate a continuation of the perpetual war between the locksmith and the thief, now raised to a higher technological level that denies "employment" to the less skilled among white-collar thieves.

Another common concern is that the quality of life is being eroded by the depersonalization commonly associated with mechanization. Plainly, the human touch is eliminated when one deals with a machine instead of a human being. Still, many of the examples of mechanization I have described are those where the human touch simply cannot be afforded, so that the real choice is between mechanized service or less service. Overall, therefore, human values may be served better by accepting mechanization. In some situations, such as when services are needed by people who cannot speak the language of the country they are in, mechanized systems can be designed that are easier to deal with than a person who speaks only the native language. Furthermore, the younger generation quite obviously is not intimidated by the mechanized interface between consumers and electronic systems. The worst concerns probably arise for workers. Many mechanized systems (semiautomatic order-picking devices, for example) tend to isolate individual workers and break up normal social patterns.

More broadly, mechanized systems may actually result in overefficiency. Efficiency is almost always bought at a cost in flexibility and resilience. For example, the US suffered from the Middle East oil crisis because its distribution system for petroleum was so efficient that inventories had been kept at a minimum. This offered little protection against an unanticipated interruption.

Finally, there are the fundamental concerns about displacement of labor. Who is to be displaced, and how is the displacement to be handled? When the transition to a mechanized system is slow (as in containerization) or is in a growing industry (such as air transportation or banking), the direct impact is seldom severe; indeed, employment will often increase. When the industry is a mature one and the pace of change is rapid, the problem is not so easily solved. Efforts to retrain the displaced workers generally have a poor record even when the economy is growing rapidly. Most of the workers will feel a justifiable anguish at being severed from their accustomed livelihood. The least skilled workers will suffer disproportionately and are the least able to fend for themselves. This is a broad social problem whose solution will require education reform and a new emphasis on encouraging the early development by young people of skills that are needed in an increasingly mechanized world. In this regard it is worth noting that future mechanization will probably broaden the range of those affected. Levels of middle management have already been eliminated by some forms of mechanization, and the spread of artificial intelligence may eventually affect even skilled professionals.

So far, demand for new services has been almost open-ended. At some point a fundamental restructuring of business and social life will no doubt come; at present it appears to be more than a decade away. The restructuring will be hard on many people, particularly those whose training and skills limit the types of employment they can seek. The immediate challenge, and a hard one, is to manage the restructuring in a more humane way than was typical of the original Industrial Revolution.

New Technology in Banking and Shopping

John Marti and Anthony Zeilinger

The new technology will have a profound effect on how we shop and conduct our financial affairs. But there is much uncertainty about what will happen, and when. John Marti and Anthony Zeilinger of the Policy Studies Institute (PSI) in London explore some of the social implications of the revolution in banking and retailing. This is taken from their study, Micros and Money *(PSI, London, 1982).*

What impact will the "Microelectronics Revolution" have on banking and shopping? Will we soon be using plastic cards with microprocessors built into them, paying our bills by sending electronic messages over the wires, doing our shopping from an armchair in the living room? Or will we still be handling cash and writing cheques and going down to the shops in the High Street much as we do today?

Automation in banking and retail distribution has been proceeding steadily over the past two decades, mainly in the form of large mainframe computers to handle the back office operations of the banks and the stock control and general head office administration in the major retailing groups. This has brought advantages (and also some disadvantages), but it has not altered fundamentally the way the banks and shops do their business, and its effects are not for the most part directly evident to the customers.

Microelectronics, however, is quite another matter. It offers the possibility of storing and processing information in ways that can be much more compact, flexible, dispersed and, above all, cheap. It therefore gives scope for a far wider range of applications than the large central computer installations used hitherto. And when combined with other new technologies, such as laser scanning, fiber-optic cabling, satellite transmissions, and two-way television systems, it opens up possibilities for major changes with far-reaching implications.

This gives rise to a number of important questions. Just how revolutionary are the prospective changes, and to what extent are they in fact technically feasible? What are the issues that will determine whether they are actually undertaken, and, if they are, in what form, on what scale, and how soon? How important are the likely advantages, and who will get the benefit of them? How high are the likely costs, and who will have to pay for them? What are the possible disadvantages and problems, and who will be at risk from them? What will be the effects on the way that banking and retailing are organized, on the jobs of those who work in them, and on the service provided to their customers?

With a view to finding answers to these questions, we undertook a study of microelectronics and other new technologies in banking and retail distribution. This took the form mainly of interviews with representatives of over 150 relevant organizations. The majority of the interviews were carried out in the course of 1981 and were conducted on a confidential and free-ranging basis. The aim was to get hard facts on technical feasibility, economic viability, and intended developments, but also more qualitative information on issues, expectations, and general attitudes. The study was part of a wider Policy Studies Institute (PSI) programme of research into microelectronics.

Effect on consumers

In principle, the ultimate purpose of bankers and retailers is to provide services that meet the needs of consumers. However, it does not necessarily follow that changes in payments systems that seem to meet the needs of bankers or retailers will inevitably be equally beneficial to consumers or, indeed, beneficial to them at all. The changes in progress and in prospect vary greatly in the certainty and extent of the advantages they offer consumers and in the risks they pose of bringing offsetting disadvantages.

In general, interbank and back office automation schemes have little direct impact on individual customers, and if they improve the efficiency of the banks they are likely to benefit their customers. Similarly, the introduction by the banks of automated teller machines (ATMs) seems to offer them positive advantages, in the form of a slightly cheaper service and greater choice of time and place for drawing out their money. However, their usefulness would be increased if they were all designed to accept a standard form of plastic card so that people could use the auto-teller machines of other banks, just as at present they use the branches.

The simpler point-of-sale systems will usually have an impact on consumers that is only marginal, but also favorable, in the form of a slightly quicker checkout. Those with automatic scanning systems will usually bring the disadvantage of there being no longer separate price labels on individual items, but this is likely to be more than offset by the advantages of appreciably faster checkouts, more accurate billing, and fuller printed receipts.

The indirect advantages—in the form of better stock control and greater efficiency generally—will be less visible to the consumer, but probably more important if, as is likely in many areas of retailing, they are passed on in the form of lower prices. The main drawbacks for the consumer are likely to be a tendency for a smaller range of goods to be stocked and, eventually perhaps, some reduction in competition if the stronger groups use these advantages to achieve dominant positions in their own particular areas.

The innovation that gives rise to the greatest doubts, from a consumer's point of view, is the proposal for direct electronic transfer of funds from the point of sale (EFT–POS). This arrangement would appear to offer the consumer few direct and obvious benefits over the present choice of using cash, check, or credit card. It might be thought that, if there were so few benefits, the consumer would simply not use the new facility and, even if no better off, would at any rate be no worse off. The concern is that, once the system had been installed, probably at considerable expense, the banks would very much want to have it used, and would in effect oblige their customers to use it by making the alternatives prohibitively expensive, or even by withdrawing them altogether on the grounds that they were uneconomic and no longer needed.

While for consumers the potential benefits are uncertain, some of the possible disadvantages are more clear. The expected need for remembering and keying in personal identification numbers could be more nuisance than signing a cheque or credit card voucher. There could be problems if a terminal develops faults, or the operator makes errors, or if a customer is refused a transaction because his account is erroneously reported as unable to support the amount required. There could also be problems if any errors made cannot easily be located afterwards and corrected and, where appropriate, compensated for. There would be disadvantages if the existing consumer protection legislation does not cover the new form of payment. There would be excessive risks in using the new system unless the system itself is fully secure from criminal attack and unless the mass of personal and commercial information assembled is given full and effective legal protection against misuse.

Impact on jobs

The effects on jobs of automation in payments systems is not easy to quantify and no comprehensive estimates have been published. The main trade unions in banking and retailing are not in principle opposed to the introduction of new technology, but they are concerned to be consulted at an early stage, and would like the opportunity to conclude technology agreements to ensure that any changes are not detrimental to their members. They foresee and accept the likely need for some redeployment and retraining, but are uneasy lest there should also be actual redundancies.

The banks have sought to allay these anxieties. For example, a senior Midland Bank executive, Mr W. D. Jarman, said in their house magazine of October 1980, that there was "No fear of jobs being lost to microchips," and that for the next 20 years, "Staff will be presented with a challenging and . . . more interesting job with as much as possible of the tedious, repetitive work being done by machines."

The unions, however, are still concerned for job security, and recently both the Banking, Insurance, and Finance Union (BIFU) and the Association of Scientific, Technical, and Managerial Staffs (ASTMS) expressed their fears over possible job losses. ASTMS quote "various studies" which predict that up to 40 per cent of staff in insurance, banking, and finance could be displaced by new technology. Their concern is understandable in that there is pressure on the banks to reduce costs, and about two-thirds of their costs are labour costs. The unions' fears are exacerbated by what they see as a complete lack of consultation between themselves and the banks on developments in new technology.

The ATMs that have recently been introduced must be assumed to have reduced the need for counter staff, and automated bank terminals could have similar effects, although when they were introduced by the Trustee Savings Bank they were used not to reduce staff, but to extend opening hours with the same number of staff.

Potentially far greater numbers of jobs could be affected if electronics funds transfer is introduced at the point of sale. If it is used instead of cash, it will mean fewer people than otherwise needed to deal with cash transactions, but extra people to handle the electronic transfers, and the net effect on employment in banking may not be great. But to the extent that the new system is used instead of checks and credit cards, it should cut out most of the present paper work and reduce considerably the staff time needed for handling a transaction.

However, it must be remembered that the introduction of new technology so far has been in the context of a rapid rise in the number of check and credit card transactions, and in the number of staff needed to handle them. And it must also be remembered that the installation of automated tellers and counter terminals will be spread over a number of years and is unlikely to displace counter staff altogether, and that the spread of EFT–POS is likely to take much longer still. Thus, for some years only a proportion of retail outlets will have the new electronic terminals, only a proportion of their customers will use them, and only a proportion of those who do use them will use them instead of checks or credit cards.

All in all the prospect would seem to be for a substantial drop in employment in banking, but spread over a very long period. Given adequate consultation and careful planning, there is not at present much reason to doubt that it could be absorbed by natural wastage and by retraining and redeployment for which, with only five main banks, there is obviously more scope than in an industry fragmented between a large number of small independent units.

In retailing the situation is somewhat different. Electronic funds transfer is unlikely to make much difference to the number of people employed in the shops. If the new facility is used instead of checks and credit cards, checkouts will probably be slightly faster, with less staff needed; but if it is used instead of cash, checkouts will be slower and more staff will be needed; overall, the net effect on the numbers employed is unlikely to be great.

The use of electronic cash registers and the simpler kinds of point-of-sale system are also likely to make only a very limited difference to staff requirements. The main impact on jobs in retailing may be expected to come from the more elaborate electronic point-of-sale systems which include automatic scanning devices and integrated stock control systems. These should reduce the staff required at checkouts by speeding up the operations involved. More important, however, will be the staff savings as a result of not having to label products individually, in not having so often to check stocks physically, in having reduced levels of stocks to handle and keep safe, and in largely automated systems for analysing sales and re-ordering stock.

However, it needs to be remembered that the most labor-saving systems will be suitable for only a limited proportion of outlets, and will be adopted by them only over a period of years. The total loss of jobs in any one year in retailing as a whole is thus likely to be far less than the normal rate of natural wastage.

The bankers

In recent years the banks have come under increasing criticism for providing a poor service to retailers (for example, in not being open when they are most wanted, in the evenings and on Saturdays), and for providing a poor service to their private account customers (again, on account of their limited opening hours and also their unwillingness to pay interest on current accounts). They have been criticized for providing a service that just under one-third of the population do not find worth using and just under half the population are reluctant to receive their pay through. And they have been variously criticized, in earlier years, for their relatively low profitability and, most recently, for making excessive profits.

At the same time, the five main clearing banks have been under increasing competition from other financial institutions. The National Girobank turned out to be less of a threat than was once expected, but the minor banks have been expanding their business, some of them helped by the enterprising use of new technology. And over a long period the building societies have been expanding their share of the home loans market; apparently the fact that one-third of home loans business is reportedly done on Saturdays was a major factor in the recent decision of Barclays Bank to resume, after 13 years, limited opening on Saturdays. A more serious threat is the way people have increasingly been attracted by the building societies' longer opening hours and payment of interest on accounts to make use of building society share accounts for their day-to-day needs instead of current accounts with the banks.

Meanwhile, the banks have been up against a number of problems in their operations. Security has been an increasing worry, both in the scale of robberies of cash and in the extent of fraud with credit cards and check guarantee cards. The increasing volume of check and credit card transactions has generated a massive increase in the number of pieces of paper to be handled, and this is placing increasing strains on the arrangements for processing them and is leading to ever-increasing costs. And these problems can only get worse if the banks are successful in their aim of attracting the other third of the population who at present do not have a bank account.

Over a long period the banks have been using automation in interbank and back office operations, and they are probably more advanced in this than the banks in most other countries. However, this has been largely on the basis of large mainframe computer installations, and the attraction of the new microelectronics technology is that it lends itself to a variety of kinds of dispersed applications which seem particularly relevant to the problems of the main clearing banks.

The most widely adopted application at present is in automated teller machines. These usefully extend both the hours and the locations at which cash can be dispensed, thereby answering criticisms, countering competition, and to some degree also reducing the pressure on staff. The banks have been rather slower to recognize the potential of automated counter terminals and of truncation of checks, but these could offer scope for improved customer service and savings in staff.

The banks are also interested in the use of transaction telephones for cutting out the fraud on credit cards and check guarantee cards, but the application that most excites their enthusiasm is the possibility of introducing direct electronic funds transfer from the point of sale in the shop (or pub, restaurant, railway station, travel agent, or wherever else a purchase is made). This would, hopefully, reduce the amount of cash in circulation, with its attendant security risks, and reduce the number of transactions by check and credit card, thereby reducing the extent of fraud. It would also reduce the amount of paper work and the numbers of staff needed for dealing with it. It would have the further advantage, from their point of view, of introducing a new and cheaper method of payment from which other competing institutions might be excluded. At a single stroke, therefore, it could make a major contribution to dealing with several of their most awkward problems.

There are, however, a number of practical issues to be settled before it can be introduced, such as the most suitable means of identification, the type of system and technical standards to be adopted, and the arrangements for ensuring security of data. There is also the more basic problem of getting the support of the other parties whose cooperation would be needed, the retailers and the consumers. The advantages for the banks are clear enough, but the advantages for the others remain to be demonstrated. The banks are believed at present to favor a decentralized, on-line system, which would be very expensive, but the retailers argue that different kinds of system would be needed for different kinds of retailing.

These differences have been made more difficult to resolve by the secrecy that the banks have maintained about their plans and, in particular, about the likely costs involved. It would seem that a successful partnership is likely to be achieved only on the basis of more open relationships and a fuller sharing of information and intentions.

Electronic funds transfer, together with the increasing use of automated tellers and counter terminals, is likely to result in local bank branches tending to become smaller and to have fewer people manning them. They might eventually become smaller and fewer still if facilities for carrying out banking transactions from the home or office by means of viewdata links become well established and widely used. However, it is still much too early to know whether, if such services can be made more convenient than the present arrangements, they will also be cheaper; or, if they can be made cheaper, whether they will necessarily be more convenient.

The retailers

In retailing, the scope for applications of new technology, the requirements for success with them, the likely scale of adoption, and the likely consequence if they are adopted all vary greatly from one kind of business to another. For retailing is not a single, homogeneous activity, but many different kinds—the range of products is very wide, and the ways in which they are sold varies almost as much as the ways in which they are made. There are several distinct types of organization—department stores, multiples, cooperatives, independents, and so on—and individual businesses vary greatly in a number of ways, including size, location, product type, product range, and market conditions.

Consequently, any generalizations about automation in retailing are inevitably hazardous and subject to many exceptions. Nevertheless, it is probably broadly true to say that the applications of new technology that interest and affect retailers most are those for automation at the point of sale. For small outlets with relatively simple operations this is often a matter of getting an electronic cash register. This will do much the same job as its electromechanical predecessor, but rather better, with some extra functions, and will be cheaper. Its advantages are fairly obvious, but not as great as the gains that can be made by the use of more elaborate systems in larger organizations.

For larger outlets, and those with more complex operations, the interest is usually in more elaborate electronic systems, especially those with automatic scanning facilities and computer links. They are much more expensive, but offer scope for much greater improvements in checkout efficiency. Usually, however, the main potential gain and reason for interest is nothing to do with the payments systems as such, but with the capacity for providing much fuller management information in general and much better sales and stock control data in particular. These can often make possible a significant improvement in the overall efficiency of the operation, although naturally the scope varies greatly from one business to another; and in retailing, as in any other field of application, new technology is likely to be successful only if built on an efficient existing system—if the existing organization and management is weak, the introduction of automated systems, far from magically rescuing it, may well serve to compound the problems.

In general, retailers' interest in point-of-sale systems tends to be concentrated on those that affect their internal operations. They are usually much less interested in, or not even fully aware of, external links, such as transaction telephones, which they tend to see mainly as a device for cutting down fraud on credit cards, and hence as an item of interest mainly to the credit card companies; and electronic funds transfer systems, which they see as offering advantages primarily to the banks. They are therefore likely to be interested in adopting these systems, which can take space and raise problems at checkouts, only to the extent that the credit card companies and banks are prepared to pay for them, or can manage to persuade the retailers that they will get benefits commensurate with whatever costs they are expected to pay.

The extent to which they adopt any new systems will be colored by the changes that have taken place in retailing in recent years. Following the ending of resale price maintenance in 1964, competition has become sharper, particularly in the past few years of recession when retailers have been struggling to increase their share of a market that is growing very slowly. In consequence, price competition has been fierce, with price wars in gasoline, a rapid growth in low-cost warehouse selling and specialist cut-price shops, and the extension of low-margin multiples into new lines of activity. The result has been a general lowering of margins and of profitability. Consequently many retailers, if they have the funds for investment at all, are inclined to check very closely whether any particular piece of new technology really will pay off before they rush to spend money to buy it.

There have been other developments in retailing that are more favourable to the adoption of new technology. The average size of outlet has been increasing steadily (and it is the larger shops that tend to offer the greater scope for applications); many of the more enterprising retailers have been diversifying into new lines (resulting in a more complex operation with a need for more sophisticated control); and an increasing share of the market has been taken by the stronger multiples (who have the scope, the finance and the organizing ability).

All in all, it seems likely that automation in point-of-sale systems will proceed cautiously, yet fairly quickly; but adoption of a full electronic funds transfer system is a much more uncertain matter. Relatively few retailers know much about it, and even fewer are eager to adopt it. There is virtually no prospect of the early adoption by them of a full on-line electronic funds transfer system, covering all the main retail outlets, and introduced all at one time—certainly not if the retailers are expected to pay for their end of it. Much more likely, it will be a matter of a number of different kinds of system to meet the needs of different kinds of retailer being introduced gradually over quite a number of years, and only after a succession of major trials to clarify the costs and the problems and to indicate more precisely the benefits to be expected for the retailers.

The effects of new technology on retailing may well turn out to fall short of the "revolution" predicted by some, but may nonetheless be important, less in its direct impact at the point of sale than in its wider impact on efficiency in retailing more generally.

It will also have an effect on manufacturers. The new systems will enable retailers to identify more quickly and accurately which lines are less popular, or are moving more slowly, or in other ways less profitable, and to modify their ordering policy accordingly. Suppliers of best selling lines will therefore tend to improve their market share, but the suppliers of less successful lines, like the operators of less successful retail outlets, will face increasing risk of going under.

Eventually, the greatest changes of all in retailing could come from the development of tele-shopping. This could lead to the reduced importance of town center shops, increased business by warehouse sales organizations and by organizations specializing in tele-sales. There could also be a new development of home delivery services to enable full value to be obtained from the long-distance ordering facilities. However, this still looks many years away, and it is by no means clear yet whether it will ever account for more than a tiny percentage of total sales.

Micros in Medicine

James D. Meindl

Microcomputers are now widely used in medical research, medical practice, and medical decision-making, says the author, who is professor of electrical engineering and director of the Stanford Laboratories at Stanford University. Diagnostic tests and laboratory analyses are quicker and easier. The new technology offers new hope to the deaf, the blind, and the disabled. Reprinted from Science, *February 12, 1982.*

The number of transistors commonly fabricated in a single silicon chip, now less than 1 by 1 by 0.1 centimeter in dimensions, has increased from one in 1960 to approximately 100,000 in 1982. Moreover, chip cost has remained relatively constant and reliability has improved during this period. A comparable rate of progress is projected for the current decade. Single chips including 1,000,000,000 transistors appear to be feasible about the year 2000.[1]

The impact on society of this growth in capability of microelectronics and computers has been accurately described as both profound and pervasive.[2] Microelectronics and computers have now come into use in virtually every aspect of modern medicine. In fact, the computer system, with data acquisition and display added to its computational and memory power, now epitomizes the application of high technology to diagnosis and treatment of disease.[3]

Medical research

Computers have been adopted, adapted, and absorbed into medical research. The crux of the creative process, idea conception, has been affected in a fundamental fashion. The kinds of ideas that are conceived, what they deal with, and the terms in which they are considered have been profoundly affected by the computer. The usable level of complexity of ideas has shifted upward dramatically. Computer models of biological systems (such as DNA) that involve many elements, interrelated in specific but analytically opaque ways, can now be evaluated, manipulated, and explored in quest of confirmation, insight, and narrowing of uncertainties.[4] Computer simulation of disease processes, physiological mechanisms, and pharmacological interactions, as well as computer analysis of data, has made many important contributions to medical research.

Indeed, the success of the computer in storing and manipulating biomedical research data has now shifted the emphasis, in many instances, to data acquisition. That is, the principal need in many modern biomedical research projects is the design of appropriate experiments and the application of novel microelectronic sensors to acquire the database for those keen insights that epitomize major advances.

A fundamental requirement in data acquisition is that the method or instrument used to perform a scientific measurement does not disturb the essential state of the system under test. In the case of animal models of human disease, for example, this may dictate that during data collection the animal be not anesthetized or physically restrained. In addition, it may be necessary to obtain data for periods of more than one year of normal life for both healthy and diseased animals, including 24-hour monitoring periods to account for circadian rhythms. One promising means of solving such problems is totally implantable micro-electronic sensors that include radio-frequency transmitters for telemetering data to recording stations remote from the research animal. Totally implantable sensors for measurement of blood flow and pressure, organ dimensions, strain, and bioelectric potentials such as the electroencephalogram, electrocardiogram, and electromyogram as well as temperature have been described.[5] Such sensors are closely akin to the artificial implantable cardiac pacemaker, a now common prosthetic device which is often vital to the life of a recipient. (In fact, special-purpose microcomputers on a single silicon chip have been designed for use in implantable cardiac pacemakers[6] and programmed insulin delivery systems.[7])

Innovative applications of microelectronics in new implantable telemetry systems offer a singular opportunity for advances in biomedical research for two reasons:

1 More than all other types of biomedical instruments, implants exploit fully the inherent technical advantages—complex functional capability, high reliability, low power drain, small size and weight—of microelectronics.

2 Implants bring microelectronics into intimate association with biological systems.

For these reasons, implants allow otherwise impossible experiments to be conducted to improve the quality of health care through advances in research.

Medical data collection

Medical data are remarkably soft in character. They are described, collected, and interpreted with a degree of variability and inaccuracy that falls disturbingly short of typical engineering standards. In order to provide data to a physician, a patient must first perceive himself as ill. The patient–physician dialogue then entails two types of data. The first is expressed by the patient and includes past medical history, a description of current illness, and statements in response to examination. The second type is directly obtained data, which do not pass through the patient's intellectual and emotional filters. These data include some aspects of the physical examination and diagnostic tests that do not require patient responses for their measured values. A patient's description of his medical history or illness may be compromised by intellectual, informational, emotional, and language barriers. Electrocardiograms, X-rays, hematology laboratory tests, and similar directly obtained data are more objective.[8]

Variability in medical history can be addressed by use of standardized questionnaires that present the same questions phrased in the same way. Computer-obtained histories appear to be generally valid when compared with physicians' interviews. Computer-based medical information systems provide a memory of previous events that is less error-prone than human memories and that enhances the completeness and retrievability of all information. These systems may also reduce data variability and inaccuracy in a more subtle way by forcing consideration of how data should be defined and in some instances quantified. The process of data collection cannot be separated from data interpretation in connection with a patient's description of his illness. Indeed, in deciding what expressed and directly obtained data to collect, a physician must select questions on the basis of tentative decisions about the probable diagnosis.

In addition to their use in data collection, computers greatly facilitate data storage, retrieval, and manipulation for medical record-keeping, report-writing, and patient or third-party billing. Important features of such a system include confidentiality, ready accessibility, unrestricted space for entries, as well as editing and backup capabilities. Once data are entered, they should be available to be manipulated for numerous research screening and accounting purposes. Graphic displays on a television screen have the advantages of clarity in showing trends, for example, in disease progression and treatment response.

Only when accurate data can be collected, stored, manipulated effectively, displayed in an appropriate context, and then communicated properly do they become information. Computers make these functions possible and thereby offer substantial potential improvements in the effectiveness of clinical practice, although successes in the field have been limited to date.[9]

Medical decision-making

Decisions are central to the diagnosis and treatment of disease, and many attempts have been made to use computer assistance in medical decision-making. To provide optimal assistance, computers should be used to augment or amplify other decision-making tools—principally the judgment of an experienced clinician. Computers are used in clinical decision-making mainly to (1) improve diagnostic accuracy by using systematic approaches and explicit decision criteria; (2) enhance diagnostic reliability by precluding unwarranted influences of nonidentical but similar cases; (3) increase cost efficiency; (4) advance understanding of the structure of medical knowledge; and (5) gain insight into clinical decision-making itself.[10]

Three paradigms of clinical reasoning have been suggested.[11] These may be described as decision-making, clinical judgment, and problem-solving. Decision-making is the process whereby the clinician selects one action from a number of alternatives. Frequently, these actions are either diagnostic or therapeutic. The decisions about what question to ask while collecting a medical history and what drug to prescribe are specific examples. In clinical medicine, decisions must be made under conditions of uncertainty. Therefore, in seeking an optimal decision, the clinician strives to formulate an appropriate set of alternatives and a consistent set of selection rules. Bayes' theorem from probability theory has been used to assist clinical diagnosis with computers. The question addressed is the following: given a conditional distribution of diseases along with their characteristic signs and symptoms and a particular clinical syndrome in a patient, what is the probable disease state of the patient?

Decision analysis and other more recent analytic approaches have led to more comprehensive techniques for decision-making by including in the process more complex value statements and selection mechanisms for specific actions. Decision analysis provides an approach to a decision problem under conditions of uncertainty. This approach requires that preferences for various consequences be numerically scaled according to the value of their utility, and similarly that judgments of uncertainties be numerically scaled according to their probability. In clinical practice, the analysis attempts to determine the optimal or best medical action for a patient.

Physicians often refer to their clinical reasoning process as judgment. Somewhat more narrowly, judgment is a term applied to the study of the degree to which information provided about a problem is used in making a decision. Multiple regression and discriminant function analysis are useful tools in the study of judgment. Within this particular paradigm, emphasis is placed on the way available information is used by the clinician and not on whether the chosen treatment is correct in some objective sense.

The problem-solving paradigm views the physician as an information-processing system that operates with the limited capabilities of the human mind. Careful analysis of the approach to solving a problem yields insight into the task environment and the strategies at work. These insights are then incorporated in a computer program with the goal of reproducing the problem-solving approach of the physician, given the original task environment. Here the focus is on the programs to be used by the problem-solver in guiding sequences of behavior. A problem-solving theory should first predict the performance of a physician handling specified tasks, including what processes are used and what mechanisms perform them. Next, it should predict incidental events that accompany problem-solving and show how changes in surrounding conditions affect problem-solving. Finally, it should explain how problem-solving skills are developed and what the problem-solver has learned when he has them. Computer-based clinical decision aids now account for a large portion of clinical computing efforts, but the acceptance of these aids by the professional health care community has been limited to date.[12]

Computed tomography

The advent of computer tomography (CT) precipitated a revolution in the field of diagnostic radiology because of the markedly enhanced resolution and reduced ambiguity it affords compared with common projection systems of x-ray imaging. Computed tomographic scanning consists of two sequential processes. First, x-ray absorption data from different views within a single cross-sectional plane must be acquired. Second, these data must be used as input to a computer, which calculates and then displays an image. This image is derived, for example, from a fan beam of x-rays passed through the body from many different angles. The x-rays are partially absorbed or attenuated as they pass through various tissues, and the variations so produced are recorded by x-ray detectors. The electrical signal outputs from these detectors are amplified, converted from analog to digital form, and stored in the computer memory. From these stored data an image of a single slice of tissue is reconstructed by the computer through the solution of a series of complex equations. The brightness of each portion of an image is proportional to the degree to which the tissue absorbs x-rays. The image is displayed on a television monitor in shades of gray or colors and is photographed for permanent recording. The image data can be manipulated to provide quantitative attenuation measurements of all regions.[13]

Computed tomography images a transverse plane through the body similar to a section that might be made by a knife cut through a cadaver. In contrast with conventional x-ray imaging, which projects all anatomy between the x-ray source and the film plate detector into a single image, CT imaging presents a new vision of anatomic detail during life. Conventional x-ray imaging reliably detects differences in density in the range of 5 to 10 percent. Computed tomography can detect differences as small as 0.5 percent and therefore resolve heretofore unobtainable detail. For example, CT readily discriminates between normal and infarcted myocardium. It is also noninvasive and applicable to outpatients. Because the information CT provides is both quantitative and immediately available for computer manipulation and storage, it opens vast new opportunities for the computer as a diagnostic tool.[14]

Ultrasonic imaging

In comparison to x-rays, ultrasound is a noninvasive diagnostic modality that has superior characteristics in some circumstances and inferior ones in others. Applications in which ultrasound usually provides advantages include (1) imaging moving targets, such as the heart or fetus; (2) imaging the pregnant uterus, where minimal x-ray exposure is advisable; and (3) imaging soft tissues, such as muscles and blood, with similar densities but different acoustic properties. Applications in which x-rays are often more useful include imaging the lungs or the brain, where air passages and the skull, respectively, severely attenuate ultrasound.[15]

The basic principle of operation of most ultrasonic imaging systems is similiar to that of sonar or radar. A train of short pulses of high-frequency ultrasound is emitted by a piezoelectric transducer contained in a hand-held probe in contact with the patient's skin. As a packet of ultrasonic energy propagates through the body, most of it is lost through absorption by tissue, but a small amount is reflected at boundaries between tissues of differing acoustic impedance. It is this reflected or scattered component, which returns to the transducer, that provides useful information. Arrays of microelectronic acoustic transducers and associated circuits are used in a computer-controlled time sequence for both electronic focusing and scanning in a sector pattern to produce a real-time cross-sectional image. High resolution is achieved at all ranges since "electronic zoom" is used dynamically to adjust the focus of each depth. In addition, a large sector angle of 70° to 90° provides a panoramic field of view[16] displayed on a television screen.

Ultrasonic instruments under computer control are now widely used for noninvasive imaging of the heart, kidneys, spleen, pancreas, liver, abdomen, gallbladder, fetus, brain, and eye. Their performance, compared with that of early instruments, has been greatly enhanced by the use of computers and microelectronics. Specially designed ultrasonic systems can measure the flow velocity of moving structures, such as blood cells in blood vessels, by comparing the frequency of the transmitted sound with the Doppler-shifted frequency of the returning sound. The frequency shift is proportional to the velocity of flow. Circuitry for distinguishing positive from negative Doppler frequency shifts allows the differentiation of forward or backward flow. Cross-sectional imaging systems and Doppler systems have been combined to allow the diagnostician to improve the accuracy of positioning the ultrasonic beam for measurement of the Doppler frequency shift. In contrast to x-rays, ultrasonic radiation is nonionizing and, when applied at the low power levels sufficient for diagnosis, causes no harmful effects. Indeed, the absence of patient discomfort, apparent safety, ease of performance, and measurement accuracy make ultrasound nearly ideal for use in human patients.[17]

Clinical laboratories

In modern medicine the clinical laboratory occupies a position of great importance. Few hospitalized patients escape undergoing a host of laboratory studies. Both mechanical equipment and electronic equipment are used in modern clinical support laboratories in response to the demand for rapid execution of large numbers of tests. This equipment is interfaced with computers in order to reduce response time, minimize errors, and control costs. Since the tests performed provide numerical results, an excellent opportunity is presented for computer data processing.[18]

During normal operation of a computer-managed clinical laboratory, patient files can be kept current on a minute-by-minute basis by transferring automatically from a central hospital computer system information regarding all discharges, admissions, and transfers. This eliminates the need for routine keyboard entry of census data into the laboratory computer system. In addition, it permits results of completed tests to be transferred automatically to the correct nursing station as patient transfers occur. As patient specimens with attached requisitions arrive at the laboratory, they are entered by keyboard into the computer, which assigns accession numbers. This process provides accurate specimen processing labels and initiates computer communication of pertinent test information. Selected test results can be printed promptly at key clinical locations. Simultaneously, the test results are stored in the computer memory for later retrieval from terminals located in patient care areas. Automatic billing for laboratory tests is facilitated by a central hospital computer system on receipt of test data from the clinical laboratory system. An itemized listing of all laboratory tests is included as an end result in the patient's computer-generated hospital statement.

Cost per procedure in computerized clinical laboratories tends to increase more slowly than costs for noncomputerized operations. In addition, the capacity to share the clinical laboratory is facilitated by the speed and improved communication of results. The clerical workload of laboratory technicians is markedly reduced, and the capacity to expand the number of tests performed without expansion of staff or space is improved. These benefits are precisely the ones computers must offer in medicine if they are to be practicable.[19] Many of the benefits cited for clinical laboratory computerization have been equally well demonstrated in other ancillary services such as radiology and pharmacy.

Although not yet in wide use, perhaps the most powerful analytical instrument used in the clinical laboratory is the gas chromatograph–mass spectrometer (GC–MS). The use of a minicomputer data acquisition and processing system for this analytical instrument greatly enhances its effectiveness in a clinical laboratory. Using the gas chromatograph as the input device to introduce separated sample constituents to the spectrometer greatly increases the qualitative information obtainable from the complex mixtures usually represented by biological samples. In fact, almost automatic identification of all components of the mixture is achieved. A key feature of the mass spectrometer, compared with most other analytical instruments, is its very high sensitivity. With built-in computers and on-line recording, processing of the output of a GC–MS has become common, and much more information is obtained from tests than was previously thought possible. On-line recording of large amounts of original data makes it possible to obtain qualitative information on complex samples in very short periods of time with a single test as well as quantitative measurements with unprecedented accuracy. Thus, the diagnostic value of the GC–MS in the clinical laboratory has been substantially enhanced.[20]

Adjuncts for diagnostic tests

Study and interpretation of the results of diagnostic tests may occupy a substantial fraction of the clinician's time in certain medical specialities such as cardiology. A large percentage of these tests may yield results within the normal range, and the data may be quite repetitive. A physician may be released from the burden of examining large numbers of test results by using the computer as a substitute. Frequently, a highly abnormal result detected by the computer is referred to the physician for additional review. Utilization of computers in interpreting results can increase the speed of entering results on the patient's records, improve follow-up by feeding back results to consulting physicians, and relieve some of the burden of interpretation by the physician.[21]

Electrocardiography was one of the first areas in which computers were applied as adjuncts for diagnostic tests. Automated cardiac dysrhythmia analysis is now the most common application of computer processing and interpretation systems. Compared with most biological signals, the surface electrocardiogram (ECG) is rather well behaved, and its clinical interpretation has been studied extensively. Briefly, the result of a sequence of electrical depolarizations coupled with contractions of the muscular chambers of the heart produces the ECG. These depolarizations can be detected by using electrodes in contact with the skin, and they manifest themselves as a series of deflections in a time-varying electrical wave form. The shapes of the wave form vary with the precise depolarization sequences and locations of electrodes on the skin. The underlying rhythm of the heart is indicated by both the wave-form timing and shape. Premature ventricular contractions are a particularly important form of dysrhythmia.

Automated computer analysis of abnormal cardiac rhythms is now well established for use in real-time ECG monitoring in both long-term ambulatory care and hospital intensive care units. However, many difficulties persist. Many facets of ECG signal acquisition have not been well standardized, with the result that signal characterization continues to be troublesome. Although advances have been made in other domains, the analysis algorithms rely largely on correlation techniques or time-domain feature extraction. The accuracy of these algorithms is improving, but none is free of weaknesses. Consequently, human interaction is necessary to compensate for deficiencies in computer analysis with most systems. Furthermore, the evaluation of system performance demands extensive effort, and to date such assessments have been hampered by the absence of standards and lack of a database that is widely accepted. Efforts to establish a firmer basis for system performance evaluation are continuing. Moreover, research in this field is being addressed to all of the foregoing issues because of the promising clinical utility of automating dysrhythmia analysis.

Automated cardiac dysrhythmia analysis presents a formidable challenge to computer technology. Serious pitfalls exist owing to the vagaries of the signal itself as well as the many sources of artifacts. Nevertheless, a number of practical systems for both patient care and clinical research have evolved over the past two decades. Moreover, the extensive effort that has been expended in pursuit of the perfect dysrhythmia analyzer has underscored the need to comprehend ECG signal analysis fully. The need to develop explicit processing algorithms has forced a more searching examination of signal content. Electrocardiography and the patients it serves will benefit from these efforts.[22] In addition to ECG analysis, computers have been applied to spirographic and blood gas analysis with rather broad success.

Monitors of critically ill patients

Computer monitoring of critically ill patients has been explored intensively for more than a decade. During this period hospital intensive care units have proliferated. For maximum effectiveness, intensive care units require prompt and accurate information about the patient's condition. Although a patient may suffer ventricular fibrillation only once in several days, the occurence of this event cannot be overlooked. Computers offer tireless observation for such infrequent events whose detection is vital. In addition, computers allow recall of data already collected in order to improve understanding of changes in a patient's condition. This is especially important when a medical team is following several patients with diverse problems.[23]

A computer-based system for monitoring critically ill patients typically includes:

1 sensors such as ECG electrodes for signal acquisition;
2 electronic modules for signal amplification and preprocessing;
3 a computer system with input and output interfaces, data manipulation and storage hardware, and a set of software programs;
4 television displays for wave forms, text, charts, and so on, and a keyboard.

In a thoracic surgical unit, for instance, both bedside and central displays are often used for presentation of the ECG, arterial pressures, and respiratory wave forms. The fidelity of the ECG signal is dependent on careful preparation of the skin and placement of the electrode to ensure acceptable electrical impedance at the electrode–skin interface. Amplified and filtered ECG wave forms are transmitted to the computer for digitizing, analysis, and display. Frequently, pressures are obtained from indwelling intra-vascular and intracardiac fluid-filled catheters. The catheter fluid column transmits hemodynamic pressures to piezoelectric or piezoresistive transducers for conversion to electrical signals. Respiratory rate estimates are derived from thoracic impedance measurements, which require careful electrode placement. Many other clinical measurements are performed both automatically and manually, and the resulting data are stored in the computer monitoring system.[24] One of the most interesting of these measurements currently being made in critical care is on-line mass spectrographic analysis of gas components from the patient's exhalations.

Computer automation of physiological measurements and attendant record-keeping relieve nurses of time-consuming tasks, enabling them to devote more time to direct patient care. Some intensive care units have gone further and use computers to assist in clinical decision-making. A few systems even perform closed-loop feedback control of a limited number of physiological variables through infusion pumps which are controlled subject to rigidly structured algorithms. Overall, experience with computer monitoring of critically ill patients suggests that the additional associated costs may be recovered through enhanced personnel performance as well as more rapid stabilization of patients, who can then be transferred sooner from intensive care to less costly rooms. However, indisputable evidence that computer-based intensive care systems are cost-effective is not yet available. Quantitating the economic value of the benefits is a major obstacle in establishing cost effectiveness. The applicability of computer monitors depends a great deal on the nature of the clinical environment. Critical care units in which patient management is highly structured offer the most promising opportunities for the future use of computers.[25]

The handicapped

The quality of life and job opportunities for many people are substantially diminished by loss of some natural function. Many suffer from afflictions such as loss of limbs, paralysis, speech impediments, deafness, and blindness. The compact size and powerful sensory, computational, memory, and display capabilities of microelectronics and computers offer opportunities to relieve these functional deficiencies. Computers especially have substantially increased the variety and quality of job opportunities available to handicapped and home-bound individuals. The intellectual character of the work that they can now perform is more challenging. Both personal and economic rewards are greater. The handicapped person can now engage in computer science and engineering, data processing, accounting, and bookkeeping, fields of gainful professional employment in which there is considerable outside contact through computer terminal, telefacsimile, and telephone.[26]

Computers have been incorporated in prosthetic devices usable by individuals with motor dysfunctions so severe that they can neither produce intelligible speech nor use any part of the body to write effectively. The operation of these devices is dependent on the availability of biological signals that are under control of the user. For example, a neck movement could be interpreted by the device as a command to produce a particular phrase previously synthesized and stored in computer memory.[27]

Mobility is one of the most acute requirements of all handicapped persons, for without it their capabilities are severely limited. This necessitates the development of lightweight, maneuverable, and reliable electrically powered wheelchairs. Microprocessors in particular offer vast improvements in wheelchair control compared with current systems. Automatic speed limiting, programmed acceleration and deceleration, obstacle avoidance, battery monitoring, and component temperature monitoring are both needed and feasible in microprocessor-based controllers.[28]

Microelectronics has been used to achieve a portable reading aid for the blind. The Optacon is a direct translation device that converts an optical image of a printed character on a page of an ordinary book, magazine, or newspaper to a vibrating tactile facsimile. By means of a simple optical system contained in a hand-held camera, an image of a printed character is focused on a microelectronic array of light sensors. Electrical output signals from the sensor array are processed in a simple computer system and then used to control a corresponding array of piezoelectric tactile stimulators or bimorphs. Tiny pins attached to the tips of the bimorphs protrude slightly through perforations in a rectangular plastic plate. A blind reader whose fingertip is resting over the perforations can then feel a tactile facsimile or vibrating image of the original printed character.[29]

Because the signals that travel the human nervous system are electrical in character, it should in principle be possible to replace defective nervous tissue with an electronic substitute. The most striking possibilities for neural prostheses are sensory prostheses for hearing and vision. In both cases the information flow is high, and the stimulation pattern that must be generated is complex in both space and time. A proposed auditory prosthesis for the profoundly deaf is based on electrical stimulation of the auditory nerve in cases where the transduction machinery of the cochlea is totally defective. This stimulation produces the sensation of sound if the auditory nerve itself remains viable. Considerations based on both auditory physiology and speech theory indicate that the independent stimulation of eight to ten subsets of auditory nerve fibers is required to convey the information content of speech to the brain. A formidable problem remains with regard to developing suitable algorithms for recognizing and mapping spoken speech into a pattern of stimulation for the electrode array.[30]

Conclusion

Microelectronics and computers have been used in nearly every phase of medical research. In medical practice, they have been effectively applied to collect, store, retrieve, and manipulate medical data and to augment, amplify, and assist the judgment of experienced clinicians in medical decision-making. Powerful new noninvasive diagnostic instruments such as the computer x-ray tomographic scanner and ultrasonic imaging systems with arrays of acoustic sensors are based on dedicated microelectronics and computers. The utility of clinical laboratories and sensitive analytical instruments such as the GC–MS has been increased through the application of computers.

Computers are now widely used as adjuncts for diagnostic tests such as the ECG and as monitors of critically ill patients. Morale-building accomplishments, social contacts, and job opportunities available to handicapped and homebound individuals have been substantially enhanced by the computer. Prosthetic aids for those suffering afflictions such as loss of limbs, paralysis, speech impediments, deafness, and blindness have advanced markedly through incorporation of microcomputers in their design. Health care represents one of the most promising opportunities for improving the quality of life in society through microelectronics and computers.

Notes

1 J. D. Meindl, K. N. Ratnakumar, L. Gerzberg and K. C. Saraswat, in *Digest of Technical Papers, IEEE International Solid-State Circuits Conference, 1981* (IEEE, New York, 1981), pp. 36–7.
2 P. H. Abelson and A. L. Hammond, *Science*, **195**, 1087 (1977).
3 R. B. Friedman and D. H. Gustafson, *Comput. Biomed. Res*, **10**, 199 (1977); S. A. Glantz, *IEEE Comput. Mag.* **11**, 68 (1978); C. J. McDonald, *Ann. Intern. Med.* **84**, 162 (1976); W. V. Slack and L. J. Van Cura, *Comput. Biomed. Res.* **1**, 527 (1968); T. L. Lincoln and R. A. Korpman, *Science*, **210**. 257 (1980).
4 W. O. Baker, W. S. Brown, M. V. Mathews, S. P. Morgan, H. O. Pollak, R. C. Prim and S. Sternberg, *Science*, **195**, 1134 (1977).
5 J. D. Meindl, *Science*, **210**, 263 (1980); R. L. White and J. D. Meindl, *Science*, **195**, 1119 (1977).
6 J. A. Berkman and J. W. L. Prak, in *Digest of Technical Papers, IEEE International Solid-State Circuits Conference, 1981* (IEEE, New York, 1981), pp. 168–9.
7 W. J. Spencer, *IEEE Trans. Biomed. Eng.* **BME-28**, 237 (1981).
8 A. L. Komaroff, *Proc. IEEE*, **67**, 1196 (1979).
9 S. R. Garfield, *Sci. Am.*, **222**, 15 (April 1970); B. J. Blum, R. Lenhard, Jr, H. Braine and A. Kammer, in *Proceedings of the First Annual Symposium on Computer Application in Medical Care* (IEEE, New York, 1977), pp. 131–8; E. Osserman, L. Katz, W. Sherman and F. Putney, *J. Am. Med. Assoc.*, **239**, 1772 (1978); R. A. Seltzer, G. W. Reimer, L. R. Cooperman and S. B. Rossiter,

Am. J. Roentgenol., **128**, 825 (1977); M. J. Meldman, D. Harris, R. J. Pellicore and E. L. Johnson, *Am. J. Psychiatry*, **134**, 38 (1977).

10 E.H. Shortliffe, B. G. Buchanan and E. A. Feigenbaum, *Proc. IEEE*, **67**, 1207 (1979).

11 C. Bordage, A. Elstein, J. Vinsonhaler and C. Wagner, in *Proceedings of the First Annual Symposium on Computer Application in Medical Care* (IEEE, New York, 1977), pp. 204–10.

12 P. Szolovits and S. G. Pauker, *Proc. IEEE*, **67**, 1224 (1979); G. L. Irvin and R. Zeppa, *Ann. Surg.*, **183**, 594 (1976); D. H. Gustafson, J. H. Greist, F. F. Strauss, H. Erdman and T. Laughren, *Comput. Biomed. Res.*, **10**, 83 (1977); J. Conklin, J. Munzenrider, P. W. Neurath and W. M. Ross, *Radiology*, **123**, 441 (1977).

13 H. L. Abrams and B. J. McNeil, *N. Engl. J. Med.*, **298**, 255 (1978).

14 C. H. Marshall, *Postgrad. Med.*, **60**, 105 (1976); R. S. Ledley, G. Di Chiro, A. J. Luessenhop and H. L. Twigg, *Science*, **186**, 207 (1974); H. K. Huang, M. Cerroni and J. C. Mazziotta, in *Proceedings of the First Annual Symposium on Computer Application in Medical Care* (IEEE, New York, 1977), pp. 356–63; L. E. CLoe, *Am. J. Roentgenol.*, **127**, 187 (1976); A. C. Kak, *Proc. IEEE*, **67**, 1245 (1979); E. L. Ritman, J. H. Kinsey, R. A. Robb, B. K. Gilbert, L. D. Harris and E. H. Wood, *Science*, **210**, 273 (1980).

15 White and Meindl, *Science*, **195**.

16 R. D. Melen, A. Macovski and J. D. Meindl, *Proc. IEEE*, **67**, 1274 (1979); R. L. Popp and A. Macovski, *Science*, **210**, 268 (1980).

17 Ibid.

18 V. Sondak, H. Schwartz, N. Sondak (eds), *Computers and Medicine* (Artech, Dedham, Mass., 1979), p. 205.

19 B. Nussbaum, T. Minckler, R. Roby and E. Ackerman, *Am. J. CLin. Pathol.*, **67**, 149 (1977); D. A. Wycoff and J. R. Wagner, *Am. J. Clin. Pathol.*, **70**, 390 (1978); B. Drewinko, B. Wallace, R. W. Crawford and J. M. Trujillo, *Am. J. Clin. Pathol.*, **67**, 64 (1977); J. P. Horwitz, E. B. Hills, D. Andrzejewski, J. Peñkala and S. Albert, *J. Am. Med. Assoc.*, **235**, 1708 (1976).

20 K. Biemann, *Proc. IEEE*, **67**, 1287 (1979); R. J. Elin, *Science*, **210**, 286 (1980).

21 Sondak et al., *Computers and Medicine*, p. 93.

22 M. Ariet and L. Crevasse, *J. Am. Med. Assoc.*, **239**, 1201 (1978); W. Chu, P. A. N. Chandraraina, D. E. Raeside, R. E. Brown and H. Poehlman, *Radiology*, **123**, 795 (1977); T. W. Sykes, R. L. Haynes and E. R. McFadden, *Am. Rev. Respir. Dis.*, **115**, 581 (1977); L. J. Thomas, Jr, K. W. Clark, C. N. Mead, K. L. Ripley, B. F. Spenner and G. C. Oliver, Jr., *Proc. IEEE*, **67**, 1322 (1979).

23 Sondak et al., *Computers and Medicine*, p. 107.

24 L. C. Sheppard and N. T. Kouchoukos, *Anesthesiology*, **45**, 250 (1978).

25 M. Hilberman, B. Kamm, M. Tarter and J. J. Osborn, *Comput. Biomed. Res.*, **8**, 447 (1975); M. E. Wade, P. J. Coleman, S. C. White, *Obstet. Gynecol.*, **48**, 287 (1976); J. J. Almasi and L. J. Heitlinger, in *Proceedings of the First Annual Symposium on Computer Application in Medical Care* (IEEE, New York, 1977), pp. 290–6; L. C. Sheppard, *Proc. IEEE*, **67**, 1300 (1979).·

26 M. J. Giannini, *Computer*, **14**, (no. 1), 12 (1981); W. Schneider and W. Seamone, *Computer*, **74**, (no. 1), R. J. Leneway and B. R. Montgomery, *Computer*, **74**, (no. 1), G. C. Vanderheiden, *Computer*, **74**, (no. 1), 54.

27 M. A. Rahimi, *Computer*, **74**, (no. 1), 19; A. Thomas, *Computer*, **74** (no. 1), 25.

28 J. H. Aylor, B. W. Johnson and R. L. Ramey, *Computer*, **74**, (no. 1), 35.

29 White and Meindl, *Science*, **195**.

30 Ibid.

Guide to Further Reading

Banking

Kent W. Colton and Kenneth L. Kraemer, *Computers and Banking: Electronic Funds Transfer Systems and Public Policy* (Plenum, New York, 1980). Charts slow progress of EFT and shift of emphasis from radical revision to modification of the existing system.

Martin Mayer, "Here Comes the Smart Card," *Fortune,* August 8, 1983.

The *Financial Times* (London) also discussed the smart card on August 17, 1982. The *FT* ran general articles on electronics in banking on July 16, 1981, October 23, 1981, June 26, 1982, December 14, 1982; January 17, 1983 and March 22, 1983; pieces on automatic validation of credit cards July 23, 1981, March 17, 1982, and February 24, 1983; and major surveys on computers in banking and finance on March 30, 1983 and October 22, 1984. The launch of CHAPS (Clearing House Automated Payments System) was covered on October 8, 1983 and on February 6, 9, and 14, 1984.

Amin Rajan, *New Technology and Employment in Insurance, Banking and Building Societies* (Gower Press, Aldershot, England, 1984).

Desmond Smith, "The Wiring of Wall Street," *The New York Times Magazine,* October 23, 1983.

R. T. Clark, "Electronic Funds Transfer: The Creeping Revolution," *Telecommunications Policy,* March 1984.

Union-inspired reports include:

Bank Workers and New Technology (International Federation of Commercial, Clerical, Professional and Technical Employees (FIET), Geneva, 1980).

New Technology in Banking, Insurance and Finance (Banking, Insurance and Finance Union (BIFU), London, 1982).

Retailing

Peter Marsh, "Will Britain Buy Electronic Shopping?" *New Scientist,* April 29, 1982.

Peter Nulty, "The Bar-Coding of America," *Fortune,* December 27, 1982.

Technology: The Issues for the Distributive Trades (NEDO, London, 1982).

"How Giant Food Harnesses High Tech To Fatten Supermarket Profits," *Business Week*, December 5, 1983.

Alan Cane, "Britain's Shopping Revolution: The Challenge to the Cheque," *Financial Times*, December 8, 1983.

Stephen B. Weinstein, "Smart Credit Cards: The Answer to Cashless Shopping," *IEEE Spectrum*, February 1984.

Health care

Thomas L. Lincoln and Ralph A. Korpman, "Computers, Health Care and Medical Information Science," *Science*, October 17, 1980. The convergence of information science and medical science has created a new field of medical information science.

"Calling 'Dr SUMEX,'" *Time*, May 17, 1982.

John Fox, "Computers Learn the Bedside Manner," *New Scientist*, July 29, 1982.

N. F. Kimber, *An Introduction to Computer Applications in Medicine* (Edward Arnold, London, 1982).

James S. Turner, "Consumers, Computers and Pharmaceuticals," *Health Affairs*, Spring 1983. Speculative article which discusses "smart" health cards, home computer doctors, and the "hospital on the wrist."

Peter Marsh, "Technology for the Disabled: The Great Bureaucratic Hold-up," *New Scientist*, November 26, 1981.

"Power to the Disabled," *Time*, December 13, 1982.

Technology and Handicapped People, (Office of Technology Assessment report, US Congress, Washington DC, 1982).

David Dery, *Computers in Welfare: The MIS-Match* (Sage, Beverly Hills, 1981). Highlights problems of management information systems for those looking for a quick high-technology cure to current chaos.

Miscellaneous

Efrem Sigel et al., *Books, Libraries and Electronics: Essays on the Future of Written Communication* (Knowledge Industry Publications, White Plains, NY, 1982).

Humberto Gerola and Ralph E. Gomory, "Computers in Science and Technology: Early Indications," *Science*, July 6, 1984.

New technology in newspapers is a very under-reported subject, but there is:

Ian Hamilton Fazey, "Making News in the Regions," *Financial Times*, May 11, 1981.

Michael Cross, "New Scientist Says Goodbye to Gutenberg," *New Scientist*, January 27, 1983.

Finally, another interesting application of microelectronics is described in Hoo-min D. Toong and Amar Gupta, "Automating Air-Traffic Control," *Technology Review*, April 1982.

Part Three: The Impact on Work

9 Employment: the Quantity of Work

Automation is Good for Us

Patrick Jenkin

This statement by a British Cabinet minister ably expresses the current orthodoxy in government thinking on the employment impact of new technology. It is not automation but the failure to automate that risks jobs, he says, making it all sound delightfully simple. The statement appeared in New Scientist, *February 24, 1983, when Jenkin was Secretary of State for Industry.*

Fears that automation will inevitably lead to higher unemployment are not new. In 1811, the Luddites rioted and destroyed the textile machinery which they saw as a direct threat to their jobs. Yet employment in the textile industry proceeded to grow during most of the nineteenth century. Nor was this an isolated example. In the same century, the fastest growing industries, in terms of employment, were those based on new technology. Railway employment rose from 29,000 in 1851 to 320,000 in 1901; in chemicals, employment over the same period trebled; and in metal manufacture employment nearly trebled.

The last century demonstrates clearly that, despite the fears to which new technology gave rise, technology promoted employment. More recent history repeats the lesson. Considerable concern was expressed in the 1950s and early 1960s about the impact of automation on jobs. Nonetheless, employment in the UK reached record levels towards the latter part of the 1960s. Despite the evident lack of competitiveness of much of the British economy, and the impact of the first oil crisis, total employment showed no overall fall during the 1970s, a decade that witnessed the widespread diffusion of computers. Even today, Britain has the highest proportion in Europe of people of working age actually in jobs.

It is easy to understand why the fears of technology-generated unemployment should enjoy a renaissance during a world recession. Undoubtedly, concern about the impact of robots and other microelectronic-based production equipment has been exacerbated by the depressed world economy and the high levels of unemployment in nearly all the major industrialized countries. But the facts do not support any causal relationship between automation, higher productivity, and unemployment. It cannot be emphasized too strongly that our unemployment problems do not stem from the installation of such equipment.

In the first place, the penetration of robots, numerically controlled machine tools, and other microelectronic-based equipment in British industry is as yet very limited. By the end of 1982, for example, there was a total of 1152 robots installed in the UK. Estimates suggest that little more than 5 percent of the stock of machine tools was numerically controlled. Quite simply, there is too little equipment installed to account for the levels of unemployment.

Second, even when such equipment has been installed it would be wrong to conclude that the overall impact on jobs has been negative. The countries with the highest level of robots per employee, Japan and Sweden, both have very low rates of unemployment. Far from high usage of robots being associated with high levels of unemployment, the opposite is true. This may at first appear paradoxical because robots are often—perhaps most often— thought of as mechanical work people, displacing humans. It is hard to associate that picture with an analysis that shows the use of robots promoting employment. To understand the true picture, it is necessary to consider how any new technology incorporating high productivity affects jobs.

The installation of a robot, for example, has both a "direct" and a "compensating" effect on employment. The direct effect is the net change in the number of jobs before account is taken of output changes. It is sometimes, although by no means always, negative. However, the installation and operation of robots and related equipment improves output, albeit not immediately, and this helps to increase jobs.

How then does output increase? The primary reason for introducing new technology such as robots is to reduce costs and improve product quality. Lower costs mean lower prices. With the improvement in product quality, this results in increased demand for our goods and services, which in turn generates higher output and employment. Profits also increase, inducing higher investment and R&D expenditure, thus creating jobs. Finally, investment in robots and other microelectronic-based equipment provides opportunities for domestic producers of such capital goods to increase output and employment.

Some specific examples show how the argument works out in practice; how automation, far from destroying jobs, can create or safeguard employment.

Tallent Engineering at Aycliffe, County Durham, last October began to operate a computer-operated press system linked to a robot welding line. The system helped the company to win a £5 million a year contract to supply rear suspension arms for the new Ford Sierra. It not only secured jobs that might otherwise have been lost but also created 80 more. Without investment in new technology, it is thought that the work would have gone abroad.

Without the robots the company could not have coped with the Ford contract. The pressure on the welders would have involved a very high rejection rate—a problem overcome by robot welding.

Flexible manufacturing systems (FMS) provide the economics of mass production to small batch manufacture. Savings come not so much from savings in manpower as from the reductions in working capital, higher production quality, better control, and much faster switching from one product to the next, thus avoiding the need to carry excess stock. Normalair-Garrett (NGL) at Crewkerne was among the first companies to introduce flexible manufacturing in the UK. NGL found that with FMS stock is now being turned over 24 times a year as opposed to the previous 3·3 times. It has also cut manufacturing lead times from 17 weeks to 2 and trebled output per operator.

A more traditional industry gives a similar account. J & J Cash of Coventry has been weaving name tapes on narrow measures of cloth since 1902. The industry's traditional labor-intensive Jacquard process had by the 1960s begun to threaten Cash's position because of rising prices and inability to respond to peak demands. A measure of automation based on a computer and a punched-tape system was introduced, and this helped cut costs and speed production. But, more recent, it became necessary to introduce an improved system based more fully on microelectronics. The investment has made Cash strongly competitive, by keeping prices down and improving turn-round. Although a few jobs have been lost, the workforce today is nearly 200, whose jobs are now more secure because of automation.

Most recently, consider the direct effect at Telford in Shropshire through the expansion of the robot manufacturing activities of Unimation (Europe). Nearly 250 new jobs have been created in the company, and as many more in its suppliers. The important point is that increased use of advanced production technology represents a growth area for the supply of the equipment involved. In robotics, for instance, a number of British companies have begun to develop and make robots, so creating jobs directly.

Measuring the wider, compensating effects of automation is, of course, difficult. However, it is clear from a variety of European studies that improved competitiveness helps jobs. The Institute of Employment Research at Warwick University recently carried out a number of macroeconomic simulations to discover what would happen if Britain installed sufficient microelectronic technology to improve its rate of productivity growth by 1 percent relative to its competitors. The answer was that such a technological acceleration would create more jobs than it displaced. Similarly, both the Rathenau Advisory Group in the Netherlands and the Industrial Institute of Economic and Social Research in Stockholm, carried out simulations that indicated that slow introduction of microelectronic technology would have a worse effect on employment than keeping up with other countries.

It is not automation but the failure to automate that risks jobs. Although microelectronic production technology can increase the total number of jobs, the occupational structure of employment is likely to change significantly. In particular, the demand for engineers, technicians, computer programmers, and software experts can be expected to increase sharply. (The NEDC says that already 16,000 more programmers are needed.) Fewer machine-tool operators, welders and production fitters are likely to be required. In particular, the new demand is likely to be for those concerned with software.

Already there is evidence that some firms that have introduced microelectronic production technology have experienced difficulties as a result of a shortage of engineers, technicians, and programmers. What is needed is the right training and retraining, and the government therefore makes available a wide range of training schemes and courses.

The overall message is clear. We must make use of microelectronic technology. We must be prepared to adapt as some industries and occupations expand and others contract. Secure employment will come from embracing new technology, not from pretending that it does not exist.

This is the way to build a high-wage, high-productivity country. We can then become more competitive, sell more goods, and employ more people. We can also afford the better social services which are the hallmark of a civilized state.

Job Losses Now, Maybe Some Later

Tim Brady and Sonia Liff

Jobs may well be created through new technology, say the authors, but it will be a long time before the gains offset the losses from traditional industries. That is the broad conclusion of this British government-backed study, which outlines some of the complexities of the job creation and displacement process, with special reference to the West Midlands, one of Britain's industrial heartlands. Tim Brady is with the Science Policy Research Unit, Sussex University, England and Sonia Liff is with the Technology Policy Unit, University of Aston in Birmingham. Taken from Monitoring New Technology and Employment *(Manpower Services Commission, Sheffield, England, June 1983).*

In this section, we discuss the implications of technological change for employment in manufacturing bearing in mind the effects of the recession. To start with, we will examine job displacement effects. This is a complex issue. First, there are the direct displacement effects associated with the introduction of a specific piece of plant. Second, there may also be significant job losses resulting from the failure of firms to adopt new technology in products or processes.

The recession has heightened competition in many sectors, and this has increased pressures on firms to cut costs and innovate. This competition, based around attempts to retain a share of static or contracting markets, may mean that the direct employment effects on successfully innovative firms are relatively small compared with losses in non-innovative firms. The latter are much harder to identify since they occur at sites other than those at which the technological change is taking place.

Unless markets expand or new markets are found, the number of firms that can gain competitive advantage through technical change is limited. We found that a common effect of the introduction of labor-saving new technology has been to increase capacity, leading to a concentration of production. This could have serious implications for the number of firms that can survive and hence for long-term job opportunities.

Industrial restructuring often accompanies radical technical change. In this way new technology can indirectly affect employment away from its point of introduction. For example, technical change may affect the raw materials or components used in a production process. This may affect employment in firms that supply these materials. Similarly, the changing patterns of production discussed above will have indirect employment effects often resulting in job gains at the point of introduction of new technology—but job losses elsewhere.

We also examine the evidence of job creation flowing from new technology. These jobs involve producing new goods or services either directly for the consumer market or for industrial markets. In both cases we need to recognize that job creation will not necessarily lead to net job gains in a sector. For example, in the case of robots, jobs will be gained in producing and maintaining them, whereas jobs will be lost at the point at which they are introduced. We also briefly examine some of the schemes that have been adopted to promote the creation of new jobs based on new technology.

Job displacement effects

One of our main concerns has been to try to identify the ways in which technological change can affect job loss. The magnitude of job losses during the recession has made it particularly difficult to identify those directly attributable to changes in technology. However, the number of such job losses is likely to be small in comparison to the total. Table 9.1 below shows net employment in selected sectors in June 1970, June 1980, and June 1982, and the actual and percentage change in employment between June 1970 and June 1980 and between June 1980 and June 1982.

The long-term decline of the manufacturing industries has continued and in many instances has accelerated. The depth of the recession is evident in that even those service sectors that exhibited substantial employment growth during the 1970s have shed labor between June 1980 and June 1982.

Over the last year, we have monitored the press for job losses and gains with the aim of identifying sectors where technological change has been a significant factor. Unfortunately, the press reports do not always include the reasons for the job losses. Where an explanation is given, the predominant cause of job losses advanced is lack of demand, caused by the recession. Government spending cuts were the cause of many anticipated job losses in some service sectors. The effects of technological change are rarely reported in the press, and more detailed examination of sectors is necessary to provide suitable material for analysis of the role of technological change.

The introduction of a specific type of equipment may have different effects on employment in different firms within a sector, depending on work organization, existing degree of capitalization, availability of skilled man-

Table 9.1 Numbers in employment, June 1970, June 1980, and June 1982

	June 1970	June 1980	June 1982* (thousands)	Change 1970–80	Change 1980–82*	% change 1970–80 (%)	% change 1980–82 (%)
Total: All industries and services	21,993	22,355	20,651 (20,771)	+362	−1704 (−1584)	+1.6	−7.6 (−7.1)
Agriculture, forestry, fishing	454	351	345	−103	−6	−22.7	−1.7
Mining and quarrying	407	347	325	−60	−22	−14.7	−6.3
Manufacturing industries	8164	6711	5655	−1453	−1056	−17.8	−15.7
of which: Food, drink, and tobacco	767	669	605	−98	−64	−12.8	−9.6
Chemicals and allied products	440	436	388	−4	−48	−0.9	−11.0
Mechanical engineering	1092	877	722	−215	−155	−19.7	−17.7
Instrument engineering	161	147	129	−14	−18	−8.7	−12.2
Electrical engineering	815	739	642	−76	−97	−9.3	−13.1
Vehicles	830	699	551	−131	−148	−15.8	−21.2
Metal goods n.e.s.	591	518	430	−73	−88	−12.4	−17.0
Textiles	633	382	299	−251	−83	−39.7	−21.7
Clothing	430	319	260	−111	−59	−25.8	−18.5
Paper, printing, and publishing	619	539	493	−80	−46	−12.9	−8.5
Construction	1294	1242	1024	−52	−218	−4.0	−17.6
Gas, electricity, and water	382	342	331	−40	−11	−10.5	−3.2
Transportation and communication	1549	1483	1363	−66	−120	−4.3	−8.1
Distributive trades	2617	2821	2656	+204	−165	+7.8	−5.8
Insurance, banking, finance, and business services	943	1292	1300	+349	+8	+37.0	+0.6
Professional and scientific services	2830	3658	3660	+828	+2	+29.3	+0.1
Miscellaneous services	1908	2571	2496	+663	−75	+34.7	−2.9
Public administration	1446	1539	1496	+93	−43	+6.4	−2.8

* Estimates of employees in employment are provisional for June 1982 and understate the level of employment mainly in the service industries. Supplementary series that include an allowance for underestimation are shown in brackets.

Sources: Department of Employment Gazette, March 1975 and September 1983

power, industrial relations, etc., within individual firms. The introduction of complete systems, for example, in the context of a green-field site, may result in greater labor savings than piecemeal or incremental introduction of the same technology in existing sites.

The job-displacing effects of a particular technology may vary from sector to sector depending on its application. The extent to which job losses occur is not predictable from the nature of the technology alone. It depends, among other things, on the existing stage of development of the technology, and the level of demand in the sector. If such equipment is introduced when demand is rising, the "direct" job-displacing effects may not be apparent. If demand is static or falling, then the introduction of such equipment is likely to lead to job loss.

The "direct" effects of automation

Production automation

Severe competitive pressures in many sectors brought about by contracting markets have made automation attractive to many firms as a means of cutting costs to help maintain profit margins. Examples of this sort of technological change are fairly widespread in several of the sectors that we examined. The introduction of automatic component insertion equipment for the manufacture of printed circuit boards in the TV industry has led to a significant decrease in the number of people involved in manual assembly operations. Automation of the controls on machinery in the paper packaging industry has also led to the direct displacement of labour, with fewer operators required for certain machines. Automation in some of the large iron foundries has reached such a level that the labour force required is as little as a tenth of that needed in foundries with more traditional forms of production.

The development of microprocessor technology has opened up new areas for automation, so that even in sectors where the process has always been highly automated there is scope for automation in the more labour-intensive parts of the production process. Thus, robots are being used to unload diecastings in the aluminum foundry sector. The scope for this sort of automation will increase as the technology improves and is reduced in price.

There have been various estimates of the numbers of people that are directly displaced by the introduction of robots. It has been suggested that there is a widely held view that a robot installation replaces on average five people. With just over 700 robots installed at the end of 1981, this would mean that some 3500 jobs had been lost in Britain as a direct result. In contrast to this "widely held view," Peter Davey, coordinator of the Science and Engineering Research Council's robotics research program, believes that the maximum number of people that a robot can displace is one or two, depending on the application. He further suggests that, assuming a robot population of 20,000 by 1990 (an optimistic assumption according to Davey), the maximum number of jobs lost would be 40,000, against which would have to be balanced some 20,000 jobs created to maintain, supervise and control this number of robots.

Empirical work at the Technology Policy Unit (TPU) at the University of Aston found a job displacement rate of 0–3 per shift or 0–6 taking into account shift working, and averages of 1.6 and 2.9 respectively. Research has been carried out in West Germany which also includes actual case-study data about the job displacement effects of robot introduction. This broadly agrees with the TPU findings, suggesting that between 0.8 and 6.2 workers had been displaced by each robot installed. The exact number depended on the number of shifts being worked and the application.

There were more staff reductions when robots were being used in parts-handling than in tool-handling applications. Furthermore, the maintenance and manufacture of industrial robots created considerably fewer jobs than were lost directly through the introduction of robots. TPU research on job creation suggests a figure of 0.25–1.0 people per robot over, say, eight years of robot life, depending on the development effort involved. The job displacement effects are likely to be felt more keenly when assembly robots are fully developed and new areas of potential penetration are opened up.

Design automation
The introduction of CAD in engineering firms has had little effect on employment levels so far. The proportion of draftsmen in total engineering industry employment declined steadily from the mid-1960s to the late 1970s, but this was nothing to do with CAD. The impact on draftsmen's employment has been small so far because most user firms are in the learning and experimentation stage. It is only since the arrival of relatively cheap American "packaged" CAD systems from about 1977 that scope for extensive CAD use has occurred.

Job loss may have taken place, but it has not often been apparent: fewer contract staff have been employed; vacancies have been written off the books; the design workload has been increased using existing staff. The stage has now been reached where a number of firms are about to make a transition from the experimental to systematic use of CAD, so past experience of the effects on employment are unlikely to be reliable indicators for the future.

However, it is probable that there will be job reductions in mechanical engineering, in electrical engineering (where a significant part of the design workload is mechanical), and in aerospace and vehicles. CAD can also play a part in firms' competitiveness—by facilitating the design of new, more competitive products. Non-users may find themselves in a less competitive position which could lead to job losses. The effect of this sort of job loss would vary from sector to sector according to the nature of the competition within a sector.

Indirect effects of technological change

Firms that adopt successful innovations in products and production proces-ses may enhance their competitiveness and thereby reduce job loss. Even relatively successful innovators are likely to suffer job loss in some occupations, but in the process they may gain greater security for the jobs that remain.

Technical change cannot guarantee an end to job loss. The recession has seen the closure or rationalization of many firms that have introduced new technology. The TV industry, for example, is a sector where all the firms have invested in new technology in both products and in production processes, but nevertheless some have been forced to close down factories or to carry out redundancy programmes.

More usually however it is firms that fail to innovate that are likely to suffer the most job losses. Competition within a static market is based on capturing part of that market from other firms, and in such a context losers are inevitable. In this situation some firms may be in a better position to succeed than others.

For example, the option to innovate may not be open to many small firms. The high cost of capital investment to automate processes or to introduce equipment that will raise the quality of the products prevents some firms from adopting this strategy. If they are to survive they have to cultivate specialist markets and rely on their flexibility and low overheads to maintain these markets. Firms unable to find such market niches and unable to innovate are likely to go out of business, or at least to suffer very substantial job losses.

There are several examples of this kind of market segmentation with the larger firms being able to innovate and the smaller ones having to rely on a particular niche. In the West Midlands, it was found that overcapacity, owing to a reduction in demand from car companies, had led to many small spring-making firms becoming unprofitable because they relied on traditional technology. The larger firms had been able to improve the consistency and quality of their products through high spending on R&D and investment in new technology, and had suffered fewer job losses as a result.

Similarly, in the iron foundry sector a handful of large firms had been able to automate and were winning all the volume orders. The small foundries, unable to invest in much equipment, either had been forced to close or had to rely on their lower overheads and their ability to meet short lead times as a basis for obtaining business.

In the fiberboard packing case sector, the larger firms were investing in microprocessor-based new technology for the control of their machines and in automation of materials handling. The smaller firms, and there were many of them, were concentrating on maintaining flexibility to take on varied work at very short notice. They were helped in this by the tendency for customers to demand shorter lead times and order smaller batches than had been the practice some years ago. Here, it was often the medium-size firms, which were not large enough to spend vast amounts on investment but were too large to achieve the sort of flexibility of the smaller firms, that were being squeezed.

Smaller firms in the TV industry concentrated on particular markets—the black and white portable market or small-screen color sets—rather than offering a complete range. The larger firms usually carried a complete range of sets and included features such as teletext. They also spent large amounts on R&D and invested in automated production and test equipment.

Longer-term employment effects

Technologies that result in large productivity increases have implications for the number of firms that can survive, even if they invest in new technology. Automated machinery often needs to be fully utilized to justify the large capital expenditure incurred.

While many firms may have introduced new technology mainly as a means of cutting costs through labor reductions, one of the effects has been to increase capacity. This is because new equipment is made in discrete units which run much faster than older equipment. Each new unit may offer production capacity several times higher than previously installed machinery. In some cases, firms may install one machine to replace several old ones, but in other firms the installation of only one new unit may increase the firm's total capacity.

Interviews in the color television industry confirm that output could be increased readily from present levels without very much increase in employment. At one firm, for example, a 20 percent increase in output could be achieved without any increase in employment, while at another a 50 percent rise would require only a 10 percent rise in employment. This is a sector where demand is relatively buoyant. Several new entrants have come into the industry in recent years and some of the existing firms are expanding capacity with the intention of entering European markets. Excess capacity is likely to continue unless new markets can be found or production capacity is transferred to new products such as video recorders. This may mean that the full-productivity potential of new technology cannot be realized.

Demand in other sectors is not so buoyant so there is less likelihood of new firms entering, but the problem of overcapacity still exists. In these sectors it is likely that firms will continue to be under pressure to cut costs. In this environment labour-intensive areas of production such as materials handling are likely to be subject to increased automation, so further job losses can be expected.

It seems probable that there will be continued automation of production processes in British industry. As yet there are few examples outside the electronics industry of fully integrated systems linking design to manufacture (CAD/CAM systems). It is likely that developments in particular areas such as assembly and materials handling will have more effect on employment. Unless there is a substantial increase in the level of demand, it seems doubtful that there will be significant expansions of employment in mature industries.

Job creation effects

In view of the depth of the recession (and its persistence), it is not surprising that evidence of job creation was difficult to come by. However, it seems that new technology is a strong influence in those sectors where jobs are being created. The employment monitor identified significant job gains in only two sectors—electrical engineering and distribution.

In the electrical engineering sector, 23 firms reported plans for new jobs in the future (sometimes the future was some years away). All were involved in electronics in some way. Five were microchip producers, four were telecommunications equipment manufacturers, eight were manufacturers of computers or related hardware equipment, five were producing consumer electronic goods (including four TV manufacturers), and one firm was producing radar equipment. Half of this number were Japanese or American-owned. The other job gains in manufacturing were all in firms producing high-technology products, or using high-technology equipment.

The vast majority of the jobs created in the distribution sector were in large supermarket chains. They were concentrating on the superstore type of outlet which usually requires a staff of between 200 and 300. It is this type of store that has most to gain from the use of electronic point-of-sale equipment, and it is likely that many of these new stores will incorporate such equipment. It is important to recognize that distribution is suffering net job losses. Job losses are less evident since they often occur in smaller establishments. What we are seeing is a restructuring of the industry, in part as a response to technical change.

It is quite probable that the employment monitor failed to identify many examples of small-scale job creation. The study of the West Midlands suggests that many of the small businesses started there are not producing distinctly new products or utilizing much new technology in their production processes. Choice of activity seemed rather to be based on already existing markets, and firms were often started by people who brought their contacts and "know-how" from their previous employment. It is unlikely that the creation of jobs from such small businesses will do much to counteract the massive job losses that have taken place in the region recently.

Other schemes to encourage job creation have been adopted. "Science parks" are thought to speed up the innovation process by providing a context that allows a more direct flow of ideas between inventor and innovator, and by providing venture capital and support services. The relevance to job creation is that inventions that might otherwise never have been developed, or that would have developed at a much slower rate, can be helped to reach commercial viability.

Various models of science parks exist, but most involve the provision of premises (generally close to a university) and of shared services for small firms that are concentrating their efforts on the R&D stage of high-technology ventures. The proximity to a university is intended to provide a source of ideas, advice, and skilled manpower which, it is hoped, will speed up the innovation process. There is currently considerable interest in science parks, and several are being established. One such scheme, Aston Industrial Science Park, is seen by Birmingham City Council as a way of providing the foundation for longer-term changes in the technological base and direction of West Midlands industry.

Attempts to change the industrial structure of an area based on attracting and supporting new high-technology firms is a long-term strategy, since firms take time to grow from the R&D state, however successful they are eventually. There will be few jobs created in the short term, and these will be high-technology-based and thus unlikely to provide many opportunities for skilled and unskilled people being made redundant from manufacturing.

New products, if they are successful, offer better long term prospects for job creation. Overall, the number of jobs based on new technology may well increase as the firms grow, but it is likely to take a long time for these to offset job losses from traditional industries.

High Tech is Low on Jobs

Business Week

Forecasts by the US Bureau of Labor Statistics and Data Resources Inc. show clearly that high tech can't come near to replacing jobs lost in manufacturing because the sector is small and productivity is rising fast. Nearly every state and city in the US has therefore embarked on a largely fruitless quest to create large numbers of high-tech jobs in their own versions of Silicon Valley. But some good is coming out of it. This report first appeared in Business Week, *March 28, 1983.*

From the halls of Congress to the gritty industrial cities of Buffalo and St Louis, high technology is being invoked as the cure for the nation's economic ills. Like a talisman, high tech is being called on to spawn new industries to meet ever-stiffening foreign competition, revitalize decaying smokestack industries, and put 11 million unemployed US workers back on the job.

Nearly every city and state is convinced that, as new industries replace old, survival will depend on attracting and nurturing high-tech industry. "It's a great bandwagon business," concedes April Young, executive director of the Fairfax, Va, Economic Development Agency. "There are 4500 economic development agencies in this country, and it's fair to say that every one of them is after high tech." Even Akron, one of the last US cities that would be considered high tech, is attempting to change its image by calling the area "Polymer Valley," a major move to capitalize on its traditional base in rubber and plastics.

The demands being made on high-tech industries add up to a tall order—so much so, in fact, that most localities will come away empty-handed over the next decade. In spite of their spectacular growth, the new industries will account for only a fraction of total US employment by the mid-1990s. At the same time, years of neglect in the US educational system have already put a serious crimp in the supply of scientists and engineers needed to drive high technology.

Meanwhile, Congress is rushing to make a national priority out of science education and high technology by proposing more than 200 pieces of new legislation. But "the idea that we are going to have giant Hewlett-Packards growing up in every town is really a bit of a misperception," acknowledges George A. Keyworth II, President Reagan's science adviser.

"There is not enough high technology to go around," declares Gerhard O. Mensch, a management professor at Case Western Reserve University. "Although there is a whole basketful of new technologies that people are developing, they will be distributed very unevenly. A few places will get the lion's share; most places, hardly a ripple."

One reason for the exaggerated expectations for high technology is that there is virtually no agreement on exactly what high tech is. "Everyone talks about high technology, but ask them to define it. Most can't," says Eugene Strull, general manager of advanced technology at Westinghouse Electric Corp.'s Defense & Electronic Systems Center. And even when they think they can, state and city officials, technology experts, and government labor analysts all come up with different terminology. It was only recently, in fact, that the Bureau of Labor Statistics (BLS) even formulated a definition.

Most local officials seeking high-tech industry go along with James P. Fenton III, executive director of Connecticut's new High Technology Council, who calls high tech "any industry that is going to create jobs in the 1980s and 1990s." But such experts as Westinghouse's Strull call high tech any "technology with a high rate of change." That includes the cutting-edge technologies of information processing, communications, semiconductors, robots, and biotechnology.

For statistical purposes, the BLS has tried to define high tech by saying that 36 of the 977 industries assigned standard industrial codes qualify because their R&D expenditures and number of technical employees run twice as high as the average for all US manufacturing. This includes makers of drugs, computers, electronic components, aircraft, and laboratory equipment. Also qualifying are service industries—including computer programming, data processing, and research laboratories. The BLS defines an additional 56 industries as "high-tech-intensive," which means their R&D spending and technical employment are above the national average. This includes most of the chemical industry, petroleum refining, and the makers of such products as textile, printing, electrical, and medical equipment.

Even when the broader BLS definitions are used, the number of jobs that will be created in high-tech industries in the next ten years is disappointing. Forecasts made by the BLS and for *Business Week* by Data Resources Inc., in fact, show that the number of high-tech jobs created over the next decade will be less than half of the 2 million jobs lost in manufacturing in the past three years (see figure 9.1). While high-tech industries, as defined by the BLS, will generate ten times the number of jobs expected from the rest of industry, it will still amount to only 730,000–1 million jobs. And most of those will be in traditional occupations, not technical ones. Fewer than one-third will be for

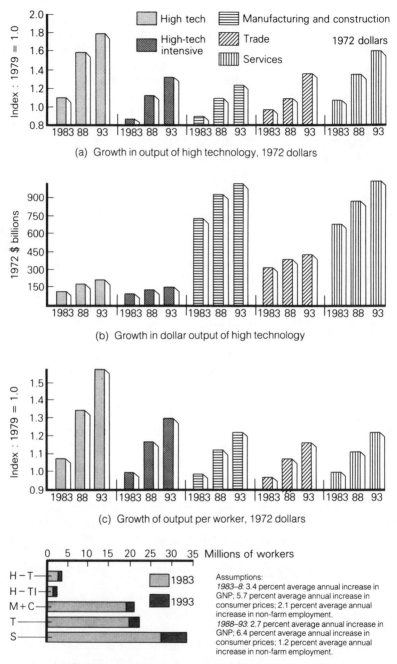

(a) Growth in output of high technology, 1972 dollars

(b) Growth in dollar output of high technology

(c) Growth of output per worker, 1972 dollars

Assumptions:
1983–8: 3.4 percent average annual increase in GNP; 5.7 percent average annual increase in consumer prices; 2.1 percent average annual increase in non-farm employment.
1988–93: 2.7 percent average annual increase in GNP; 6.4 percent average annual increase in consumer prices; 1.2 percent average annual increase in non-farm employment.

(d) Projected numbers employed in high-tech sector, 1983–93.

Figure 9.1 High tech's output will grow rapidly compared with other industries, but because the high-tech sector is small and its productivity is increasing, it will create only a modest number of jobs.
Source: Data Resources Inc.

Table 9.2 Most new jobs in high tech will not be technical

Occupation	No. of jobs	Growth, 1980–93 (%)
Operatives	226,183	22
Managers & clerical workers	149,650	22
Engineers	110,930	40
Craft & related workers	98,121	21
Engineering & science technicians	70,693	37
Professionals, except managers & clericals	33,305	22

Source: Data Resources Inc.

engineers and technicians, according to DRI, and the remainder will be managers, clerical workers, operators, and other factory workers (see table 9.2).

Few jobs will be created in high-tech industries because these industries have a relatively small base. The entire output of the high-tech sector is less than twice that of the auto industry. The high-tech industries employed only about 3 million workers in 1979 (the latest data available), or about 3 per cent of the nation's nonagricultural workforce. This total will climb to about 4 percent by 1993, DRI says.

Rapidly rising productivity is another factor that detracts from high tech's ability to create jobs. Output per high-tech worker in 1972 dollars is expected by DRI to rise by 46 percent between 1983 and 1993, nearly double the projected increase of 24 percent for manufacturing and 23 percent for services in the same period. This means that, while dollar output in high-tech industries will grow 87 percent over the next decade—from 7 percent of gross national product to 10 percent or $206 billion—the number of workers needed to produce this increase will need to rise by only 29 per cent.

On top of that, the growth of high-tech jobs will slow as those industries automate production. By 1993 they will employ 3.2 machine operators for every engineer, down 11 percent from the ratio of 3.6 to 1 in 1980.

The skilled workforce in the semiconductor and electronics industries, for example, will grow faster than the operators and assemblers during the next decade, according to DRI projections, even though less skilled people will account for the bulk of new jobs. By 1993 the number of engineers employed in those industries will jump by 60 percent to about 56,000, while the number of operators will move up only 42 per cent to 377,000.

"We will be upgrading jobs through automation," concedes Michael W. Hart, director of personnel relations at National Semiconductor Corp. "I think we will be changing all our collars to white over the next few years." The California chip maker added automation and increased the diameter of the silicon wafers from which it makes integrated circuits from 3 in. to 5 in. In that batch process, it increased the output of chips by 66 percent, "using the same number of operators with the same skills," says Hart.

Many high-tech companies are using the very products they make to replace labor in their plants. Hewlett-Packard Co., for one, estimates that several thousand of its computers are now being used throughout the company. "It's a major effort," says David E. Sanders, HP's director of corporate development.

The products of high-tech companies will also change the workforce in other industries. Robots, for example, are beginning to displace jobs in traditional industries such as steel and autos. "There are many unmanned factories in the future," says Richard M. Cyert, president of Carnegie-Mellon University.

In the auto industry, for example, the number of robots working will grow from 2400 now to between 15,000 and 25,000 by the end of the decade, according to a recent study by the W. E. Upjohn Institute for Employment Research. While these robots are expected to create 3000–5000 jobs, they will replace up to 50,000 auto workers.

Where high tech will cause growth in jobs is in those industries that build their products with high-tech components. "We don't see a lot of jobs being created by the core industries like electronic components or the manufacture of basic genetic materials. But there will be substantial multiplier effects downstream," predicts Victor C. Walling, manager of the business futures program at SRI International. High-tech companies, for example, will add only 27,000 new computer programmers and systems analysts over the next decade, DRI estimates. But an additional 225,000 programmers and analysts will be added throughout the economy.

At the same time, unskilled jobs in high tech, such as assembly workers, will face continued competition from lower-cost foreign labor. Atari Inc. recently stunned many of those who look to high tech for new employment when it moved 1,700 jobs overseas, and that kind of move is bound to continue. Hewlett-Packard, for one, predicts that its overseas workforce will grow faster than that in the US. "We are currently building a couple of major plants in Mexico and the US," says HP's Sanders.

In their all-out dash to lure high tech from its bastions in California's Silicon Valley and Boston's Route 128, states and cities have not taken the time to analyze employment patterns. "In the past six months, every place in the country has suddenly gotten the idea of high technology," says Thomas A. Vanderslice, president of GTE Corp. Everyone believes the US is in the midst of an economic transformation on the order of the Industrial Revolution. "We are moving from the bump, funk, and clunk era to a world in which everything whirs gently," says George T. Rehfeldt, a group vice-president at Cincinnati Milacron Inc.

The great jobs race

Every state and city plans to make it big. Georgia is a good example. The Georgia Institute of Technology figures, unrealistically, that the number of high-tech jobs in the state will soar from 28,000 to 200,000 by the turn of the century. And most states have even less chance than Georgia. "Unless they build up a critical mass, they're not going to make it," predicts Vanderslice. "There's too much competition now."

"The war between the states is heating up," declares Trygve Vigmostad, deputy director of Michigan's Office of Economic Development. Californians joke that the major beneficiaries of the hunt for high tech are the hotels in Silicon Valley, which are often filled with hundreds of eager recruiters from the industrial heartland.

Cities and states are setting up a raft of innovative programs to attract and nurture high-tech industry. Cincinnati, Syracuse, and New Haven, for example, are offering inducements such as tax breaks, research parks, and venture capital. North Carolina's highly successful Research Triangle Park has been overwhelmed by more than 100 state and city delegations over the past 12 months, all wanting to know how they can duplicate the park back home.

A growing number of local governments, from Chicago to Georgia, are trying to "grow their own" high-tech companies. Startup companies, for example, can lease space and share services in a ten-story "incubator" building that anchors a new research park just set up by Illinois in Chicago's West Side Medical Center. Chicago officials are considering a proposal, too, that would establish a privately funded Chicago Science Foundation— modeled after the federal National Science Foundation—to support local research.

Another novel idea catching on that aims to attract and hold high-tech startups is the formation of local venture-capital funds to invest in new companies that agree to locate locally. Cleveland is launching a nonprofit venture-capital fund with an initial capitalization of up to $10 million, and Cincinnati is planning a $15 million fund. In Pittsburgh, bank executives were astounded by complaints from fledgling high-tech companies that they had to go to the West Coast for venture capital. Subsequently, Pittsburgh National Bank formed a venture-capital group.

Chicago is considering a plan to set up a for-profit Science & Technology Investment Corp. to provide capital to startup companies until they are able to attract the attention of venture capitalists. In return for providing the startup capital, the investment fund would get an equity position in an infant company. Initial funding would total about $15 million, enough to make 30 half-million-dollar investments. The agreement would also include a capture provision: "If you accept these dollars, you are obligated to locate your firm in Chicago," explains Louis H. Masotti, the Northwestern University professor who wrote the plan.

Last fall, Michigan's legislature voted to allow up to 5 percent of its $7.5 billion in retirement funds to be invested in small-technology-based companies as venture capital. The Connecticut legislature is considering a proposal to set up a $7 million venture fund as part of a $17 million program to improve the state climate for high technology. Similar efforts are under way in Massachusetts, Minnesota, Mississippi, and Wyoming.

Georgia is taking a different tack and importing the venture capital needed to fund new companies. In November, Georgia's Advanced Technology Development Center brought together executives from 17 companies in its "incubation" center with 294 venture capitalists. As a result, brags Jerry Birchfield, director of the Georgia center, "we got 50 percent of the companies funded. That's only half bad." The center provides startups with funding and housing for up to three years. If they cannot attract additional funding, they must leave. "It's profit or perish," says Birchfield.

But money is just one factor in attracting high-tech companies. Unlike smokestack industries that need access to raw materials, energy, and transportation, high-tech plants locate where the quality of life is high enough to draw a skilled work force. "You don't locate plants for cheap labor or even taxes," insists GTE's Vanderslice. "You locate where the people want to live." It was because of this that GTE moved one of its major computer laboratories from suburban Chicago to Phoenix. "We went through hell in Northlake trying to recruit software people," Vanderslice says. "You can locate in Boston, in spite of the high taxes, because it has a nice ambience. You can do it in parts of Dallas and Houston, but you sure can't do it out in the boonies."

Indispensable to that quality of life is a good technical university nearby. "What developers of high tech don't understand is that the university environment is the engine of technology and the driving force behind bringing technology to an area," says Thomas M. Nies, president of Cincom Systems Inc., a Cincinnati computer software company.

Any industrial recruiter worth his salt knows that the high-tech enclaves of Silicon Valley and Boston's Route 128 were built around Stanford University and the Massachusetts Institute of Technology. And the smaller high-tech concentrations— in Research Triangle Park, Minneapolis, Austin, Denver, and Philadelphia—owe their existence to superior local universities. "If you don't have the educational institutions, it's not going to happen," says Roger Ahlbrandt Jr, a research director at the University of Pittsburgh.

Those states lucky enough to have top-notch technical schools are busily trying to duplicate the climate in which high-tech industry flourishes. Pennsylvania, for one, has created what it calls the Ben Franklin Partnership—a consortium of state government, universities, and businesses to support research, turn scientists into businessmen, and retrain displaced workers for high-technology jobs. Governor Richard L. Thornburgh sees such programs as "our best hope for future growth in jobs. The boom-time level in our heavy-manufacturing base is simply not going to return."

Not to be outdone, the New Jersey legislature is expected to pass a similar program that is also named after one of its own: the Thomas Edison Foundation. "The entire package is to attract high-technology industry to the state," says Thomas J. Edwards, director of policy and planning, the New Jersey Democrat who wrote the bill.

But those localities lacking a highly rated university will undoubtedly have to overcome serious image problems if their plans to bring in high-tech companies are to succeed. Cincinnati has laid a careful plan based on a study done by SRI, but it is "perceived by Silicon Valley as a technological backwater," admits Bruce S. Crutcher, an economic developer in the city's Chamber of Commerce.

Sometimes even a leading university is not enough to lure high-technology companies. Yale University might not be able to help turn around the industrial decline of New Haven, which has lost thousands of manufacturing jobs in firearms, sporting equipment, apparel, chemicals, and tiremaking over the past five years. To arrest that trend, New Haven joined forces with Yale and Olin Corp. to found Science Park Development Corp. But the park has so far bagged only one company, despite a wealth of inducements.

The industrial Midwest may have a better shot at attracting high-technology companies than the aging cities in the East. The expansion plans of 691 high-tech companies were analyzed last summer by Congress's Joint Economic Committee, and it concluded that "the Midwest offered the best overall investment climate." The committee noted that New England, the mid-Atlantic states, and the West "are not expected to maintain their positions over the next five years." Rising costs will make it difficult for Massachusetts and California to remain competitive, it added.

Retraining workers is one way that states can build the skilled-labor pool needed to attract high-tech companies and put their displaced workers back on the job at the same time. Until recently, however, retraining has not received much emphasis from either employers or state and federal governments, according to many experts. West Germany, for example, offers up to three years of training for laid-off workers, while in the US thousands of workers whose jobs have disappeared have had nowhere to turn to learn new skills.

Now, however, states are realizing that retraining programs can be a help in attracting high-tech industry. Retraining displaced workers obviously cannot produce engineers, but it can provide a pool of technicians, such as the 70,000 such workers DRI estimates that high-tech employers will need by 1993 to service computers and related equipment. "I definitely see more and more retraining," says Larry G. Selland, Idaho's director for vocational–technical education. "Technology is affecting the whole workplace."

Programs started by 17 states are aimed at training workers specifically for high-tech jobs, according to a recent count by Congress's Office of Technology Assessment. And legislatures in Massachusetts, Maryland, Connecticut, and Washington are considering proposals for similar programs.

California, for example, started a $55-million-a-year retraining program in

January that it hopes will turn out at least 15,000 experienced workers a year for jobs in high-tech industries. Financed by revenues diverted from its overfunded unemployment-compensation system, the program demands assurances that there will be a job for each worker who completes a course. Companies can choose the workers they want to retrain—even if they are newly hired—and they can decide whether to do the training themselves or have it done at a college or vocational school.

But the unemployed in such hard-hit industries as automobiles and steel are not lining up to take advantage of retraining programs. Many auto workers are not signing up for one program funded by General Motors Corp., because they still hope to get their old jobs back. Another drawback is that high-tech employers often do not pay as well as the auto industry. "If 5 percent of GM workers placed in new jobs got wages anywhere near equivalent to their former salaries, I would be shocked," says Maurice R. Treadwell, an administrative assistant at the United Auto Workers.

Many unemployed workers from smokestack industries simply spurn high-tech jobs as menial and demeaning, while high-tech companies continue to rely on new recruits to meet staff requirements. "What we are finding is that the great majority of jobs are going to new employees,'" says Charles W. Minshall, research leader for community and regional development at Battelle Memorial Institute. "The semi-skilled worker of the high-technology industry sits with a doily on her head, listens to the soap opera on the radio, and fiddles with a piece of wire. This is very different from the auto worker who bangs on a car."

The nation's push for high technology may not solve current unemployment problems, but it might be the best thing that has ever happened to high tech. Concern that the US was losing its technological edge to Japan and other foreign competitors has been rising for nearly a decade. Now the recession has brought that worry to the forefront. "All this concern is a reaction to competition. And the only thing that's so unusual about it is that it came so late," says science adviser Keyworth.

Experts are convinced that these efforts by city, state, and federal governments are creating a fertile climate for starting up high-tech companies. That, they say, will in turn unleash a wave of significant new technology in the future. "Much of the great changes will come through startup companies—there has been a virtual explosion of these," says D. Bruce Merrifield, Assistant Commerce Secretary for productivity, technology, and innovation. A recent study by the Brookings Institution estimates that new companies contributed just over 50 percent of the new jobs created between 1976 and 1980.

The dramatic shift under way in industrial America will bring many wrenching dislocations with it before high technology revitalizes the nation's industrial complex. "A lot of the small cities that are dependent on one smokestack industry are not going to make it," says Joseph A. Steger, provost of the University of Cincinnati. "We are going to see a new urban landscape." Nonetheless, says Commerce's Merrifield: "We're looking at the most exciting period of growth the world has ever seen, and the US will be the greatest beneficiary of this."

Women: the Vulnerable Group

Diane Werneke

Women's employment is concentrated in precisely those less skilled occupations in the service sector which are most likely to be affected by microelectronics. Drawing upon case studies, International Labour Office researcher Werneke looks at the position of women in US banking, insurance, and general office work, where employment growth has not matched the growth in the volume of business. Taken from Microelectronics and Office Jobs: The Impact of the Chip on Women's Employment *(ILO, Geneva, 1983).*

During the last two decades women have entered the labor market in increasing numbers, accelerating a trend evident in most countries since World War II.[1] The majority of women have found employment in the tertiary sector of the economy (table 9.3) and in nonmanual occupations (table 9.4). Indeed, as women have entered the labor market, certain activities in the service sector and occupations have become increasingly "feminized." Today, throughout the industrialized countries, one of the fundamental characteristics of the labor markets is the marked segregation by sex. Women are concentrated in a limited range of occupations and are most likely to be found working in relatively less skilled and lower-paying jobs than their male counterparts. This concentration has important implications, because many of the occupations to which women gravitate are those that the new information technology is beginning to transform.

Microelectronic technology has the capability of profoundly affecting so-called information activities—the creation, processing, storage, and transmission of information. A large proportion of the workforce in most countries is engaged in these activities. In the US, it has been estimated that about half of the labor force work in information-related occupations;[2] in the UK the information sector is said to employ nearly half of all workers.[3]

The information sector, like the rest of the economy, is not homogeneous.

Table 9.3   Concentration of women workers by sector of activity, 1977

| Country | Sector | | | |
	Agriculture (%)	Industry (%)	Services (%)	Total number employed (thousands)
Federal Republic of Germany	7.5	29.5	61.6	9,012
France	8.6	24.4	67.0	8,077
Italy	13.2	31.2	55.6	5,266
Netherlands	1.5	14.0	83.3	1,154
Belgium	2.1	23.0	72.4	1,120
Luxembourg	0.5	12.5	80.0	40
United Kingdom	1.3	25.6	72.3	9,373
EEC 9	6.3	26.5	66.3	35,189
United States (1980)	1.6	17.9*	80.5†	41,283
Canada (1979)	3.2	15.8	81.0	4,022
Australia (1979)	3.8	17.1	79.0	2,137

*Includes manufacturing, mining, and construction.
†Includes transportation, public utilities, trade, finance, public administration, private household services, and miscellaneous services.
Source: Eurostat, *Economic and Social Position of Women in the Community 1981* (Luxembourg, Statistical Office of the European Communities, 1981), table 30; US Department of Labor, Bureau of Labor Statistics, *Employment and Unemployment: A report on 1980*, Special Labor Force Report 244 (Washington, DC, 1981), calculated from table 27, p. A-26; and OECD, *Labour Force Statistics, 1968–79* (Paris, 1981).

Table 9.4   Concentration of women workers, by occupation (percentage of female labor force)

| Country | Occupation | | | | |
	Clerical (%)	Sales (%)	Professional & technical (%)	Administrative & managerial (%)	Service workers (%)
Federal Republic of Germany	31.0	13.2	13.8	1.5	16.9
France	26.9	10.4	19.7	1.5	15.3
Italy	14.4	12.4	13.1	0.2	13.5
United Kingdom	30.8	12.2	12.2	0.9	23.3
United States	34.3	6.8	15.2	5.9	21.0
Canada	34.0	10.5	19.3	4.8	18.0
Australia	33.7	13.0	14.4	2.7	15.5

Source: ILO, *Year Book of Labour Statistics*, 37th–39th issues (Geneva, ILO, 1977–79).

Within it there are skilled occupations—those that create, analyze, coordinate, and interpret information—and less skilled—those that manipulate information.[4] The latter's tasks may be characterized as information-handling. It is in these activities that women by and large are concentrated as secretaries, typists, bookkeepers, stenographers, cashiers, and the like. In contrast, the upper echelons of information occupations are dominated by men, who comprise the vast majority of senior and middle management and professional workers. Microelectronics is conceived of fundamentally as a tool to assist the latter group in their decision-making, analysis, and communication by speeding up and broadening the flow of information to and from them. The use of microelectronics, as far as the information-handlers are concerned, is a tool to increase their productivity in delivery of that information and consequently has more significant employment implications.[5]

Looking at the information-handling occupations, the degree of sex segregation becomes evident. If, following the analysis of Joy Selby Smith,[6] a "female" occupation can be defined as one in which 50 percent or more workers engaged in that occupation are women, one finds women workers overwhelmingly concentrated in a few "female" occupations. Many of these are information-handling occupations. Selby Smith found that in Australia over 85 percent of women in the paid workforce were concentrated in 18 of 61 occupations listed by the Statistical Office, all of them "female" occupations. Half of Australian women were found to be in "female" occupations that were likely to feel a significant impact from developments in microelectronics.

Using detailed labor force data, a similar analysis can be performed for the US. About one-third of all women workers are concentrated in white-collar "female" occupations—mainly clerical—which are expected to be affected by the use of microelectronic technology (table 9.5). Only 16 percent are classified as professional and technical workers. When one looks at management occupations, the proportion is smaller again—only 6 percent. Those information-handling occupations—bookkeepers, secretaries, cashiers, typists, and so on—are clearly dominated by women: in many of these occupations more than 90 percent of these workers are women.

In the UK one-third of women work in offices and of these more than 90 percent are employed in routine clerical jobs. The top three female occupations are clerical, cashiers, and typists.[7] It is likely that similar concentrations are found elsewhere in the industrialized countries.

This analysis of occupational data does not mean that one-half of all Australian women or one-third of women in the US are being threatened by job loss as the result of the use of microelectronics in the office. But it does suggest that large numbers of women bear the burden of the adjustments that are required. Moreover, it is not just those women who are now employed that will be affected. It may cause problems among women who are seeking to re-enter the workforce after a spell in domestic activity and among new entrants hoping to follow in the tradition of the women before them. For example, in France, 50 percent of girls are being trained in typing, shorthand, and accounting, and in Australia the same percentage are seeking clerical jobs.[8]

The use of microelectronic technology in the office will result in some *labour displacement*. Some women may lose jobs, and the use of new technology is likely to cause a curtailment of growth of the office jobs

Table 9.5   Female white-collar occupations likely to be affected by microelectronics in the US, 1980

Occupation	Percentage of women in total workforce	Number of women employed (thousands)
Bookkeepers	90.5	1,723
Cashiers	86.6	1,346
Secretaries	99.1	3,841
Typists	96.9	991
Bank tellers	92.7 ·	515
Billing clerks	90.2	147
Clerical supervisors	70.5	169
Collectors	56.4	44
Counter clerks	73.4	257
Estimators	56.2	300
File clerks	86.4	280
Insurance adjusters	57.5	100
Office machine operators	72.6	682
Payroll & time keeping	81.0	188
Receptionists	96.3	606
Statistical clerks	78.0	302
Stenographers	89.1	57
Telephone operators	91.8	290
All other clerical	77.1	1,435
Total employed		41,283

Source: US Department of Labor, Bureau of Labor Statistics, Employment and Unemployment: A Report on 1980, Special Labor Force Report 244 (Washington, DC, 1981), calculated from table 23, p. A-22.

generally held by women. And because of certain factors affecting women workers, one can postulate that adjustment to change may be quite difficult in some cases. For one thing, if existing jobs are lost, women generally have more limited access to alternative job possibilities than men because of their domestic/family responsibilities. Many of their traditional skills such as speed typing and shorthand may be outdated as firms search for personnel who are acquainted with word processing and other new technology-based office equipment.

Those who can quickly grasp the capabilities of a word processor or can efficiently perform a computer search of microfiche files will be in demand. This may mean that some older workers will find it difficult to adjust to change. Older persons also tend to lack extended formal education and the general office skills that provide a useful basis for adjusting to change.[9]

Women may also experience difficulties in finding new jobs because of inadequate information. Many who are part-time workers or who withdraw from the labor force do not benefit from the social aspects of work, which are important elements in the informal information networks that are widely recognized as a key job search tool. Because of these part-time and intermittent characteristics in their labor force participation, women are seen as secondary workers, participating in a separate and unequal sector of the labor market. Jobs in the secondary sector are characterized by low pay, poorer working conditions, little chance of advancement, and considerable instability.[10] The information sector is also generally characterized by lack of unionization, which means that women are in a relatively poor position to bargain about the introduction of new technology.

If, then, the use of new technology curtails the growth of those traditional clerical or information-handling occupations in which women work, what is the likelihood that the opportunities for work that new technology opens up will become a source of employment growth? Although difficult to predict the range of jobs that will probably be created by the spread of microelectronics, one can definitely point to computer applications—programming, systems analysis, data management—and electronics engineering as growth areas.

Based on evidence already available, the outlook for women in these jobs is not encouraging. For example, although in the US women hold 25 percent of computer specialist occupations and represent 20 percent of the engineering and science technicians, they tend to be concentrated at the lower end of the skill spectrum.[11] In Sweden, 90 percent of those working as data processing managers were men, 80 percent of those in data processing planning were men, while 97 percent of routine data entry positions were filled by women.[12] According to electronic equipment suppliers in Europe, the proportion of women in higher-skill-level computer courses has grown significantly over the last two decades, but women still account for only 10–25 percent of all participants in courses at this level.[13]

The reason for the underrepresentation of women in the highly skilled and expanding computer occupations lies partly in their early education.[14] It is well known that girls in secondary schools take fewer science and mathematics courses than boys. Even in today's more technologically advanced society, women remain unaware of the increasing mathematical requirements of any field of study. For example, in Europe girls are disproportionately concentrated in general subjects rather than technical studies, and they choose courses in the humanities, languages, and arts as opposed to science, mathematics, and other technical subjects.

It is not surprising, therefore, that, lacking an appropriate technical background, women are underrepresented in highly skilled computer occupations. Having once shut the door to science and mathematics, they lack the background to study for the higher professions in the computer industry. A similar situation exists with respect to general management and supervisory training. Whether given in-house or at a training center, the proportion of women represented is low. Whether this is due to the lack of desire or, owing to family responsibilities, an inability to invest time and effort in training on the part of women themselves, or is the result of employers overlooking them as candidates, the consequence is the same: the access of women to new jobs and advancement is likely to be severely limited unless their participation in relevant course work in school and training beyond school is substantially broadened.

So far, this section has focused on the labor displacement effect of the use of microelectronics in the office and on the reasons why women will find it difficult to adjust to the changes required because of their relative immobility, both occupationally and geographically, and their lack of appropriate education, training, and skills. Another aspect of the effect of new technology on jobs that is being viewed with concern is its impact on *job content* and *work organization*.[15] Some observers argue that the use of new technology will enhance jobs and upgrade skills of the information handlers. According to this view, new technology can reduce the tedium of routine tasks and the monotony of the job by allowing more varied and interesting activities to be undertaken. It may also result in workers becoming more highly specialized and less susceptible to easy substitution and replacement than they are at present.[16]

On the other hand, many argue that the introduction of new technology will cause the jobs of information handlers to become more routine and less skilled. Following the principles of scientific management, jobs will be broken down into a series of steps that will then be reordered and divided among different groups of workers to save time and increase productivity. As a result, jobs will become fragmented and deskilled. The mechanization of office work occurs when office machines "subordinate the work of the clerk to the tempo of the machine, which takes over the larger part of the discretion involved in the operation, and which requires full-time specialised attendants . . ."[17]—not unlike the factory assembly line.

Support for both the favorable and unfavorable scenarios exists in the literature on the impact of new technology. The outcome depends on how work is reorganized when new technology is introduced. For this reason, the importance of participation and consultation over the changes brought about by the introduction of new equipment has been emphasized, particularly by the European trade unions.

Along with the issues of potential displacement and changes in work organization, the health aspects of the new office equipment have received a great deal of attention.[18] The trade unions have highlighted the health aspects of visual display units and the lack of attention paid to environmental or ergonomic factors associated with their positioning as part of computer systems. Health hazards cited include headache, eye strain, backache, and possible exposure to radiation. The last has been unsupported by medical studies to date, which show that the radiation emitted is far less than that resulting from an average day's television viewing. But some observers maintain that the new office equipment is introducing to office jobs problems of stress that are associated with the mechanization of the factory.[19]

The North American experience

In contrast to the well publicized debate on technology and employment in many European countries, microelectronics and its possible effects on jobs has not been a highly visible public policy issue in North America. The tremendous growth in employment over the past two decades has lulled many North Americans into a sense of security. Fears about rising unemployment, while certainly a concern, are related to inflation, contractionary public policies, and the swings of the business cycle.

Part of the reason for this lack of attention is past experience. In the early 1960s the fear of widespread unemployment as a result of the introduction of computers became a major public issue. In the US a presidential commission was appointed that took the complacent view that, once technology arrived, it would create new industries and jobs. While job loss would occur, "job displacement is the price of a dynamic economy."[20] In fact, the commission's conclusions were borne out by experience. New jobs were created; massive employment did not result.

There are, however, a number of observers in both the US and Canada who are wondering if the new wave of microelectronics might not produce a different and more troubling outcome because of its technical and economic attributes. Although much of the focus has been on robotics and their invasion of the factory production process, office work is gaining increasing attention because this is the sector where the spread of new technology has become the most apparent. But only two North American reports, by Working Women in the United States[21] and by the Institute for Research on Public Policy[22] in Canada, have addressed themselves to the particular situation of women. However, as clerical occupations in both countries are highly feminized, most studies dealing with office work can be considered to have implications for working women in the office.

The fear that workers will lose their jobs stems from the recognition that the use of microelectronic technology affords tremendous increases in office productivity. If one can do a job it used to take two to do, it would seem that only one would be employed, and the other displaced, if the cost of buying or renting new equipment were less than the cost (salary) of a second typist. In North America, the costs of word processors and some minicomputers have reached this point. Consequently, one would expect to find evidence of jobs being lost in companies engaged in information processing.

Banking and insurance are activities particularly well suited to the introduction of microelectronics, because one of their main functions is handling large quantities of simple information. For example, in the US it is estimated that about 81 percent of the banking industry's output is used in providing information services. For insurance, 83 percent of the industry's costs originate from information activities like actuarial work necessary to sell its services.[23] Indeed, evidence available from trade publications indicates growing awareness of and introduction of new technology.

Banking is also a key sector of employment for women. In Canada, 70 percent of total bank employees were women in 1980, and, while they were making significant inroads into the sales and managements ranks, 80 percent of women workers were in clerical positions.[24] In the US a similar picture emerges: women accounted for one-third of the bank official and manager positions and 92.7 percent of tellers.[25]

Banking
In the banking sector, the application of computers to banking activities has been widespread and long-standing in North America. During the 1960s labor-saving technology was introduced during a period of increasing demand for banking services. Computers automated in-house functions such as checking and savings accounts and improved information retrieval. Consequently, the number of jobs in banks grew, but not as quickly as the volume of work. For example, in the US banking transactions rose 8.3 percent per year while employment grew 4.3 percent annually between 1960 and 1973.[26] The activities of clerical banking personnel also changed, with proofreaders and other backroom clerks moving behind the counter to service customers.

Beginning in the early 1970s, branch banks increasingly acquired on-line computer/communications networks which allowed the branches to offer computerized services. Teller terminals allowed employees in branch banks to consult the central computer directly for the current position of an account and immediately to bring the account up to date with the appropriate debit or credit.[27] This speeded up branch service considerably and brought to the teller counter a number of services previously performed in centralized data processing. Branch banking expanded rapidly during this period. In the US transactions rose 7.2 percent per year while employment slowed down somewhat, to 3.2 percent a year, between 1973 and 1976.[28]

A Canadian case study of the banking industry[29] provides some interesting insights as to the employment impact of the increasing use of information technology. The on-line teller terminals truncated the job functions required to complete a banking transaction. The system reduced the amount of after-hours preparation of daily transactions records by the teller. Consequently the teller could absorb the data entry clerk's function in spare time. Indeed, the case study found that clerical employment lagged behind overall bank employment growth during the 1975–80 period, whereas earlier it had grown faster.

The study found that, as the on-line systems were introduced, the traditional work of a teller was reduced substantially: one official estimated by one-half. However, this had not yet resulted in a reduction in teller personnel. Instead, the amount and type of work they did had been expanded. New job descriptions changed from the information handling traditionally involved in teller work to information marketing, where tellers were engaged in handling customer complaints, explaining bank services, and promoting new services.

The advent of microelectronics is making several new developments feasible in the banking sector. The most noticeable is the increasing appearance of automatic teller machines (ATMs). In addition to paying out money like cash-dispensing machines, automatic tellers can accept deposits, transfer money from one account to another, and make other transactions. The system enables average bank customers to do about 80 percent of their banking business at a time convenient to them. In 1980 there were more than 20,000 automatic tellers in use in the US[30] and about 100 in Canada,[31] and banking journals indicate that they are being installed at an increasing rate. A study in the US by Arthur D. Little Inc. projected that the number of these machines would rise to 54,200 in 1985 and 71,000 by 1990. Although there is a wide spectrum of views among banks as to whether the machines are cost-effective, the increase in their numbers is driven by market forces, because in urban areas the bank without an automatic teller is at a market disadvantage.

What are the implications for the employment of bank tellers? One American banking industry report[32] suggested a levelling-off of employment growth if it was assumed that the number of machines would increase along the lines forecast by Arthur D. Little Inc. and that the growth in the number of bank offices would slow down, even assuming that many teller functions could not be taken by automatic machines, particularly the more time-consuming ones. Whether the growth in other bank services will allow any displaced teller to be reabsorbed in connection with other bank activities is an open question that depends in part on the use of microelectronics elsewhere in banks.

One area of banking that is likely to use microelectronic technology increasingly is the trust department where stock transfers take place. There, speed of transfer and quality control are the keys to customer services. For example, at a Citicorp subsidiary, stock transfers require the processing of 12,000 certificates daily for more than 1000 different issues. In moving from central processing of these stock transfers to a decentralized work station operation, the number of staff was reduced from 147 by one-third. Those displaced were transferred to other parts of the bank's operations.[33] A recent Dun and Bradstreet survey found that two-thirds of the largest banks in the US had not yet automated their securities activities.[34] Pressure to do so is likely to mount as security trading activity increases during the 1980s. Labor displacement may be partly offset, as was the case with Citicorp, however, by higher levels of service in the future, such as cash management, cost accounting, and portfolio management.[35]

Citibank is also cited in another case study which involved automation of processing letters of credit.[36] In June 1975 it took 14 people three days to process a letter of credit; in 1979 one person working with a minicomputer and filing records could do the same job electronically. Within three months staff were reduced from 142 to 100. A small number were reclassified as professionals; others were transferred to other operations. Overall Citibank, one of the first to automate, had reduced its clerical staff from 10,000 to 6000 during the 1970s. By extensive use of retraining, and because of the growth of business, the bank had been able to place these clerks in other bank operations.[37]

This is of course encouraging, but will other banks be able to expand their services as fast and maintain employment? Many bank analysts believe the answer could be yes, because "as the power of the computer expands . . . we are going to move inexorably into new uses of information"[38] such as tax counseling, budget counseling, and retail "what-if?" services (e.g., should one purchase a house given an expected income growth?). In addition, bankers believe that the use of new technology will give banks the flexibility and quick response time that allows them to compete more effectively in the market place.

Smaller banks are also beginning to use new technology. By using a mini-computer, many banks have shifted their data processing requirements from a service bureau to in-house operations with significant cost savings. This has transformed some clerical jobs of coding and sorting into computer-based activities. In one bank[39] there were two clerks who each spent half of their time preparing the general ledger of accounts. With a minicomputer system one operator could deliver all the information in an hour. This, however, had little impact on staff requirements. "We have the same personnel in the bookkeeping department as we did before—two clerks and a supervisor. Under the new system two not only do the work of three, but they can produce all of the new management reports as well. The supervisor is freed for other duties."[40]

The banking industry is complex, and its need for and use of employees varies substantially by size of bank and type of services in which it is engaging. The case studies and examples presented cannot therefore be generalized. Nevertheless, the information reviewed does permit several summarizing observations.

First, the growth in both teller and clerical work is likely to slow down substantially as ATMs and minicomputers are introduced. Second, the extent to which workers will be redeployed to other parts of the bank's operations rather than displaced will depend on the bank's ability to increase its services to customers. Finally, whether retaining their jobs or finding new ones, it is clear that those engaged in these occupations will have to learn new skills in information marketing or computer operating. Retraining by the employer should become an increasingly important issue.

Insurance
As in banking, the insurance industry has substantial accounting and statistical requirements, and the need to store vast amounts of data and

transmit information. Consequently, its business operations have been well suited to computer applications. In fact, the industry was among the first to use computers to store and process records in the 1960s and to apply them to an increasing range of activities during the 1970s. Nonetheless, employment expanded during the period as business increased rapidly. In the US the number of jobs grew about 2 percent a year over the 1960s and 1970s, with no slowdown in evidence from the most recent data for 1981.[41] In Canada employment grew about 20 percent during the 1960s but slowed down in the 1970s to about 14 percent.[42] In both countries the insurance industry employs a large number of women: they accounted for about 60 percent of all insurance jobs in 1980. Most women are found in the lower skilled clerical occupations—in Canada 90 percent of the clerical jobs in the industry are held by women.[43]

Looking first at the insurance industry in the US, it is expected that the increasing use of microelectronics will slow down the growth of clerical employment in the 1980s, and several occupations are anticipated to decline because of the need to reduce costs: "as operating expenses continue to climb, the cost to create and maintain paperwork also increases in terms of labor, space, and productivity."[44] Consequently, one of the chief goals of using new technology is to reduce the unit labor requirements of each business operation.

Computers have already been applied to a wide range of activities including billing and collection, underwriting, and claims. This has generally resulted in reduced unit labor requirements for many operations.[45] For example, the use of computers and microfilm storage enabled Blue Cross and Blue Shield to keep constant the number of clerical employees engaged in processing health and disability claims despite a large increase in volume. Five years ago the 127 employees of Medicare B were processing 830,000 claims annually; in 1981 the same number of workers processed 1.4 million claims annually.[46] More generally, large declines have been recorded for those clerical jobs involving record keeping, transcribing premium payments, assembling records, calculating policy values, and handling remittances. Other positions dealing with transactions and communications, such as accounting and correspondence, have remained stable, but transaction volumes have risen considerably. Stenographer and typist occupations generally declined. Women are by and large concentrated in these occupations.

Recent advances in microelectronics will foster two developments in the insurance industry: the automation of an increasing number of work functions, and the increased use of computers in branch and agent offices. Looking at two activities, actuarial research and underwriting, the use of new technology will reduce the necessary labor. For example, by allowing the rapid retrieval of data and their manipulation, studies and projections based on mortality and morbidity rates can be done many times more quickly than in the past and with lower clerical requirements. Some higher skilled work may also be affected.

In underwriting, for example, computers can be used to code, rate, and issue policies. One property and liability company was able to reduce policy-issuing personnel by one-third by computer participation in the underwriting process.[47] The use of optical character recognition technology, which allows information to be fed directly into the computer from the original document, has already reduced the number of data entry clerks required. And the use of microfilm/fiche technology, particularly computer-assisted retrieval of documents stored on microfilm/fiche, will reduce the number of file clerks and typists required.[48] These more efficient and streamlined operations are expected to develop further with the increasing use of integrated communications systems, which allow consolidation of the different records held on each policy on a central computer.

On balance, the US Department of Labor finds that computerization of practically every mechanical operation in the insurance industry is expected to reduce the availability of low-skilled, entry-level clerical positions in the 1980s. The demand for secretaries, typists, key punch, and other machine operators, bookkeepers, and file clerks is likely to decline. However, employment in managerial and technical occupations is expected to increase, especially those related to electronics technology.[49] To date, the combination of the high turnover among clerical personnel and industry growth has resulted in few displacements. In the future it is likely to depend on the extent that clerical workers are able to move into computer-based activities.

In the Canadian case study of a major insurance company, similar conclusions emerge: "there has been a consistent reduction in clerical job functions associated with different insurance transactions."[50] The industry itself was found to be less labor-intensive as routine work such as underwriting and policy issuing was automated, reducing substantially the clerical component of traditional support functions.

Another interesting observation was that new technology reduced the need for management employees as the computer's monitoring capabilities lessened the need for direct supervision. Also, computer-aided instruction reduced the teaching work that supervisors used to perform. Consequently, first-line managers as well as clerical workers have been playing a diminishing role in the Canadian insurance industry. It is into these positions that women have recently made their breakthrough into management, and hence the impact of new technology may be to reduce these opportunities for upward mobility.[51]

Moreover, the Canadian study of the insurance company found not only a reduction in the proportion of clerical workers employed but also a widening skill gap that does not bode well for future opportunities. During the three-year period that new technology was applied, employment of specialists and professionals increased substantially—10 percent in the first nine months of 1979 alone. However, few of these new positions were filled by clerical personnel whose jobs were becoming less in demand: "Of the job openings posted internally during the first six months of 1980, only half resulted in internal candidates being chosen."[52] The reason given was that these new positions required previous related experience. Consequently, without access to training, clerical employees will be cut off from the path of upward mobility.

Thus, in the insurance industry in North America the outlook for many women is not promising. The automation of many insurance activities to date has not been accompanied by widespread displacement of workers because the growing number of business transactions has allowed many to be reabsorbed elsewhere in companies and because high turnover has meant that reductions in the clerical workforce would be accomplished by not filling vacancies. However, more widespread use of new technology is likely to affect many workers, who, without access to the training and skills required to move into new jobs, may find themselves displaced in greater numbers than in the past.

General office work

Information creation, handling, and communication are the key activities of most offices in every industry, and hence one would expect the impact of microelectronics to be spread widely in offices throughout the economy. However, actual job displacement of office workers outside the banking and insurance sectors appears to be sparse, despite the tremendous savings in costs and increases in productivity that use of new technology in the office seems to afford.

A model for office automation was developed by IBM, the essentials of which include the elimination of the one-to-one manager–secretary relationship. The secretarial function would be split between a centralized word processing pool and an administrative assistance section which would take over all but the typing tasks, for example, the routing of incoming telephone calls, messages, and so on, most of which would be done electronically through the office communications network.

To date, it appears that few offices resemble this model, as all but the most advanced have opted for a piecemeal approach that generally involves use of word processing equipment as a first step toward an automated office. Among those companies that have developed an advanced system, information is available on a few. A case study of a major corporate head office in Canada that substantially automated its operations found a major impact on employment following the introduction of new technology in the information services department.[53] There was a substantial reduction in the clerical workforce despite a major expansion in the department's activities. The additional demand for labor occurred solely in the professional and managerial ranks. Only 2 of the 130 clerks who were displaced (100 of them were women) moved up to meet this increased demand: neither of these was a woman. Although none of the clerical workers displaced from their jobs was made redundant by the company, some had to accept demotions to fit into other departments that had openings for clerical personnel.

In the US, one major West Coast bank is said to have saved $500,000 through the use of word processing pools. The number of secretaries was reduced from 200 to 100, and a word processing center of 40 typists was established resulting in a net reduction of 60 secretarial posts.[54] At Xerox, the application of advanced equipment did not result in the actual loss of jobs, but few new jobs were created despite an increase of 200 percent in output, which was of "500 per cent higher quality."[55]

Other anecdotal reports from business magazines also suggest some reductions in secretaries and clerical employees with the introduction of new office equipment. Nonetheless, there has not been a widespread trend toward overt labor displacement. Even with increasing use of the word processors in the US, secretaries have been one of the fastest growing occupational groups in the last eight years, and the US Department of Labor expects that secretaries, typists, and general office clerks will be the source of the largest number of new jobs in the 1980s.[56]

What accounts for this demand in spite of potentially labor-saving equipment in the office? Several industry analysts, focusing on the secretary, found widespread resistance on the part of managers to losing their secretaries. Seen as executive perks, no professional wanted to be without them.[57] Moreover, the reduced costs of word processing equipment make it feasible to take a decentralized approach to the equipment rather than establishing pools of typists centrally to ensure maximum utilization of the equipment. Thus, in many offices word processors are seen as the next generation of typewriters and are used as such.

In addition, increasing emphasis is put on the need to increase managerial/professional productivity as this is the most costly element in an office budget.[58] Consequently, rather than reducing the amount of paperwork, word processing systems have been used to increase the amount of text output per professional. "The extra work done by authors tends to be more self-initiated and proactive . . . just the sort of work one would like to encourage in management."[59] The potential benefits derived are estimated at $8500–$13,000 per professional served.[60] Seen from this viewpoint, an analyst says that "managers have found that at least in the first year they must be prepared to add to total secretarial and clerical staff."[61]

In a similar vein, some offices are finding that, with the decreasing cost of new equipment, they can use it to improve their services to customers and clients. For example, one trade association had contracted out to a service bureau to do electronically some of its increasingly complex information activities performed for members. This involved a supervisor and five clerks coding sheets and proofing the key punching. However, when it became economically feasible to install a minicomputer, these activities were undertaken in-house. The minicomputer was reported to be able to process in four days what would have taken four-and-a-half months previously.[62]

Thus, in summarizing this section on labor displacement in offices several observations can be made. First, the potential for job loss is clearly evident as advanced equipment like automatic teller machines, computerized filing, storage and retrieval, and word processors can replace many functions performed by clerks, typists, and secretaries. To date, however, there has been little visible evidence of increasing job loss. To be sure, labor displacement has occurred in a more subtle way, as the examples point to both a reduction in the clerical content of each service provided and a reduction in the labor-intensity of clerical work in general. Thus, employment growth has not paralleled the rise in the volume of business that would have occurred in the absence of new technology. Developments in new technology have not been generally translated in wholesale job loss, in part because of the growth in demand for information-based services over the past two decades and in part because of the uneven pace of diffusion of new technology as some companies have advanced further down the road than others.

Moreover, the existence of the technology does not determine the outcome: the latter depends also on how the technology is used. In some cases, technology may be introduced to enable the better use of information or to supply new services rather than to reduce personnel. In this view, the benefits of using new technology may result from better information content rather than faster information handling.[63] Finally, it is clear that, whatever the outcome, in each department, company, or industry that introduces new technology, it will be the lower-skilled clerical and, in some cases, first-line management that are affected. The majority of these lower-level occupations are held by women.

Notes

1 ILO, *Problems of Women Non-Manual Workers: Work Organisation, Vocational Training, Equality of Treatment at the Workplace, Job Opportunities*, Report III, Advisory Committee on Salaried Employees and Professional Workers, Eighth Session (Geneva, ILO, 1981), pp. 4–6.

2 M. Porat, *The Information Economy: Definition and Measurement*, Publication 77–12(1), (Washington, DC, US Department of Commerce, Office of Telecommunications, 1977).

3 Quoted in E. Bird, *Information Technology in the Office: The Impact on Women's Jobs* (Manchester, Equal Opportunities Commission, 1980), p. 9.

4 Porat, *The Information Economy*, p. 106.

5 This is not to say that management will not be affected by new office technology.

6 J. Selby Smith, "Implications of Developments in Microelectronic Technology on Women in the Paid Workforce," unpublished paper to OECD Working Party on Informatics, Computer and Communications Policy, March 1980.

7 Bird, *Information Technology in the Office*, p. 16.

8 M. Hult, *Technological Change and Women Workers: The Development of Microelectronics*, report submitted to the World Conference of the United Nations Decade for Women on Equality, Development and Peace, Copenhagen, July 14–30, 1980 (Copenhagen, doc. A/CONF 94/6; mimeographed), p. 22.

9 ILO, *Problems of Women Non-Manual Workers*, pp. 53–70.
10 P. B. Doeringer and M. J. Piore, *Internal Labor Markets and Manpower Analysis* (Lexington, Mass., D. C. Heath and Co., 1971).
11 US Department of Labor, Bureau of Labor Statistics: *Employment and Unemployment: A Report on 1980*, Special Labor Force Report 244 (Washington, DC, April 1981), pp. A-22–3.
12 Hult, *Technological Change and Women Workers*, p. 24.
13 ILO, *Problems of Women Non-Manual Workers*, p. 63.
14 Ibid., pp. 53–70; and V. Hasty, "Women and their Science-related Education—or Lack Thereof" (ILO, unpublished research, July 1981).
15 K. Schramm, "Business education for the future," in *International Labour Review* (Geneva, ILO), January–February 1980, pp. 115–27.
16 J. Sleigh et al., *Manpower Implications of Microelectronic Technology* (London, HMSO, 1979).
17 D. Lockwood, *The Black Coated Worker* (London, George Allen and Unwin, 1958), p. 89.
18 Sleigh et al., *Manpower Implications of Microelectronic Technology*, pp. 100–2.
19 S. L. Sauter et al., *VDT-computer Automation of Work Practices as a Stressor in Information Processing Jobs* (Madison, Wisconsin, University of Wisconsin Department of Preventive Medicine, unpublished and undated research).
20 H. R. Bowen and G. L. Mangum (eds), *Automation and Economic Progress* (Englewood Cliffs, NJ, Prentice-Hall, 1966), p. 5.
21 Working Women, *Race against Time: Automation of the Office* (Cleveland, Ohio, 1980).
22 H. Menzies, *Women and the Chip: Case Studies of the Effects of Informatics on Employment in Canada* (Montreal, Institute for Research on Public Policy, 1981). In addition, see S. G. Peitchinis, *The Effect of Technological Changes on Educational and Skill Requirements of Industry* (Ottawa, Department of Industry, Trade and Commerce, Technology Branch, 1978) for an overview of technology and the Canadian economy.
23 Porat, *The Information Economy*, p. 30.
24 Menzies, *Women and the Chip*, p. 42.
25 *Employment and Unemployment: A Report on 1980*, p. A-22.
26 Working Women: *Race against Time*, p. 16.
27 ILO, *The Effects of Technological and Structural Changes on the Employment and Working Conditions of Non-manual Workers* (Geneva, ILO, 1981), pp. 21–3.
28 Working Women, *Race against Time*, p. 16.
29 Menzies, *Women and the Chip*, pp. 41–9.
30 E. B. Cox, "Prospects for Automated Tellers, 1981–1990," *American Banker*, June 2, 1981, p. 16.
31 Menzies, *Women and the Chip*, p. 43.
32 Cox, "Prospects for Automated Tellers," p. 28.
33 M. Jaffe, "Trimmed Staff and Equipment Make Citibank Bullish on DDP," *Bank Systems and Equipment*, October 1980, pp. 58–60.
34 *ABA Banking Journal*, May 1981, p. 63.
35 Ibid., p. 58.
36 R. Matteis, "The New Back Office on Customer Service," *Harvard Business Review*, March–April 1979.
37 *Chicago Tribune*, September 19, 1979, p. 16.

38 R. D. Edwards, "Distributed Data Processing," *United States Banker*, April 1981, p. 16.
39 Cited in "Minicomputers: One Bank's 'Big' System Approach," *Savings Bank Journal*, November 1978, pp. 55ff.
40 Ibid., p. 52.
41 Unpublished industry employment data produced by the US Department of Labor, Bureau of Labor Statistics, Washington, DC.
42 Menzies, *Women and the Chip*, p. 34.
43 Ibid., p. 33.
44 M. L. Wright, "Where We Are Today: The State of Automation in Selected Insurance Companies," *Best's Review Life/Health Edition*, September 1980, p. 62.
45 US Department of Labor, Bureau of Labor Statistics, *Technology and Labor in Five Industries*, Bulletin 2033 (Washington DC, 1979), pp. 41–9.
46 R. Polis, "Micrographics in Insurance: Reducing the Paper Burden," *Best's Review Life/Health Edition*, January 1981, p. 36.
47 US Department of Labor, *Technology and Labor in Five Industries*, p. 43.
48 Polis, "Micrographics in Insurance," p. 30.
49 US Department of Labor, *Technology and Labor in Five Industries*, p. 48.
50 Menzies, *Women and the Chip*, p. 37.
51 Ibid.
52 Ibid., p. 39.
53 Ibid., pp. 25–31.
54 Z. Zerman, *The Impacts of Computer/Communications on Employment in Canada* (Montreal, Institute for Research on Public Policy, 1979), p. 189.
55 *Business Week*, August 3, 1981, p. 62.
56 "Occupational Employment Growth through 1990," *Monthly Labor Review*, August 1981, p. 46.
57 S. Johnson, "Word Processors Spell Out a New Role for Clericals," *New York Times National Recruitment Survey*, special supplement to the *New York Times*, October 11, 1981, p. 30.
58 L. D. McCartney, "Putting It All Together," *Dun's Review*, August 1979, pp. 70–2.
59 T. M. Lodahl and N. D. Meyer, "Six Pathways to Office Automation," *Administrative Management*, March 1980, p. 74.
60 Ibid.
61 Ibid.
62 G. Murphy, "From Service Bureau to In-house Minicomputer," *Association Management*, July 1979, pp. 74–5.
63 J. T. Dyer, "Data: An Allocable Resource," *Best's Review Life/Health Edition*, May 1980, pp. 78–84.

Guide to Further Reading

United States

Colin Norman, "The New Industrial Revolution," *The Futurist*, February 1981; reprinted in *Dialogue*, no. 54, 1981. A very optimistic view of future job prospects.
Eli Ginzberg and George J. Vojta, "The Service Sector of the US Economy," *Scientific American*, March 1981. Deals with shift of employment to the Sunbelt.

The growth of the service sector is also outlined in "America's Restructured Economy," a special issue of *Business Week*, June 1, 1981.

Advances in Automation Prompt Concern Over Increased US Unemployment, General Accounting Office, May 1982. See also *Occupational Outlook Handbook*, issued by the Bureau of Labor Statistics every Spring.
Martin Anderson, "Shake-Out in Detroit: New Technology, New Problems," *Technology Review*, August–September 1982.

Henry M. Levin and Russell W. Rumberger of Stanford University argue, contrary to popular belief, that high tech creates more low-skill jobs than high-skill ones in "High Tech Requires Few Brains," *Washington Post*, January 30, 1983.
Richard McGahey, "High Tech, Low Hopes," *New York Times*, May 15, 1983.
A. F. Ehrbar, "Grasping the New Employment," *Fortune*, May 16, 1983.
"Detroit's Jobs That Will Never Come Back," *Business Week*, May 23, 1983.
"The New Economy," *Time* cover story, May 30, 1983.
"A Productivity Revolution in the Service Sector," *Business Week*, September 5, 1983.
"Gains in Employment Will Start Coming Harder," *Business Week*, December 19, 1983.
Edward Cornish (ed.), *Careers Tomorrow: The Outlook for Work in a Changing World* (World Future Society, Bethesda, Maryland, 1983). Articles from *The Futurist*.
S. Norman Feingold, "Emerging Careers: Occupations for Post-industrial Society," *The Futurist*, February 1984.

Henry M. Levin and Russell W. Rumberger, *Forecasting the Impact of New Technology on the Future Job Market* (School of Education, Stanford University, 1984).

Britain

Ken Green and Rod Coombs, "Slow March of the Microchip," *New Scientist*, August 7, 1980; "Employment and New Technology in Tameside," *Futures*, February 1981; and (with K. Holroyd) *The Effects of Microelectronic Technologies on Employment Prospects* (Gower Press, Aldershot, 1981).

Verner Wheelock (ed.), *Careers in an Information Society* (University of Bradford, 1982).

J. D. Whitley and R. A. Wilson, "Quantifying the Employment Effects of Microelectronics," *Futures*, December 1982.

The Impact of New Technology in the London and Home Counties Area (Association of Professional, Executive, Clerical, and Computer Staff (APEX), London, 1982).

Arthur Francis, "The Social Effects of Computer-aided Engineering in Britain," *Electronics and Power*, January 1983. Excellent overview: makes six controversial propositions, the first being that higher unemployment is inevitable.

Virginia Williams, "Employment Implications of New Technology," *Employment Gazette*, May 1984 (HMSO, London). Review of literature and some recent studies.

Women

Emma Bird, *Information Technology in the Office: The Impact on Women's Jobs* (Equal Opportunities Commission, Manchester, England, 1980).

Heather Menzies, *Women and the Chip: Case Studies of the Effects of Informatics on Employment in Canada* (Institute for Research on Public Policy, Toronto, Canada, 1981).

Microelectronics and Women's Employment in Britain, a report by members of the Science Policy Research Unit (University of Sussex, England, 1982).

Ursula Huws, *Your Job in the Eighties: a Woman's Guide to New Technology* (Pluto Press, London, 1982).

Jan Zimmerman (ed.), *The Technological Woman: Interfacing with Tomorrow* (Praeger, New York, 1983).

Women and New Technology (Trades Union Congress, London, 1984).

Cynthia Cockburn, *Brothers: Male Dominance and Technological Change* (Pluto Press, London, 1983).

Rosemary Crompton and Gareth Jones, *White-collar Proletariat: Deskilling and Gender in Clerical Work* (Macmillan, London, 1984).

Amy Dru Stanley, "High-Tech Will Hurt Women," *The New York Times*, September 19, 1983.

10   Skills: the Quality of Work

The Quality of Employment in Services

John Child, Ray Loveridge, Janet Harvey, and Anne Spencer

Jobs in health care, banking, and retailing are being changed rapidly by the new technology. But the picture emerging is a varied one: human skills are being replaced in such tasks as cash transactions, with a consequent loss in job satisfaction. In highly professionalized, high-risk services like medicine, there is producer and consumer resistance. This analysis by researchers at the University of Aston Management Centre was presented to the annual meeting of the British Association for the Advancement of Science, at the University of Sussex, England, August 1983, and first appeared in the Symposium's proceedings, New Technology and The Future of Work and Skills *(Frances Pinter, London, 1984).*

This paper inquires into how the jobs of service providers are changing with the introduction of new technologies. Examples drawn from banks, hospitals, and retail stores show considerable variation in the changes that are underway. An analysis of factors contributing to this variation is then developed. The paper focuses on changes in the quality of employment rather than in the absolute level of employment. Quality of employment refers to the skill content of jobs and to the control that job-holders have over working practices. Studies in Western countries have indicated that the use of skills is associated with job satisfaction and psychological wellbeing among a majority of workers (cf. O'Brien, 1980); some degree of control over working practices is a necessary condition for skill utilization.

It has been argued that there is a long-term trend in manufacturing industry toward the reduction of workers' opportunities to use skills and to exercise control over their work. This is seen to result from management's attempts to tighten its grip over working practices in order to raise the level and predictability of performance (Braverman, 1974; Gordon, Edwards, and Reich, 1982). It is also claimed that even highly qualified professional service providers are experiencing a comparable erosion in their quality of employment. Most are now employed in organizations where they are subject to managerial efficiency drives stimulated by public expenditure crises (as in the case of health care) or by commercial competition (Oppenheimer, 1973).

Technology has been regarded as an important instrument of erosion in the quality of employment. Merton commented in 1947 that "labor-saving technology" creates the "forced obsolescence of skills." Others have provided recent examples of the deskilling effects of new electronics-based technology on office and shopfloor work (e.g. Cooley, 1981) and have seen in management's application of this technology an attempt to win control over work processes from the worker (e.g. Feickert, 1979). Haug (1977) has argued that in the services the ability of computers to apply algorithms to stores of data with speed and precision makes it possible to substitute workers of lower skill for professionals—for example in computer medical diagnosis.

This line of argument assumes that managements both seek and are able to apply new technology in ways that degrade the quality of employment. Common though it may be, a degradation policy is not the only one in evidence, for reasons that include (1) a need for workers to have sufficient skill to cope with work processes that are variable and uncertain; (2) the desire to develop human capital; and (3) a belief in the value of enlisting workers' commitment to managerial objectives (Sorge et al., 1983; Wilkinson, 1983).

Also, it has become apparent from studies of manufacturing that some groups of workers may be able to resist pressures toward a degradation of their work on the basis of (1) occupying a strategic position within the production process; (2) drawing strength from cohesive collective organization; and (3) enjoying a scarcity value in the labor market (cf. Wood, 1982). Claims of an ideological nature by the members of given occupations to have the exclusive right to conduct certain activities according to their expert discretion may also inhibit the threat to the quality of employment posed by new technology, even though the work is of a kind that could be automated (cf. Loveridge, 1972).

In the provision of services directly to the public, consumer ideologies may constitute another conservative force. The progress of technology as a vehicle for degrading such service work will depend on the acceptance by customers that direct access to the judgment of an expert and the availability of personal attention are no longer necessary requirements (Child and Fulk, 1982).

Evidence on the accommodation of jobs and work organization to new technology, emanating mainly from manufacturing, therefore points to the influence of ideological and political forces in addition to that of a technical logic. There is no reason, then, to expect a simple uniform relationship between the introduction of new technology and changes in the quality of employment. Studies from manufacturing industry do not indicate uniformity, and the debate over how far the services represent a radical departure from traditional industrial production (cf. Bell, 1973; Gershuny, 1978) has drawn attention to additional factors in service provision that are likely to increase variation in the technology–employment relationship. These include the application in some services of public welfare rather than commercial criteria, and the wide variety in the organization of services which in some cases is grounded on professionalism. The services constitute sharply contrasting contexts for the introduction of new technology.

The selection of banks, hospitals, and retail stores as research locations is in recognition of the contrasts between three sectors that share a significant role in modern society as providers of services directly to the public (Gershuny and Miles, 1983). The particular new technologies being investigated all contain potentialities for degrading the quality of existing employment. Our research is designed to examine whether these potentialities are realized in different contexts.

Hospitals

New technologies in British hospitals tend to be lodged within departments and attuned to the special needs of those departments as defined by medical consultants. In recognition of this diversity, three examples are considered: (1) computer medical diagnosis, (2) patient monitoring, and (3) laboratory automation.

Computer medical diagnosis

In areas of medical diagnosis where there is only a limited range of symptoms that can be readily identified and measured, and where their relation to underlying causes is relatively well-known, computer diagnosis is today as predictively accurate and more consistent than that of most doctors (cf. Rogers et al., 1979). On the grounds of task performance, the new technology could be said to substitute for the doctor's diagnostic skills.

A specific example arises in a hypertension clinic located within a major hospital. The consultant in charge of the clinic developed, with the initial help of computer programmers, a 22-stage interactive system which stores basic patient information, can provide a full patient history, indicates the physical examination required for the patient, and when the results are inputted produces a recommended plan of treatment including subsequent investigations, drug doses, and frequency of future appointments. The system operates via a VDU terminal on a desk in the clinic deliberately positioned so that the screen is visible to the patient seated at one side of the desk.

This system captures all the necessary data requirements and diagnostic routines. As the consultant put it, "The doctor is now in the computer and a nurse could easily take its orders." Indeed, it would be relatively straightforward to allow for patients to interact directly with the system if everyday language were substituted for medical terminology. The substitution of direct patient interaction with the system for the doctor as intermediary might even have positive benefits, for the consultant indicated that the plan of treatment coming from the computer carried greater authority and conviction than if it came from a doctor. This was the reason for ensuring that the terminal was visible to the patient. The system substitutes for the medical service provider in other respects as well. It provides a discipline by making sure tests are taken, information is recorded, and a systematic treatment plan is devised. A measure of the improvement this offers can be judged by the consultant's claim that in 10 percent of hypertension consultations the doctor omits to take the patient's blood pressure, while many doctors fail to set out a specific or systematic plan of treatment. In terms of task requirements, the system has created redundancies in certain areas of personal medical skill and opened up the possibility of delegating the performance of work to more junior staff if not to patients themselves.

In practice, no such degradation in the medical role has taken place, for reasons that are very clear. The consultant in charge had designed and introduced the system. He maintained total control over the process of applying the new technology. It was his decision—"a very conscious decision"—not to incorporate everyday language in the system. He insisted that junior doctors use the system. One junior doctor took the view that the system impinged on his professional autonomy and started to enter false information and identity codes in protest: he was found out and had to leave the unit. From this position of control, the consultant had also resisted the admitted possibility of delegating patient interaction to nurses, let alone any move towards self-administration of the service by patients.

The reasons advanced by the consultant for his resistance to any change in the medical role within the clinic should be considered for their ideological function in legitimating an established structure of medical control as well as for their relevance to performance of the task at hand. He claimed that expert medical judgment is still required for decisions on the truthfulness of patients' responses, and on whether to override the recommended plan of treatment in the light of knowledge about the patient's reliability. In his view, direct interaction of patients with the system would be too time-consuming and would increase the risk of input errors. None of these problems would appear to militate against delegation of the medical role to a trained nurse. Indeed, the consultant had hoped to connect the system to satellite users so that *non-specialist* doctors in local clinics could employ it to provide treatment.

This case identifies control over the innovation process within oganizations, and the ability to legitimate the maintenance of a central role in service provision by reference to professional ideology, as factors that can maintain the integrity of service employment in the face of a contrary task logic that new technology imports into the work situation. Whether this defence of the quality of employment can be sustained in the long term remains an open question.

Patient monitoring
New technology for the monitoring and recording of patient systems is particularly apparent in intensive care units. Microelectronic equipment will typically monitor ECG, blood pressure, venous pressure, and respiration. It is possible to attach microprocessors to each of a series of monitors in order to record data automatically, analyze them in terms of trend changes, cross-compare with other trends, and then send alarm signals if appropriate. This facility has been used in some intensive care units to set up a central nursing station which takes the nurse away from continuous close contact with the bedside and is potentially labor-saving.

A central nursing station involves a degree of substitution by new technology for the nurse's monitoring role. It is possible to go further and to import technology that "closes the loop." This replaces human intervention in response to signals of a changed patient condition, and the administration of corrective action (for example, administration of drugs) is then activated automatically according to a computer program. A closed-loop system could take over a considerable element in skilled nursing discretion and leave the nursing role confined more or less to a tendering to patients' basic physical needs, the correcting of monitors, and the alerting of medical staff to emergencies.

In the case of one intensive care unit we have studied, these technological possibilities have been consciously rejected. The concept of a central nursing station was rejected by both the consultant in charge and the senior nursing officer, in the belief that nurses have a role by the bedside in the intensive care context. On the other hand, the automatic recording of patient data was seen as a valuable development insofar as it relieved nurses of interruptions to the time they could give to patients. Even so, nurses tended to regard these data more in the light of information that was of research interest to doctors than as inputs to which they would respond in the course of their work. Observation indicated that greater reliance was placed on signals such as changes in the patient's facial complexion.

The concept of the closed-loop system, automating the cycle of monitoring–corrective action--monitoring, was decisively rejected as being too dangerous. It is difficult to evaluate the degree to which this rejection was based on a systematic technical evaluation rather than an ideological belief. By way of justification, the story was told of a closed-loop situation in the US when the nurse in attendance happened to be wearing a nylon slip. The static electricity this produced sent the monitor haywire; the patient was administered a vast oversupply of the drug, with predictable results.

A very consciously articulated philosophy of intensive personal nursing was entrenched in the unit studied. The consultant anaesthetist in charge stated his belief that a sustained pressure of nursing activity at the bedside was in the patient's best interest. This philosophy is clearly antithetical to the application of new technology for any form of labor-saving that distances the nurse from the patient. Since the consultant had built up the unit and controlled the process of selecting and bidding for new equipment (most funded by charitable donations), he was in a position to apply his philosophy to the unit's new technology. Nurses in the unit shared this approach.

This case again illustrates how, in the hospital context, professional and semi-professional staff with strong occupational traditions and ideologies, and performing a critical role in patients' survival, are able to determine the use of new technology.

Laboratory automation

Since the late 1950s, equipment has been available for the automation of standard biochemical tests. Automatic analyzers can divide specimens into measured samples each one of which will be subject to a different test; they administer an appropriate quantity of reagent to each sample, and they record the results. Prior to automation, all these operations would have been carried out manually at the bench by laboratory technicians.

The present example is a small district general hospital biochemistry laboratory of nine technicians, a biochemist, and a consultant pathologist in charge. In 1980 this laboratory introduced a pair of discrete centrifugal automatic analyzers, which now perform approximately 80 percent of the tests requested from the laboratory. Unusually, the laboratory also acquired a computer at almost the same time for storing, rapidly accessing, and analyzing patient test records, and for printing items such as specimen identification labels and test result documents. The intention, now achieved, was to input test results directly from the automatic analyzers to the computer.

Laboratory automation *per se* replaces the much slower but generally more satisfying manual conduct of tests by technicians. Computerization relieves technicians of the manual recording of test results, which they generally regard as a chore that interferes with their core activities. In this particular laboratory, the two new automatic analyzers were replacing five older machines which were less productive and required greater attention from the technicians operating them. It was therefore expected that both the investments in new technology (analyzers and computer) would allow technicians more time to carry out interesting non-routine work of a kind that calls more substantially upon skills acquired during their training of approximately four years.

As in the other examples of new hospital technology, the process of deciding to bid for the analyzers and the computer, the subsequent lobbying and development of a case, the selection of the equipment, and its introduction into the laboratory's work organization were all closely controlled by the consultant in charge. The biochemist conducted an appraisal of alternative models of analyzer applying criteria agreed with the consultant.

Participation by laboratory technicians in the choice of analyzer was not encouraged on the grounds that the consultant wished to impose his work norms with the new equipment and not allow technicians to develop their own.

The introduction of new automatic analyzers and of the computer has affected the content of the laboratory technicians' job in different ways. The computer removed lower-level chores and also greatly enhanced the service that technicians and the biochemist can provide in response to inquiries inside and outside the hospital. Computerization has also been a vehicle for the promotion of two male technicians. Their promotion resulted from selective sponsorship by the consultant in charge and has led to a feeling of grievance on the part of female technicians, who perceived that they were deliberately excluded from this opportunity.

The further automation of testing, on the other hand, has contributed to a degradation in technicians' work. The machines require a set sequence of minor routine tasks, and they require the close physical attendance of the operator, which considerably inhibits social interaction. Their introduction has not released more time for non-routine work because of the very considerable increase in the laboratory's workload that has occurred (some 30 percent in two years). It is important to bear in mind the context of growing workload, since it is this, in combination with investment in new technology, that has led to a degradation in the technicians' work. The combination of workload and new technology has also led to shift-working and to the delegation of automated testing to a sub-technician grade being contemplated as necessary future adaptations in work organization.

This laboratory case points again to a high degree of control by senior medical staff over the choice of new technology and its introduction into their departments. The technical staff, whose jobs were the most immediately affected by the new equipment, did not exercise any appreciable influence in these decisions. Their exclusion from discussions on the introduction of the automatic analyzers eroded their ability to regulate jointly with management the intensity of work, so that an intensification accompanied the new technology.

Technicians have been able to defend their right to rotate through the full range of laboratory work on the basis of an established occupational role model that is grounded on a hierarchy of formal qualifications. There was considerable discontent among the technicians we studied because of the growing disparity between the model as a normative point of reference and the reality of diminishing prospects for upgrading, lack of time to conduct non-routine work, and the threat of shiftwork. This discontent had triggered off unionization.

Banking

The focus of our research is retail branch banking, in which services are provided to personal consumers and small to medium businesses. The main areas of new technological development in retail bank branches are (1) machine room on-line data transmission; (2) automated telling; and (3) systems centred around customer data files. The first two areas have now gained widespread application, while the third is still at an early stage in British banks. The conclusions reported in this section derive from research covering three branches of a large British clearing bank: two branches had been recently furbished with new technology, while the third operated with older equipment.

Banks operate very large centralized information processing networks in which paper (checks, printouts) remains an important medium. Transmissions through the network are standardized both in content and timing. They are therefore well suited to electronic automation, and banks have been pioneers in the advanced use of computer technology although constraints exist in, for example, the legal requirement that individual check entries and signatures be verified. The machine room or "back office" of a branch provides its link with the network. New technology in branch machine rooms typically means on-line terminals and check encoding machines.

There were 3510 automated teller machines (ATMs) in use within the UK by the end of 1982, and some 4300 were projected to be in service by end-1983. The earlier type of ATM merely dispensed cash on insertion of a magnetically coded plastic card and the customer's personal identification number (PIN). Current models can offer a range of additional services; they can display and print out statements, initiate Giro payments, establish or amend standing orders, order checks, and provide information about fixed-term deposits. Certain American ATMs can even report weather forecasts and football scores!

Most ATMs in Britain are "through-the-wall" and are accessed by the customer outside the branch, but there is a move now to install ATMs inside the branch to improve the opportunity of marketing other services to customers. In the two more advanced branches studied, new technology had also been introduced to speed up over-the-counter cash transactions at the same time as providing sufficient security to permit customer-friendly screenless counters. The technology consisted of cash dispensers operated by cashiers through a keyboard. Each dispenser was located in a safe controlled by a time lock, and served two cashiers. Cash paid in was rapidly transmitted by pneumatic tubes to a secure area. This cashier service was available only for transactions up to £200, and any other customer inquiries or needs were referred to other staff.

In this particular bank, systems centred around personal customer data files are still at the development stage. These would provide a profile of relevant information for each customer, including items such as creditworthiness, loan repayment history, services the bank had attempted to sell to the customer, and personal background, rather than the present structure of information, which is account-based. These files would replace present paper-based systems and could much more readily be kept up to date. They would be available to branch staff via VDU terminals and could be used in connection with decisions on loans, as a basis for marketing initiatives, and to eliminate office work connected with amendments and the like.

These technological initiatives are planned centrally through three committees which concentrate on the bank's main market segments. The managers in charge of its management services division play a key role in these committees, and the division has its own new technology development establishment. In keeping with the somewhat paternalistic ethos in the bank, and with its community ideology, the process of enlisting staff acceptance of technological changes has involved consultation rather than participation. Contact with trade unions has been kept informal and any negotiation studiously avoided. The two new technology sites studied were actually new branches (one replacing two older branches) and staff were asked if they were willing to work in them. Potential union opposition was muted by discussions and visits before their opening, while a system of staff meetings is seen by management to provide a channel for identifying any subsequent discontent.

Machine rooms in the new technology branches processed their work via on-line terminals, whereas in the "old technology" branch most work was batched up to be sent by a messenger to a computer located in a city central accounting unit. The use of newer on-line technology improved the quality of staff employment both as assessed by observation and as expressed by the employees. Processing inputs and outputs at an on-line terminal not only required the new skill of terminal operation but also constituted a more rounded task in contrast to simply sending out work to be processed in another unit.

On the whole, the new technology also permitted staff to have a greater control over the pace and timing of their work. Although there were still phases to the working day in the new technology machine rooms when different tasks tended to predominate, the on-line terminals permitted staff to put work through to the computer over the course of a whole day and to take advantage of a lull to clear miscellaneous tasks such as amendments. The new technology allowed staff to have greater flexibility in organizing the timing of their work. The main disadvantage lay in the risk of machine downtime, which can lead to a need to work overtime since transactions have to be processed on the day they are made. By contrast, in the old technology machine room the rhythm and pressure of work was to a large extent dictated by the need to assemble the day's work to go out for processing and to deal with the day's input, which has to be distributed without delay.

Automated telling via "through-the-wall" machines did not impinge to any degree on the content of bank teller's jobs except for those delegated to service the machines, though their introduction has saved on increases in staff and reduced the cost of supplying money from 30p to 5p per transaction.

The main change to job content had come about through the introduction of new technology to open-plan screenless counters within the broad context of a policy initiative by the bank to differentiate small cash transactions from larger transactions and other business within the new branches. This creates a new and narrower role of cashier, to which the new technology of secure tills and automatic cash dispensers has been applied.

The new arrangement divides the traditional job of bank teller into routine and less routine components. The less routine have been absorbed into the new job of lobby clerk, which deals with all inquiries except account balances and is also responsible for marketing services. Cashiers concentrate on small cash transactions not involving large amounts of coin and on balance enquiries. Both the jobs of lobby clerk and cashier are normally carried out by female staff.

Full-time cashiers expressed dissatisfaction with their role and the new technology accompanying it. They had lost the variety that came from dealing with customers' queries, including tasks such as ringing other bank branches whose customers wished to cash a check. The new shared tills were balanced at the end of the day for the branch as a whole, and cashiers had therefore lost the use of this skill. They particularly resented the fact that, despite losing this responsibility, transactions were still traceable to individual cashiers and they were still held individually accountable for any errors that might come to light.

There were also complaints that the new technology made the job boring and routine. The absence of screens improved their ability to talk with customers, though it was observed that, possibly owing to the tedium of the job, most talk was between cashiers themselves. The part-time "auxiliary" cashiers took a different view, however, expressing a preference for the convenience of the new equipment and clearly enjoying the ability to chat to customers. This contrast serves as a reminder that the quality of employment that workers perceive their jobs to offer may depend significantly on what they seek from employment in the first place.

The prospect of customer data files accessible via VDU terminals presents the senior management of the bank with a choice that has direct implications for the intrinsic quality of branch jobs (particularly the lobby clerk). On the one hand, by providing a full customer profile in terms of background, financial history, and the like, such files could be used as a basis for a lobby clerk or junior loans officer to exercise greater initiative and discretion over matters such as overdrafts and loans. On the other hand, it is recognized that the new technology could be used to support a policy of degradation. By programming the system to set out data requirements on a step-by-step basis and then to indicate appropriate decisions within given parameters, provision can be made for what a senior manager in the management services area called "idiot-proof systems." This latter option, "enabling you to use cheaper labor and still do a reasonable job at that level," was being seriously considered.

The centrally planned introduction of new technology in this bank appears to be characteristic of the sector as a whole. It clearly involves choices that have direct consequences for the quality of employment and of which senior managers appear to be aware. They are determined, however, to retain their control of new technology development, rejecting any idea of joint regulation with unions and relying instead on internal communication channels to signal any serious staff discontent. To date the impact of new banking technology has fallen primarily on lower-level and mostly female staff, many of whom have not in the past sought a long-term career in the bank. Senior bank officials freely admit that they have been introducing new technology within an unpublicized dual employment policy in which a stereotype of the female worker as a non-career-seeking short-term employee has justified a lower tier of routine jobs. The application of new technology to facilitate the degradation of jobs such as cashier, and prospectively to place tighter limits on the lobby clerk, has been informed by this ideological conception and not simply by the demands of the task. With an increasing number of women now wishing to continue their employment in banking, and a certain proportion seeking career advancement, the traditional stereotype is likely to be rejected by these employees with increasing insistence, a rejection that may fuel a more forceful collective resistance to new technology policy.

Retailing

Two companies illustrate the introduction of new technology in retailing. One is a group of 12 department stores, and the other is a multiple supermarket chain. Both companies have installed electronic-point-of-sale (EPOS) equipment incorporating scanning and linked to computer-controlled systems. A store in each company has been studied in detail, one of them being the pioneering British supermarket application of EPOS.

EPOS terminals potentially can form the "front end" of totally integrated computer-based systems which provide for inventory management, the reordering cycle of open-to-buy decisions and purchase order management, accounting, credit management, and sales analysis on the basis of precise itemized operating data from the store. This high level of integrated control has generally still to be realized. EPOS systems of the kind operating in the companies studied did, however, capture itemized sales information via wands attached to terminals that read magnetic tickets (in the department stores) and via laser bar-code scanning in the supermarkets. The supermarket system also incorporated automatic price look-up, and in both cases the EPOS terminals fed itemized sales data into a computer for analysis.

The benefits claimed by equipment manufacturers to derive from the introduction of EPOS involve the enhancement of managerial control, including some features that impinge on the jobs of in-store staff. Automatic price look-up removes the scope for "fiddling" by checkout staff. The availability of sales data precisely identified by time and item provides the information to deploy staff flexibly in line with the changing incidence of workload. The use of bar-coding instead of individual item pricing affects staff economies but at the same time reduces the number of employees who have personal contact with customers through answering queries and the like. The use of new technology to lessen contact between service providers and the customer forms part of a longer-term policy to reduce direct labor costs in the retailing of branded or standardized commodities by shifting the process toward self-service.

There are several other significant characteristics of the context into which new retailing technology is being introduced. Pressures on margins from high and increasing levels of competition, together with the key nature of buying and location decisions, have sustained a culture of entrepreneurial trading and the centralized initiative that accompanies it. Multiple retailers have always tended to maintain a centralized buying operation, controlling sources of supply, product range, pricing, and qualities purchased. Although individual department stores were for some time usually permitted to operate and buy autonomously, their owning groups were also turning to centralized buying by the early 1970s. Unpublished findings from the Oxford "Growth of Firms" study (cf. Francis, 1980) showed how decisions on new investments in retailing companies relied heavily on the approval and judgment of the chief executive, and further indicate the centralized entrepreneurial ethos of these companies. Finally, the structure of retail employment is relevant. Women account for approximately two-thirds of their retailing workforce, and retail firms have been increasing their ratios of part-time to full-time staff. It is not surprising, therefore, that trade union membership is low and that links between full-time union officials and members in stores are usually tenuous.

The involvement of staff representatives or union officials in the process of introducing new technology was virtually non-existent. While at national level there is talk of consultation and the establishment of new technology agreements (e.g. Distributive Trades EDC, 1982), at local level change was introduced unilaterally. For instance, the first that local store staff below the store manager heard of the scheme to install EPOS in the department store company was an announcement in the national press. Staff were then told by local management only that new tills were being put in. In fact, managers in both companies looked to the introduction of EPOS as a means to enhance their control over store operations and staff behavior.

The increase in managerial control over the work process was more marked in the supermarket. With EPOS, every checkout operator's efficiency can be monitored and their working hours scheduled to suit the peaks and troughs of activity. Operator performance is also reinforced by the feedback that EPOS now enables supervisors to give. "Leakage" of stock has been reduced both by the direct point-of-sale controls and by the changes that the new technology has allowed management to make to the stocking of shelves. Since individual items no longer need to be priced, more attention is now directed towards accurate shelf-placing, which necessitates the elimination of "merchandise drift" along shelves.

In the department store, a control room routinely checks on erroneous entries made by EPOS operators. A printout indicates cause or type of error. It also shows total takings and number of transactions per hour, which has provided the basis for a more precise deployment of staff at different times of day. Some restriction has been placed on management's ability to check on individual operators in the department store by the growing practice of keying in one operator number for substantial periods of time rather than keying in the individual operator number for each transaction.

The application of EPOS also reduced middle management autonomy. For example, the department supervisors in the supermarket are no longer required to exercise discretion in the display and quantity of stocks on the shelf, since these are dictated by the system. Within the department store, EPOS had already permitted senior management to have specific information on each department's performance, which eliminated what the general manager described as "guessing games" by departmental managers.

With the introduction of computerization, the price labelling of goods is now carried out by a recently formed pre-retail department, which applies a selling price recommended by head office. Previously, departmental managers had enjoyed considerable discretion in deciding the final selling price, which they adjusted to meet their own assessments of local demand. Their discretion to reorder stock has also become subject to increasing regulation, and there has been a move to use data provided by EPOS to introduce automatic reordering of repeat "essential merchandise" under the administration of head office buyers. It is worth noting that departmental managers were not involved at any stage in the planning of EPOS-based information system or in discussions of alternative technical options. They saw the installation of EPOS terminals as having created "extra work," which was seen as being of benefit only to buyers and others at head office.

Changes to the skill content of other jobs present a complex picture which can only be summarized here. In the supermarket, the point-of-sale checkout job now incorporates somewhat greater variety, a higher level of training to deal with malfunctions, and the potential for more customer contact. On the other hand, it has become subject to a greater level of direct control and pressure for fast throughput.

In the department store, EPOS has thus come to the sales floor as a reinforcement of an already present trend away from the personal sales service formerly offered to the customer. Sales staff are now located at a central point in each department, this location being determined by the EPOS installation. Their role had effectively been narrowed to little more than a cashier. They did, however, experience conflict between offering advice to customers on their purchases in the manner previously given, and still expected by the customer, and their new concern for the accurate recording of information into the terminal. More time is taken in processing transactions, because the EPOS terminals in the department store require a more complex sequence of operations, interruption of which by customers can be irritating. The terminals require that a formal set of procedures be followed to process a transaction, and they impose their own speed which slows down when the store is busiest.

In an attempt to cope with these problems and pressures, one branch of the department store group had introduced a new division of labor which separated the specialized duties of cashier from those of salesperson. In order to reduce interruptions to the cashier's duties and to facilitate customer service, women (though not men) in each role were dressed in distinguishing uniforms. There is a parallel between this essentially *ad hoc* adjustment and the more systematically planned division in the bank between cashier and "lobby clerk" roles. In each case the cashier role was the one most closely attached to and circumscribed by new technology, and both bank and retail cashiers complained that the narrowing of their jobs led to boredom.

Discussion

The cases described point to considerable variation in the ability of different service providers to influence the way new technology impinges on the content of their jobs. In the case of the hypertension clinic consultant, the new technology was developed at his volition and as an extension of his conception of service provision. The process remained under his control throughout. The use of monitoring systems in intensive care also remained under the control of the consultant and the nursing staff. However in other cases, such as the laboratory technicians, bank cashiers, and retail sales staff, service providers did not participate effectively in the selection, development, or application of new technology. They were observed to have lost elements of skill and/or control as a result of the way technology was applied, and on the whole they expressed dissatisfaction with the changes that had taken place.

The extent to which new technology threatens the quality of service providers' employment appears to depend on the position they occupy in the decision-making structure of their organizations, and in particular on the power they have to define "appropriate" working practices and job content. Three relevant factors have emerged. First, the nature of the tasks being performed clearly differed among the cases described, and this prompts the question of whether task is a factor underlying service providers' position and influence. Second, there were also indications that some service providers enlist claims of an ideological nature in defense of their present roles and working practices. The credibility of such claims will be enhanced if the public valuation of their services is high. Third, labor market conditions appear relevant to the ability of service providers to defend the integrity of their jobs against pressure from threats such as the substitution of alternative labor. Highly professionalized groups have been in a position to act as gatekeepers to a specialist labor market, with state recognition.

A recognition of the interdependence between task, ideology, and market power in determination of workplace control draws upon both Weberian and Marxist traditions within sociology. For example, Bendix (1956), working very much in the former tradition, demonstrated that command over new techniques (task) and the development of an ideological basis for public acceptance were twin foundations for the growth of managerial authority in industry. Loveridge (1972) similarly identified the emergence of skills to perform new significant tasks, coupled with ideological justifications for the exclusive right to such performance, as key factors in the long-term development of white-collar occupational groups.

Jamous and Peloille (1970) have formulated this type of analysis in terms that are particularly helpful to the subject of this paper. In analyzing the historical transformations undergone by the French university-hospital medical elite, Jamous and Peloille concluded that these resulted from a combination of task and ideological factors. They suggested that the critical task factor was the ratio between "indetermination" and "technicality."

Indetermination (I) refers to the content of task performance that cannot be codified and expressed in the form of rules (methods and procedures). It consists of those performance aspects that rely upon the special qualities of each worker, and which of course readily form a mystique that outsiders cannot easily penetrate, be they managers, experts in new technology, or customers.

Technicality (T) refers to the part played in the work process by means that can be learned and communicated in the form of rules and hence are potentially translatable into information technology programs. The higher the I/T ratio, the greater is likely to be the worker's intrinsic quality of employment and his or her potential for defending it against inroads via new technology. Moreover, Jamous and Peloille recognized that the I/T ratio is in effect publicly expressed in specific historical settings. In other words, indeterminacy is claimed through ideologies, and the I/T ratio that prevails in jobs will therefore depend not just on the technical content of tasks but also on the public acceptance of ideological claims that these tasks are appropriately conducted by experts.

Writers in the Marxist tradition have given particular attention to attempts by employers to enhance managerial control over work in order to extract a greater surplus from the labor process. New technology fits into this scenario for its potential to intensify work and/or to substitute for reliance on the worker. The extent to which employers can apply new technology to these ends is seen to depend on the worker's ability to resist, which is affected, *inter alia*, by their market position. Thus Friedman (1977) has pointed to the relevance of product and labor market conditions for the relative power that labor and management can exercise effectively in the workplace.

Task dimensions

There is some consensus among organizational theorists that the most significant task dimensions for an understanding of how work is organized are those relating to *uncertainty* and *complexity* (cf. Perrow, 1970; Van de Ven and Ferry, 1980). The number of exceptions encountered in performing the task and its general variability, a lack of clarity about what is required and about cause–effect relationships, are all factors contributing to uncertainty. Complexity is increased by factors such as the amount of relevant information to be absorbed in carrying out the task, the number of steps involved, and the number of contributions required from different sources. Since we have in view the direct provision of services to members of the public, a further significant attribute of the task is the *cost of making an error* or providing a substandard service.

Uncertainty in the service task may also derive from difficulties in understanding and specifying the customer's requirements in addition to the factors just mentioned. The greater the uncertainty, the more difficult it will be to transfer the provision of the service to a technological system, because of the presence of exceptions and because the definition of exceptional requirements necessitates direct exploratory interaction with the customer. New technology in its present state of development is suited to processing data within *given* parameters and known cause-and-effect rules.

When present in a service where the cost of error to a customer is also high, task uncertainty increases the risk facing the customer. High risk is likely to lower the public acceptance of the absence of a service provider with certified competence or his/her substitution by automation. Apart from the fear that automation can break down, in risky transactions members of the public wish to be able to identify a person who is responsible should failures arise, as the rising number of court cases indicates. Task analysis therefore predicts that, under conditions of uncertainty and risk, new technology is less likely to threaten the integrity of the service provider's job and is more likely to be used in a complementary relationship, for instance to perform analyses and store information.

The presence of uncertainty and risk in service tasks would appear to constitute a far more significant constraint on the encroachment of new technology than that arising from complexity. In fact, new information technology can readily handle complex data so long as these are in a structured form. In these circumstances, new technology can be more consistent than the human being. Tasks where the degree of uncertainty and level of risk are relatively low but the amount of data to be processed is high are in principle ideal cases for new technology to substitute for personal service provision.

A task analysis in terms of uncertainty, cost of error, and complexity of data transactions does appear to discriminate between the extent to which new technology had impinged on the quality of service providers' employment in many of the examples given earlier. Thus, bank cashier and retail point-of-sale jobs could be said to represent tasks with low uncertainty, low risk to the customer, and relatively high complexity of data processing (at least over the working day as a whole). Jobs concerned with sanctioning overdrafts and loans in the bank, and with departmental supervision in the department store, involved higher levels of uncertainty and accompanying judgment. The risk to the customer from refusals (loans, refunding purchases) and/or poor advice was also somewhat greater. These are jobs that new technology is beginning or threatening to degrade, but where the outcome remains uncertain. At the other end of the spectrum—in intensive care—uncertainty, risk to the patient, and complexity are all high. New technology has been introduced but is used in a complementary manner that has enhanced rather than taken away from the skill and control exercised in the work process by the specialist nurses.

Two of the cases, however, do not so readily fit the predictions that can be derived from task analysis. The new technology-based hypertension system was applied to a service where cost of error could be high, but where on the other hand uncertainty was low enough for a computer program to handle the data requirements (complexity). The other anomaly is more apparent than real, namely the bank machine room, where new technology had somewhat enhanced use of skills and control. For the basic structure of work even in the new machine rooms remained governed by a highly mechanistic set of procedures. This kind of work should be amenable to further automation, being low in uncertainty and risk and high in complexity. Management in the bank does indeed envisage future moves in that direction.

Ideology and market power

The hypertension case furnishes an instructive example of a so-far successful defense of a service provider's job territory in circumstances where the introduction of new technology is threatening. The consultant's ability to control the terms on which the technology was developed and applied doubtless reflected his organizational position as a member of the hospital's professional elite, a position that has become institutionalized and publicly recognized in terms of the role model of hospital consultant. Nevertheless, it is significant that he defended his role by claims to exclusive expertise in making judgments about patients, which to an observer appeared to be as ideological as they were scientific in nature. This case illustrates the potential of ideology as a means for occupational groups to preserve a social acceptance that their work contains important elements of indeterminacy. A similar example is provided by bank managers in the sphere of making loans. However, while the hospital consultant is likely to continue to succeed in his defense of indeterminacy for some while to come, the position of bank manager appears far less secure.

As the risk to the public diminishes, and as the service provided is more routine and more readily evaluated by the public, so the defense of indeterminacy in face of technological change becomes more difficult. Compared with doctors, bank officers have a weaker public identity as professional experts. Loans, for example, are being offered by many sources today, and it is relatively easy for the customer to weigh up alternatives and decide on the best method of financing. It does not make much difference whether he or she obtains an offer from a bank manager or from a junior member of bank staff following the directions of a computer program. This is even more the case with highly standardized services such as food retailing and cash dispensing. In short, while service providers may for a while be able to defend indeterminacy with ideological claims, these will not remain credible if they no longer reflect the nature of the task, including any changes in the way it can be performed made possible by new technology.

There is also a reciprocity between the scope that service providers have to develop an ideological basis for the creation and defense of indetermination and the extent to which the providers of capital funding seek operational control over investments in new technology. Thus, the hierarchy of authority in the National Health Service still rests on the professional elitism of medicine, though it is strongly challenged by the new administrative ethos of managers recruited since the 1974 reorganization. Banks are centralized bureaucracies in which paternalism has been the dominant managerial ethos. In the past, this tradition has provided ideological encouragement for a degree of indeterminacy in the role of local branch management. In the bank studied, however, the center had undertaken strategic assessments that may lead to a progressive withdrawal of discretionary functions from local branches. In retailing, the choice and application of new technology resides primarily with the chief executive and his immediate associates involved in buying and locational decisions.

The part that indeterminacy plays in the defense against technological encroachment is associated with the market position of service providers. It helps to preserve their role in the organizational internal labor market against substitution by less "expert" employees and to provide a rationale for the control of recruitment through occupationally administered qualifications. Indeterminacy preserves the services provider's ability to move within the external labor market. This can be seen clearly to operate in medicine and nursing, though market forces are now moving adversely in regard to an excess supply of doctors which may in turn weaken their defense of indeterminacy (Child and Fulk, 1982).

In banking, the structure of qualifications offered by the Institute of Bankers has never provided a basis for this degree of market power, because the banks have always controlled the volume of new entrants into training and have maintained the dependence of staff on the employer for career progression by refusing to recruit from each other. Moreover, the predominantly female staff in jobs such as cashier and machine room operator are effectively excluded from this career path and confined to work where the "technicality" component is virtually total.

In retailing, indeterminacy has traditionally resided in skills such as buying. This element is under threat from the systematic knowledge now available through EPOS systems and, being a role out of contact with the public, it is not protected by customers' expectations. The market position of sales staff in the mass turnover non-bespoke areas of retailing is extremely weak, since replacements can readily be found and the period of required training is minimal. Customer acceptance of self-service and now of EPOS removes any ideological support for the preservation of that small element of indeterminacy that used to reside in the personal advice-giving and persuasive selling component of the salesperson's job.

This paper has explored some examples of how the introduction of new technology has affected the quality of service providers' employment. It has argued that this has to be understood as a political process which centers on the extent to which service providers have an ability to influence or resist technological change. The nature of the service task being performed and its consequence for the public appear to be significant determinants, but account has also to be taken of the effects of public, occupational, and managerial ideology directed to the nature of service provision and to the labor market position of the workers concerned. The substantive changes in tasks brought about by the introduction of new technology do not sufficiently explain the manner in which these are organized and distributed between the roles of existing incumbents.

References

Bell, D. (1973), *The Coming of Post-Industrial Society*. New York: Basic Books.

Bendix, R. (1956), *Work and Authority in Industry*. New York: John Wiley.

Braverman, H. (1974), *Labor and Monopoly Capital*. New York: Monthly Review Press.

Channon, D.F. (1979), *The Service Industries*. London: Macmillan.

Child, J. and Fulk, J. (1982), "Maintenance of Occupational Control," *Work and Occupations*, 9, 155–92.

Cooley, M. (1981), *Architect or Bee?* Slough: Langley Technical Services.

Coombs, R. and Green, K. (1981), "Microelectronics and the Future of Service Employment," *Service Industries Review*, 1, 4–21.

Distributive Trades EDC (1982), *Technology: The Issues for the Distributive Trades*. London: National Economic Development Office.

Feickert, D. (1979), "Of Men and Minos," *Computing Europe*, 22 November, 31.

Francis, A. (1980), "Families, Firms and Finance Capitals," *Sociology*, 14, 1–27.

Friedman, A. (1977), *Industry and Labour*. London: Macmillan.

Gershuny, J. (1978), *After Industrial Society?* London: Macmillan.

Gershuny, J. and Miles, I. (1983), *The New Service Economy*. London: Frances Pinter.

Gordon, D.M. Edwards, R. and Reich, M. (1982), *Segmented Work, Divided Workers*. Cambridge: Cambridge University Press

Haug, M.R. (1977), "Computer Technology and the Obsolescence of the Concept of Profession," in M.R. Haug and J. Dofny (eds), *Work and Technology*. London: Sage, chapter 14.

Jamous, H. and Peloille, B. (1970), "Changes in the French University–Hospital System" in J.A. Jackson (ed.) *Professions and Professionalisation.* Cambridge: Cambridge University Press.

Loveridge, R. (1972), "Occupational Change and the Development of Interest Groups among White-Collar Workers in the UK: A Long-Term Model," *British Journal of Industrial Relations,* **10**, 340–65.

Merton, R.K. (1947), "The Machine, The Worker and The Engineer," *Science,* **105**, 79–81.

O'Brien, G.E. (1980), "The Centrality of Skill-Utilization for Job Design," in K.D. Duncan, M.M. Gruneberg and D. Wallis (eds), *Changes in Working Life,* New York: John Wiley, chapter 11.1

Oppenheimer, M. (1973), "The Proletarianization of the Professional," *Sociological Review Monograph,* **20**, 213–27.

Perrow, C. (1970), *Organizational Analysis: A Sociological View.* London: Tavistock Press.

Rogers, W., Ryack, B. and Moeller, G. (1979), "Computer-aided Medical Diagnosis: Literature Review," *International Journal of Biomedical Computing,* **10**, 267–89.

Sorge, A., Hartmann, G., Warner, M. and Nicholas, I. (1983), *Microelectronics and Manpower in Manufacturing.* Aldershot: Gower Press.

Van de Ven, A.H. and Ferry, D.L. (1980), *Measuring and Assessing Organizations.* New York: John Wiley.

Wilkinson, B. (1983), *The Shopfloor Politics of New Technology.* London: Heinemann.

Wood, S. (ed.) (1982), *The Degradation of Work?* London: Hutchinson.

The Politics of Technical Change

Barry Wilkinson

All the things that make up the day-to-day experience of the shopfloor worker are matters of social choice. Essentially political decisions—sometimes concealed—are being taken today that will determine whether the craftsman of tomorrow is to be computer-aided or computer-degraded, says the author, a lecturer in business administration at the National University of Singapore. This article comes from the Industrial Relations Journal, *Summer 1983, and a more detailed account of the research appears in the author's book,* The Shopfloor Politics of New Technology *(Heinemann, London, 1983).*

The increasing use of the new microelectronics technology on factory floors has radical implications for the modern organization and experience of work. Payment systems, patterns of shiftworking, styles of supervision and control, and the structure and distribution of skills—all the things that determine the experience of the shopfloor worker—are being redesigned around the latest equipment and machinery. This paper reports on case studies concerned with the introduction of such new technology in British industry, and the implications for work organization. Here I shall elaborate on three aspects of the case studies.

1 *The economic rationales in terms of which technical change is generally conceived.* It is vital to understand their practical usage, not only because productivity, efficiency, etc., are major motivations behind technical change, but also because economic rationales can serve to *conceal* the political processes and social choices that lie behind innovative efforts and the manner in which the new technology is introduced.

2 *The shopfloor politics, which influence the precise direction that technical changes take.* These are important in any understanding of technical change, not simply because they might provide "resistance" to the introduction of

new technology, but because the very logic of technical change is dependent on these detailed political processes.

3 *The alternative uses of technology and the alternative ways of organizing work*, which were reported by managers and workers. In particular, I shall focus on the possibility of the development of a *computer-aided craftsman*, as an alternative to the *computer-degraded craftsman*, and argue that trade union and worker organizations might more effectively take up this possibility as a serious issue.

Most of the cases on which this paper is based have already been reported in detail elsewhere.[1] This paper will concentrate on and develop the themes described above arising from the case studies, which were undertaken in 1980 and 1981 in the West Midlands engineering industry.

Innovations in most firms visited, included the four upon which most attention was concentrated, included the introduction of microelectronic control devices for small- and medium-batch production processes. Of course, there is nothing new about automatic control, but the advent of microelectronics has led to a great increase in the range of processes amenable to automation. This includes many batch production processes, which have traditionally proved difficult and costly to automate because of the frequent need to reset machines with small and medium batches of component. Microelectronics has made this economically feasible by allowing the development of more powerful, and indeed cheaper, programmable control devices.

The centrally important aspect of these devices for our purposes is that their use allows the movements of production machinery to be *predetermined* through the programming or setting of the new controls, thus eliminating the need for manual intervention at the machine itself. This provides the opportunity to remove the skilled or semi-skilled machine operator from his or her central position in the production process, and to create a new skill elite in the form of programmers or process controllers. Thus one might expect a *deskilling* of the blue-collar labor force, and a *shift of control* from the shopfloor to the new white-collar staffs; and certainly the expectation was expressed by most (though not all) managements during my research. As we shall see, however, the real picture was neither so simple nor so unproblematic.

The new control devices are held to confer several economic advantages. For instance, predetermining the job can make the process easier, and the machinery itself can be in more constant use; this has the desired effect of increasing utilization rates, a major concern for engineers, particularly when production machinery is expensive and pay-back periods critical. Second, the program, often in the form of a paper or magnetic tape, can be filed and recalled whenever needed; thus the programming effort for very small batches or "one-offs" may be worth the effort. Third, manning levels may be reduced on some machinery, contributing to higher productivity. Finally higher accuracy and predictability of production is sometimes held to improve product quality and reduce the scrap rate. At least some of these economic advantages were cited as justifications for the introduction of the new technology in all the case study firms. However, perceived political and social benefits were often added to the economic advantages, and the economic rationales themselves could be used to conceal the social and political dimensions. That is, the economic rationales made up an ideology.

The four major case studies were of a plating company, a rubber molding company, a machine tool manufacturer, and an optical company. All were introducing, or had recently introduced, control devices for batch production processes. These will now be discussed, together with some illustrations of similar social and political processes in other industrial sectors.

Automatic plating

In one case study, a plating company was attempting to automate a plating line. Basically, plating involves dunking metal components into various solutions in order to coat them with a plate of zinc, chrome, or aluminum, etc. In this factory, in the old system there were three men per line, and their work involved loading, controlling, and unloading the machinery. Workers would control the movements of jigs or barrels that held the components by use of control levers, moving them in and out of different vats of solutions for varying amounts of time—depending on the type of component, the finished required, etc. (figure 10.1).

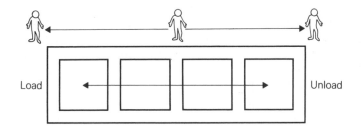

Figure 10.1   The old plating system (simplified view).

With the *new* system (figure 10.2), the movements of jigs or barrels were to be controlled *automatically*, by using an electronic control device to *predetermine* the carriage movements. These new lines would now need only two workers per line, and this increase in productivity would mean that the control system would pay for itself within a year or two, and increase profits thereafter. The operators shed would be lost through natural wastage. (In this company there were a total of five lines, and in the first instance two were automated.)

In terms of the conventional view of industrial innovation, the change appears at first sight unproblematic. There is a straightforward economic

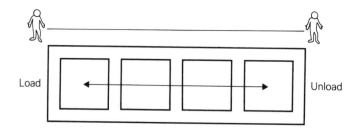

Figure 10.2   The new plating system.

motivation in increasing productivity, and the consequences are straightforward and inevitable. Some workers will lose their jobs (or, if lucky, will be redeployed), and there is an ("unfortunate") loss of job satisfaction and control for those remaining—only the tasks of loading and unloading are left.

But during interviews, without any prompting, managers pointed out that there was *more* to their motivations. For with the old system some workers did not wholly conform to management demands. They were described as "lazy" and "unreliable." They would take a lot of time off work, take long breaks, and sometimes work slowly, leaving components in vats for longer than necessary. For managers this meant a waste of time and plating materials, and constant detailed supervision was relied upon. *This*, managers insisted, was one of the spurs to innovation—automatic plating allowed greater management control over the pace and quality of output. The aim was "automatic control" in a wider sense than is normally implied by the term.

Further, managers tried to *make certain* of their control, as was discovered after asking how the new computer control worked. (After all, I thought, workers might still adjust this control in order to continue exerting their own influence over machine pace.) The manager took me on to the shopfloor, past the plating lines, out of a back door and into a room at the back of the factory. Here were the electronic control panels. "The platers can't get at them here," he said. In fact, only managers and one or two engineers are allowed in the room.

However, the outcome was dependent on the response from shopfloor workers. For obvious technical reasons, before full-scale automatic production could go ahead, frequent use had to be made by the platers of a manual override. This was necessary while new computer programs were being "proved," and while the whole system was debugged. Platers *continued* to use this manual override after debugging was complete, and thus managed to retain some of their traditional control over output. Old supervisory styles were thus still relied upon, and the effort extracted from the workers was still up for negotiation. The works manager put it as follows: "Some operators still do not adjust to the automatic lines. They use the manual override if you don't watch them. . . . It's a bad habit. It's difficult to get it into their heads that automatic is the best way. They think they can do a better job manually."

To make it clear that the view of platers as an "unreliable" lot of "bolshie" workers was peculiar to management, it should be pointed out that, although restraining effort was an important objective (and anyone who has worked on a plating line will find this understandable), some of the platers did take a pride in their work. Indeed, they explained that one of their justifications for using the manual override was that in this way they could get a better-quality finish on the product, as well as reduce the scrap rate. In *their* view, it seems, *management* standards were shoddy.

Although still unsuccessful on my last visit to the company, managers' intentions were clearly to take control of production away from the shopfloor worker. Platers were now supposed to "stand and watch" an automatic process, intervening only if things went wrong. The platers *could*, of course, have been purposely left in control if the control panels had been kept on the machinery itself. And innovative effort *could* have been directed toward the loading and unloading of the machinery—the more mundane aspects of the plating process. So if we continue to insist that economic motivations are central to technical change, we must at least recognize that there may be *competing definitions* of what is "efficient" (some of the platers had a clear alternative view in their emphasis on quality), and that these definitions may have a basis in shopfloor politics (the definitions begin to make some sense in this context).

In another instance, this time in a machine shop, managers were divided on whether to retain skills and control on the shopfloor, or to transfer them into the Manufacturing Facilities Office, where white-collar workers would take control. Over a period of several years the company had introduced nine CNC machine tools for the production of small batches of relatively complex components, and blue-collar workers had so far successfully defended their traditional position of shopfloor control.

This was achieved by taking advantage of the fact that CNC machines (unlike NC) have a tape editing facility at the machine itself. When program tapes were brought from Manufacturing Facilities to the shop-floor, many operators insisted on carrying out their own tape "proving" (editing out program faults during the production of the first of a batch), and some operators had even learned, for some of the simpler components, to carry out their own *programming*, using the tape editing facility. At the insistence of shop stewards, existing skilled machinists had been given the opportunity to work on the new computer-controlled machines, and this may partly explain the operators' desire to exercise and develop their skills, though it should be stressed that it was only through individual operators' ingenuity that this was possible. They had been given only two weeks' formal training in CNC, and the assumption was that programmers would take virtually all responsibility for programming and editing.

While most recognized the remarkable achievements of the machinists, some senior managers and engineers were unhappy that blue-collar workers remained in control. However, the superintendents and foremen who worked on the factory floor, many having been skilled machinists themselves in the past, sympathized with the operators and thus helped shopfloor workers retain control. The programmers, for their part, were unhappy about the situation. One suggested to me that "CNC was designed in the first place because semi-skilled labor is cheaper. . . . Management should decide exactly what the operators' tasks are, and give us full authority for the programs."

The programmers' desire was that a key that locked the tape editing facility should be in their own hands, and taken away after each new batch was set up, so that machinists would be left only with peripheral tasks (loading, inspection, etc.). The keys, in a sense, symbolized the battle for control on the shopfloor, and so far had never left the machines. Other researchers have recorded this contest during the introduction of CNC,[2] and we can assume that comparable events are occurring in many machine shops today. These working practices developed quite informally, and had become "institutionalized" in a set of unwritten rules. The way the control of production develops, however, is problematic. Unwritten rules may be a precarious basis for workers' control.

If workers' skills were unambiguously recognized and accepted by managers and engineers, then we could see the emergence of manual data input (MDI) CNC machines, which are designed for operator programming. Encouragingly, a team of engineers and social scientists at UMIST are currently working on this.[3] But perhaps a more likely option is the development of control systems that centralize control more completely, and certainly many of the senior managers in the firm discussed here were interested in the latter. Bryn Jones has shown how this might work. Direct numerical control (DNC) and flexible manufacturing systems (FMS) could be used in future years to override the operator–programmer conflict—for instance by using one central computer for the control of several machine tools simultaneously. This would eliminate the current contest between part-programmer and machinist by undermining the main tasks of both parties.[4]

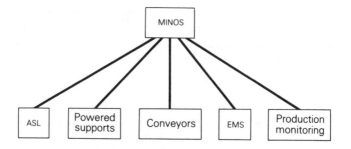

Figure 10.3   The new Mine Operating System (MINOS).

Events in machine shops are more complex than indicated, but the central political question has been raised: should the informal control of work exercised by machinists be developed as a viable alternative, representing a new tradition of skill enhancement, or should it be seen as a subversive threat and eliminated at the next technological opportunity?

Automatic coal cutting

Most processes in the engineering industry, when compared with coal getting, are relatively simple to automate. One of the reasons for this is because the workpiece—the piece of metal to be cut or shaped, plated, or formed—can be made to a consistent standard. Thus speeds, feeds, and movements of machinery can be the same for the whole of a batch of components. With mining, the "workpiece" is the coal face itself, and variations in underground conditions vary enormoulsy—even within the same pit. Thus, precise movements of, for instance, coal cutting machines may have to be different for each cut, and computer control is a far more difficult technical accomplishment.

After failing with ROLF (Remotely Operated Longwall Face) in the 1960s, the NCB has again taken up an interest since the mid-1970s, pouring millions of pounds into research and development in automation, but this time using the more sophisticated microelectronics technology. The new system being developed is called MINOS (Mine Operating System) and would, if successful, control and coordinate the movements of a variety of underground machinery, in turn fitted with microcomputers of their own (see figure 10.3).

On the coal face the center of attention is the Anderton Shearer Loader (ASL)—the coal cutting machine. Presently this and other machines (e.g. conveyors and hydraulic roof supports) are steered and controlled *manually*. The aim is to make the machines automatic. Experiments to control individual machines automatically are going on in pits at the moment. The ultimate aim of the integration of the whole system via MINOS may, however, be many years away yet.[5]

Motivations cited by the NCB in engineering journals fall into three main categories. First, dramatic improvements in labor productivity are expected. A 1979 Department of Employment study predicts jobs at the face will be cut from an average of 22 to 15 per shift,[6] though some engineers would consider this a conservative estimate. Second, improvements in working conditions and safety are stressed.[7] For instance, there will be fewer men underground; the ASL driver will no longer have to steer at the head of a machine in a constant spray of dangerous dust and water; and the MINOS package includes an Environmental Monitoring Subsystem (EMS), which can constantly and more accurately measure, for example, levels of methane gas and dust. Third, increased management control over production is expected. Managers at the pit top would now have detailed information on performance at their fingertips, and machinery movements would be remotely controlled from the

MINOS control room.[8] For remaining underground workers, then, it might be more a matter of keeping up with the machines, rather than using them as tools.

Presently face workers command control over the typical mechanized technology, and indeed the shift from shortwall to mechanized longwall techniques of coal getting in the 1940s and 1950s probably *increased* managers' dependence on the skills and knowledge of face teams.[9] With colliers currently trained in the operation of a variety of equipment, and production depending on the experience of those workers with at least several years' knowledge, "management control" comes mainly in the direct form of shift deputies. With automation under MINOS, the threat to workers' job control is obvious, particularly since the NCB has suggested that deputies will man the control rooms. I do not have the evidence to demonstrate that undermining the central position of face workers in coal getting is a central motivation for MINOS, but certainly this could be the consequence. Currently the majority of officials in the NUM were previously face workers, and this probably partly explains their solidarity and militancy—thus MINOS could have implications more far-reaching than job control and work satisfaction.[10]

So far there is little in the way of coordinated strategy on job control from the NUM, though at pits where the new MINOS technology is being introduced some local negotiations have already gone on.[11] For instance, it has been agreed at many pits that computerized production monitoring will not be used for disciplinary purposes or in calculations for bonus pay. Also, there have been agreements that, if and when deputies man the new surface control rooms, they must transfer, within six months, from NACODS (the supervisors' union) to the NUM.

Thus, from the limited discussions I have had with Nottinghamshire NUM branch officials, it is already obvious that job control is to be an important agenda item in bargaining as the new technology is introduced. This contrasts with the engineering industry, where the conflicts I have documented were more likely to go on at an unofficial, even "subterranean," level.

Automatic rubber molding

Another striking case of the tactic of *isolation* was explained to me by managers of a rubber molding company. Here managers had found unwanted working practices developing on new rubber molding equipment during its "productionization" (their phrase) on the shopfloor. As in the plating company, workers were able to become familar enough with the new machinery and control systems during their debugging to "work the system" to their own advantage. In this case managers were alarmed by this partly because the operators were at least as expert in machine control and methods as the time-study men, and could thus negotiate relatively high piecework rates for new jobs. Bargaining over this and related issues went on between management and trade union for many months before a system of work organization was finally thrashed out.

But, managers told me, "we've learned . . . a lesson" from this experience, and the latest wave of new production machinery, which was in preparation during my visits, was to be debugged in a newly built development area away from the shopfloor. This time, when machines went on to the shopfloor, they were to be ready straightaway for full-scale production, working practices already having been designed in detail by development and process control staffs. Here, then, is a case where the design of the factory itself was informed by existing shopfloor conflicts.[12]

Automatic manufacture of glasses

An instance that stood out as something of an exception in terms of management strategy was a small optical company. Here, a new managing director, bringing a strong philosophy of participation and involvement, had introduced a range of electronically controlled equipment and machinery. Instead of taking advantage of possibilities for deskilling, he instigated a system of job rotation, and ensured that the skills and knowledge involved in production remained on the shopfloor.[13] Managers justified this in terms of eliminating boredom and allowing work satisfaction, but also by reference to the productivity advantages inherent in maintaining a skilled and flexible workforce. New skills were certainly different, more resembling a "science" than a "black art," but this did not mean a new elite of technicians as in the previous case. The determination to maintain or upgrade the skills of shopfloor workers was confirmed by a recent recruitment of four apprentices—remarkable in a firm with less than 60 workers—and by the fact that the reason why one group of workers (the "surfacers") was not rotated was that they had so far refused to cooperate. They had been partly deskilled because of computerization, but they wished to maintain the remnants of traditional "craft" control of their own area of work.

Economic rationales

There still exists a tendency among social scientists to treat new technology as a wholly economic phenomenon, as if technology were somehow autonomous: because of its economic superiority, new machinery and techniques *demand* adoption, and the consequences are inevitable. Within this tendency there is a debate as to whether the so-called *impacts* of automation will bring an enrichment or a degradation of work,[14] but the logic is the same: managers are treated as the intermediaries or *messengers* in a rational process of technical change, and any attempts by workers to influence the new work organization are simply considered opposition or *resistance*.[15]

The most important aspect of this pervasive ideology for the purposes of this paper is its use within firms as a legitimation of actions and choices that are essentially political, often in motivation as well as consequence. In the plating company, for instance, it was questionable whether management's claim that a lower-quality product and higher scrap rate was tolerable merely *coincided* with the fact that automatic (management) control meant a loss of control for the platers. When managers were trying "to get it into their heads that automatic is the best way," they were using an apparently objective argument ("the best way") to wrest control from what they considered a problematic workforce. Prior to the introduction of new technology in the rubber molding company, management gained agreement from the workforce that "to compete and survive we must utilize the most efficient techniques," before using the technology to centralize control over production in the hands of white-collar staffs. Again, whether this effect is really "most efficient" is questionable.

It should now be obvious that what constitutes an "efficient" technology and an "efficient" work organization is debatable. Occasionally, indeed, there were competing definitions of efficiency, which began to make some sense in the context of the politics of the workplace. One instance was provided by the platers' emphasis on quality as opposed to managers' emphasis on quantity despite a high scrap rate. Another instance was in the machine shop described above, where both programmers and operators had developed elaborate arguments about the efficiency of the preferred methods during their battle for control over machinery.

It is vital to recognize that alternative arguments about efficient production organization were often generated by workers, and reflected their own specific interests. It is also important, however, to recognize here certain limitations. Most importantly, workers became involved only at the implementation stage of innovation, often after some constraints on the organization of work had already been imposed by management in the design and choice of technology. Even at the implementation stage, it was managers who were more likely to use the economic justification, and certainly they were the most expert in its deployment. Workers' use of the efficiency argument could thus be precarious, since faith in it might play into management hands. This is especially true where workers' responses to management designs are individualistic or "subterranean," as is often the case. I came across one instance where workers had obviously been taken in by the economic rhetoric and had been defeated. This was a group of progress-chasers in a machine shop, whose job control had been dismantled by use of a new computer-based system of shop scheduling. One resignedly summed it up by saying, "that's modernization, isn't it?"

In sum, arguments about "efficiency" can be used as a legitimating device for politically informed decisions on the technical and social organization of work. While this hardly made for a consensus on the shopfloor when it came to establishing practical working arrangements, it did help managers keep job control off the agenda in bargaining over change. With the remarkable exception of automatic coal cutting, in all the cases mentioned above official negotiations were limited to questions of payment, gradings, job protection, health and safety, etc. That is, they were limited to the conventional topics for collective bargaining, the fate of skills and work organization remaining management prerogatives—shrouded in the ideology of technical progress and efficiency. This general finding squares with other available evidence on official bargaining over technology, even where special procedures have been set up in the form of New Technology Agreements.[16]

The negotiated order

Students of industrial relations will be familiar with the notion that work organizations constitute a "negotiated order,"[17] in that working practices, and relations between workers, supervisors, managers, etc., become embodied in a set of unwritten rules. These unwritten rules—the custom and practice of the factory floor—influence both the type of technology that managers seek to introduce and the social consequences of that technology. Often managers feel that current custom and practice represents a lack of control on their own part, and try to remedy this by taking advantage of the opportunities lent by the new microelectronic technology.

In the cases described in this paper, this largely meant that managers would attempt to gain greater control over production by isolating and centralizing the skilled elements of work away from the shopfloor. Other less subtle—and more politically visible—means are occasionally employed. In two engineering factories I visited, for instance, electronic machine monitoring devices were used to secure management control over the pace and quality of production. In one firm, die-casting machines were fitted with digital readouts which displayed the speeds at which machines were running. Managers said that workers would previously run machines more slowly than the optimum rate, but now foremen could see any "shirking" at a glance while walking around the factory floor. Clearly, the electronic devices conferred *no technological advantage whatsoever*—it was only through an increase in *input* (increased worker effort) that an increase in output was gained. Similarly, a tire manufacturer was considering the installation of production monitoring devices to its tire-building machines, so that a range of information on machine speeds, downtime frequency, etc., would be relayed to a senior manager's office. However, workers and their trade unions had recognized the "big brother" implications, and managers were sceptical of the possibility of an unconditional acceptance. As in the case of the NCB described above, the firm would probably have to agree that *disciplinary* uses of computerized production monitoring would not be allowed.

In the retailing sector, where workers are relatively unorganized, electronic Point of Sale (POS) terminals are being used by the larger chain stores, with no opposition from workers. As with machine-monitoring devices, they mean that a closer check, for instance on the throughput of "checkout girls," can be made on workers. They also mean (incidentally?) that "sweethearting"—charging lower prices to family or friends—can be eliminated by passing magnetically coded labels attached to goods across a laser beam which "reads" the code and automatically records the correct price. As in the other cases, then, managers are attempting to build a "bias" into machines that change custom and practice to increase their own control.[18]

Where very large capital investments are involved, existing custom and practice can be *side-stepped* by introducing new equipment and machinery to "greenfield sites." An example of this was illustrated in a recent study where a food processing company set up a new factory several miles from the existing plant. Managers said they were using the new site as a "test bed" for their new ideas on work organization, and "green labor" without working practice traditions in the food industry were deliberately selected.[19] Similarly it has been suggested that a separate factory for the new BL Metro line at Longbridge in Birmingham was an important factor in the ability of managers to impose new working practices of their own design.[20]

All this goes to show that, in practice, industrial innovation is a political process, carried on by people with specific interests and in particular positions of power. Inevitably, it is intended to serve the perceived interests of specific groups of people—albeit not always successfully. From what I have described it is obvious that any attempt by *managers* to humanize work, or to maintain and extend workers' skills and control over production, will require a major shift in their political perceptions and actions. Beryl Tipton has summarized the political qualities of job restructuring:

A significant policy to improve the quality of working life even at workplace level is inevitably a *complicated social and political undertaking*: one that could not be accomplished without a *commitment to the value of equality and a greater belief that talent is quite widely distributed* among the population than is apparent from current practice.[21]

She is pessimistic, suggesting that it is "hard to believe that substantial changes involving the erosion of management's own position . . . would be initiated by management itself."[22] Thus, the difference between managements' recent experiments with new "humanistic" forms of work organization and those that could be demanded by workers might be distinguished by the concepts of workers' control and industrial democracy.[23] Certainly many of the managers I met treated work organization as a question of who controls production, and only rarely did they recognize and accept workers' rights to fulfill their own aspirations for control.

Craftsmen: computer-aided or computer-degraded?

Within the broad traditions of the deskilling and degradation of work, among the details of working practices, we occasionally found alternative methods. Sometimes they were becoming "institutionalized" to represent a different tradition of skill enhancement and worker control, an example being the case of operator programming of machine tools. Redirecting technology along more human lines is a matter of nurturing these new traditions, rather than allowing them to be eliminated at the next technological opportunity.

Enid Mumford, among many others, has pointed out that skilled machinists could be formally trained in programming, so that the whole machining process would remain with them.[24] A similar argument would hold true for most other production processes. Howard Rosenbrock has demonstrated that the old and new skills are compatible, and he sums up the possibilities in his image of the "computer-aided craftsman."[25] Indeed, the case studies above provided instances where individual machine operators trained themselves in programming and process control and the recognition and development of these practices could be a simple matter.

The problem is that trade unions, and particularly local stewards' organizations, would have to begin to demand certain rights to workers normally seen as management's prerogative: to exercise and develop skills; to define what constitutes a "good job"; and to formally participate in the choice and use of new technology. As we saw, the mineworkers' union is having a significant influence over the use of new technology, though even here the influence is mainly a "rearguard action," coming after major technological decisions have already been taken. Most other British trade unions, on the available evidence, have even more ground to make up.

In the meantime, workers will continue to seek to exercise their skills and maintain a degree of control over their work, albeit largely in subterranean and unorganized ways.

Notes

1 B. Wilkinson, *The Shopfloor Politics of New Technology*, London, Heinemann, 1983.

2 See, for instance, H. Shaiken, *Computer Technology and the Relations of Power in the Workplace*, Berlin, International Institute for Comparative Social Research, October 1980. Shaiken comments on a case where operators would seek control in order to adjust feeds and speeds. This could increase bonus pay. In the case discussed here, in contrast, craft workers justified maintaining control in terms of "work satisfaction," "self respect," etc.

3 An economic justification of MDI machine tools is provided by the UMIST researchers. See J. Boon, et al., "The Development of Operator Programmable NC Lathes," unpublished paper, University of Manchester Institute of Science and Technology, 1981.

4 B. Jones, "Technical, Organisational and Political Constraints on System Redesign for Machinist Programming of NC Machine Tools," paper presented to IFIP conference on "*System Design—For, With and By the User*," Riva del Sole, Italy, September 20–24, 1982.

5 For an overview of these developments see V.M. Thomas, and K.W. Chandler, "Monitoring and Remote Control—Progress Towards Automation in British Coal Mines, 1977," *Mining Engineer*, October 1978, pp. 251–64.

6 Department of Employment, *Manpower Implications of Microelectronic Technology*, London, 1979.

7 See, for instance, P.G. Tregelles, "A Mining Engineer Looks at Sixty Years of Change—and at Where do we go from Here?" *Mining Technology*, October 1980.

8 See P.G. Tregelles, "The Contribution of Measurement and Computation to Improved Control and Management of Mechanised Mining Systems," *Mining Engineer*, October 1981, pp. 199–206.

9 For an account of the social dimensions of this transition, see E.L. Trist and K.W. Bamforth, "Technicism: Some Effects of Material Technology on Managerial Methods and on Work Situation and Relationships," in T. Burns, (ed.), *Industrial Man*, Harmondsworth, Penguin, 1969, pp. 331–58.

10 At least one commentator does see MINOS as explicitly designed to attack the industrial and political strength of the mineworkers. See Conference of Socialist Economists, Microelectronics Group, *Microelectronics: Capitalist Technology and the Working Class*, London, CSE Books, 1980, chapter 10. Detailed research in the mining industry is needed to test this thesis.

11 These negotiations were related to me during seminars for Nottinghamshire branch officials organized by the NUM. Further study will doubtless throw up many more instances.

12 Stephen Marglin has gone so far as to suggest that the factories arose in the first place out of considerations of control over workers. They conferred no *technological* advantages, he argues, but meant that tighter supervision and discipline could be employed, and that embezzlement—a major problem for employers with the putting-out system—could be virtually eliminated. See S. Marglin, "The Origins and Functions of Hierarchy in Capitalist Production," in T. Nichols (ed.), *Capital and Labour: Studies in the Capitalist Labour Process*, London, Fontana, 1980, pp. 237–54.

13 This case raises the interesting question of whether smaller firms in general are more likely to go for a "decentralized" rather than a "centralized" labor control strategy. My own case studies of small firms were too limited in number to test this hypothesis, and instances of both strategies were uncovered.

14 For an optimistic view see R. Blauner, *Alienation and Freedom*, Chicago, University of Chicago Press, 1964. Davis and Taylor have been even more optimistic in referring to automation as "post-industrial technology"! See L.W. Davis and J.C. Taylor, "Technology, Organisation and Job Structure," in R. Dubin (ed.), *Handbook of Work, Organisation and Society*, Chicago, Rand-McNally, 1976. For a pessimistic treatment see J.R. Bright, *Automation and Management*, Boston, Harvard Business School, 1958.

15 For a fuller discussion of this approach to technical change, see Wilkinson, *Shopfloor Politics*, pp. 9–12 and 17–19.

16 For an overview of the official union response to new technology in Britain, see G. Bamber and P. Willman, "Technological Change and Industrial Relations in Britain," *Bulletin of Comparative Labour Relations*, Summer 1983. An account of the contents of Technology Agreements can be found in K. Robins and F. Webster, "New Technology: a Survey of Trade Union Response in Britain," *Industrial Relations Journal*, Spring 1982, pp. 7–26.

17 The concept of the negotiated order is elaborated in A. Strauss et al., "The Hospital and its Negotiated Order," in G. Salaman and K. Thompson (eds,), *People and Organisations*, London, Longman, 1973, pp. 303–20.

18 A fuller account of the social and political dimensions of POS can be found in J. Cosyns, R. Loveridge and J. Child, "Microelectronics, Organisation and the

Structuring of Employment in Retailing," paper presented to the EGOS/SSRC Colloquium on "*Organisational Innovations in the 1980s*," Strathclyde Business School, Glasgow, March 29—April 2, 1981.

19 More details of this case study can be found in B. Wilkinson and S. Smith, "Management Strategies for Technical Change: Two Case Studies from the Food Processing Industry," *Science and Public Policy*, April 1983.

20 A. Francis, M. Snell, P. Willman and G. Winch, "Management, Industrial Relations and New Technology for the BL Metro," Department of Social and Economic Studies. Imperial College, Mimeo, November 1982.

21 B. Tipton, "The Quality of Training and the Design of Work," *Industrial Relations Journal*, Spring 1982, p. 36; emphasis added.

22 Ibid., p. 36.

23 For a recent discussion see H. Levie and R. Williams, "User Involvement and Industrial Democracy: Problems and Strategies in Britain," paper presented to IFIP *Conference on "Systems Design—For, By and With the Users*," Riva del Sole, Italy, September 20–24, 1982.

24 E. Mumford, "The Design of Work: New Approaches and New Needs," in J.E. Rijnsdorp (ed.), *Case Studies in Automation Related to the Humanisation of Work*, Oxford, Pergamon, 1979, pp. 9–17.

25 See H.H. Rosenbrock, "Automation and Society," paper given at the Workshop on "*Feedback and Synthesis of Linear and Non-Linear Systems*," Bielefeld/Rome, June 22, July 3, 1981.

Using the New Technology

David A. Buchanan

You can't predict the impact of new technology on the quality of work simply from a knowledge of its technical features. So much depends on why and how it is used. Management has a choice of means and ends, which should be clearly identified, says the author, who is a lecturer in organizational behaviour at the University of Glasgow, Scotland. This article appeared in the European Management Journal, *vol.1, no. 2, 1982. A full report on the research appears in David A. Buchanan and David Boddy,* Organisations in the Computer Age (*Gower Press, Aldershot, England, 1983*).

In his article for the first issue of the *European Management Journal*, Roy Jenkins discussed the challenge of innovation and the lack of success of Britain and Europe in introducing new technologies. The economics of world trade, he argued, are such that our attitudes toward preserving industries in decline and our reluctance to change are no longer appropriate. Innovation, Jenkins claimed, is "unwelcome because uncomfortable."[1]

It is widely assumed that we must employ new information and computing technologies to remain competitive in world markets. It is also generally accepted that, because microprocessor technologies are widely applicable, faster, smaller, cheaper, and more reliable, their application will be inevitable, rapid, and beneficial. A report by the British Advisory Council for Applied Research and Development (ACARD) concluded in 1979 that, "The rate of technological innovation in United Kingdom industry will need to increase if its products and manufacturing processes are to match those of our major competitors. This is a necessary condition of our future survival as a trading nation."[2]

Developments in electronics and computing have thus introduced a new technological imperative into the climate of innovation in industry for the 1980s. Survival depends on rapid and effective technical change.

Success is not guaranteed simply by adopting the latest electronic gadgetry. The research shows that popular predictions about the radical and automatic advantages of new technologies are exaggerated. The aim of this article is therefore to identify the factors that contribute to the effective application of information and computing technologies.

The article aims to show that:

1   it is not possible to predict the organizational consequences of technical change from a knowledge of technological features and capabilities;
2   the impact of technical change depends on why and on how it is used;
3   management has a choice of means and ends in the use of new technology;
4   these choices can and should be identified and evaluated in advance of change.

These conclusions are based on research concerning the introduction of seven different types of new technology by six companies in central Scotland (see table 10.1).[3]

Table 10.1 The introduction of new technology in Scotland

Company	New technology
Caterpillar Tractor Company	Numerically controlled machine tools
	Computer coordinate measuring machines
Reiach and Hall	Computer-aided architectural drafting
Govan Shipbuilders	Computer-aided lofting
Y-ARD	Word processing
United Biscuits	Computerized equipment controls
Ciba-Geigy	Computer-aided process control

The managers who introduced these technologies were interviewed to find out why and how they had done so. The opinions recorded in this article thus reflect the thinking of the managers on the research sites at the time of the study and do not necessarily reflect the current policies of these companies. Workers who used the new technologies were also interviewed and observed to find out how their work experience had been affected.

The reasons for change

The research showed that different levels and functions within management had different expectations of the new technology. Their reasons for using it fell into three broad categories.

Strategic objectives

Caterpillar wanted to upgrade its machining and inspection equipment to meet changing market conditions. Govan Shipbuilders wanted to reduce the lead time between completing a ship design and starting production. Reiach and Hall wanted to attract customers by being market leaders in the use of new technology in architecture. United Biscuits has a company policy of maintaining investment in modern equipment to reduce costs and improve product quality. Ciba-Geigy wanted to increase capacity and give customers improved product consistency, quality, and price.

These are the external, economic, market, and customer-orientated objectives to which Roy Jenkins and ACARD refer.

Operating objectives
Caterpillar, United Biscuits, Reiach and Hall, and Ciba-Geigy wanted to improve the consistency and quality of their products and to reduce costs. Caterpillar and Y-ARD were replacing obsolescent equipment. Caterpillar also wanted to overcome bottlenecks in workflow created by other changes. Govan Shipbuilders wanted to computerize one part of the production process to be able fully to exploit other technical changes in the yard. Ciba-Geigy wanted to improve labor productivity, reduce production bottlenecks, and control energy use to cut plant running costs. Reiach and Hall also wanted to increase capacity, and improve the flexibility of their drawing process. Y-ARD was concerned with reducing numbers and costs of support staff. United Biscuits wanted to cut product waste.

These are internal, technical, performance-orientated objectives, clearly related to the achievement of strategic objectives.

Control objectives
Caterpillar, Reiach and Hall, United Biscuits, and Ciba-Geigy wanted to reduce human intervention and errors in the workflow. Caterpillar and Y-ARD wanted greater management control over the workflow. Reiach and Hall wanted to reduce staff coordination problems. Reiach and Hall and Govan Shipbuilders wanted to improve consistency in the transfer of information around the company. United Biscuits and Ciba-Geigy wanted full machine monitoring and control over key process variables. United Biscuits was working toward centralized, computerized production control, and wanted better and faster information about key aspects of production. Ciba-Geigy wanted to improve the recording and analysis of process performance information.

Control objectives are thus orientated towards:

1   reducing human intervention;
2   reducing dependence on operator control;
3   increasing the amount of production performance information;
4   increasing the speed at which information becomes available.

Strategic thinking was characteristic of senior management. Operating concerns were characteristic of middle-line and financial management. Control was the main concern of middle and junior-line management.

It is usually middle and junior-line management that determine *how* the technology is used. They decide how work is organized around the technology. The pursuit of strategic and operating objectives influences the initial investment decision. But the pursuit of control objectives influences the effects of technical change on the organization of work, operator skills, and performance.

The research demonstrated, however, that the pursuit of control objectives is not necessarily consistent with the achievement of strategic and operating objectives.

Technology, organization, skills, and performance

Technology can either *replace* human effort and skill, or it can *complement* it. The domestic washing machine and the industrial robot are examples of replacement. Musical instruments and conventional machine tools are examples of complementarity. Many devices do both; the car replaces the need to walk or ride, but makes demands on the driver to steer a route safely through numerous obstacles at speed.

Information technologies replace human manual skills. Computer coordinate measuring machines replace the conventional surface table on which machined components were inspected with gauges, callipers, and other manually operated devices. Computer-aided drafting replaces manual drawing skills.

Information technologies also, however, complement human information processing skills such as problem-solving and decision-making. But the organization of work around the technology must allow operators to develop and apply new knowledge and skills to use the technology effectively.

The following cases illustrate the relationships between technology, work organization, skills, and performance, and highlight the importance of choices of work organization. These and the other studies from this research are described in detail in the full report of the project.[4]

Case 1: Process control
Technology   Ciba-Geigy is a multinational manufacturer of specialist chemical products. The company's site near Glasgow had eight plants manufacturing pigments, and this study concerned one plant where modern computer controls had recently been introduced.

The plant made a range of pigment types on four continuous process lines. Two of these lines were water-based and reliable. The other two were based on alcohol and the product was corrosive, thick, greasy, hard to pump, and difficult to clear when it blocked. This equipment thus suffered more breakdowns than the water-based process equipment.

The process was run by one control room and one plant operator. Information on the state of the process was presented in the control room in various ways, including a video display screen, trend recorders, equipment status lights, digital displays, and a printer. The computer control loops were mostly automatic, but worked up to preset limits. If a process variable went beyond its set limit, an audible alarm sounded in the control room, and process control passed to the control room operator. If the operator's action was not effective, the computer shut down sections of the plant or the entire process, depending on the severity of the fault.

The control room operator adjusted the process through a keyboard, or by giving the plant operator appropriate instructions. The plant operator patrolled the plant and could visually inspect the process. The computer sensors could become fouled with the product and give inaccurate alarms. No alarm was given when items of equipment or sections of the process that were not directly monitored became faulty.

Work organization   The control room and plant operators were not supervised. The site management regarded the lack of supervision as an experiment and this was considered to have been successful. The operators rotated jobs daily.

Skills   The minimum training time for these operators was one year. They had responsibility and discretion, and their actions had a significant impact on the plant output. They retained an integral controlling role in the process and required a high degree of skill and knowledge to perform that role effectively. The computer controls were tools that helped them in their work rather than replacing them.

The operators required an understanding of the process, the product, the equipment, and the computer controls. The operators used their experience and judgment to override the computer controls when they knew they were not functioning correctly. Control was thus delegated, rather than allocated, to the computer. The operators also had to be able to explain past and future needs to management and maintenance staffs.

The number of interdependencies in the process was high. The speed of events was faster than with manually operated batch production. The operators were under pressure to respond quickly to problems. They rarely dealt with one process variance at a time. They had to interpret the *pattern* of information from the computer and from the visible and audible condition of the process, to be able to trace the source and cause of deviations and to decide what action was necessary.

The computer could not effectively control the process without skilled human intervention. The organization of work gave the operators experience that allowed them to develop their knowledge of the process and to use that knowledge to control the process with the help of the computer system. This is an example of *complementarity* between operators and computing technology.

Performance   The labor productivity of this plant was about twice that of comparable plants on the site that did not use computer controls.

Case 2: Word processing[5]
Technology   Text typed into a word processor is held in a computer storage medium (normally a magnetic disc) before it is committed to paper. Once stored, the text can be edited to correct spelling and presentation before it is typed. The same text can be printed repeatedly, perhaps with minor adjustments, from the stored copy without retyping.

Y-ARD is a marine engineering consultancy whose main product is technical reports for clients. These reports are typically lengthy, complex, have more than one author, and go through several revisions before release to the customer. This is an ideal application for word processing.

Work organization   When the word processing system was installed, the company grouped their 11 video typists in two word processing centers, taking them away from the small groups of report authors with whom they previously worked. There are no aspects of word processing technology that demand this organizational arrangement. Management wanted to be able to control more closely the flow of work to and from the typists. Each group had a supervisor, and a coordinator to oversee the group's activities.

Skills   The video typist has to learn computer file management procedures, and the editing codes for manipulating stored text. The job is therefore more skilled and demanding than conventional typing. The typist can experiment with alternative page layouts to produce documents that look better and communicate more effectively, and can demonstrate presentation styles to authors on the visual display screen before printing.

Performance   The output of the typists in pages per day was six times higher with word processing than with conventional typing. The number of typists employed was reduced. But the authors were dissatisfied with the system. They spent more time reading proofs than before, and the waiting time for reports had not been reduced. The typists liked the system as it produced better-quality output, but they were also unhappy with the reorganization. The proportion of the company's payroll that went on typing support staff did not fall. More managers had been appointed, each with a secretary, and the video typists were given a salary increment.

The typists did not know which authors they were working for, and so could not develop and apply their knowledge of individual authors' styles and preferences. Rather than check problems with authors, the typists would "have a go," as correction was easy. The typists also were not able to show authors alternative layouts for documents on the system. The authors never saw the typists, were not fully aware of the capabilities and constraints of the system, made unrealistic demands on the typists, and did not make full use of the system.

The reorganization that accompanied the introduction of word processing in this case gave management the control over typing that they sought. But it created a typing service that was not as effective as expected. How else could the work of the typists have been organized? The typists could have remained with the author groups to which they were already attached. But the pooling arrangement provided some training advantages, helped to cope with peaks in workload, and enabled them to share expensive printers.

Typists could have been allocated to specific authors or groups of authors while still located in the pool. This would have retained the contact between typists and authors and would have enabled typists within the pool to cover for each other and share peak loads.

This is an example of *complementarity frustrated* by work organization. A more effective service might have been created with a work organization that enabled typists to develop and exploit their knowledge of the capabilities of the word processing system, and of the needs of authors. At the time of the study, management was considering putting the typists back with their author groups.

Case 3: Biscuit packet weighing[6]

This and the case that follows are from the same company. Together, they demonstrate how some of the disadvantages of computing technology may be overcome, and how the complementarity effect may be enhanced.

Technology　Packets of biscuits have to be as close as possible to their stated weight to comply with consumer protection legislation, and to reduce the costs of packing excess biscuit. In 1979 United Biscuits began to replace electromechanical packet-weighing machines with microprocessor-controlled systems. The electromechanical checkweighter simply accepted packets that were at or above the required weight and rejected those below it, but could keep no record of packets weighed. The electronic checkweigher recorded the weight of every packet, removing the need for a half-hourly manual check on actual packet weights.

The weight of every packet was displayed on the wrapping machine, and the operator could, within limits, adjust the number of biscuits put into each packet to compensate when necessary. The packet weight information was also displayed graphically on a video display unit for the ovensman, and was updated every two minutes.

Work organization　The organization of work on the line was not changed when this device was installed.

Skills　The wrapping machine operator and ovensman now had more information than they previously had, provided rapidly, which enabled them to control the process more effectively. When packet weights strayed beyond the ability of the wrapping machine to compensate, the ovensman had to make appropriate adjustments to the oven controls. This is a complex task, as action to correct one feature of a biscuit, such as its thickness or "bulk," affects other features, such as moisture content and color.

The electronic checkweigher thus *complemented* the craft skills of the ovensman, and created a job in which he:

1 retained discretion to monitor and control the process;
2 got rapid feedback on performance;
3 had a good understanding of relationships between process stages;
4 was able to control the process more effectively;
5 had a visible goal that he could influence;
6 felt that the job had more interest and challenge.

Performance   The proportion of excess biscuit wrapped was halved.

The computer system produced production performance analyses which the ovensman did not get. Any errors by the ovensman appeared within two minutes on his own video display, which passing managers could see, and on analyses of production over each shift. The work of the ovensman thus become more "visible," as more people could see the effects of his actions faster.

The introduction of computerized biscuit packet weighing did not lead to any reorganization of work. Each line was manned by a dough mixer operator, a machineman who supervised the cutting of biscuits from the dough, an ovensman, a wrapping machine operator, and a packing team. There was some job rotation within each of these occupational groups, to cover for absence and breaks. The potential effects of changing this organization of work will be considered after the next case, as this would affect all of the operators on each line.

Case 4: Biscuit Dough Mixing
Technology   United Biscuits began to computerize their biscuit dough-mixing operation in the early 1970s. The computer controls replaced the mixing of dough in open vats supervised by craftsmen bakers or "dough-men." The recipe for each biscuit type was stored on punched paper tape. A recipe change required the programming and punching of a new tape. When the tape was read into the computer, the automatic controls fed the correct amounts of flour and water to the mixing vessels. The mixer operator (the job title that management preferred to "doughman") started the mixing machine on instruction from the computer, and emptied the mix into a hopper 20 minutes later when the process was complete. The mixing machine was fully enclosed. The mix was interrupted two or three times for the operator to add small quantities of other ingredients or "sundries" such as salt and other chemicals that were difficult to insert automatically.

In 1980, a microprocessor-based device called a "recipe desk" was being installed to replace the old computer system. Recipe adjustments could be made quickly and easily using small thumb wheels on the desk panel, and programming skills were no longer needed.

Work organization   When the mixing control computer was first installed, a computer room group was set up to program and operate it. Mixing was thus centrally controlled by this group from this room. The new recipe desks were also installed in the computer room.

The doughman used to hold a position of status; he supervised other manual workers and had considerable mobility in the factory. The mixer operator was now paid less than the ovensman, and worked alone at one location.

Skills   The computer controls deskilled the doughman's job, although the need for human intervention had not been removed. Previously, he could see, hear, and feel the dough as it mixed and could when necessary add ingredients to adjust the quality. This was no longer possible. The mixer operators:

1   had a poor understanding of the process and equipment;
2   could not visualize the consequences of their actions;
3   could not trace sources or diagnose causes of process faults;
4   developed no knowledge or skills that would make them promotable.

These job characteristics may be typical of work in "nearly automated" production systems, where traditional skills are replaced, and where the work experience does not equip the operator with the capabilities to carry out the residual but key functions effectively.

Performance   Management felt that dough output was more consistent with the computerized controls, but there were disadvantages. As the doughman's crafts skills were replaced:

1   specialist maintenance was required;
2   mixer operators became bored, apathetic, and careless;
3   mixer operators rejected responsibility for breakdowns;
4   management lost a source of supervisory recruitment.

The mixer operators sometimes forgot to add sundries, and bad mixes had to be dug manually from the dough hopper, a tedious and time-consuming task.

The computer thus reduced human control over the mixing operation, but did not replace the mixer operator. How could that job have been organized to overcome the disadvantages, while retaining the benefits of the new technology?

This could be achieved by redesigning the mixing job alone. The mixer operators could have been responsible for the computer controls. The introduction of the smaller, more flexible, and easy-to-use recipe desks made this option easier than with the conventional large computer.

But this could be achieved more effectively by redesigning the organization of work on the line as a whole. The mixer operators could have become part of a "line team," rotating jobs with the other operators on the line, including the ovensman. This would have enabled them to retain and use some of their craft skills and knowledge, develop knowledge of other stages in the process and of the functioning of computer controls, and overcome the boredom and isolation of the mixing job. Management in this way would have gained a more skilled, flexible, and motivated workforce from which to recruit supervisory management. But this option had not been considered.

It is interesting in this case to note that the range of management choice was constrained by organizational and physical structures created by previous decisions. The factory layout and the pay differentials would have made it difficult, but not impractical, to create line teams.

Using the new technology effectively

Information and computing technologies are not magic productivity cures. Performance depends on how work is organized around the technology. These cases demonstrate that there are choices of work organization that are not determined or constrained by the technology. Other current European research supports this finding.[7]

Work can and should be organized to take account of the complementary relationship between information technology and its users. But in these and other applications studied, management choices were constrained by a preoccupation with control objectives. The pursuit of control objectives inhibits the design of effective forms of work organization.

The implications of this research for the management of technological change therefore seem to be as follows.

1 Identify the reasons for technical change

This research has suggested that different management objectives affect different facets of the process of technical change. Investment in new technology may be supported by plausible strategic and operational aims. But in practise the technology is often used to achieve management control objectives, which this research has also indicated to be inconsistent with strategic and operating goals.

Change should be problem-pulled, not technology-driven. The introduction of new technology should be aimed at clearly identified strategic and operating goals. It may be useful in this respect to consider the various possible answers to the question, "who benefits?" from a particular technical change, and to consider in particular how the customer will benefit. It may also be advisable to consider in advance the performance criteria on which the effectiveness of the application will be judged.

Technological change is particularly uncomfortable where management feel that they are losing control over what they perceive to be their sphere of influence in the organization. The only case of resistance to technological change encountered in this research concerned management, who felt that they would lose control over the workflow in their sections with the introduction of new computerized machinery. The effective use of information and computing technology may, however, require that management reconsider or relinquish control objectives.

2 Identify and evaluate work organization options
Productivity is not dependent on technology alone. It requires people with the necessary skills, and with the incentive and motivation to develop and apply those skills effectively. Productivity is thus dependent on the organization of work.

A preoccupation with the replacement of operators and skills is inadequate. The ways in which new technologies complement human capabilities must also be considered. Technical change creates new opportunities for work reorganization, and the range of options should be identified and evaluated in terms of the reasons for the change in technology.

The optimum form of work organization is one that gives operators:

1   more and faster information feedback on performance;
2   meaningful, interesting, and challenging goals;
3   control over the workflow;
4   discretion over methods and task allocations;
5   opportunities to develop skill and knowledge through the work.

In a recent article in *The Economist*,[8] Norman Macrae presents a similar argument for setting up "intrapreneurial groups" within large organizations. These are autonomous groups with around ten members each which work in competition with each other. Word processing departments, for example, would act as "minifirms," providing services for their "customers" within the organization. Computing and information technologies make these kinds of organizational arrangements possible, through their abilities to capture, store, analyze, and distribute information.

3 Accept the changing role of management
Control of workflow, and workers, is traditionally the task of junior and middle-line management. The research evidence presented here suggests that workflow control should become the task of the operators of the new technology, if the capabilities of that technology are to be fully used. These technologies may therefore replace more managerial than operating jobs.

The main conclusion from this research is that management must relinquish control over workflow, in office and manufacturing operations, to create work systems that are strategically and operationally effective. This advice runs counter to traditional management attitudes about the reliability of employees, the organization of work, and the rights and duties of management. With information and computing technologies, it may not be effective to design work systems on the assumptions that employees are untrustworthy, that task fragmentation is more productive, and that managerial decision-making prerogatives must be maintained. The optimum form of work organization outlined above is based on opposite assumptions.

Management may hold on to traditional values and use the new technology ineffectively, or accept the change in attitudes and role required to exploit its capabilities. But the pressures to change on the attitudes and role of management may be irresistible, for three reasons.

First, competitive pressures will place higher premiums on the successful achievement of strategic and operating goals. The choices now concern *how* to use information and computing technologies, not *whether* to use them. The organizations that use these technologies successfully will survive.

Second, information technologies open up access to information in organizations. Information is a source of power and authority, and more people at all organizational levels will be able to get more of it. These technologies will therefore make inroads into areas of decision-making where management has traditionally claimed rights and prerogatives, and may lead to the development of new forms of worker participation.

Third, expertise is also a source of power. Users of information and computing technologies develop expertise and a "feel" for their functioning that supervisors who merely observe do not have. Often there are no visible operations to supervise. Many managers resist the pressure to learn basic keyboard and computer terminal skills. Management thus becomes dependent on the knowledge and skills of operators.

A concentration in management decision-making with new technology and its capabilities is narrow and inadequate. Technical change must be considered in the wider context of organizational change as a whole. The organizational changes that are required to make effective use of technical developments will be of much greater significance to organizational performance and the future role of management.

Notes

1 Roy Jenkins, "Welcome to the European Management Journal," *European Management Journal*, 1(1), 1982, 5–7.
2 ACARD, *Technological Change: Threats and Opportunities for the United Kingdom*, HMSO, London, 1979.
3 D. A. Buchanan and D. Boddy, "Working with Information Technology," *Management Research News*, 5(1), 1982, 12–13.
4 D. A. Buchanan and D. Boddy, *Organizations in the Computer Age: Technological Imperatives and Strategic Choice*, Gower Press, Aldershot, 1982.
5 D. Boddy and D. A. Buchanan, "Word Processing in a Marine Engineering Consultancy: Y-ARD," *Effective Computer Applications*, 1(1), 1981, 36–59.
6 D. Boddy, and D. A. Buchanan, "Information Technology and the Experience of Work," in L. Bannon, U. Barry, and O. Holst (eds), *Information Technology: Impact on the Way of Life*, Tycooly International Publishing, Dublin, 1982, 144–57.
7 J. Evans, "The Worker and the Workplace," in G. Friedrichs and A. Schaff (eds), *Mircoelectronics and Society, for Better or for Worse: A Report to the Club of Rome*, Pergamon Press, Oxford, 1982, pp. 157–87.
8 N. Macrae, "Intrapreneurial Now," *The Economist*, April 17, 1982, 47–52.

Guide to Further Reading

Apart from the books and articles already referred to in the four contributions above, readers will find interesting discussions of the deskilling and "labour process" arguments in the writings of Howard Rosenbrock (for example in *New Technology: Society, Employment and Skill*, Council for Science and Society, London, 1981) and Mike Cooley (especially *Architect or Bee?* Langley Technical Services, Slough, England, 1980).

Useful material also appears in:

D. F. Noble, "Social Choice in Machine Design: the Case of Automatically Controlled Machine Tools," in A. Zimbalist (ed.), *Case Studies in the Labour Process* (Monthly Review Press, New York, 1979).

Thomas B. Sheridan, "Computer Control and Human Alienation," *Technology Review*, October 1980.

Harley Shaiken, "Computer Technology and the Relations of Power in the Workplace," Discussion Paper no. 80–217 (International Institute for Comparative Social Research, Berlin, 1980).

Stephen Wood (ed.), *The Degradation of Work?* (Hutchinson, London, 1982).

Craig, R. Littler, *The Development of the Labour Process in Capitalist Societies* (Heinemann, London, 1982).

Arthur Francis, Mandy Snell, Paul Willman and Graham Winch, "The Impact of Information Technology at Work" in Liam Bannon, Ursula Barry and Olav Holst (eds), *Information Technology: Impact on the Way of Life* (Tycooly, Dublin, Eire, 1982).

Erik Arnold and Peter Senker, *Designing the Future—The Implications of CAD Interactive Graphics for Employment and Skills in the British Engineering Industry* (Engineering Industry Training Board, London, 1982).

David A. Buchanan and David Boddy wrote up two of their case studies (video typists and biscuit-making) in the *Journal of Occupational Psychology*, vols 55 (1982), pp. 1–11 and 56 (1983) pp. 109–19.

Barry Wilkinson and Steven Smith, "Management Strategies for Technical Change," *Science and Public Policy*, April 1983.

E. Fossum et al., *The Computerisation of Working Life* (Ellis Horwood/John Wiley, Chichester, England, 1983). On the work of the Norwegian Computing Center.

Automation and the Workplace: Selected Labour, Education and Training Issues (Office of Technology Assessment, US Congress, Washington DC, 1983).

The Impact of Advanced Information Systems (National Economic Development Office, London, 1983).

"Robotics: Future Factories, Future Workers," special issue of *The Annals of the American Academy of Political and Social Science*, edited by Robert J. Miller, November 1983.

Cynthia Cockburn, *Brothers: Male Dominance and Technological Change* (Pluto Press, London, 1983). On skills in the UK printing industry.

Rosemary Crompton and Gareth Jones, *White-Collar Proletariat: Deskilling and Gender in Clerical Work* (Macmillan, London, 1984).

11 Management and Labor: the Organization of Work

Robots at Work

Fred K. Foulkes and Jeffrey L. Hirsch

Managers who successfully introduce robots carefully select their sites, move slowly, retrain displaced workers, and educate and keep informed both line managers and the unions. These are the common-sense conclusions to emerge from the authors' case studies of US companies going over to robots. Foulkes and Hirsch are, respectively, professor and junior fellow at the Human Resources Policy Institute, Boston University, and this article first appeared in the Harvard Business Review, *January–February 1984.*

In Japan, where companies now use more than 30,000 robots, workers are beginning to seek protection from the advance of automation. In the US, where roughly 7000 robots are in use, we have yet to see much resistance. Most US workers have not been sorry to relinquish the hot, heavy, hazardous, and frequently boring tasks that robots perform. But the advent of more and "smarter" robots, capable of performing sophisticated, interesting tasks, may introduce a new era of conflict between employees and management. In at least one US auto manufacturing plant, sabotage of a robot installation has already occurred, and a study on the use of robots by a large corporate leader in robotics predicts that "workforce unrest and sabotage will be characteristic as people attempt to prove the new technology will not work."[1]

Fear of loss of jobs is at the root of workers' objections to robots. Estimates vary on the extent of displacement of workers by robots; at present, one robot does the work of anywhere from 1.7 to 6 employees. Some researchers estimate that the new generation of "smart" robots, equipped with rudimentary vision or tactile sense, could displace as many as 3.8 million workers.[2]

In the long run, normal attrition and a shift of job opportunities to nonmanufacturing industries may offset the unemployment such automation causes. In Japan, for example, from 1970 to 1981 employment in manufacturing grew by only 70,000 workers, while employees in the retail, wholesale, and service industries increased by 5.2 million. Many experts believe that with planning and training by both the federal government and industry, the transition to new jobs will be manageable. By 2025, they say, most current operators will have left their positions, the robotics industry itself will provide some new jobs, and people entering the labor force will acquire skills in jobs that won't be automated.[3]

But this transition is still in the future. For now, managers will have to alleviate employees' fear and mistrust of technology that brings the fully automated factory closer. Thoughtful consideration of the issues that robots raise and careful advance planning of policies concerning them are critical to the integration of robots into factories.

During the past three years we visited many US companies that use robots. Their experiences as they progressed from decision to installation yield valuable lessons for human resources managers who are helping to plan the automated factories of tomorrow. In this article, we explore the personnel issues these companies faced when introducing robots and some of the approaches they adopted. Although we focus on robots, many of the issues we raise and recommendations we make are applicable to other forms of automation.

Deciding to use robots

In late 1981, General Motors Chairman Robert B. Smith noted: "Every time the cost of labor goes up $1 an hour, 1000 more robots become economical."[4] Ninety-five percent of the managers we visited cited the desire to reduce labor costs as their primary reason to use robots.

Robots are especially appealing to management in some heavily unionized industries where hourly wage rates are very high. In the auto industry, for example, while an employee costs the employer between $23 and $24 per hour, including benefits, an industrial robot performing the same job costs approximately $6 per hour, including purchase and maintenance. Under these circumstances, managers understandably are attracted to robots. GM alone expects to invest $1 billion to install some 20,000 robots in its facilities by 1990.

Overemphasis on short-term profitability can, however, jeopardize a robotic project from the start. Pressure to show a satisfactory return to headquarters may cause managers to overlook workers' interests. In a Carnegie-Mellon study, one manager observed that, although he had wanted to install robots in settings where dust and fumes endangered workers, the application he was considering did not show the direct labor savings that corporate headquarters was seeking.[5]

Most companies demonstrate concern for workers by assigning the first robots used to the three Ds—dull, dirty, or dangerous tasks—or the three Hs—hot, heavy, and hazardous jobs. One manufacturing company we visited developed robots for the express purpose of solving problems related to difficult tasks resulting in injuries to employees' wrists. Now, after having used these applications in its production facilities for two years, worker compensation claims have decreased dramatically.

In addition to reducing labor costs and eliminating hazardous jobs, robots can increase both the quality and quantity of output. At General Electric's Louisville plant, computer-aided engineering and robotic equipment enable the company to produce 20 percent more dishwashers in 20 percent less floor space. At the GE locomotive facility in Erie, 68 machine operators formerly took 16 days to produce a locomotive frame. Now, with the help of several robots, 8 unskilled workers do the job in 1 day. At some warehouses of Walgreen's, the drug chain, where packers handled 110 less-than-case-load shipments per hour, robots now load 900 per hour.

Managers who have successfully adopted robots have been careful to select a simple first application. "It's critical to establish a positive track record and build employee confidence with a first installation. Be sure it's going to work," advises one manager. In locations where robots didn't work and had to be dismantled, constant breakdowns undermined morale and destroyed the employee enthusiasm that characterizes successful projects.

The automation installation at one printing company demonstrates the effects a technical failure can have on employees. This company operates a plant with 400 employees in Dunkirk, New York. In late 1981, the company put in a robot that was supposed to grab loose newsprint sheets as they came off a printing press. When the robot arrived, employees were curious. They expressed concern over potential job loss and openly predicted that the robot "would never be able to do this work as well as a man."

They were right. For more than six months the company tried to debug the robot on the factory floor. In all that time no employee suggested any way to make it work. Finally, in August 1982, the company removed the robot. According to local management, employees, feeling that their mistrust of the robot was vindicated, are unlikely to welcome further experiments. Had the company chosen a simpler task for its first robot, workers might have responded with enthusiasm.

Today, between 70 and 80 percent of robots are retrofitted and integrated into existing production lines. Yet evidence indicates that employees are more likely to welcome this technology if it is dedicated to a new product or is a part of a new process—in short, when the robots do not threaten their jobs. Whereas workers who man robots in new processes are generally volunteers, employees in existing production lines, who are retrained or transferred, may feel that they are being forced out.

Even if no workers are displaced, retrofitting robots can cause morale problems. A study by Westinghouse Electric Corporation of its robot installations found that, in plants where a robot was installed alongside outdated manual equipment, its struggle for survival was uphill and frequently unsuccessful. The message the company sent to the workers and their supervisors was clear: "We have not listened to your pleas to repair our equipment—we have spent our money on a robot instead."[6]

Ideally, automation should be introduced as a part of plans to increase capacity in response to growth in product demand. The experience at IBM's Dayton, New Jersey, plant is a case in point. In 1981 the company decided to use robots to manufacture a product for which it had formerly subcontracted. Management announced six additional job openings requiring experience in multi-station production with robot involvement. Because no employee would be displaced, workers viewed the arrival of the assembly robot as an opportunity and 40 employees expressed interest in the new positions. Today the robotic operation produces 300 products per hour, compared with the 80 to 100 per hour the company acquired under the subcontract, and employees haven't reported any dissatisfaction with the process.

One domestic auto plant also chose to apply robots in a new product line. When the company began automated manufacturing of a new engine in a plant with 15,000 employees, workers were eager to participate. "Instead of feeling threatened, people felt challenged," a personnel director remarked. He added, "The robots weren't replacing workers but creating a need for more human talent." TRW, GE, and Westinghouse have also reported positive employee responses when introducing robots while expanding capacity.

Clearing the way

After managers have selected a site, they need to make a smooth introduction of the robots. Success often depends on paying special attention to the problems of displaced workers, gaining line management support for the change, and educating employees in the use of the equipment before installation.

Developing a displacement policy

Prudent managers take time to make long-term forecasts of job requirements before widely introducing robotics into their operations. It is important to try to predict where and what types of jobs, skills, and experience will be required. Long before implementation, managers need to identify which jobs robots will perform.

A no-layoff policy obviously contributes enormously to workers' acceptance of the machines, and in most companies we visited, attrition and retraining have sufficed to prevent layoffs. For auto manufacturers and others whose robotic projects are part of a fight for survival, however, attrition and retraining may not be enough. To help make a no-layoff policy practical, some companies anticipating many robotics applications offer voluntary early retirement to employees and limit the number of new permanent employees by hiring part-time and temporary employees, retirees, and former employees as necessary.

If layoffs do become unavoidable, a company needs to determine what severance or salary continuation benefits to offer. Well in advance of the installation, managers at all levels must be prepared to articulate, defend, and discuss layoff policies. We saw one very successful robotic installation that caused the layoff of 14 workers. Enthusiasm for the robots nevertheless remained high for several reasons: the United Auto Workers local had never opposed automation; the company operated in a depressed area where workers saw money spent on modernizing equipment as a commitment to keep existing plants open; and the robots were an engineering triumph, with virtually no technical problems.

Job analysis is essential to planning the transition from manual to automated operations. We found that companies with successful installations analyzed both the skills of the employees involved in the existing procedure and those required to run a robot. Often the job requirements are much different.

One aeronautical manufacturer found that the manual drilling and routing job the company was automating required the driller to have excellent perceptual and motor coordination, whereas a robot operator would need cognitive and communications skills. As monitor and controller, the operator would have to dictate the robot's processes and communicate well with programmers, maintenance staff, and others in the system.[7] And, of course, the operator is not the only worker whose job requirements change as a result of automation.

The need for robot operators, technicians, engineers, and programmers will also create opportunities for retraining. Robert U. Ayres of Carnegie-Mellon University predicts that "in the next decade most robot technicians will be recycled workers previously working in the same production area where the robots are now working."[8] GM and other companies are now training many of these specialists at company expense. Many companies are also cooperating with local institutions to educate a new generation of engineers, computer scientists, and technicians.

The work of maintenance personnel, in particular, changes when robots are used, for it often requires both electrical and mechanical skills traditionally considered separate crafts.

Gaining line management support

A GE study of worker acceptance of robots found management a greater roadblock than hourly workers. Westinghouse also reported that line management was sometimes the neglected stake-holder and that the supervisor was often overlooked, received no training, and was not kept abreast of events. At one successful installation, one such supervisor confided that he had felt left out when he was not included in discussions about how the robot would affect his operations. More prudent management at another location sent all its supervisors to the robot manufacturer for training.[9]

Because every department is likely to be affected in a facility where robots are put in, management needs to identify the critical interfaces ahead of time. For example, preventive maintenance of robots' sensitive electronic and hydraulic systems is the most effective way to reduce breakdowns. In one installation where managers neglected to include the maintenance supervisor in planning discussions, the supervisor felt he had no stake in the project's success and allowed the maintenance schedule to slip. Eventually, the manual backup system took over production.

Such an oversight is less likely in a company that often changes its manufacturing processes. Here, the people who need to work together and to be informed about robots' operational needs are already in frequent contact.

Educating employees

Most managers we talked with said that the education programs preceding a robot installation should be thorough and involve the largest number of employees possible. In most cases, line management casually informed employees that a robot would be purchased well in advance of installation. After word spread, managers confirmed the news by holding a general meeting and, when necessary, advising the union. (In some cases, companies informed the union earlier.) Workers who would be directly involved received training and were often kept informed of developments.

Westinghouse has perhaps the most elaborate and extensive employee education program. According to the company, the probability of success of a robot installation is directly proportionate to the number of affected people who are involved in the discussions and planning of the project.[10] Management at Westinghouse has found it essential to develop what it calls a "sense of ownership" in all employees who will be involved with robots.

One Westinghouse division used an ideal education and information plan. After spreading word of the coming robot through informal channels, local company officials met with the union leadership and formally advised it of the company's decision to install a welding robot. At this first meeting, division officials stressed the company's need to remain competitive and maintain market share. Local management described the union response as "curious and concerned." Union officials wanted to know how many employees the machine would displace. Company officials stressed their commitment to their policy of not displacing employees because of technological change and said that any reduction in the work force would come through attrition. (The company and the local union had a mature and constructive bargaining relationship, with no strikes since 1950 and no layoffs since 1972.)

The next step was to show union leadership videotapes obtained from the manufacturers of working robots. Management informed union officials that the welding robot was being tested at the corporate R&D laboratory prior to installation and suggested that the company be permitted to choose one employee to learn how the equipment functioned by working with R&D at corporate headquarters. The union approved the idea, and a young welder spent six weeks working with the robot while it was being tested.

What is a Robot?

Robots are combinations of sophisticated microelectronic technology, usually involving a computer, and various mechanical devices. The Robot Institute of America defines them as reprogrammable, multifunctional manipulators. Others describe them as automatic devices that perform functions ordinarily ascribed to human beings and that operate with what appears to be almost human intelligence.

Robots can be programmed to perform numerous tasks such as welding, assembly spray-painting, lifting, dipping, inspecting, and carrying parts. The robot frame is a substitute for the human arm and its microprocessor takes the place of the human brain. Robots help make flexible manufacturing systems possible.

Robots may be equipped with different kinds of arms. One design has the equivalent of a shoulder, which is able to bend and swivel, and an elbow capable of bending. Although flexible, this robot arm lacks the ability to handle heavy loads. Other designs extend out from a central pillar and can swivel around while moving up and down the pillar. Most of the robots being used in the US have appendages that can grab, hold, twist, and turn whatever product is being manufactured.

Some robots have the ability to "feel," which enables them to grab a part and, merely by touching it, distinguish it from other parts. Other more advanced robots are able to "see" and distinguish objects. Some assembly robots have as many as 22 finger "hands."

Many different types of robots are in use, and many more are being developed in research labs. These vary greatly in cost. Generally, the least expensive robots are pneumatically driven. These are light and move quickly but have little strength. They usually sell for $8000 to $20,000. Hydraulically driven robots, which use fluids to move the limbs, are more powerful and more costly. Electronically driven robots are the most expensive of all, costing $80,000 or more.

Throughout this period, company officials answered employee questions and listened to employee suggestions concerning the robot. Management tried to keep the union and employees informed of all important developments and stressed the positive results of the technology for workers.

The final step in this Westinghouse division's plan is the most impressive. Senior local management and senior union officials flew to Japan for a two-week tour to view robots at work in plants. One purpose of the trip was to show union leaders the competition facing US industry in general and Westinghouse in particular. According to the company, the trip reinforced the union's understanding that the company's future depended on an increase in productivity through robots and other new technology.

Managers at a different Westinghouse plant came up with another novel and, to the company's mind, effective way to introduce a robot. When the machinery was ready to be moved from the on-site research room to the shopfloor, management decided to put on a show. An engineer programmed the robot to demonstrate its abilities, including some production activities performed in the plant as well as other skills such as pouring water out of a bottle into three glasses. The trainer spiced the presentation with bits of humor.

During the last of the 11 performances, which was put on for the union committee, what could have been a public relations disaster turned out to be a blessing. The robot malfunctioned; its heavy arm smashed onto the table, knocking over and breaking all the glasses and other props. Everyone laughed. The manufacturing manager's embarrassment gave way to gratification. "The employees saw that it was only a stupid machine and could make mistakes," he said later. "If I had to do it again, I would run that mistake right into the robot's program."

Like Westinghouse, some domestic auto makers educate their employees extensively. At one auto plant we visited, which employs 5500 people, management informed the UAW bargaining committee some eight months in advance that the company would begin to install 11 materials-handling, painting, and assembly robots in September 1981 on a new production line. The company showed several films on robotics to union officials first and later held a meeting for all plant employees. The workers' immediate reaction was concern over job loss, which later dissipated when only three robots were actually installed. And some displaced workers expressed satisfaction with their new jobs: one ex-machine-loader enjoys feeding parts to the robot instead of being tied into the machine cycle.

Local management at the auto plant estimates that, since the first installation, 70 hourly and salaried employees have attended technical courses about robots on company time. These programs aim to educate employees and to encourage them to feel involved in the success of the robotics operation. Executives who seek to instill a sense of ownership of robots in their employees argue that involved employees often come up with valuable suggestions. Westinghouse cites the case of a newly assigned robot operator who pointed out that the robot was being installed directly over a joint in the concrete floor. Moving the location a few feet avoided a possible problem later as the building settled.[11]

Some companies, however, have introduced robots successfully without formal announcements or meetings. One large electronics producer, which operates a nonunion plant, relied wholly on informal channels of communication to educate employees. Six months before the first robot was assembled in May 1979, line management informally advised the 500 employees of the decision to use a robot. The company proceeded with its plans and a year later installed three more robots. In September 1981, the company added four more. Local management reports that employees have not voiced any concerns about displacement or job security. One company spokesman said that the company's growth (in the two-year period, the number of employees had risen by 60 percent) and the fact that employees saw no need to worry about losing their jobs explained the ease with which robots were accepted.

Managers who do not use formal announcements or elaborate education programs think that there is no reason to exaggerate the importance of robots; they see them as just a part of ongoing automation. One transportation company we visited, which has developed its own robotic systems for the sole purpose of eliminating hazardous jobs, is undertaking a remarkable test of this approach. As a rule, foremen informally advise only the employees affected by a robot some three months in advance of installation. Later, personnel and engineering staff show slides, explain why the company is automating, and discuss safety precautions with those employees who will work with the robots.

Management of this company is now installing robots in four different locations at roughly the same time, and at three of the plants will inform employees in the usual manner. At the fourth plant, however, managers have told no one, not even the foreman, about the machine. At present, an unexplained hole is in the floor where the system will be placed, and rumors of job loss are rife. The company wants to find out how serious worker unrest will become and whether workers will file a grievance. While such an experiment might be interesting, the risks are high and seem hardly worth taking.

Despite the success of those companies that have chosen to educate only a handful of employees in an informal manner, it is in the best interest of management anticipating future large-scale use of robots to educate as many employees as possible and to sell the idea of robots before they arrive.

Working with the union
In their planning for robots, unionized companies are often constrained by contract clauses. A recent analysis of a representative sample of 400 union contracts shows that 21 percent restrict technological change in some way.[12] More than half of the restrictive contracts contain clauses obliging the company to give the union advance notice of plans to use new technology.

In the interest of maintaining a cooperative bargaining relationship, the conscientious employer should not wait for the union to request advance notice of new technology but should voluntarily extend this courtesy. Moreover, union officials could be embarrassed if management suddenly installed robots and they had no knowledge of such plans.

Unions seek to protect their members from displacement. Of the contracts that restrict technological change, 13 percent oblige the company to make an effort to retrain displaced workers, 18 percent require retraining, and 10 percent require retraining only if the employee qualifies for the new job.[13] Although such provisions reduce employee anxiety and fear, they restrict management flexibility and are unlikely to gain wide acceptance in the near future.

Unions often urge management to form a joint committee on technological change. In such proposals, union officials usually include provisions requiring union involvement in designing and planning the adoption of new technology as well as taking part in decisions regarding its introduction. This requirement places obvious constraints on management, which explains why most companies have not accepted such proposals.

A joint advisory committee on technological change does seem, however, to function well at several companies. It meets regularly to discuss the latest developments in technology and how the company plans to use automation. Such a committee allows labor and management to learn the issues and problems each considers important. In 1979, GM and the UAW formed a joint national committee on technological progress and continued to employ it in arriving at their 1982 agreement.

One important union concern is the potential reduction in bargaining unit size. Union officials often question the bargaining unit status of employees who will program or maintain robots. They want to maintain the integrity of the bargaining unit and expand it to include these employees, while employers may want to exclude them. (In the 1982 contract negotiations with GE, the International Union of Electrical, Radio, and Machine Workers made this type of demand.) To gain workers' support of the use of robots, some employers agree to include routine programmers and maintenance workers in the bargaining unit. Such action, however, can also increase the union's power to bargain and strike.

Working with robots

In most companies the introduction of robots is gradual. Usually, plant managers decide to introduce only one or two robots at a time. A gradual introduction is beneficial for technological reasons because it gives engineering people time to discover the strong and weak points of the operation and to make appropriate adjustments. It is also wise human resource management because it allays employees' fears.

The speed with which a robot operates at full capacity depends partly on whether the company has developed and tested it in a research lab or on the shopfloor. Since startup problems are inevitable even after development and testing, some companies take advantage of the debugging process to involve their employees in the project and demonstrate that the robot is a piece of machinery like any other.

At one Westinghouse plant, for example, management had the new equipment delivered directly to the plant and set up on the site where engineers worked out the bugs before integrating the robot into the production system. Westinghouse's rationale for on-site debugging was simple: local management wanted to make sure the employees understood that the robot posed no threat to them. From all indications, the employees have adjusted very well and are working smoothly with the machine.

Safety has been a special concern with robot installations. Injuries are most likely to occur during the setup and when the robot malfunctions or is reprogrammed. Companies may need to install safety devices such as light-beam detection systems that shut off the robot when the beam is interrupted and pressure-sensitive pads on the floor that stop the robot when stepped on. Many of the installations we visited are roped off from general traffic, with the operators' controls located outside the barrier.

In companies where the integration of robots has been successful, management continually solicits employees for ideas on how to improve their operation. For example, IBM pays any employee whose suggestion is accepted 25 percent of the resulting cost savings in the first two years. At the company's Dayton, New Jersey, plant, three of the top ten employee suggestors were robot operators; several employees have earned more than $1000 for their money-saving ideas.

An incident at a Westinghouse plant, where a robot was installed to spray-paint electric motors, also shows the value of encouraging employee ideas. At first, the robot could barely cover the motor with paint. After much tinkering, the production people decided that, to enable it to paint the entire product properly, they should equip the robot with two spray guns. Even the addition of a second gun was insufficient. Production and engineering people were beside themselves over the failure of the robot to do the job. Hearing of the difficulty, a young employee, himself a painter, volunteered to program the equipment. With his supervisor's approval, and after a great deal of trial and error leading it through the movements step by step, the employee succeeded in programming the robot to paint electric motors satisfactorily.

'By any other name'

Regularly, in plants throughout the world, employees name robots and sometimes even give them nameplates. Robots have been named after retired employees and foremen; in Japan, they are often given the names of rock-and-roll or baseball stars. In one US auto plant, because both were deemed unreliable, employees named their robot after the supplier's salesman.

The process of naming a robot "humanizes" it, and this has concerned some managers. On one hand, they see the humanizing effect as good because employees tend to feel more comfortable working with a robot with a name. On the other hand, some company officials are concerned that the naming process could make the robots too human, and this could eventually create employee relations problems. At least one major US corporation has a corporate policy that robots are not to be named. Ultimately, we think the answer to the naming issue is best left to employees. If they feel comfortable naming robots, management should allow it.

There may be as many as 150,000 robots at work in the US by 1990.[14] The input of managers will help determine whether workers accept this high-tech equipment as a means to a more efficient production, with ultimate benefits to all in the form of a higher standard of living. Along the way, managers will make many critical decisions. They can choose to reduce exhausting and hazardous work. They can choose to use robots in new processes, new plants, or new products, and thereby create challenging new jobs for qualified people. They can decide to participate in long-range training and educational programs, both within the company and in collaboration with federal, state, and local agencies as well as colleges and universities. In the end, it is people who make robots work. Effective decision-making on the use of robotics could make the difference between success and failure to both the company and its employees.

Notes

1 *Changing Technology: Managing an Organization for Successful Robot Installation*, a report by the Human Resource Council of the Construction Group (Westinghouse Electric Corporation, Pittsburgh, PA, 1982), p. 16.
2 Testimony during joint economic hearing on robotics and unemployment before the Joint Economic Committee of Congress on March 18, 1983 as reported in *1983 Daily Labor Report* no. 54, F-1, (Washington, DC, Bureau of National Affairs, March 18, 1983).
3 Robert Ayres and Steve Miller, "Industrial Robots on the Line," *Technology Review*, May–June 1982, p. 43.
4 "G.M. Shift Outside Suppliers," *New York Times*, October 14, 1981.
5 "L" Corporation, April 21, 1981 interview, "The Impacts of Robotics on the Work Force and Workplace," Appendix B (Carnegie-Mellon University, Pittsburgh, PA, June 14, 1981).
6 *Changing Technology*, p. 8.
7 *Human Factors Affecting ICAM Implementation*, (Honeywell Systems and Research Center, Minneapolis, MN, September 1981), p. 129.
8 "Behind Every Successful Robot There is a Technician," *New York Times*, October 17, 1982.
9 *Changing Technology*, p. 10.
10 Ibid., p. 9.
11 Ibid.
12 "Basic Patterns in Management and Union Rights," *1983 Daily Labor Report*, no. 29, D-1 (Washington, DC, Bureau of National Affairs, February 10, 1983).
13 Ibid.
14 *1983 Daily Labor Report*, no. 54, F-1.

Chips and Crisps:
Labor Faces a Crunch

Bernard Leach and John Shutt

This intriguing study from Manchester Polytechnic shows the uphill struggle that trade unions now face in trying to resist technical change and preserve employment in a labor-intensive sector that suffers from overcapacity and poor growth prospects. In this instance it is the cereal and snacks industry in north west England, and the new technology takes the form of the Ishida computer-weigher. This paper was first read to the annual meeting of the British Association for the Advancement of Science at the University of Sussex, England, August 1983, and first appeared in the Symposium's proceedings, New Technology and the Future of Work and Skills *(Frances Pinter, London, 1984).*

> Ultimately, men and women together, we have to ensure that technical knowledge becomes common knowledge. Decisions over what to produce and how to produce it should be made by all who are affected. [Cynthia Cockburn, *Brothers*, 1983]

Since 1979 Britain's food industries have been in a relatively healthy state, compared with many other industries. Figures from the Food and Drink Industries Council showed profit margins in 1981 higher than at any time since 1973. While nationally 37.8 redundancies per thousand employees were recorded in manufacturing as a whole in the period 1977–82, this compared with a figure of 29.4 per thousand for the food, drink, and tobacco sector, with metal manufacturing, textiles, and engineering being hit much more severely.

In September 1981, of the 629,000 workers employed in the entire food manufacturing sector, over 97,000 were employed in the Northwest, one of Britain's major manufacturing regions. In this region only 8000 jobs were lost in the space of three years (1978–81), a decline of 7.7 percent.

But since then the position has worsened dramatically. The rate of closure and job loss has accelerated as the giant multinational food companies have announced decisions to rationalize production, frequently coupled with decisions about major capital investment programs in new technologies.

Table 11.1 Shares in ready-to-eat cereals market (percent)

Kelloggs*	48.2	Quaker	5.0
Weetabix	19.7	Own-label biscuits	3.5
Nabisco	7.9	Own-label muesli	4.0
Own-label cornflakes	6.0	Others	5.7

*Kelloggs' share had gone up to 56 percent in 1981, according to the *Financial Times*, May 11, 1981.
Source: Super Marketing, December 19/26, 1980.

Cereals and snacks are both sectors dominated by a small number of large multinational companies. As table 11.1 reveals, Kelloggs, Weetabix, and Nabisco have between them 75.0 percent of the ready-to-eat cereals market, while in crisps [potato chips] Nabisco (Smiths, Walkers) and United Biscuits (KP) have a combined market share of about 70 percent (see *Super Marketing*, 1980; Monopolies Commission, 1982). Both industries have been characterized by increasing demand for their products throughout the last decade. For example, the average intake of breakfast cereals in the UK, boosted by the decline in cooked breakfasts, rose by 66 percent from 1966 to 1979 (National Food Survey, 1980). Similarly, crisp consumption increased by 25 percent in the 1975–82 period, with savory snacks showing a spectacular growth. Savory snacks were virtually unknown before 1971 (see MINTEL, 1982).

However, household spending on food is declining and the "recession-proof" tag has disappeared (NEDC, 1983). With no further significant market growth expected, competition is intensifying, particularly from supermarket "own-brand" products. It is not surprising, therefore, that maintaining and increasing market shares preoccupies the leading companies and accounts for the vast sums they spend on advertising. In 1981, for example, in an effort to halt market decline the Smiths Food Group spent as much on advertising and promoting its products as it paid in wages to direct factory employees; while in 1982, Kelloggs UK budgeted to spend £15 million on advertising alone in order to prevent the erosion of market share to "own-brands" products.

In 1982 Nabisco Brands, who already owned Walkers' Crisps, took over Huntley Palmer PLC and thereby acquired the Smiths Food Group. Nabisco is the fourth-largest food processing company in the US and eighth in the world league. Nabisco and Kelloggs are two of the leading American multinational companies and see themselves as world leaders in the food industry. Through extensive advertising on the one hand and a crash capital investment program on the other, both companies have embarked on a new phase in the battle to expand and maintain market shares.

Technical change: the Ishida computer-weigher

Technical change is a vital component in this drive for profitability and elimination of competition. The growth of giant multinational companies spread across the globe is accompanied by increased capital concentration as their investment resources for new technology leave smaller competitors behind. The end result is that competition can *decline* as the weak go to the wall or are taken over, and cartels led by a dominant market leader develop. As the *Guardian* said at the time of the Nabisco takeover of Huntley Palmer, "The takeover of Huntley Palmer creates a considerable monopoly at a time when the present government, most prominently through privatisation, is trying to break them up" (*Guardian*, October 21, 1982).

In many respects Kelloggs and Nabisco are quite different companies in different sectors of food manufacturing, but common processes are at work in the drive to cut production costs through cutting labor costs and introducing waste-saving equipment. Since 1981 the arrival of the Ishida computer-weigher has transformed product weighing and packing lines for both manufacturers.

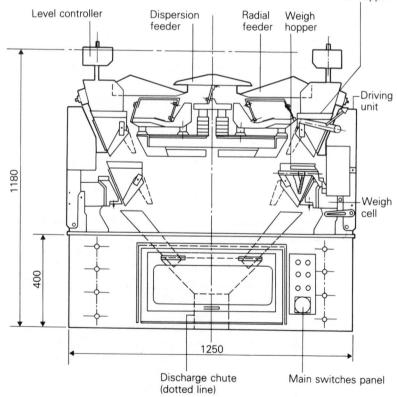

Figure 11.1 The Ishida computer-weigher.

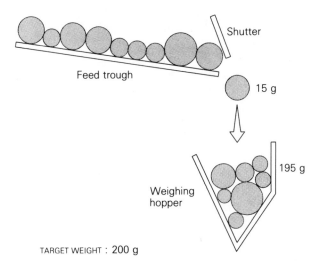

Figure 11.2 Example of conventional weighing system. Against the target weight of 200 g, the weight of the products in the weighing hopper is 195 g. If the weight of the product falling into the weighing hopper is 15 g, giveaway becomes 10 g (195 + 15 − 200 = 10).
Source: Driver Southall Ltd, Walsall. The authors wish to thank this firm for permission to reproduce this promotional material.

The Ishidas (see figure 11.1) are imported whole from Japan by Driver Southall (part of Avery GEC) of Walsall. Although Driver Southall produces weighing/filling machines for powders and granules (e.g. tea, sugar) they were late coming into the field for the weighing of piece goods, like cereals. Enter the Ishida.

Ishida Scales Manufacturing Company Ltd of Kyoto, Japan, are a fourth-generation family firm employing over 500 people. They developed their first computer-weigher in-house in 1973, but for six years its export potential went unrealized, since the firm traditionally never exported. Driver Southall first saw Ishida by accident in 1980. Realizing its potential, they negotiated to become its sole West European distributors and are now selling it to over 300 UK customers for everything from crisps to dry pet food.

Ishida developed their computer-weigher in response to a unique problem. The Japanese use a lot of small green peppers in cooking. The problem came in trying to automate their packing, since they came in a variety of shapes and sizes. This is a problem with many small piece goods, from crisps to frozen prawns.

The conventional system of weighing is to have a gravity feed trough controlled by a shutter which allows a certain amount into a weighing hopper. When the target weight (e.g. 200 g) is reached, the shutter is automatically closed. This system is not too wasteful with standardized piece products (e.g. grains and powders), but where piece weights are irregular, the piece "in flight" that tips the weighing hopper above the target weight can lead to unacceptable "give-away"—in figure 11.2 that means the consumer will get a bag of 210 g weight when the stated weight is 200 g, i.e. 10 g "free."

For crisps, one way of overcoming the weighing problem was to remove raw material variation. Hence the popularity of extruded snacks based on powdered maize [corn], and Smiths' development and promotion of the square crisp (based on powdered potato). Short of genetic engineering to produce identical peppers, Ishida did not have the same option open to them. Instead, they approached the problem from a different angle altogether. Their machine uses a series of weigh-heads fed by a vibratory system which randomly disperses the product into one of the ten weigh hoppers (14 in the latest machine) arrange concentrically. The computer compares the static weights in the ten hoppers, and out of 1023 possible combinations (according

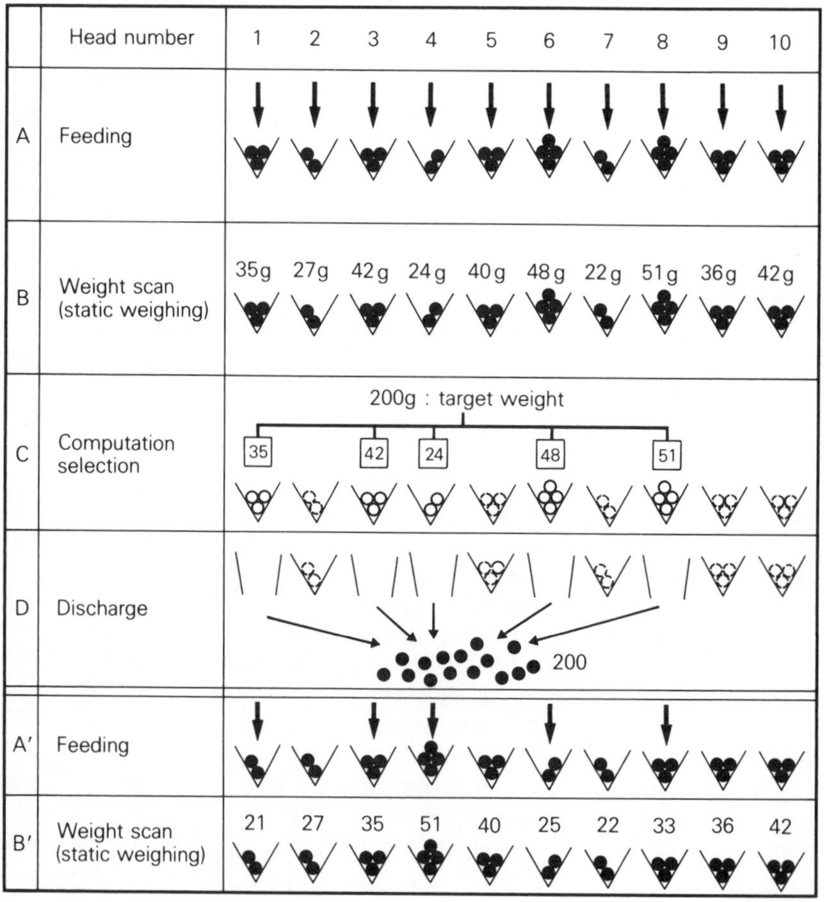

Figure 11.3 Computer combination weighing system.
Source: Driver Southall Ltd.

to Driver Southall) selects that combination of hoppers closest to the target weight and discharges them to a packing machine below (figure 11.3). This system is much faster than previous systems, and it can cut "giveaway" by up to a factor of ten. The cost of the Ishida is relatively small in relation to the increase in output, reduction of waste, and productivity.

In a period of rapid technical change, the actual scale of job loss resulting from the introduction of a new machine such as the Ishida will depend on a wide variety of factors. In particular, the circumstances of the individual company, its position in the market place, and the bargaining strength of the trade unions concerned will determine the conditions under which the capital investment is introduced. The introduction of the Ishida into production in our two case studies occurred in firms operating in different organizational contexts. The Smiths Food Group's Snack Food Division, for example, operated in eight factories spread throughout the UK prior to its takeover by Nabisco in 1982. By contrast, Kelloggs' main breakfast cereals are manufactured in just one plant in Trafford Park at the core of the Greater Manchester conurbation.

Smiths Crisps—new workerless flavor crisps

Crisps—Thin slices of potato fried in oil, usually sold in packets made from thin packing film.
Extruded snacks—A powder or paste is extruded through a die-plate and cut into single-bite portions and then fried or baked. [Monopolies Commission, 1982]

Crisp production was started in the UK in the early 1920s by Frank Smith in London, who used a recipe for a French speciality table dish. Until 1963 there was no real competition in the crisps market—Smiths *was* crisps. Indeed, in 1963 Smiths took over Tudor Food Products, their main competitor in the Northeast. Golden Wonder Crisps (Imperial Group) came into the market in 1963 with the major innovation of cellophane film bags, which increased shelf life up to six weeks. Since then, Smiths' market share of both snacks and crisps has declined, first under General Mills (US) management and from 1979 under Huntley Palmer (formerly Associated Biscuits).

The distinguishing feature of the two rising competitors in the crisp market, Walkers and United Biscuits, has been their early and heavy investment in new technology. Smiths UK Food Group, on the other hand, did not receive sufficient investment in the 1970s. The parent company, Huntley Palmer, preferred to embark on a series of foreign acquisitions, using biscuit and snack food profits. By 1981 Smiths Food Group employed approximately 5000 full-time and part-time workers in eight factories (plus depots and offices), many of which were outdated. Walkers, with a similar market share, had only two factories. The Smiths Group profits were increasingly squeezed by its competitors' ability to maintain profits while the real value of crisps sold dropped.

Smiths' strategy for the 1980s
By the 1980s, in order to maintain competitiveness, Smiths Food Group were obliged simultaneously to invest in new machinery, to close down old factories, and to reduce labor costs. The first public hint of a changing strategy came in the *Sun* newspaper (July 16, 1981) with the headline "2000 Crisp Men to Go in Shock Closures." Subsequently it turned out that only three of the eight factories were "safe," and that even they stood to lose a high proportion (up to 60 percent) of their workforce. For example, Paulsgrove, the first factory in the Smiths Food Group to install Ishidas, reduced the workforce from over 400 to 180 even before autopackers were introduced. At the same time, production capacity increased by an estimated 44 percent.

Twenty-five gram bags are cheap (10–12p) items, but as packing film is an expensive item, the bag must be as small as possible. With erratic bag-filling this can lead to problems with bag seal, leading to a high proportion of rejects. With Ishidas, 25 g bags of crisps are a maximum of half a gramme overweight compared with a previous 6 g maximum (table 11.2). Although a ten-head Ishida costs £28,000 and a 14-head £39,000, more than double the previous weighers, customers generally expect the machines to pay for themselves in nine months. A few weeks after the introduction of Ishidas at Paulsgrove, the National Joint Shop Stewards' Committee was told:

Table 11.2 Crisp packing lines—old and new

	Old	New	New
Weigher	Wrights (UK)	Ishida 10-head (Japan)	Ishida 14-head (Japan)
Operation	Gravity feed, mechanical weighing and discharge	Vibration feed, static computer-combination weighing and discharge	Same as Ishida 10-head
Film wastage	17%, since many bags overweight and rejected	0.3% computer scan ensures no discharge unless target weight achieved, therefore no rejects	Same as Ishida 10-head
Product "give-away" (25 g bags)	6 g	0.5 g	Same as Ishida 10-head
Bags per minute	46	55	80–85 single tube packer 110–115 twin tube packer
Cost	n/a	£28,000	£39,000
Quality control	Manual sampling	Auto printout on weight in each hopper before it goes into the bag	Same as Ishida 10-head

Source: Driver Southall Ltd.

Management now have definite knowledge that our competititors wanted to get hold of Ishida machines. It was now necessary to place the order for another 20 Ishida machines. This has to be done by next Monday to avoid losing our place in the queue. [NJSSC Minutes, December 1981]

Since that time, Smiths Food Group has accelerated its program of closures and redundancies. Four factories have closed and 1500 jobs have been lost in the space of two years.

Toward the automated crisp factory
The spate of new capital investment is leading to massive overcapacity within the crisp industry, since output is increasing at a time when the recession is biting into household expenditure. Even if expenditure on crisps were to grow, it would appear that crisp production capacity already far exceeds likely demand. The Smiths shop stewards in Stockport put it like this:

If all the new 40 Ishidas were installed at Peterlee, they could produce over 3000 million bags of crisps per year, compared with Stockport's 40 million a year. [Stockport Factory Closure Bulletin, 1982]

The introduction of the Ishida is only one part of the development of a mechanized process that will ensure the creation of the automated crisp factory. With the introduction of the Ishida, the technology race is now on and the logical outcome can only be further concentration of capital in the crisp sector. Golden Wonder, for example, is behind both Walkers and Smiths in packaging technology, having concentrated on improving crisp quality, and it has only just started to install Ishidas at its Corby factory. Golden Wonder's market share in crisps has already fallen to an estimated 18 percent from a 1976 figure of 35 percent.

The UK production of crisps in 1981 was over 3700 million packets per year, with more recent estimates in the industry of over 3900 million (Ministry of Agriculture and Fisheries, March 1983). To put it another way, about 7080 are produced per minute. Given the potential speed of 120 bags per minute of a 14-head twin tube Ishida, a simple calculation suggests that, in theory at least, 59 Ishidas working 24 hours a day for 365 days of the year would be sufficient to cope with the existing UK demand for crisps (an additional 24 would cope with the savory snack market!) Even if practical problems of machine downtime are included, not many more would be needed. Figure 11.4 illustrates the difficulties that crisp firms now face, while table 11.3 summarizes the direction of change in the Smiths Food Group in August 1983.

The shopfloor response
It is the massive overcapacity within the sector that already exists that has thwarted the plans of trade unionists to develop alternative proposals to closures. For example, it was widely acknowledged by shop stewards that the best crisps were produced at Smiths' Great Yarmouth factory. In fact, Marks

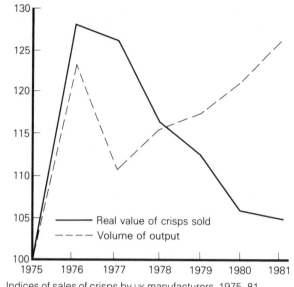

Figure 11.4 Indices of sales of crisps by UK manufacturers, 1975–81.
Source: Business Statistics Office, MMC Study.

and Spencer are said to have insisted on being supplied with their own brand of crisps from this factory. This ensured that the factory was profitable, yet the factory was closed. Investment and technical change means that even profitable factories will be closed, simply because they are surplus to capacity. Moreover, in this case Smiths were also determined to ensure that the factory plant was completely scrapped. Despite an offer from Adams and Dack, a Norwich-based company that planned to produce "Canary" crisps (nickname of the local football team, Norwich City), Smiths insisted on placing a covenant on the sale of the property prohibiting any future purchaser from manufacturing crisps there.

Initial investigation would seem to confirm that the prime locational choices influencing Smiths capital expenditure program were influenced by the age of existing plants, their "green-field" status, and distributional potential. Factories that closed were older conurbation plants with cramped conditions and little room for expansion (Stockport and Park Royal), or were badly placed for national distribution networks (Great Yarmouth).

With the installation of the Ishidas at Paulsgrove, the unions at Smiths Food Group were fully aware of the threat to jobs. As the Great Yarmouth factory convenor said at a national combine meeting on May 10, 1982, following the announcement of the closure of his factory, "It's the Ishidas that are closing factories; in one shift they will produce as much as Yarmouth in a week." Management were not assured of being able to manage the closure and redundancy program while maintaining production. The Smiths Food Group combine, the National Joint Shop Stewards' Committee,

Table 11.3 Smiths Food Group—closing down and moving out

Factory	Location	No. of employees					Comments
		1981*				1982†	
		f/t equivalent	f/t	p/t	Total	f/t equivalent	
Lincoln		540	95	546	641	367	Main snack site, £15m investment planned (but in doubt since Nabisco takeover).
Paulsgrove	Portsmouth	400†	65	246	311	188	Crisp producer. First to have Ishida; workforce reduced by 100+ in first phase.
Swansea	South Wales	130	41	262	303	172	Makes Monster-Munch, Sizzles, and Frazzles. Future doubtful.
Fleetwood	Lancashire	180	22	216	238	130	Makes square crisps. Future doubtful.
Great Yarmouth	Norfolk	188	26	186	212	119	Closed January 1983, despite firm offer from Adams Food.
Stockport	Greater Manchester	358	43	419	462	252	Closed March 1983.
Peterlee	County Durham	480	97	694	791	444	Biggest crisp producer. Ishidas installed 1982–3.
Cardiff	South Wales	110	28	165	193	110	Nut producer. Closure announced.
Park Royal	West London	197	0	0	0	0	Closed 1980 after battle by workforce for better redundancy terms.

*Union estimates.
†Source: Nabisco Brands.

planned a campaign of opposition to the closures, led by the main union, the Transport and General Workers' Union. As well as the combine there was also the "muscle" of what was known as the Trunker Group. Unlike many other crisp manufacturers, Smiths employed its own fleet of trucks. If the truck depots (the Trunker Group) went on strike, they had the ability to bring the group to a halt in a few hours (as manufacturers and retailers were unable to carry large stocks of crisps). The union's main strategy was to try to negotiate a new technology and job security agreement, while in the meantime maintaining the status quo (i.e. no introduction of new equipment without mutual agreement).

Why did the trade unions fail?
Our research indicates a number of causes of the failure of the trade unions. First, they did not develop a perspective on the introduction of the Ishidas that transcended individual plants and allowed them to combat management control over the introduction of the new equipment. Information was released piecemeal to the unions, making it difficult for them to plan long-term strategy. For example, management talked of the "continued refurbishment" of the Stockport plant in January 1982, and put forward the possibility of Ishidas being installed at Stockport and Swansea in March 1982. Two months later, on May 5, 1982, management announced the closure of the Yarmouth and Stockport factories. Second, the trade unions neglected basic organization. By the effort of a few individuals, an effective national Smiths Food Group combine had been established, but decisions were made that were frequently ignored or overturned at branch level. Members were often given no information by the unions about what was going on, leaving management free to develop its own viewpoint to the workforce. Third, while 80 percent of Smiths Food Group workforce were women, many part-time, the union combine negotiators were overwhelmingly full-time males, with an overrepresentation of the depots as opposed to the factories.

For the unions the crunch came following the May 5, 1982 closure announcement. They were either to take action then, or to all intents and purposes abandon the struggle. A national mandating committee met on May 10, 1982 and passed a series of motions which included: opposition to the closure of Yarmouth and Stockport factories; refusal to accept the need for further loss of jobs; refusal to cooperate with management in any transfer of work until satisfactory negotiations had been completed; refusal to handle any new technology including computers; and opposition to a transfer of work within the group outside normal quotas. These strong proposals were passed overwhelmingly by the mandating committee. The sites voted 4 to 1 for industrial action, but lack of basic organization at factory level and disunity over forms of action led to the crumbling of union resistance. By July 28, 1982, the opposition to closures had collapsed and sanctions were called off. Three months later the Smiths Food Group convenor was sacked,

> The trade unions are demanding:
>
> (a) Investment in all Smiths factories, thereby keeping jobs in all the areas
> (b) Full and accurate information from the Company on their future plans
> (c) The introduction of new technology only after full consultation
> (d) No compulsory redundancies
> (e) Workforce to share in the productivity benefits through a shorter working week

Figure 11.5   Source: Factory Closure Bulletin, "Save our Stockport Factory," 1982.

and in 1983 the factories closed without a struggle and without any of the Stockport workers' demands being achieved (see figure 11.5).

Most of the rationalization of the Smiths Food Group had thus been done before the completion of the 1983 Nabisco Brands takeover of Huntley Palmer. Yet the fact is that job losses in the Group have only just begun, as the recent announcement of the Cardiff factory closure and the following comment from the *Financial Times* shows: "Nabisco reckons it will have to pump some £95 million into the business over the next five years and lay off as many as 4000 employees to bring Huntley's operating efficiency up to scratch" (*Financial Times*, October 30, 1982).

Cornflakes—the crunch to come?

Kelloggs operates in 29 locations throughout the world, supplying products to more than 130 nations. Its main product is ready-to-eat cereal, but since 1979 it has been seeking to diversify, particularly in the UK. Backed by aggressive advertising, R&D in Kelloggs is carried out at Battle Creek, Michigan. In the UK, Kelloggs has two main manufacturing plants, one at Wrexham, opened in 1979, which produces Super Noodles, bran products, and muesli; and its main breakfast cereal manufacturing plant at Trafford Park, Manchester, the second largest such factory in the world, producing 10.25 million packets of Corn Flakes and Rice Krispies per week. These products have been consistent money spinners since Kelloggs first started producing at Trafford Park in 1938.

Kelloggs' strategy for the 1980s

In 1980 the Kelloggs Company of Great Britain commissioned a study by business consultants McKinsey & Co., in order to examine the case for re-equipment of their manufacturing units. This followed a prolonged dispute at the Trafford Park plant in 1979, which led management to doubt the wisdom of a strategy that concentrated too much production capacity in one particular location. While the consultants' report was never made available to the trade unions, there can be little doubt that it led to the restructuring of working practices at the plant. As table 11.4 shows, over five years the workforce will have been reduced by approximately 585, the regular weekend shifts will have been abolished, the packing lines will have been re-equipped with Ishidas, and the concept of the disposable temporary worker will have been introduced for the first time since the early 1970s.

Both unions and management agreed on the need for new technology and fresh capital investments. Many of the machines at Trafford Park were originally second-hand when installed in the 1930s. In January, 1981 the company briefed the trade unions on the reasons for the introduction of new equipment:

The current market position is not unsatisfactory, and Corn and Rice in particular have had a successful 1980. Recent studies, however, have indicated that our competitors are closing the gap through technological change and more efficient working practices. To protect our market share and ensure continued success, we have looked closely and urgently at our manufacturing costs and efficiencies. Only through an improvement plan of the scope outlined will we feel confident of meeting our targets for the remainder of the 1980s. [Kelloggs Briefing Notes, 6 January, 1981]

Kelloggs is also under pressure from supermarket "own-label" breakfast cereals, and has experienced a 14 percent drop in its share of the cereals market in the past decade.

Table 11.4 Job loss and technical change at Kelloggs—Trafford Park, 1981–5

Year	No. employees	Comment
1980	2085	
1981	2079	360 redundancies declared with change of shift patterns (from 7 to 5-day cover).
1982	1774	Packing Line Requirement Plan unveiled: job loss of 277 by end of 1984 predicted.
1983	1783	Includes 200 temporary workers (contracts end October 1983).
1984	1650	6 Bag-in-Box lines installed.
1985	1494	Possibly include up to 400 temporary/seasonal workers.

Source: 1983: trade union estimates.

Even before the new investment programme was announced, Kelloggs, disturbed by the effects of the ten-week strike, had already decided to take a tougher stance with the trade unions. The Company Public Affairs Manager told the *Financial Times* on September 24, 1980:

In the past the Company's hourly paid workers got their information either on the grapevine or through their union. We are now asserting our right and duty to communicate. Unions still have a function in representing their members, but that does not include communicating Company information. In future shop stewards will get such information on the basis of their function as employees rather than as officials of a trade union.

The trade unions' response to the Ishida redundancies

At Trafford Park, the main trade union is USDAW, who organize the process workers, with 1400 members (nearly a quarter of whom are women). In addition, there are 300 craft workers (all of whom are men), and these are organized primarily by the majority craft union, the AUEW. Despite the significant number of women within the factory, out of 15 USDAW shop stewards only 2 are women, and of 9 on the negotiating committee, only 1 is a woman.

When the company first announced their packing line re-equipment programme, USDAW's main concern centered on the 277 future redundancies for their members. Realizing the potential disruption that might occur, and bearing in mind the lessons of the bitter 1979 dispute, management has followed a policy of periods of pressure on the unions, eased for a while if there are any signs of industrial action. Together with this "hard-and-soft" approach, management has made full use of the new policy of direct communications to the workforce through supervisors, the in-house journal, and specially commissioned video tapes (viewed in company time). As in the crisps case, they have also been successful in playing off sections of the workforce against one another. Hence, the emphasis on the "uprated skills of the line operators" (all men and well-represented in USDAW positions) as opposed to the redundancies among packers (of whom over 60 percent are women) in the negotiations surrounding the introduction of the Ishidas.

The craft unions have traditionally been well organized at Kelloggs. Through a year-long overtime ban they eventually succeeded in June 1982 in negotiating a job security agreement, the key clause of which states: "The company does not intend to make any Kellogg craft area employee compulsorily redundant as a result of the imminent changes in new technology." While this agreement is by no means watertight from the union point of view, it is interesting to note that the management has refused a similar agreement for the USDAW workers. The reasoning behind this is quite simple. The craft workers had an overtime ban for over a year, until the company negotiated the job security agreement. The craft workers were determined and well organized, and the company needed their cooperation in installing and maintaining the new equipment.

Kelloggs' aim has been to reduce the work hours required to produce 1000 cases of breakfast cereals from approximately 73 to around 26. Already on the lines installed, average product "giveaway" has been reduced from 20 g to an estimated 2 g. The company's plans, as outlined to USDAW in November 1981, called for the loss of 277 jobs, all but a handful of which were packers' jobs. Over 45 percent of packers' jobs were to go.

USDAW's difficulty in formulating an effective response, as judged by the acceptance of redundancies and re-introduction of temporary labor, was due partly to their complex structure. At factory branch level this made it almost impossible to develop a strong shop steward system. In particular, the key negotiating committee was not accountable to the shop stewards, since they were elected by a mixture of secret ballot (branch officials) and show of hands at the annual general meeting (three others). Thus, the shop stewards were without real power. Many members were unaware of who their shop steward was, and if they needed advice often went direct to negotiating committee members.

There have been other results of the introduction of new packing lines at Trafford Park apart from the loss of jobs for women. The attempt to move to a workforce composed of possibly 1000 full-time workers, plus up to 400 temporary workers for the busy summer months, marks a significant deterioration in the terms and conditions of employment of the process workers. Many who previously had security of employment will now become seasonal workers, hired the next season depending on demand and their previous work and sickness records, and without even the minimal protection afforded by the Employment Protection Acts.

Conclusion

We have focused on the automation of packaging lines because it was one of the most labor-intensive areas in food processing and ripe for automation. Crisps and breakfast cereals are not alone in being affected by the automation of packaging. For example, United Biscuits' largest biscuit [cookie] factory at Harlesden, north London, has cut its workforce by a third to 1130 since 1977. The workforce on each packaging line has been cut from fourteen people to two (*Economist*, March 13, 1982). There can be little doubt that developments in packing technology are drastically reducing job opportunities, and that women are bearing the brunt of job losses. It is not yet clear whether part-time work, so important to many women, will increase in relation to full-time work. However, the indication is that secure part-time employment conditions will be harder to negotiate, since temporary part-time and seasonal jobs are on the increase.

The case studies examined here illustrate the uphill struggle facing organized labor in relation to technological developments. Not only has management a monopoly of the detailed economic and technical information, but also it can appeal to the dominant ideology continually drummed in by the media and all political parties that technological progress is inevitable and essential to the firm's/country's survival. To combat this all-pervasive world view, it is not enough for the trade unions to suggest alternatives that lack credibility among their own members. It is not surprising therefore that trade unionists often prepare for the fight against redundancies caused by new technology with an attitude that is already resigned to eventual defeat. Our two case studies show that firms have been able to carry out a program of redundancies and closures on their own terms with few concessions to the trade unions.

A factor mitigating against effective union action has been the failure of union organization to adapt to changing circumstances and to overcome sectional interests and rivalries. The unions need to overcome the lack of effective organization between plants in the same group, and divisions within the workforce inside the factories—craft vs process, men vs women, full-time vs part-time, permanent vs temporary.

The two sectors investigated in this paper are both characterized by high rates of growth of output, fuelled by new technology which has also led to large numbers of redundancies. Both are reaching a stage where capacity will far outstrip demand. The counter-cyclical strengths of the food industry (underperforming the market during more buoyant times, and outperforming the market in recession) are shown clearly in the case of both crisps and breakfast cereals, but the signs are that volume growth is reaching its limit and that even economic recovery would not significantly affect this.

For manufacturers to maintain or increase profits this has meant a new technology race to become the lowest-cost producer and to pressurize competitors out of the market, a policy adopted by Kelloggs in cereals, Nabisco in crisps, United Biscuits in biscuits, and Associated British Foods in wrapped bread. These manufacturers have to spend vast sums of money on advertising to persuade consumers to eat more expensively and to eat products that carry better profit margins—extruded snacks such as Monster Munch rather than crisps, Crunchy Nut Cornflakes rather than plain Cornflakes.

A new technology that started off with a problem of packing green peppers seems to be ending up in large sectors of food manufacturing with a capacity that far outstrips demand, and large numbers of redundant workers joining the 16 percent of the population known in the trade as "subsistence consumers" (Henley Centre, 1982).

References

Cockburn, Cynthia (1983), *Brothers*, London, Pluto Press.
Henley Centre (1982), *Manufacturing and Retailing in the 80s—a Zero Sum Game?*
Labour Research (1981), *Food Industry, Fatter Profits*, London, Labour Research.
Ministry of Agriculture and Fisheries (1983), *Statistics 1962/83 Potato Crisps and Snack Foods*.
MINTEL (1982), *Consumer Spending: 10 Year Trends.*
Monopolies Commission (1982), *Report on Nabisco Takeover of Huntley Palmer.*
NEDC (1983), *Review for the Food and Drink Manufacturing Industry.*
Smiths Joint Shop Stewards' Committee (1982), "Save Our Stockport Factory," anti-closure bulletin.
Super Marketing (1980), December 19.
The Economist (1982), March 13.

The Making of the Metro

Paul Willman and Graham Winch

The British motor industry has performed so badly in the past that the UK *now imports 1 million cars per year. One bright spot, however, is the new range of cars from state-owned* BL, *which began with the Austin Metro, launched in 1980. The Metro plant is now the most productive in Europe. The authors show how management overcame Britain's Byzantine industrial relations system to introduce the new technology that made success possible. Paul Willman is with the Cranfield School of Management, Bedford, England and Graham Winch is at Imperial College, London. This is a summary of their book,* Making the Metro: Technological Change, Management Strategy and Industrial Relations at BL Cars *(Cambridge University Press, England, 1984).*

The British Leyland Motor Corporation, now BL plc, was formed in 1968, and is unique in the British car industry in being the product of a series of postwar mergers and rationalizations. Vauxhall and Ford have expanded through growth rather than amalgamation, and the Rootes Group, now owned by Peugeot Citroen, was largely formed before the war. The two companies that came together in 1968—British Motor Holdings and Leyland— were themselves loosely integrated, and the new corporate structure exacerbated this fragmentation.

During the 1950s, the British motor industry enjoyed an unprecedented boom in which companies could sell all that they could produce. Production at full capacity in an easy market, coupled with full employment and hence a tight labor market, meant that the employers' concession costs were low and resistance costs high in collective bargaining. The postwar shop steward movement easily established itself in the motor industry, and managements failed to invest adequately in the modernization of their production processes.

The crisis broke in the early 1970s as the world market moved into recession and import penetration grew. In these respects, the British car industry presents a case study in the more general phenomenon of the process of "deindustrialization," and the most widely discussed element in this process has been BL because of its total dependence upon state finance and its role as the only remaining major British-owned car producer (see Bhaskar, 1979; Dunnett, 1980; Turner, 1973 for details).

During 1974 the financial position of BL worsened rapidly as it ran up to the limit of its overdraft facilities with little prospect of further funds being made available. The government responded by appointing the Ryder Committee in December 1974 to analyze the "present position and future prospects" of the company. The Report was presented to the Secretary of State for Industry in March, 1975. It covered all aspects of the company's operations, and concluded that BL should remain a producer of cars and commercial vehicles throughout the product range. In order that this be successfully accomplished, the injection of a large amount of state funds for capital investment was recommended, to be supported by extensive rationalizations of the product range, operating divisions, and engineering resources. In the area of industrial relations, recommendations were made in three areas—payments systems, industrial democracy, and collective bargaining. This paper will analyze the interaction between the developments associated with the implementation of these recommendations, and those associated with the development of the product range and modernization of manufacturing processes.

Change at Longbridge 1974–80

The major investment of the period 1974–80 was the retooling of the body assembly facilities at the Longbridge site. The complex at Longbridge near Birmingham—the old "Austin" works—is one of the two major manufacturing sites in (now) Austin Rover, the other being at Cowley, near Oxford.[1] These two sites have received the greater part of investment for new products, the £275 million expenditure for the Metro involving £106 million spent on the body assembly line in the new West Works, while the Maestro project at Cowley has latterly cost £210 million.

The new equipment at Longbridge consists primarily of large dedicated multiwelders for spotwelding, together with two parallel lines of robots to perform finishing needs: in addition, some robots are used for handling operations, giving a total of 40 robots in the new West Works. This is rather lower than typical or subsequent automation exercises in the industry, and reflects the inflexibility of this single-model line. In all, over 70 percent of welds are performed on multiwelders, under 10 percent by robots, and about 20 percent, particularly seam welds, manually.

Ten unions represent hourly paid workers at the Longbridge site, their activities being coordinated by a Joint Shop Stewards Committee (JSSC), which in turn elects a Works Committee consisting of seven members plus a minutes secretary. Four places go to the Transport and General Workers Union (TGWU), three to the Amalgamated Union of Engineering Workers (AUEW), and the remaining place represents the other unions. The TGWU and AUEW are the dominant unions on the site, with the TGWU having the largest membership owing to its organization of around three-quarters of the direct production workers. The JSSC also elects a chairman and secretary, and the secretary is a designated works convener.

In 1975 the Longbridge Operations Direct Workers and Indirect Workers Agreements were reached. These agreed a two-grade structure for direct workers, and a seven-grade structure for the indirects. A notable feature of the agreements is that they embodied the principle of mutuality—for instance, the application of industrial engineering techniques was to be preceded by full discussion with shop stewards and the membership, and mobility was based upon the 1968 plant mobility procedure agreement which specified discussions between the supervisor and the members and shop stewards concerned as the first stage. These two agreements formed the basis of negotiations at Longbridge on pay and conditions until they were superseded by the 1979 corporate document. The Longbridge procedure agreement was negotiated in 1976. It provided for union recognition, the role of shop stewards and the Works Committee, and procedures for negotiation and the settlement of disputes.

This formal system of plant agreements was supplemented by a large number of local agreements throughout the plant. These were negotiated at sectional level, yet were written agreements that could be "held up against" management. Most of these agreements were for special payments of one kind or another, but many covered working practices—such as the 6A agreement, which specified that, if there were to be layoffs on the night shift, they had to occur before 10 o'clock. There were also around 50 groups of workers listed in the Direct Workers Agreement who qualified for some form of fixed supplementary payment on the main grade rate. Such fragmentation led to a very complex and time-consuming bargaining situation—one line manager estimated that he used to spend 45 percent of his time on industrial relations issues. It also militated against mobility of labor, because the agreements were not compatible between different areas. Again, perhaps the crucial feature of this informal system was the emphasis upon mutuality and the role of the shop stewards—management would always approach the workforce through the shop steward.

The "principle of mutuality" is long established in the British engineering industry. Payments by results began to be introduced into the industry towards the end of the nineteenth century with the aim of enabling employers to gain greater control over the effort and output of their workforce. In order to exclude trade union officials from the setting of piecework prices, the engineering employers imposed the principle of mutuality after the 1898 lockout (Jeffreys, 1946, p. 48). The essence of the principle is that "piecework prices and bonus or basic times shall be fixed by mutual agreement between the employer and the worker who is to perform the work" (CSEU, 1979, S. 5.1). However, the other workers in the plant clearly have an interest in the prices agreed by individuals, and so the setting of prices quickly became the subject of workplace negotiations; indeed, this bargaining was the origin of many of the shop stewards' committees that arose during World War I (Goodrich, 1975, chapter 12). Thus, in times of full employment and workplace trade union strength, the principle of mutuality became a major source of workplace bargaining. Any changes in working practises, particularly measurement of the job, have an impact upon piece earnings, and so mutuality became enshrined in formal agreements as the power of workplace organizations grew during the 1950s and 1960s.

It is often argued that the installation of automatic machinery means that payment-by-results systems are no longer viable, because the operator has no control over output, and that the job requirement is consistency of effort rather than maximization; this point has been made with reference to the motor industry by Turner and his colleagues (1967, p. 97). The company had begun to implement a corporate policy of moving away from incentive systems to measured daywork (MDW) in the early 1970s, and Longbridge had switched in 1972. The principle of mutuality survived this change to MDW to become mutuality over the effort bargain, in that standard times were bargained over on the shopfloor.[3]

Following upon the policies advocated by the incoming Labour government in 1974, there were a number of experiments in industrial democracy in

BL CARS PARTICIPATION SCHEME

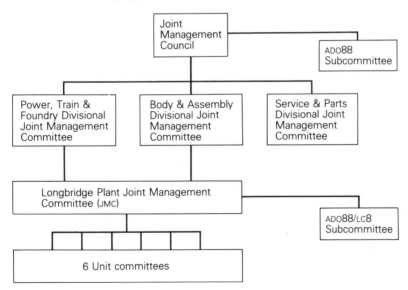

Figure 11.6   BL cars participation scheme, January 1976–September 1979.

Britain (see Elliott, 1978, for an overview). The participation scheme at BL Cars was one of the more comprehensive examples. The scheme started in January 1976 and continued until September 1979, when the unions withdrew—the overall structure is shown in figure 11.6. The terms of reference of the scheme were to review corporate performance, and to examine ways of improving the competitiveness of the company; issues that were normally discussed in the collective bargaining forum were excluded. The appropriate functional management served on one side, while the workforce was represented by elected lay union officials. The two most important levels for the Metro project were those of the Council, and the Longbridge Joint Management Committee (JMC). At each level, a Metro[4] subcommittee was formed which handled the discussions. In early 1976 the Council approved a presentation of the overall concept of the Metro, which included an evaluation of the production technology to be used. It was felt that high technology was where the future lay, and that all the other car companies were going down the same road. Although there was no overt attempt by management to increase control over the "refractory hand of labor" (c.f. Braverman, 1974), there was a definite feeling among management that reducing the number of workers yielded intangible benefits beyond productivity gains.

The company had been placed under the control of the newly formed National Enterprise Board (NEB), a state holding company, in 1975. When approached for finance for the project in 1976, the NEB expressed grave reservations about its viability, and the company promised to provide more detailed figures. To gain union involvement in providing these figures, the Metro subcommittee was set up; and owing to the very limited amount of time available, the union side was allowed complete access to the manufacturing planning meetings. The outcome of this access was that the union side fully supported the management recommendations on the targets for the new car; and in particular, the targets that were laid down for productivity.

At the end of 1976, the discussions were handed down to the plants that were to produce the car. At Longbridge a special committee was also set up. The early meetings were dominated by discussion of the working practises that would be required to meet the productivity targets specified. The key areas were industrial engineering, mobility of labor, and restrictive practises, particularly among the maintenance trades. A rest allowance target of 12 percent was also stressed by management. The major proposal on work organization was the implementation of team working. The idea was to combine flexibility of labor with increased job satisfaction—teams made up of material handlers, on-line maintenance workers, operators, and quality controllers reporting to team supervisors were to cover production zones. The concept also involved the amalgamation of the existing maintenance trades into two—electrical and mechanical. These proposals from management met the approval in principle of the trade union side, although the maintenance trades' representatives were not involved because they were not cooperating as part of their campaign for toolroom autonomy (see Scullion, 1981, for details of this campaign). Once agreement had been reached on these issues, the discussions settled down to a routine of the presentation by management of the various aspects of the production facilities, followed by union questioning.

The participation scheme at BL Cars, unlike many others of that era, did have genuine commitment from the management involved, at least for the first two years, and the trade union leadership was also supportive. However, line management and the trade union membership became increasingly hostile to the scheme. In October 1977 Sir Michael Edwardes was appointed chairman, and he turned first to a reappraisal of the Ryder plan in the light of continuing failure to meet its targets. The Edwardes plan, announced in February 1978, reorganized the company, reduced sales targets, and cut parts of the expansion and modernization programme (Bhaskar, 1979, chapter 7). This new policy included redundancies and closures, and meant that the trade union side was put into an increasingly difficult position: the scheme began to break down. However, all levels of managers and trade unionists interviewed agreed that the Metro subcommittees did some of the best work under the auspices of the scheme.

It has been argued that the 1970s have witnessed a shift toward the formalization of bargaining, the spread of job evaluation, and a move to pay bargaining at the level of the single employer (Brown and Terry, 1978; Brown, 1981). In these respects, BL Cars represents a classic case study in the problems implicit in these developments when starting from a highly fragmented base that was the result of amalgamations over a long period of time. So far as the Metro project was concerned, the developments associated with the centralization of bargaining within BL Cars acted as an important constraint upon bargaining over the new technology; but at the same time, that bargaining made a significant contribution to the final shape of the corporate agreement.

In October 1976 the Longbridge JMC was asked to commit itself to the recommendations of the Council subcommittee. The negotiating committee was brought into the discussions on this commitment, and the hourly paid representatives expressed particular concern over the productivity recommendation. They tried to write assurances on mutuality into the commitment, but this was rejected by management, who reported to the NEB a failure to agree. The response was "no commitment—no ADO88." The JMC met again, after balloting the workforce on management's proposed form of the commitment; the result was in favor. The manual unions backed down and gave their support to this form of the commitment, while stressing that they would be seeking payment for the changes in working practices proposed. A notable feature of these negotiations is that management, driven by the NEB, wanted a commitment to the achievement of productivity targets from the unions before they, and therefore the unions, knew how these were going to be achieved in any detail.

Early in 1978, management made a presentation to the trade union negotiators of the new working practises that they were seeking—again, the issues were industrial engineering techniques, team working, mobility, and two-trades maintenance. Later in that year management tried to open negotiations for the areas affected by the Metro project; the unions however replied by stressing the importance of payment for change and the negotiations covering the whole site. Management's proposals would have meant two bargaining units on the Longbridge site, a move that was against the company policy of the centralization of bargaining; it would also have opened them up to comparability claims. The problems associated with a separate agreement on working practises for the Metro, and the union pressure for a single-site agreement, meant that management changed its mind, and by the summer of 1978 had drawn up a draft agreement covering the whole site.

The 1978 draft marked a radical departure from the 1975 agreements. The protective clauses on mutuality in the earlier documents were removed, and the range of industrial engineering techniques was fully specified while the extent of mobility was broadened. Clauses on the method of team working and the operation of two-trade maintenance were also included. No negotiations took place on this draft, but the "Management of Change" negotiations that finally took place in May 1979 were based upon a similar document. The main issues at these negotiations were payment for change, and the grading of team workers. Discussions on both issues were constrained by the negotiations on the centralization of bargaining. Management's proposals for payment were based upon the incentive scheme then being negotiated centrally, and on the completion of the job evaluation exercise for the establishment of the corporate grade structure. This meant that no money was payable immediately. To avoid this problem, the union side demanded payment in a lump sum upon the reaching of agreement, but this gave management no means of ensuring the achievement of the agreed productivity targets.

The "Management of Change" negotiations were inconclusive, but in many respects they were just a holding operation while the corporate negotiations proceeded. As a matter of policy, management was trying to get away from the notion of payment for change; such a concept is inherent in mutuality, but is incompatible with a job-evaluated grade structure unless the actual job content changes. Moreover, team working was not supported by many managers at Longbridge—it was seen as something imposed from head office smacking of behavioral science and not relevant to those who considered themselves as "hard men who knew how to make motors."

In drawing up a draft of the agreement for the 1979 corporate negotiations, management invited submissions from all the plants concerned; Longbridge had spent the last three years thinking about new working practises in relation to the Metro, and the ideas developed in that forum were largely incorporated into the draft that was presented to the trade unions in the form of the so-called "92 Page Document." Although the clauses on team working and two-trades maintenance were greatly toned down, those on mobility and industrial engineering remained as forceful as before. Although amendments were made to the draft during the ensuing negotiations that readmitted an element of mutuality, these were little more than cosmetic in effect; the central place that mutuality had held in 1975 agreements was completely overturned.

The negotiations on the "92 Page Document" dragged on for five months until management imposed the draft, together with those amendments that had been agreed, in April 1980. In doing this, the company was greatly aided by the weakening of the trade union strength consequent upon the sacking of the Longbridge convener, Derek Robinson, in November 1979. The formal reason for the sacking was that Robinson had led the campaign against the BL Cars Recovery Plan announced in September 1979, which involved substantial redundancies and plant closures; although one manager argued that the sacking "reflected a completely new determination on the part of management to manage the business in the only way it could be successful." In terms of new working practises, management concerned with the Metro project argue that they got all they wanted from the successful imposition of the "92 Page Document."

While these developments were taking place, the issues of demarcation in maintenance working were being resolved. In July 1977 a Unimate robot was installed on the Allegro line to test the reaction of the workforce. As expected, this provoked a demarcation dispute, and the use of "programmable logic controllers" (PLCs) for maintenance diagnostics caused similar problems. The dispute was between the electricians and the two mechanical trades over programming and the accessing of information. The issue finally went to a national conference under the engineering industry procedure, and an interim agreement lasting one year was reached in February 1980 which gave the electricians responsibility for both jobs. However, in practise the working has been much more fluid; the demarcation has not been maintained, and the issue has not been raised again. The maintenance trades, as part of their long-running dispute over differentials and toolroom autonomy, also refused to undergo training, and it was only in April 1980 that full training commenced. On both these issues, however, management claim that there was little impact on the overall project—they had foreseen both problems and had allowed themselves plenty of time to sort them out before they could affect production of the new car. There seems little doubt that the main reason for installing the Unimate on the Allegro line was to allow the demarcation issue to be raised and resolved in good time.

Sir Michael Edwardes has stated that "1978 was the year in which the industrial relations nettle was grasped" (cited in Chell, 1980, p. 22). In June 1976 there had been 58 bargaining units for hourly paid workers in BL Cars; in November 1979 there was one. Against these developments the unions had been unable to mount an effective opposition. The longstanding weaknesses of the Leyland Combine Trade Union Committee meant that the shop steward organization had no means of coordinating its activities above plant level, and this vacuum was filled by the CSEU. Bargaining over the new technology was constrained by management's concern to maintain the integrity of the new bargaining structure, and this meant that they were not prepared to make concessions for the Metro project. The unions fully supported the implementation of the new technology, but demanded payment for change and the retention of the traditional protection of mutuality; their weaknesses meant that they could not enforce these demands.

The positive aspect of management's strength was the credibility of the Edwardes team that took over in November 1977; the negative aspect was the desperate financial situation of the company (Chell, 1980, p. 69). For personnel managers at Longbridge, the role of Edwardes was quite fundamental. They felt that the company was now being managed from the top, and that tough decisions would get support from senior management in a way that they never had before. The belief that the company was on the brink was shared by the CSEU executive and the membership. The former's interventions were always to move the trade union side toward the acceptance of the company's proposals. The union members demonstrated their views in the series of ballots on the key aspects of the centralization proposals, and thereby supported the belief among management that the rank and file leadership did not represent the voice of the membership (Chell, 1980, p. 26). These views were strongly influenced by company propaganda, the statements of the CSEU, and the threats by the government and the NEB to withhold funds. The divisions on the union side and the twin strengths of management meant that the company had made remarkable progress in the restructuring of the industrial relations system by the time the new car went into full production.

The system at Longbridge after April 1980

The imposition of the "92 Page Document" in April 1980 had two distinct impacts. First, the main level of bargaining was moved sharply upwards. Except within closely defined areas, all bargaining now takes place at the company level. This shift in level upwards is dependent upon the removal of mutuality from the shopfloor and has meant a great reduction in the amount of time spent by shop stewards and line managers on industrial relations. One manager reckoned that the time he spent on industrial relations dropped from 45 to 5 percent, and management reduced the number of full-time shop stewards to the seven on the Works Committee.[5] Second, on working practises, management claims to have successfully removed a high proportion of restrictive practises and special payments. In Unit 1, management drew up a list of the practises it wanted to remove, informed the shop stewards in

April, and largely implemented the changes in June. However, the demarcations in production and maintenance still remain—there has been very little progress towards two-trades maintenance at plant level, and there is still the demarcation on the Mini line in the West Works No. 1.

Perhaps the major impact of the imposition, at least in the short term, has not been in the area of bargaining levels or working practises, but in management style. The key to the new style has been a reassertion of what one industrial relations manager described as "management's responsibility to manage." This can be seen most clearly over the issue of communication with the workforce. In the past, management spoke to the workforce through the shop steward, but as the centralization of bargaining developed and the crisis of 1979 was resolved, the means of direct communication with employees was developed. At plant level this means direct mailing to the employee's home; on the line it means a policy of briefings with the whole workforce where possible. As the Metro line was being manned up, there was a deliberate policy of circumventing the shop stewards in communication, and at the launch, the whole workforce received a briefing.

These changes in style were codified by a statement issued in December 1979 on "Management in BL" (see Chell, 1980, appendix VIII) which emphasized the necessity for firm and consistent leadership, backed by senior management. Longbridge management sees itself as more forceful and determined than it used to be, and more than one manager stressed the importance of the fact that they could now rely on support from senior management for firm action. The success of such a strategy depends, in the end, upon the suspension of mutuality—the shop stewards are now informed of changes when they would have been asked about them in the past. The changes were described to us as deliberate and dramatic.

In the run-up to the Metro launch in October 1980 there were no particular industrial relations problems. The main concern of industrial relations management in Unit 1 was to ensure that the disciplines established through the imposition of the "92 Page Document" and the gains made were not jeopardized under the pressure for output as the launch date approached. The pre-launch pressures were very high, and a new operations manager was brought in from another plant to ensure production targets were met.

Management saw the new body-in-white facility in West Works No. 2 as an opportunity to start with a clean slate in an area where no tradition of custom and practise had built up. On the production side, the manning of the new works took place in two stages. At first, volunteers were requested; later, compulsory transfers taking into account seniority were made. When selecting among the volunteers, management picked those whom they considered to be flexibly minded, and those with "bad records" were not selected. However, as the volume of production increased, management could not afford to be so choosy, although the seniority rule meant that younger workers tended to be transferred. Because of skill shortages, recruitment for maintenance workers took place outside the company—the result being that many of the maintenance workers have only ever worked for BL in the new West Works.

Partly by luck and more by judgment, therefore, the workforce in the new facility was not steeped in the trade union traditions of Longbridge. The established trade union hierarchy within the unit remained in the old No. 1 Works. Although the company does not recognize senior stewards within the units, the informal senior stewards for both the TGWU and the AEUW remained in the old West Works, and management recognize that the general level of experience of the stewards in the No. 2 Works is low. The stewards in the new works are also poorly organized—there are 20 production stewards for 1400 men in the new works, as against 24 for 600 men in the old works.

Five shop stewards from the West Works No. 2 were interviewed—three production workers and two tradesmen. All five complained about the lack of communication between stewards, particularly with opposite numbers on the other shift. There are no regular meetings between stewards in the new West Works, and few contacts with other stewards in the Unit. The reasons for this recognized state of poor organization were mainly laid at the door of the selection procedure by the stewards. As one put it, when they selected the volunteers they chose the worst trade unionists, and the good shop stewards who were moved in the compulsory transfer took the attitude that they were not going to help out those who had failed to back Derek Robinson. Also, as the groups were moved across, management did not at first recognize shop stewards. Management had a deliberate policy of not talking to the shop stewards, and of trying to build up the role of the supervisor as the means of communication with the workforce. Against these tactics, the trade unions found it difficult to build up a workplace organization. The shorter lunch breaks and faster pace of working were also blamed for less effective meetings by one steward.

None of the shop stewards interviewed felt that there had been a particular attack upon him in the new works—they were ready to place the problem at the level of the plant, or even the mood of the country. But management has built upon this situation to make further progress on implementing new working practises than in other parts of the plant. On the production side, the demarcation between the NUSMW and TGWU has not been established, and the Metro line is generally considered by management to present few industrial relations problems. It is on the maintenance side that working practises in the new facility have most changed. Although there is still four-trade maintenance, a compromise has been reached in the form of two-trade response.

As the equipment was being installed, maintenance management decided which two trades were most suited to the particular maintenance problems of each machine. In practise, this meant the electricians and one other trade were allocated to each machine. The pipefitters were not really involved as they tend to be allocated to site services, but each machine was allocated to either the machine tool fitters or the millwrights—the latter tended to get the heavier work, such as the conveyors. The shop stewards were then informed that, for instance, the KUKA multiwelder would be an electrician-and-machine-tool-fitter job. The electricians and fitters have also been moved out of their maintenance pens and are now permanently allocated to a machine; this has reduced maintenance response times, and was considered to be the major advantage of team working from the maintenance point of view.

Thus, management has made major gains in implementing new working practices in the Metro body-in-white facility. The opportunity to start from scratch with a new facility, coupled with the general tenor of industrial relations in the company, allowed management to break with previous working traditions. Although much of the thinking behind the new working practises was stimulated by the challenge of implementing a major capital investment in automated equipment, none of the working practises implemented is in any sense determined by the technology, with the possible exception of two-trades response. The new working practises apply equally across the Longbridge site, and it was the newness of the technology rather than its intrinsic information processing capabilities that allowed the more successful implementation of the working practises in West Works No. 2.

On two-trades response, however, the greater size and complexity of the equipment, together with diagnostic facilities provided by the PLCs, does seem to have generated a more cooperative approach independent from the benefit of the workers being largely new to Longbridge and BL. The fitters need the information generated by the PLCs in order to tackle their jobs, and the logic of those doing the actual repair work on the robot hardware also doing any reprogramming that might be necessary has also been informally accepted. The placing of the workers together on the machines, rather than together with their own trade in the maintenance pen, may also have helped generate this atmosphere. Again, the newness of the technology, with the challenge of tackling new problems, may have been an element in generating the more flexible approach.

Conclusions

Broadly, the Metro project went through two distinct phases. The unions had long supported the principle of investing in high technology to make the company competitive in world markets, and management in 1976 and 1977 displayed a genuine commitment to participation as the way of solving the company's longstanding industrial relations problems. The early discussions over the Metro were therefore conducted within a cooperative atmosphere, and a broad measure of agreement on priorities was reached. The situation may therefore be described as one of "bargained acceptance" by the workplace organization (see Francis and Willman, 1980, p. 14), or, more precisely, that of a "productivity coalition" (see Streeck and Hoff, 1981, p. 18).

From the beginning of 1978 the mood in the company changed, and the trade unions found themselves increasingly powerless and divided as management imposed its plans for restructuring the company. By the time the Metro was coming into full production, the trade unions were unable to resist the imposition of new working practises and to press for payment for change. The role of Sir Michael Edwardes in bringing about this situation is crucial—he provided direction and gave management the confidence to act at the local level, a credibility that was reinforced greatly after the sacking of Derek Robinson. The dire financial situation also greatly strengthened management's hand.

In this later confrontational period, management built upon the commitments that had been given by the unions on the Metro project during the cooperative phase. The earlier phase provided a crucial sounding board for management to establish how far the union side was prepared to go, and it is

Table 11.5   Direct labour levels: new West Works vs theoretical conventional assembly

	New West Works	Conventional
Typical manpower savings in subassembly (per shift):		
Bodyside assembly (m/welder)	10	58
Dash assembly (ASEA robot)	1	5
Underframe floor assembly (m/welder)	11	76
On framing lines:		
Manual welding vs ABF plus Unimate robots	1 operator per line	80 operators
Conventional Metro body build: 942 jobs more than present method		

Table 11.6   Metro chronology

December 1974	Ryder Committee appointed
March 1975	Ryder Report presented
January 1976	Participation Scheme inaugurated
October 1976	Longbridge JMC committed to productivity targets
October 1976	NEB approval for ADO88 project
February/March 1977	Toolmakers' strike
October 1977	Ballot of workforce accepts parity and centralisation
October 1977	Sir Michael Edwardes appointed Chairman of BL.
December 1978	Ballot of workforce accepts 5% pay offer and proposals for assimilation to the parity 5-grade structure based upon productivity
June/July 1979	Management of Change negotiations
September 1979	Participation ends
October 1979	Ballot of workforce accepts BL recovery plan
October 1979	Draft national agreement (92 Page Document) presented
November 1979	Derek Robinson, Longbridge convenor, sacked
April 1980	92 Page Document imposed
September 1980	Metro launched

doubtful that management would have been as successful in the imposition of new working practises and the removal of mutuality from the shopfloor if it had not been able to build on the base of the earlier period. This of course begs the question of whether the project would have reached its successful conclusion without the shift to confrontation. These considerations suggest that it is more sensible to talk of the impact of industrial relations on technological change than the reverse. The new technology may well have provided a lever for changes in working practises, but the overall developments in the industrial relations system in BL Cars had a much more profound impact on the project than the project had on these developments.

In any event, the changes must be seen as a success for BL. Since 1980, the number of man-hours required to produce a car at Longbridge has halved; much of the improvement occurred after the new technology came on stream, indicating changes in organization and manning. A recent survey shows the Longbridge facilities to be the most productive in Europe, with an output of 55 cars per man per year (*Engineer*, February 9, 1984). Moreover, for employees and unions, the success of the cars is highly relevant.

The new facilities are less labor-intensive than conventional assembly: some estimates of the difference are given in table 11.5. However, no redundancies were specifically associated with the new line: in fact, as a new facility giving extra capacity, it created more than 1,000 new jobs at Longbridge in 1980. The background to these relatively small employment changes was, in any case, the loss of 56,000 jobs within the company over the previous four years, when no new products were developed.

Notes

1 Longbridge employs 13,000 and has a capacity of 7500 cars per week; Cowley employs 10,000 and has a capacity of 7500 cars per week (figures for April 1982).
2 An idea of the flavor of workplace industrial relations under piecework in the car industry can be gained from Clack (1967).
3 Beynon (1975) describes workplace bargaining under MDW at Ford's Halewood plant.
4 The Metro was code named "ADO88" until 1978, when the car was reconceptualized and coded "LC8."
5 This number has now been reduced to two.

References

Bhaskar, K. (1979), *The Future of the UK Motor Industry*, London, Kogan Page.
Beynon, H. (1975), *Working for Ford*, Wakefield, EP Publishing.
Braverman, H. (1974), *Labor and Monopoly Capital*, New York, Monthly Review Press.
Brown, W. (1981), *The Changing Contours of British Industrial Relations*, Oxford, Basil Blackwell.
Brown, W., and Terry, M. (1978), "The Changing Nature of National Wage Agreements," *Scottish Journal of Political Economy*, vol. 25.

Chell, R. (1980), "BL Cars Ltd—The Frontier of Control," MA thesis, University of Warwick.

Clack, G. (1967), *Industrial Relations in a British Car Factory*, London, Cambridge University Press.

CSEU (1979), *Handbook of National Agreements*, London.

Dunnett, P.J.S. (1980), *The Decline of the British Motor Industry*, London, Croom Helm.

Elliot, J. (1978), *Conflict or Cooperation*, London, Kogan Page.

Francis, A., and Willman, P. (1980), "Microprocessors: Impact and Response," *Personnel Review*, vol. 9.

Goodrich, C. (1975), *The Frontier of Control*, London, Pluto Press.

Jeffreys, J.B. (1946), *The Story of the Engineers*, London, Lawrence and Wishart.

Ryder Committee (1975), *British Leyland: The Next Decade*, London, HMSO.

Scullion, H. (1981), "The Skilled Revolt Against General Unionism," *Industrial Relations Journal*, vol. 12.

Streeck, W., and Hoff, A. (1981), *Industrial Relations and Structural Change in the International Automobile Industry*, Berlin, International Institute for Management.

Turner, G. (1973), *The Leyland Papers*, London, Pan Books.

Turner, H.A., Clack, G. and Roberts, G. (1967), *Labour Relations in the Motor Industry*, London, Allen and Unwin.

New Technology and the Unions

Roy Moore and Hugo Levie

British labor unions have been slow to react to the introduction of new technology and have often lost the chance to influence events. These four case studies show that new technology can slice right through conventional trade union organization, exposing its weaknesses on many fronts. The authors are with trade union-orientated Ruskin College, Oxford and a fuller account of their research may be found in European Pool of Studies Information Bulletin no. 8, *European Commission, Brussels, 1982.*

This paper represents the main findings of the Ruskin College research project on "The Impact of the Introduction of New Technology on Trade Union Organization and Structure," which was sponsored by the European Commission.

The lessons to be learned from the research fall into two categories: substantive and methodological. As far as the former are concerned, the research has demonstrated that new technology does not emerge as a novel, separate, self-contained and standardized issue for unions. It is enmeshed with other bargaining preoccupations and with wider managerial strategies, but its character does tend to expose existing weaknesses of trade union structure, organization, activity and servicing more sharply than do other problems with which they deal.

At its harshest and most challenging, new technology can slice right through the hitherto seemingly rational logic of the structure and organization of trade union representation, and consequently can convert a strength into a liability. It can reveal the inadequacy and inappropriateness of the conventional range, level, and time horizons of collective bargaining activity in relation to the fundamental trade union aim of defending and advancing the interests of members. And it can expose the absence of reliable, relevant training, research, and servicing backup for workplace representatives whose function and role are changing, whose relationship with members is more tenuous, and whose strategic advantage in the current political, economic, and managerial climate is markedly diminished. A key precondition of any real improvement would appear to be the open, honest, and self-critical identification of these current weaknesses of trade union organization on the part of those representatives, both lay and full-time, who are ultimately responsible for initiating viable change.

This point leads directly onto the methodological side of the research lessons, for the project was deliberately designed and executed in close collaboration with the workplace representatives featured in each case study, in order for it to provide a learning experience for them as well as for the researchers (who all had a recent history of sympathetic assistance provided to local trade unionists). While the injection of an external stimulus and the research resource backup undoubtedly contributes to this process of self-awareness and self-appraisal, it need not be a necessary precondition. What is required for the benefits of the process to become more generally available is a recognition, first, that education and training in a trade union context should not be developed as a separate "ancillary" function to workplace organization and activity; and, second, that union research is not necessarily best executed as a centralized, desk-based head office function, but can benefit massively from being planned and executed in close collaboration with workplace and full-time representatives who are actually confronted by the problems that represent the subject of the research.

Impact of the new technology

Unemployment

The introduction of new technology is leading to job losses in all four firms studied. In the case of GEC, this is very marked; within BL it is more of a gradual process; the same is true of Alfred Herbert, but there the picture is even more dependent on the order book. Within Midland Bank the prospective job losses are enormous. However, from our studies another relationship between technical change and unemployment emerged: the present very high levels of unemployment make it extremely difficult for trade unions to develop adequate answers to technical change.

Early in 1981, official unemployment in Coventry stood at 16.7 percent and was still rising. In two years, unemployment there has trebled. The same picture emerges in Oxford, where British Leyland is by far the main private sector employer. The threat of unemployment is encouraging a much more quiescent attitude among those who are still working. In many companies the trade unions have not been particularly successful in coping with major reorganizations and mass redundancies. The unions are often fighting hard to maintain their bargaining rights, and even those fights are not especially successful in companies like BL, or for example the old Alfred Herbert's plant at Edgwick. Trade union answers to new technology must be seen against that background of high unemployment and a membership that is in many cases very scared for its own jobs.

When it comes to company reorganization, time and again management offers an ultimatum: either the proposed changes are accepted, or the future of the whole operation is endangered. With high unemployment in the local community, and members who are increasingly aware of this, it becomes very difficult for trade union representatives not to fall for such an imperative ultimatum. When it comes to the introduction of a specific piece of new technology, common practise in each of our four case studies is that the new equipment arrives without any prior negotiation, or consultation. High unemployment and its effects on the membership make it difficult for the unions to change that situation.

Collective bargaining
In none of the four cases we examined has technical change been negotiated. The effects of the changes might be the subject of collective bargaining, such as redundancies (GEC), changes in job description (Midland Bank), changes in working practises (Alfred Herbert), changes in manning levels and grading (British Leyland); however, such bargaining takes place after the technical change has occurred, after the investment in new technology has been made.

Whatever influence the trade unionists have had over the shape of technical change has come by stealth. The reason for this apparent ineffectiveness is not just that management in these four cases decides unilaterally the future shape of work in the company: it is also that the current patterns of collective bargaining are not very well geared to negotiating technical change. To explain this point, it may be useful to ask what negotiating technical change before the event would involve. The crucial assumption that active trade unionists have to make when they want to answer this point is that there are many alternative technical solutions to any one problem. Negotiating technical change in a particular situation does not mean accepting, or rejecting, new technology out of hand: it means negotiating for the kind of new technology that is acceptable to the workforce and the trade unions. Often this will involve fighting against the ideology put forward by management and accepted by many trade union members, that the proposal put forward by management is the only solution.[1]

If trade unions want to influence technical change before the event they can choose between two approaches: first, setting conditions to the use of particular pieces of equipment; second, preparing alternative proposals for technical change. An example of the former approach was found in our case study at GEC, where the staff unions, on the basis of an AUEW (TASS) initiative on health and safety, have established a code for the introduction and use of visual display units (VDUs) which gives them some control over the way VDUs can be used by the company. In our case study at British Leyland we found by comparison an example of the second, more far-reaching, trade union approach to negotiating technical change: at Longbridge, at the time of the

BL participation scheme, the shop stewards were able to influence the equipment, layout, and work organization for the car model that was then being planned, the Metro.

These examples show how new technology, the way it is used, and its effects on work organization can be influenced if there is an early trade union input. However, it must be admitted that these examples are rare. Part of the reason for that is linked to the first theme emerging from our case studies. One of the effects of high unemployment and a subsequent decrease in trade union influence at workplace level has been that principles like "mutuality" and "status quo," which would have given trade unions some bargaining power in cases of change of technology and work organization, have been eroded at BL, Edgwick, and to some extent GEC—places where the unions used to be strong. (This argument does not go for BIFU at Midland Bank, nor is it fully applicable to the staff unions at GEC.)

The rolling-back of trade union bargaining power is only part of the reason for the fact that in many cases members and trade union representatives are forced to accept new technology as a *fait accompli*. Unmistakably, the structure of collective bargaining and the way trade unions are equipped to influence the shape of industrial change both play an important role. Problems are to be found in union organization, the overloaded agenda of annual wage rounds, the mismatch between trade union and management organization and levels of collective bargaining (often chaos appears to reign, as at GEC), the shortcomings of existing negotiating procedures, and the lack of timely provision of information. As one trade unionist involved in our project said, "Trying to cope with technical changes via existing collective bargaining traditions is like fitting a six foot corpse into a five foot coffin!" The reality we found in our four case studies compares very unfavorably with the TUC book, *New Technology and Collective Bargaining*.[2] It would be useful to know whether trade unions in other European countries are suffering under a similarly overloaded state of existing negotiating arrangements—especially if the assumption is accepted that to no single problem can there be only one conceivable "new technology." The dangers of new technology to employment, skills, and democratic control over information are becoming increasingly clear. Collective negotiations appear, at present, one of the few ways by which society can control what new technology will be introduced. More knowledge about the effectiveness of collective bargaining and other trade union responses to new technology in various European countries appears a priority not just to the unions, but to society at large.

Company reorganization

Academic researchers may want to single out new technology as a subject of specific interest; for trade union activists it is different. Technical change does not come alone: in GEC it came with changes in orders from the Post Office for telephone equipment; in Alfred Herbert it came with a takeover, a total scaling-down of the company and product range, and a complete change in industrial relations; in Midland Bank it was part and parcel of a reorganization of all bank branches consequent upon a determined shift of corporate marketing strategy; and finally, in British Leyland, it came together with major changes in industrial relations and the phasing-out of old models. The introduction of new technology is embedded in many other changes, and that makes it more difficult for trade union representatives and researchers in the following ways.

1 The trade union representatives may be negotiating changes in the payment system or in work study procedures without being aware of the implications for management plans on technical change.

2 It may even happen that the shop stewards in one section unknowingly have accepted a change in work organization that will facilitate the introduction of new technology elsewhere. This can happen particularly where new technologies are integrated and cut across different groups of workers (draftsmen and toolroom workers, paint-shop workers and indirect workers at BL, machinists and programmers at Edgwick, draftsmen and clerical workers at GEC).

3 Problems in trade union structure and organization may imply that negotiations about company reorganization and new technology are held at one level, without adequate information about the effects on the union and its members at other levels. (This certainly appeared to be the case for BIFU at Midland Bank, where there was a wide gap in information about technical changes and company reorganization between the union representatives at national and local level.)

4 In many cases (Alfred Herbert springs immediately to mind, but the same is most certainly true for Midland Bank and BL), company reorganization and the introduction of new technology do not develop according to a scenario carefully planned in detail by management before the first change is made. There are too many variables, and management does not control them all. Nevertheless, management often has got a great advantage over the unions; because it is much better placed to know which variables are crucial at a given point in time and to act upon that knowledge, management will almost invariably have the initiative and will be able to make up the scenario as it goes along. Union representatives normally have not got the knowledge to know which variables are crucial in determining technical change—while management at least have a map and a torch, the union representatives have to maneuver by the light of the moon and their sense of orientation.

5 The fact that technical change is always just part of a whole series of changes at company or workplace level means that both researchers and trade union representatives have to divide their time between many important and related issues. The researcher who is interested in trade unions and technical change cannot concentrate solely on the new technology. The shop steward who is aiming at defending the members' longer—as well as short-term interests—cannot remain fixed on the day-to-day problems that may well be caused by big underlying changes in company and work organization. Both

the researcher and the shop steward will have to find a way to use their time effectively.

6 Many trade union representatives find that their existing facilities and resources are grossly inadequate to deal with company reorganization and new technology. The staff representatives at GEC found themselves in a position where they could obtain more time off from the company to be involved in our case study, which they saw as a priority. At BL the current management regime is keen to decrease the facilities to shop stewards, to make it even more difficult to develop a soundly based response to the changes at the plant.

7 A final reason why trade union representatives find it difficult to cope with the complexities of technical change is related to the points made under the previous theme. Conventional arm's length collective bargaining often does not touch upon the managerial decision-making procedures that lead to technical change.

Disclosure of information

To be able to respond to technical change in a way that is constructive, takes full account of the members' interests, and pays respect to related changes in work organization, trade union representatives will need information at a time when they still could help shape the decisions. The starting point may be a total reorganization of the main production lines in the company as at Midland Bank; the introduction of a new product, such as a new car at BL, or components for System X at GEC; or a paring down and standardization of the product range and production process, as at Edgwick. Whatever the starting point, the unions will need early information about those proposed changes and the related investment and manpower plans.

Unfortunately, the timely provision to all concerned of detailed financial, economic, and manpower data was not something the four companies we dealt with in our research considered as normal practice. The provision of information about investment alternatives, with an indication of their respective consequences for work organization and employment, was even more out of the question. This state of affairs is deplorable, because we found that trade union representatives who want to anticipate technical change have a dire need for basic information about the financial structure and investment history of their company. In particular, if they want to develop a strategy that goes beyond *ad hoc* relations to the piecemeal introduction of new technology, such information becomes vital. We are not suggesting that better provision for the disclosure of company information would automatically lead to much more developed and sophisticated trade union responses to the introduction of new technology.[3] However, a near total lack of relevant information certainly did not help the unions in the four case studies to cope with technical change. The fact that many trade union representatives lack basic information and understanding about the financial and management structure of their company exacerbated the lack of more specific data on the changes we examined.

To counter this basic problem a number of staff representatives at GEC have decided to follow union education courses on financial information. Another indication of how seriously trade unionists take the lack of company information emerged from the questionnaire survey that we undertook at BL. Overwhelmingly, both shop stewards and members answered that lack of information about management's plans for the plant was one of their most serious problems. In fact the members said that, apart from the local press, the stewards were their only source of information.

We are aware of various forms of legislation in Sweden, Norway, Germany, and Holland that give trade unions access to at least a minimum of financial and manpower information. It would be relevant to know whether such access puts those trade unions at an advantage when it comes to dealing with technical change.

Information about the industry

1 Fifteen years ago Alfred Herbert was still one of the world's leading producers of machine tools. Today the firm as such no longer exists. The only thing that is left at Edgwick, its old center, is a medium-sized machine tool plant employing 700 people. New technology played an important role in Alfred Herbert's downfall: while German and Japanese competitors were investing in new technology, both for their production processes and in their production lines, management at Alfred Herbert in the 1960s and 1970s was not investing sufficiently, or was even disinvesting. The conclusion to be drawn from this history of Alfred Herbert by the trade unions at Edgwick is that they need information about the machine tool industry overall as much, if not more than, information about the company itself if they want to assess the options for the future.

2 Midland Bank is watching organizational and technical change in the banking world in North America and France closely. Midland Bank's branch network reorganization is not growing as rapidly as management planned originally. One of the reasons for this is technical: management appears to be waiting for the most appropriate technology to appear on the market (albeit as a result of trial and error to an extent).

3 The number of people employed in GEC's Telecom Division and the work they do depends on the place of the company in the telecom industry, the orders from British Telecom and possible orders from foreign telecommunication authorities. If the Post Office (now British Telecom) decides to decrease its investment in System X, to spread it over a longer period, or to buy less from GEC and more from its competitors, this may have immediate effects for the workforce in Coventry.

4 What cars are built at Cowley, and where the components originate, depends increasingly on collaboration between BL and car firms like Honda

and Volkswagen. The way they are built and the tools that are used to build and design them are also partly dependent on systems developed by other car companies. The stewards at Cowley require, if anything, more links with other firms. Information about the industry must therefore be important to them.

In each of our case studies, in very different ways, information about the industry is vital if trade unions want to influence new technology before it arrives.

Equal opportunities

Who suffers most from the introduction of new technology? Our case studies all point in that same direction: Asian workers and women in GEC, women in Midland Bank, older workers within BL, women in the case of GEC. They are the groups that suffer most. The picture that emerges is that new technology does not enhance equal opportunities. On the contrary, things seem to be getting worse, particularly if trade unions fall back on older, defensive, craft traditions; if trade unions fail to increase the chances of their female members to become involved in trade union activities; or if trade unions allow the views that tend to gain popularity in times of depression in certain quarters to spread among their members ("women should not work, but be at home and look after the children"; "colored workers are stealing our jobs, they do not belong here"; "workers who are too old to do a proper job should not take the place of able young men").

The magnitude of the threat of new technology to equal opportunities was, in our case studies, particularly apparent for women.

1 There is both a loss of existing jobs for women and a lack of recruitment of women in all four companies. This is not just true for traditional women's jobs, such as secretaries and clerks. In the 1960s there were several hundred women on the shopfloor at Edgwick: presently there are two left. We suspect that the same is now happening at BL and GEC.

2 The employment effects of new technology hit part-time workers first; at least, this has been true at GEC, where they were the first to be made redundant. The majority of those part-timers were of course women. In more general terms, it appears that the introduction of new technology may be coupled with an increasing casualization of sections of the workforce, and the brunt of this casualization is carried by women.

3 In the case of Midland Bank, the women's jobs, e.g. the clerical and administrative jobs, are automated first and most. This implies that the remaining women have a good chance of seeing their jobs most and worst affected by new technology. Worst, because the boredom, stress, and isolation that accompany machine-minding is going to affect women in Midland Bank much more than in the past.

4 The already poor career prospects for women at Midland Bank are going to worsen, because of a restructuring of the career pattern. The general interest of the banks in creating career possibilities for women is reflected in

the directive, sent by one of the other major banks to its managers, which told them not to take on any woman with more than four O levels. In other words, a woman with any fancy idea of making a career is barred from entering employment.

The recent document of the European Commission (Com. 578, p. 15) previously quoted points to the problem of equal opportunities:

Moreover, the skills favoured by microelectronics are characteristic of posts at present mainly occupied by men, and without a special effort in this direction it may be that the diffusion of the new information technologies throughout the production process will reinforce labour market inequalities.

In direct reference to this quotation, our case studies point out the following:

1   New technology *is already reinforcing labor market inequalities.*
2   The "skills favored by microelectronics" is an unfortunate and unhelpful expression. *There is no indication at Midland Bank that employment or career opportunities for women are obstructed because of new technology that does not favor their skills.* The same is true for clerical and design work at GEC. The reasons are to be found elsewhere.
3   Finally, our case studies fully substantiate the opinion of the European Commission and other institutions, which say that a special effort should be made to provide retraining for women and other groups who are particularly affected by the introduction of new technology. However, as long as employers are allowed, in practise, to discriminate against women who want to work in jobs that are not traditional women's jobs, retraining may not be enough.

New technology and trade union organization

Divisions between unions

Clearly, different groups of employees (unskilled and skilled, technicians, designers, foremen, computer specialists, clerical workers, managers) have different interests when it comes to technical change. Some will be threatened more than others; some will feel, rightly or wrongly, that with new technology their services will become more indispensable. A first rumor about the introduction of new technology may lead to very different reactions from the various groups.

The arrival of manually instructed numerically controlled machine tools may lead to a running battle between the programmers and the machinist. This was certainly the case at Edgwick. There the machinists (organized in the AUEW-Engineering Section) gained a victory over the programmers (AUEW-TASS); a "victory" based upon the old craft traditions, which is divisive in the short run and may turn out to be temporary and relative.

At BL a lot of the energy of the manual trade unions is spent on making sure that other manual unions do not gain anything out of the changes in work organization that management is imposing. A major dividing line runs between the TGWU, representing the semi-skilled workers, and the AUEW-Engineering Section, as the major union for the skilled workers. There are many further divisions between groups of workers, sometimes organized in the same union. The answer to the threat of new technology given by many of those groups of workers is to defend "their" jobs against other sections of the workforce that may want to snatch them away. As researchers, we were frequently left with the impression that, while management changed the rules and relocated the pitch, the unions are happy to continue fighting among themselves back in their own backyard.

These real or perceived differences in interest between groups of workers are not something specific to the UK; what is specific, however, for British industrial relations is that in many companies those different groups of workers are organized in different trade unions. Consequently, the introduction of new technology may well exacerbate poor relations between unions, particularly in companies where they are already strained. Of our four case studies, Midland Bank is a positive exception, with practically a single union. In the other three, shop stewards recognize the danger described above, but only in the case of GEC do the trade unions seem ready to try to work more closely together.

Weaknesses in trade union organization

New technology exposes a number of existing weaknesses in any trade union organization. The value of the present interest in technical change may well be that it brings some of these problems home. In our case studies we found that trade union representatives are becoming aware of these problems as constraints upon their ability to influence technical change. A simple reason for this exposure could be that, for many trade union representatives, it may be the first time that they are trying to influence something as central to the organization of work as technical change. The following questions are generated by our research:

1   Are the levels at which the trade unions negotiate with the company matched to the levels at which the company takes its decisions? Or are the unions always talking with the wrong management team?
2   Do individual shop stewards feel themselves supported in their daily work as union representatives by the senior stewards and the union at large?
3   Is there sufficient contact between union representatives and members?
4   Is there a shop stewards committee that combines information from the different offices or departments where the members work?
5   Is there adequate exchange of information with other unions at the plant, or (as in the case of Midland Bank) with other areas of the union?
6   Are the various sections of the membership (men and women, clerical and technical) involved in union activities, or is the union, locally and nationally, dominated by a specific group?

Role of the union representative

New technology and the related changes in work organization may well force a trade union to rethink the role of the individual shop steward. At the British Leyland plant the introduction of new technology is in some cases linked to a total reorganization of major production areas. This is coupled with major redundancies and redeployment, and the consequence is the emergence of a smaller number of shop stewards who are relatively inexperienced. These new stewards and the older, more experienced, stewards in areas where new technology was introduced more gradually have to cope with changing production methods and an increasing integration of different areas of the production process. This last aspect especially changes the role of the stewards: they become much more dependent upon one another than in the past. The unions within Alfred Herbert, and the Banking Insurance and Finance Union (BIFU) at Midland Bank, have to redefine the role of union representatives, but for different reasons. In the first case it is because the changes in collective bargaining at the plant have affected the relations between senior stewards and other stewards, and because most of the experienced trade union representatives have been sacked anyway. In the case of Midland Bank it is because the whole concept of office representatives is relatively new within the Bank. It is to be expected that in these two cases technical and other changes in the work organization will influence the future role of union representatives.

At GEC there appear to be several influences upon the changing role of staff union representatives:

1   as senior representatives of the staff unions seek to increase their unions' contact with their membership, the role of the sectional representative necessarily becomes more pronounced;
2   the increasing number and presence of female staff representatives can positively influence the way that representatives perform their duties and perceive their role;
3   the use of health and safety procedures as a vehicle for expressing doubts about, and seeking more control over, new technology can change the staff representative's role; and
4   increasing cooperation between the various white-collar unions inevitably affects the work of individual trade union activists.

Thus, in each of the four case studies it is possible to identify pressures upon the role of the individual union representative that are associated with technological change and related changes in work organization and industrial relations. In no case are the pressures necessarily negative: rather, they may well enable the trade unions concerned to strengthen their presence at the workplace in the longer term through an enhanced role for their workplace representatives. This, however, is conditional upon unions giving priority to servicing their representatives and strengthening their internal democracy. Otherwise there remains a real danger that individual workplace representatives having to cope with technical change would feel "out on a limb."

Effects on trade union organization

Having indicated how the introduction of new technology can expose existing weaknesses in trade union organization, and how it can affect the role of an individual union representative, this report now considers the direct impact of new technology upon union organizational patterns at workplace and company levels, and its consequences for inter-union organization. Our case studies included the following examples.

1 A consequence of the introduction of new technology and branch network reorganization within Midland Bank will be that the relative significance of groups of workers will change. This brings with it implications for trade union effectiveness as well as for career and promotion patterns, as has been witnessed, for example, by the growing significance among union membership of computer staff in the banking world in the UK and other European countries.

2 Also in the case of Midland Bank, it is clear that the reorganization of the bank branches combined with the introduction of new technology will force the union, BIFU, to rethink not only the role of office representatives, but also the place of those representatives in the local and national union organization. Also, BIFU has to reconsider how it can facilitate more involvement in union affairs of its female members, who are especially threatened by technical change.

3 Three factors have combined to change department relationships at the BL Cowley body plant: the redevelopment of the site; the introduction of new production machinery and new information technology into such areas as design, stock control, and maintenance; and the introduction of new production methods. The resulting trend is for departments to become more interdependent, which inevitably calls for increased communication between different sections of a union as well as more collaboration between unions. Shop stewards become much less independent in their decision-taking than in the past. They have to rely upon information from other sections to help them understand the reasons behind changes in their own departments, or to assess whether it would constitute a precedent for them to accept new management proposals on an issue such as work study.

4 The Edgwick case study demonstrates how the union representatives have to reshape their own shop steward's organization. The position of sectional stewards will have to be strengthened, for example by ensuring that a member with a problem or grievance approaches the steward of his or her department, instead of the senior steward. Such qualitative improvements in a new shop stewards' organization are a precondition of the development of the more sophisticated trade union strategy that appears to be required.

5 Examples of changes in union organization as a response—partially at

least—to new technology are perhaps the clearest and most constructive in the GEC case study. The threats posed by the production of System x have forced the staff unions to work toward a joint union organization at plant— and possibly even at company—level.

These examples from our case studies demonstrate how technical change may bypass or even outdate existing ways in which a trade union monitors change in a company and develops strategy accordingly. It seems important to note that this point is not specific to the UK; it would be very interesting to know how unions in other countries, which are organized differently, deal with this.

New technology and public opinion
An assumption underlying our research approach is that controlling new technology cannot be reduced to an issue for a specific group of workers, and their trade union, facing the introduction of a new piece of equipment. Controlling technical change is also a wider social issue. As things stand, however, collective negotiations about the introduction of new technology appear to be one of the few possible—albeit piecemeal—means available to society to exert any control over that major social issue. The effectiveness of trade union representatives in attempting to negotiate technical change depends to a great extent on the views of their members who will be affected by this change. Such views are often already formed in general terms, and may be influenced only marginally by the specific, particular technical change under negotiation.

The media are giving a lot of attention to new technology. Its advantages and disadvantages are covered extensively. However, this leaves trade union members with very general and vague opinions, like:

"New technology is inevitable."
"The chip threatens employment."

or

"The introduction of new technology will make British industry more competitive."

Our case studies have shown, by contrast, that, at the more specific local, or company, level, the formation of the opinion of trade union members is, certainly in its early stages, often left to the employer. Midland Bank shows its employees a film about the benefits of branch network reorganization and technical change, well before the change will affect them. Leyland Cars issues a booklet to its workers about the wonders of the new Japanese model and the related new working practices. It seems undesirable to leave trade union members with just their vague, general views about new technology to cope with this much more specific employer-orientated information. It is a very positive development that BIFU is considering preparing its own specific information about the changes that management is implementing. The same would be true if the GEC unions in Coventry went ahead with the idea to start a public debate about the consequences, for a city that is heavily dependent on the telecommunications industry, of the introduction of microelectronics, both for producers and users of telecommunications equipment.

Trade union servicing

Education

Technical change does not come alone. Trade union representatives have to cope at the same time with a lot of changes, ranging from redundancies and the effects of high unemployment to changes in payment systems and job descriptions. It is not accidental that the TUC book on *New Technology and Collective Bargaining* mentions the following bargaining issues:

1   skills and work organization;
2   training;
3   work and pay;
4   information and control;
5   health at work;
6   shorter working time;
7   products and services.

Our detailed case studies of the impact of new technology on trade union organization, and any associated influence of union organization upon technical change, carry direct implications for trade union education. In the case of the Transport and General Workers' Union at BL, the need to improve educational facilities for shop stewards was the real driving force behind the senior stewards' involvement in our research work. A cornerstone of the latter was a questionnaire survey of more than 200 stewards and members, which fully substantiated this perceived need, of both union representatives and members generally, for more education. The establishment of good basic shop steward training on the role of the shop steward, work study, and health and safety at work, together with more advanced, specific courses on new technology, thus appears as a virtual precondition for the union's capability to get a better grip on technical changes at the plant.

At GEC, staff representatives are taking advantage of their unions' educational courses to increase their insight into the financial and managerial structure of their employer. They hope as a result to understand better the information that might enable them to deal more effectively with GEC's investment strategy for telecommunications and its plans for new technology. Another priority adopted by the staff representatives during our case study has been that of union membership education.

The need for a reappraisal of trade union education facilities both constitutes a major general conclusion of our research, and represents a specific priority adopted by trade unionists involved in each of our case studies. BIFU in particular has acknowledged the educational implications of our work with their representatives at Midland Bank. It is hoped that the respective case study report will itself serve as educational material on various courses for newly elected trade union representatives, female representatives, and members, and for representatives and members elsewhere in Midland Bank who have not yet experienced branch network reorganization; and on special courses on new technology. As well as providing such specific additions to existing courses, it is possible that our research findings may also help the union to rethink its overall education program.

Precisely because new technology cannot be isolated from other bargaining issues, or from existing weaknesses in trade union organization, our case studies have shown a great need for more trade union education. Existing and possibly new courses should enable trade union representatives to assess present and future changes, both at their workplace and in their industry. It should also enable them to rethink their strategies on all the issues mentioned above, and on their cooperation with other stewards in their own union and in other unions.

Research
Our research has indicated the potential of detailed case studies on technical change that are developed together with the trade union representatives in a specific company or plant. It seems particularly important that these case studies are developed from a broader working relationship between researchers and trade unionists. It has been the latter characteristic of our work that has made it possible for our studies to become an educational exercise for all involved parties, and not just for the researchers.

It is clear to most people, both within and outside the trade union movement, that a struggle for some control over technical change and any related work reorganization will be one of the most important tasks for trade unions. If trade unions fail to improve their ability to influence such changes, the penalties will be high in terms of unemployment and social dislocation as well as industrial relations disruption.

The foundations for the development of such trade union capabilities lie less in detailed, separate treatment of new technology as a separate issue on the union agenda than in an overall appraisal of the effectiveness of trade union organization, structure, and means of activity. A higher priority and increased resources for trade union education and research are integral components of such a reappraisal.

It would be encouraging to believe that many more groups of shop stewards could have the opportunity of being involved in research activity such as this project, which could enable them to develop their understanding of the changes occurring in "their" company and "their" industry. Such a stimulus could help them—perhaps even force them—to work together more closely with other representatives of their own and other unions within their workplace and the wider company, or industry, beyond. It could also improve their ability to negotiate socially acceptable technical change, in both a local and a wider sense.

Our four case studies have helped to raise the understanding of all concerned. But they have inevitably fallen short of actually helping the four groups of trade union representatives to influence technical change. For this, longer-term research on a wider basis, and even more integrated into the trade union movement, would be required. Such research would need to be both locally and nationally based, and to encourage and rely upon the direct involvement of active trade union representatives and members. It will also cost money—a lot of money. But, in its absence, many trade union representatives will continue to deal with new technology from a disadvantaged position, "out on a limb," the costs of which could be far greater and could involve more than money.

This project suggests that conventional research techniques based upon externally organized investigations may fall short of an accurate representation of practitioner's perceptions of issues. Equally, it questions the viability of continued reliance by trade unions upon education as a separate function from trade union activity, and research as a centralized desk-based undertaking.

This is not however to say that there is little contribution to be made by trade union research departments in assisting their union representatives to deal with the introduction of new technology, or to imply that resources available for the development of a research capability on behalf of trade union representatives need be statically limited. On the first count the development of case study research into education and training materials is an obvious dual function for trade union servicing, as in the lubrication and extension of channels of information, "early warning" and "best practice," both within unions and between them. Research departments can also be instrumental in demystifying new technology, its apparatus and applications, and the managerial systems and strategies that accompany its introduction. And, not least, research departments are well placed to monitor and even anticipate what is happening at the frontiers of new technology, so that at least the next generation of technical change can be handled with more confidence by trade union representatives.

On the second count, several possibilities for amplifying resources available for research on behalf of trade unions can be identified. TURU has already drawn attention to the legislative developments in Sweden, which have provided for consultants operating on behalf of unions but financed by employers.[4] A companion paper[5] considered problems and possibilities for research sympathetic to trade union needs, and TURU continues to explore the possibility of "plugging in" to the resources of research-funded agencies on behalf of trade unions, who are surprisingly reticent applicants for funds to which they have a legitimate and competitive claim. There is no reason why research-funding agencies should continue unchallenged in regarding academic institutions as the natural and exclusive depositories of their resources. Further, unions might consider negotiating for research along with the other items that comprise the agenda of their collective bargaining: provision for analysis and appraisal of the outcome of new agreements—or even the introduction of new technology itself—could become a valid and viable subject of research by the unions' own research department or by consultants or independent researchers acting on the union's behalf.

None of this can be unleashed, however, without an open, honest, and perhaps at times self-critical appraisal by trade unions of their fundamental aim of defending and advancing the interests of their members. Neither a European Commission research grant nor the external stimulus of outside researchers "creating space" for trade union representatives need be necessary preconditions of such a process (though it is hoped that they helped considerably in the case studies that formed the backbone of this research project). What is needed is initiatives by those trade union representatives, both full-time and lay, who are ultimately responsible for initiating viable change. Ironically, the adversity of dealing with new technology may yet provide the climate for such initiatives.

Notes

1 This ideology unfortunately is widespread. In a recent document by the European Commission to the European Standing Committee on Employment, "New Information Technologies and Social Change" (Com. 81, 578 final, October 12, 1981) this ideology of an unambiguous choice between "the" new technology, or decline returns in the following way: "It is undeniable that in present labor market conditions, rationalization acts to increase productivity and may in consequence increase unemployment. But from a company point of view, introduction of new technology improves competitivity and may enable them to retain or increase their share of the market. Further, the economic constraints arising from national and international competition should also be noted: a firm which did not introduce these technologies would risk the loss of its markets and put existing jobs at risk. Consequently, there is no escaping from the introduction of the new technologies in our societies."

2 *New Technology and Collective Bargaining, a Workbook for Union Representatives* (TUC Education Department, London, 1981).

3 In *The Shop Stewards' Guide to Company Information* (Spokesman, Nottingham, 1979) we identified many factors that may constrain a union's use of company information.

4 *Employee Consultants and Information Disclosure: Some Notes on the Swedish Experience*, Discussion Paper no. 21, TURU, July 1980.

5 *Research on Employees Conditions: A Swedish Trade Union View*, Discussion Paper no. 22, TURU, May 1980.

Guide to Further Reading

Management's "problem"

"The New Industrial Relations," *Business Week* Special Report, May 11, 1981.

Brian C. Twiss (ed.), *The Managerial Implications of Microelectronics* (Macmillan, London, 1981).

Roderick Martin, *New Technology and Industrial Relations in Fleet Street* (Clarendon Press, Oxford, England, 1981).

Steve Early, "Unions and Management in Europe Seek to Ease Transition to New Technology," *Transatlantic Perspectives*, February 1982.

Don Mankin, Tora K. Bikson, and Barbara Gutek, "The Office of the Future—Prison or Paradise?" *The Futurist*, June 1982. Makes the point that technology will not determine the working atmosphere in the office of the future—management will.

Yasuo Kuwahara, "The Japanese Way of Robot Life," *Employment Gazette*, London, August 1982.

Martin Anderson, "Shake-up in Detroit: New Technology, New Problems," *Technology Review*, August–September 1982.

Shoshana Zuboff, "New Worlds of Computer-Mediated Work," *Harvard Business Review*, September–October 1982. New technology offers managers the chance to enrich jobs and to develop new forms of work organization, thus transforming old-style management–labor relations.

Henry C. Lucas, Jr, *Coping With Computers, A Manager's Guide to Controlling Information Processing* (Free Press, New York, 1982).

"How Computers Remake the Manager's Job," *Business Week*, April 25, 1983.

John Thackray, "America's New Technology Trap," *Management Today*, June 1983. On new vogue for "technology management" (an elusive idea).

Peter Nulty, "How Personal Computers Change Managers' Lives," *Fortune*, September 3, 1984.

A comprehensive literature review may be found in John Child, "New Technology and Developments in Management Organisation," paper from ssrc Work Organization Research Centre, University of Aston, September 1983.

Graham Winch (ed.), *Information Technology in Manufacturing Processes, Case Studies in Technological Change* (Rossendale, London, 1983).

Lowell Steele, "Managers' Misconceptions About Technology," *Harvard Business Review*, November–December 1983. On unrealistic assumptions about new technology.

Sheila Rothwell, "Supervisors and the New Technology," *Employment Gazette*, London, January 1984.

Susan Fraker, "High-Speed Management for the High-Tech Age," *Fortune*, March 5, 1984. On the growing problem of products with a very short life.

Malcolm Warner (ed.), *Microprocessors, Manpower and Society* (Gower Press, Aldershot, England, 1984).

Jon Clark, "Management Strategies and the Introduction of New Technology," paper to British Sociological Association conference, Bradford, April 1984. Summarizes work of the New Technology Research Group, University of Southampton.

The union response

The union response and the union point of view is covered in:

Harley Shaiken, "Computers as Strikebreakers," *Technology Review*, April 1982.

Negotiating Technological Change, a Review of Trade Union Approaches to New Technology in Europe (European Trade Union Institute, Brussels, 1982).

Robin Williams and Russell Moseley, "Technology Agreements: Consensus, Control and Technical Change in the Workplace," paper to an EEC/FAST conference which appears in Nils Bjorn-Anderson et al. (eds), *Information Society: For Richer, For Poorer* (Elsevier, Amsterdam, 1982).

"Survey of New Technology," Labour Research Department *Bargaining Report*, no. 22 (London, 1982). Looks at New Technology Agreements (NTAs).

Ian Benson and John Lloyd, *New Technology and Industrial Change* (Kogan Page, London, 1983).

International Association of Machinists, "Workers' Technology Bill of Rights," in *democracy*, Spring 1983. Calls for NTAs to cover job displacement, retraining, data protection, participation, the environment, and equality.

"New Technology Special," *Labour Research*, November 1983.

Part Four: Implications for Society

Computer Crime

Leslie D. Ball

New technological developments such as the personal computer and distributed processing in organizations have made computer crime much easier—for all grades of staff. Estimates vary as to its true extent, but swift action is needed if losses are to be curtailed. The author is an assistant professor at Babson College, Wellesley, Mass., and this piece first appeared in Technology Review, *April 1982.*

Virtually every company or organization, no matter how small, is a potential victim of computer crime. The explosive growth of computers, together with the escalating value of the data handled and stored, prompts many authorities to predict that criminal computer capers will, if unchecked, increase significantly in the years ahead.

Actual losses from computer crime are difficult to estimate. Donn Parker, a senior systems consultant for SRI International and an authority on computer fraud, calculates that losses may be $5 billion per year. However, other estimates set losses at only $300 million. Even this lower figure is more than 20 times the annual take of just a decade ago.

Complicating matters, many experts believe that only 10 percent of computer crimes are made public and numerous crimes probably go undetected. What is clear, however, is that many computer crimes involve large sums of money—Parker estimates the average computer bank fraud at $500,000, while conventional bank robbers average only $2500. For example, in 1980 the Wells Fargo National Bank in San Francisco lost $21.3 million in what reportedly is the largest bank embezzlement in the nation's history. This computer-aided scheme, involving a former bank officer and several boxing promoters, caused losses totaling more than half of all US bank robberies that year. (Although the cases I will cite are well documented, industry has questioned the exact role of the computer in many "computer" crimes. See box, page 543.)

Who gets robbed

Banks often figure in computer crimes because of the increasing use of electronic funds transfer systems, which move huge sums of money among banks with electronic symbols as the only record. This system replaces the personal signatures that once accompanied every banking transaction with electronic authorization codes, a series of unique digits identifying any bank official who can authorize money transfers. Large numbers of financial transactions can be made quickly and at less expense using such a system, but trouble brews if the codes fall into the wrong hands.

For example, in 1979 Stanley Mark Rifkin, a computer consultant and former college professor, stole $10.2 million from the Security Pacific National Bank. After learning the computer's access codes during a visit to the bank's wire transfer room, he simply telephoned the bank and, posing as a branch manager, used the codes to transfer money to a New York bank—always transferring amounts less than $1 million, since such "small" transactions are subject to fewer internal bank controls. Next, he instructed the New York bank to transfer his funds to a Swiss bank. He then went to Switzerland, bought diamonds, and returned to the US. Only after boasting of the feat was he caught, and while awaiting trial he attempted a $50 million transaction from another bank. When apprehended, Rifkin told a reporter that he thought he finally had all the bugs worked out.

Banks are robbed in other ways, too. The American Institute of Certified Public Accountants (AICPA) has conducted one of the few scientific studies of the characteristics of computer crime. Of 85 bank cases studied to date, 13 involved fictitious loans, 8 involved unauthorized lines of credit, and the rest involved various forms of transaction manipulation. Most of these crimes have not been of the spectacular variety, with 70 percent yielding less than $25,000.

Computer criminals also favor insurance companies. In fact, one of the largest computer crimes so far discovered—totalling over $27 million, though actual losses may never be known—is the Equity Funding fraud. From 1965 to 1971, the company used its computer to write thousands of phony insurance policies, and then sold those policies to companies called reinsurers. According to the AICPA study, other kinds of insurance computer crime include fictitious claims, fraudulent loans against customer policies, and the switching of addresses and canceling of policies to gain premium refunds.

After theft of cash or assets, the next most common computer crime is stealing goods, since inventory items are often easily converted to cash. Falsified computer records can make it seem that goods were damaged and disposed of, shipped to a customer but returned, or simply missing.

The Jerry Schneider case is a perfect example. He was an 18-year-old college student when he started stealing from Pacific Telephone, convincing the company's computer that the thefts were actually internal transactions. Posing as a magazine reporter doing a story about Pacific Bell's parts-distribution system, Schneider learned that requests for parts come in via touch-tone telephones and that parts are delivered to any specified location. Now he was in business, entering daily orders and picking them up at various spots. He then sold the products through his company, Creative Telephone. However, an employee soon caught on, asked for a cut but was refused, and turned his boss in—but not before more than $1 million in phone equipment was stolen.

The wealth of information stored in computers, which often has considerable value to business competitors, is another tempting target. For instance, a large computer manufacturer used a computer system to store records of machines ordered but not yet shipped. Unfortunately, an employee with access to the computer system sold the information tó another manufacturer. Thus, the competitor could approach a buyer and offer a similar system at lower cost and with an earlier delivery.

Information theft is destined to grow more troublesome as word processors replace typewriters in many offices. These systems are really small computers and are often connected to a company's central computer. Easy to operate, their potential for misuse is staggering. Typewriters do not store information when a report is finished—but with word processors, data stay in the system for future recall.

Another common computer crime involves changing the information stored, often to induce a company to make decisions different from those it would ordinarily make. Such a crime occurred recently at a New York college that was automating its grade-reporting system. Students were hired for the monumental task—which required that earlier grades be key-punched onto computer cards—and some began a land-office business in making a C a B or even an A. Someone should have remembered that foxes shouldn't guard chicken coops.

Theft of computer time is also a perplexing problem. Although employees are often urged to use the computer, their use can become improper or excessive. For example, at most colleges computer time is thought of as a free good. Students and faculty often computerize mailing lists for their churches or fraternal organizations, which might be written off as good public relations. But use of the computer for private consulting projects without payment to the university is clearly improper.

The picture is similar in business—management often looks the other way when employees play computer games or generate Snoopy calendars. But if this becomes excessive, the employee is "stealing" work time. And computers can process only so many jobs at once: analysis of biorhythm charts and Uncle Harry's taxes could delay the company's payroll. Although considered less severe than other computer crimes, such activities can represent a major business loss. If a technical employee spends half an hour each day on the computer—which is not unusual—the cost in computer time is over $3000 per year, not including delays in normal processing operations.

Other events sometimes classified as computer crime include theft of programs (software is one of the fastest-growing segments of the industry), theft of computer components or even small terminals, and the destruction of an organization's computer system. For example, a University of Minnesota study indicates that, after only 4.8 days without computers, most companies cannot function properly. Computer destruction is generally due to an accidental event such as a fire or storm, but deliberate destruction has occurred.

Who does it

It is tempting to liken computer criminals to other criminals, ascribing characteristics somehow different from "normal" individuals, but that is not the case. Donn Parker of SRI believes the computer criminal "often marches to the same drum as the potential victim but follows an unanticipated path." Many otherwise trustworthy employees have committed crimes after accidentally uncovering flaws in the computer system or gaps in the controls monitoring their activities. If the payoff appears high enough, formerly honest individuals could easily succumb to such temptations.

By analyzing computer crime reports and interviewing those involved. Parker has compiled an interesting profile of the computer criminal. Such criminals tend to be relatively honest and in a position of trust; few would do anything to harm another human, and most do not consider their crime to be truly dishonest. Most are male; women have tended to be accomplices, though of late they are becoming more aggressive. Between the ages of 18 and 30, they are usually bright, eager, highly motivated, adventuresome, and willing to accept technical challenges. Actually, they sound like the type of person managers would most like to employ.

Their positions range from data entry clerk to company president. While direct access to a computer terminal is helpful, they need only a job allowing them to generate data entered into the computer. And, perhaps ironically, most computer criminals who have been caught have had no formal computer training; only a modest amount of knowledge seems necessary. They commit frauds in the areas they best understand: payroll clerks steal from payrolls, not from accounts receivable.

How they do it

Asking how many ways a computer can be burgled is much like asking how many angels can dance on the head of a pin—just as you think you have heard them all, a new method appears. But criminals use certain methods repeatedly, though with many variations. Most common is taking advantage of a lack of control on certain types of transactions. For example, dormant bank accounts receive less attention than active accounts, being checked less often by auditors. A branch manager at New York's Union Dime Savings Bank—who was also a compulsive gambler—knew this and used the computer to "borrow" from such accounts to pay gambling debts. In industry, if a data entry clerk learns that computerized inventory counts are never verified, the clerk could easily enter a lower figure and then take the items from inventory.

A similar gambit is to take advantage of an error in a computer program. A "bug" can remain undetected for years, particularly if located in a set of rarely used instructions, and often is found only by chance. For example, a benefits clerk entering incorrect data on a specific type of medical claim might uncover a program mistake. Although the clerk finds the error, the computer may not—and the clerk faces strong temptation to use the bug to file fraudulent claims.

Deliberate changes in a computer program can lead to a variety of misdeeds. Sums that are difficult to monitor or unlikely to be questioned are easiest to manipulate. Checking account fees are in this category. A bank employee could prepare a program that overcharges randomly selected accounts by a small amount (less than the cost of a postage stamp, for instance, to avoid customer complaints), netting a tidy sum for deposit in the employee's own account. Or, in an actual case, a company employee writing a program to generate annual w-2 statements included instructions to deduct $4 from everyone's federal taxes. He added the total to his own withheld tax, printed his w-2 with the incorrect amount, and received a massive refund from Uncle Sam. However, a janitor complained to the payroll department that his w-2 was wrong, and the scheme was unraveled. The programmer's error was in thinking that no one would notice such a small error.

Programs can be processed more than once to cover up schemes. Frequently, a miscreant will write two programs for a given task, one correct and one fraudulent. After the fraudulent program has generated purchase orders, checks, or other financial instruments easily converted to cash, the correct program is processed to cover up evidence of the additional transactions.

Many of these frauds can be accomplished by manipulating the computer's "operating system"—basically, a program controlling all the other programs processed by the system. Programmers who work with operating systems are called "systems programmers" and are assigned a master account number, much like a building superintendent with a key that opens everyone's apartment. If someone can gain control of a computer's operating system by obtaining a master account number, he or she can copy or change data, change programs as they are running and restore them after completion, or restrict access to various users. In the fall of 1980, for example, two teenagers programmed a home computer randomly to generate master account codes for De Paul University's computer. They reached the university computer by telephone, using a device called a "modem" that translates digital signals into analog signals suited for the phone lines. It does the reverse when the signal is returned from the computer. They succeeded after several thousand attempts (taking only a few seconds), and prevented the university from using the system for a week by changing all the master codes.

Computers can also be used as speedy "tools" to facilitate illegal or questionable activities. Several students from the California Institute of Technology wrote a program that churned out 1.2 million entries for a MacDonald's restaurant contest. Having nearly a third of the entries, they won the major prizes and many smaller prizes. What made this possible was the students' free and unlimited access to Caltech's computer and MacDonald's policy of allowing entries to be printed on plain paper. So the students' actual crime was theft of valuable computer time, although MacDonald's officials did not quite see it that way.

Another famous case depending on the speed of computers occurred at a Florida greyhound racetrack. To win a so-called "trifecta," betters must pick the exact order of the top three dogs. There are few winners, and the payoffs run to several thousand dollars. The number of possible winning combinations is enormous, so the odds are not calculated until after all bets have been placed—and then only the winning payoff is made public.

In the fraud, a computer operator began the computer calculations as usual at the start of the race. However, after the race he fraudulently recorded that several more winning tickets had been sold. Then the fraudulent tickets had to be created. Since each selling machine is inspected after the track closes, a service person could punch in the "winning" tickets at that time to be cashed later. This scheme continued for years, with the cast of characters changing several times. Only an enormous amount of detective work solved this case, and just how much was stolen was never determined.

Why is it so easy?

Nearly every bank has armed guards, video monitors, tellers with alarm buttons, and numerous other security devices. A soft-drink distributor keeps delivery trucks fenced in, with a guard dog on night sentry duty. Why is it more difficult to rob a bank of $2500 than to steal millions from its computers? Why are trucks locked up when almost anyone can enter the soft-drink distributor's computer room?

While many people may not believe that access to computer centres is often easy, my experience proves this true. During a security consulting project at an Atlantic City hotel, I spent the evening with an associate in the casino. At about 11 pm we headed for our rooms, but the elevator stopped at the floor where the computer center was located and we decided to look around. The door marked "Computer Center—No Admittance" was locked but had a bell beside it. A computer operator opened the door when we rang, letting us in without a word. For the next ten minutes we wandered through the center without speaking to the operators on duty. Finally, we said "thank you" and left. They were lucky we were not disgruntled heavy losers!

Industrial security and bank security are fairly well understood, but computer security is not. History provides data from which to estimate the probability of a robbery occurring and the likely take, and this risk analysis shows the costs and benefits of using bank guards and sentry dogs. However, computer-related crime data are not widely available, making risk analysis harder. How much is a customer mailing list worth to a competitor? How much damage would be done if the inventory file were altered? How likely is a computer crime? All good questions with no good answers. Because computer crime data are not standardized or gathered in a central repository, analysis of the statistics is difficult. Also, the complexity and variety of computer systems means that a risk-analysis method for one system might not work for another, so security programs must be designed specifically for each organization.

Managers have contributed to the relative ease of computer crime as well—though certainly some do a good job in maintaining security—and controls over computer systems are generally weaker than in other areas of an organization. When computers first became important in business, many managers did not understand them and gave free rein to the data-processing people. Computer professionals became enigmas, assumed to be doing a good job because operations proceeded faster.

Managers also transferred this attitude to company auditors, who took the approach of auditing "around the computer." Transactions were carefully followed until they went into the computer and were picked up as they exited. Too often, traditional auditing controls rigorously enforced in manual bookkeeping systems were either eliminated or weakened in favor of the computer's speed. Other standard checks, such as separation of employee duties, have also been forsaken in many firms, partly because the computer's efficiency eliminated the jobs.

Fortunately, today more auditors are working "through the computer." These auditors write programs that can trace transactions or use auditing software packages now available commercially. Such new software packages enable auditors to identify any operator who gains access to certain data during periods when those files are normally closed. The programs can also detect employees with above-average overtime payments or highlight an abnormal number of correction entries, which often signal computer fraud. Most new auditors learn such skills, but they are playing catch-up since even the largest companies have only a handful of auditors who cannot be outsmarted by computer technicians.

Some managers have resisted additional computer security control systems, considering them at odds with performance goals. Added controls might delay a computer's response by several seconds, and if 100,000 transactions are processed every day—such as at a large airline or bank—the extra delay per transaction becomes significant. A manager must maintain daily production, develop new processing systems, plan future growth, and handle personnel matters. As a result, planning for the rare possibility that a computer crime might occur—and installing the necessary security controls—does not receive much attention.

Finger-pointing often occurs, and a well-known "you're-to-blame" loop has developed in the industry. Users of computer services point to computer specialists to solve problems. Computer people point to auditors, who point to computer management, who point to manufacturers. It's time to recognize that computer security is a management problem, and that only a coordinated effort will make computer crime more difficult.

Finally, computer crime prospers because of an underlying human trait. The perpetrator of a crime can tell a questioning customer, manager, or auditor that the computer must have made a mistake. Then the criminal can correct that transaction and continue with the fraud—since nearly everyone wants to believe that the computer is at fault.

What does the future hold?

Perhaps the biggest change in business computer systems is the move to distributed data processing. Computers were once so expensive that they were installed at a central site where all processing took place. Now the hardware is much cheaper and smaller, and computer terminals are distributed throughout an organization's various departments, electronically linked together and to a large central computer. Indeed, the number of terminals in the US has more than doubled in the past five years to over 3 million, and operators can gain access to central computer files through most remote terminals.

With more processing sites, the problem of controlling computer access is faced many times rather than just once. Also, users are physically closer to the equipment, offering greater opportunity for uncontrolled use. And each site needs qualified managers, operators, and programmers—professionals who are in short supply—so companies sometimes settle for less qualified individuals who do not understand or employ normal operating procedures.

To control user access, the most common step has been to issue secret passwords. But this is far from foolproof: for example, once individuals have been granted access to the computer, they often have total access. A rapidly growing market is now developing for access-control software systems that close such loopholes. This software restricts terminal users (identified by passwords) to only those files they are authorized to use, and users can perform only certain functions. What's needed is a system of checks and balances through the entire system: no programmer should test his or her own work; programmers should not operate equipment; and operators should not write programs.

The rapid increase in word processors and other components of the automated office, such as electronic mail and facsimile-transfer devices, also represent a security headache, since many of these devices are linked directly to a company's central computer. As chairperson of the Association of Computing Machinery's Special Interest Group on Security, Auditing, and Control, I have appointed a committee to organize a conference on office automation security issues. The conference is expected to draw participants presenting solutions to word processing security problems.

Another technological development prompting security worries is the anticipated growth of teleworking systems, in which employees work at home or at neighborhood centers using terminals hooked electronically to a firm's central computer. Then of course there is the rapid growth of home computers, which can be used to tap into commercial data transmission lines. Five years ago just 1500 personal computers had been sold in the US: today that number is 500,000, and by the mid-1980s the total should hit 3 million. Wiretapping was once considered a remote possibility because of equipment costs, but Motorola now gives bankers a demonstration of how data lines can be tapped using simple equipment bought at any computer store.

There is also the worry of accidental data-transmission "cross-talk." Think of a telephone conversation in which you can hear others talking on the line: the same thing sometimes happens in data communication, and in this way one party can gain information belonging to another company without deliberate criminal effort. The communications industry has been working on this problem for years, but no solution is in sight.

To solve some of the security worries raised by distributed data processing and the transmission of electronic information, more emphasis is being placed on encryption or coding techniques. The National Bureau of Standards and IBM have developed the Data Encryption Standard (DES), a mathematical model that can be implemented (via software or a microcomputer chip) into a coding device. Each DES user has a "key"—a list of 56 zeros and ones—that enables a message or data to be coded, and anyone who knows that key can decipher the information. So in banking, for example, a financial transaction is coded and transmitted to another bank's computer, which uses the key to produce the original message. Perhaps $400 billion is transferred each day around the nation using such systems.

All encryption schemes can be broken, however, and the work required to break a code depends largely on the length of the key. Supporters of DES claim that the effort required to learn the key, using a computer trial-and-error approach, is so great that no one would ever succeed. But some critics claim that the workload is not so great, and the payoff in cracking the encryption schemes for electronic funds transfer systems makes the effort worthwhile.

One possible solution, suggested by Leonard Adelman of the University of Southern California, is to use a newly developed "keyless" encryption method. Although safer, the system requires that the message be transmitted three times, greatly increasing the work load associated with data transmission.

It seems likely that DES is adequate for virtually all the information transmitted electronically today. The problem is that few companies actually use such techniques because of what they perceive as high cost and difficult operation. But in reality DES hardware units cost less than $200, and coding a transaction requires no extra work since the unit does it automatically. In fact, all automated bank teller machines use these devices.

Should there be a law?

A computer criminal seldom goes to court and even less often goes to jail. Companies often do not prosecute, feeling that customers might react unfavorably to perceived security lapses and take their business elsewhere. Also, the current legal definition of certain crimes may not accurately cover computer fraud. For example, in conventional theft the item is physically removed from the victim. However, theft of computer information often involves copying the data but leaving the original tape or disk. Even when a fraud is suspected, gathering evidence is often difficult, and that evidence is not easily understood by a judge or jury.

Because current laws are inadequate, much activity has been occurring in both state and federal legislatures. For example, Congress has been considering for several years the proposed Federal Computer Systems Protection Act, which calls for stiff penalties for computer crime. Also, just before the Foreign Corrupt Practices Act was passed in 1978 to prevent American companies from bribing foreign officials, the Senate added a brief section requiring adequate controls on company accounting records. The Securities and Exchange Commission (SEC) has interpreted this legislation to mean that all SEC companies must have internal financial controls. If such controls are missing, top corporate officials are subject to prosecution and prison sentences. (Any company with publicly traded stock is subject to SEC review at any time.) Since most companies use computers to keep their accounting records, this law seems directed primarily toward computer security.

Other legislation protects computer data from criminal activities in certain industries or branches of the federal government. For example, the Food and Drug Administration regulates "how computer systems used in manufacturing, packing, and storing drug products should be operated and controlled." This regulation, directed at a single industry, has generally been effective.

Computer professionals must take the lead in the drive to improve computer security. As each computerized system replaces a manual system, the potential criminal is provided another opportunity. Programmers, systems analysts, and others developing new computer systems must be trained in company-sponsored or industry-sponsored programs on security control measures. More auditors with computer training are needed. Just as programmers need to understand controls, auditors need to understand programming, allowing them to participate in the development of new security systems. A healthy working relationship between programmers and auditors will go a long way toward solving the problem.

Computer users, soon to be nearly everyone in many organizations, must be instilled with a healthy respect for computers. They must understand the computer's potential and proper use, and training programs should stress that stealing with computers is no different from robbing the vault in person.

Educational institutions should take similar steps, giving more attention to computer ethics, especially since more than 400,000 college students graduate each year with computer experience. Some teachers now encourage students to break into the school's computer data base, under the misguided belief that this helps them better understand how the system operates. However, teachers can achieve this goal through more positive action. In my own classes, for example, I teach students how to monitor the flow of transactions and why controls are important.

The overriding point is that the tools to stop most computer crimes are available. Encryption devices can be purchased, and security measures can be adapted from other industrial situations. Legislation is being improved, though slowly. Missing, however, is support from top-level management. Along with strategic planning and new product development, adequate security must be a management priority. That kind of attention is essential if our nation's growing reliance on computers is to be safeguarded.

How Much Computer Crime?
Vico E. Henriques

Nearly anyone who fancies detective stories and has had even a fleeting involvement with computers can recount such legendary "computer crimes" as the Wells Fargo, Security Pacific, and Equity Funding cases. Computer-related crime has become a hot item in the news media. This is not surprising, given its many elements of a good yarn: fraud, theft, vandalism, sabotage, and stealth set against a "high-tech" backdrop. The apparent outwitting of computers—sophisticated machines blamed for uncounted billing mistakes and other snafus that usually end up being human error—is titillating.

There also has been a recent spate of computer-crime doomsday stories involving nuclear accidents or war, international incidents, potential collapse of financial markets and institutions, and terrorist attacks. Scary stuff worthy of serious attention.

But there is one thing wrong with the unexamined regurgitation of all this crime and apocalypse: we at the Computer and Business Equipment Manufacturers Association (CBEMA) think most of it is highly speculative, sensationalized, or erroneous. Thus, the myth of computer-related crime and abuse has become so distorted and exaggerated that the real problems are not being properly addressed.

No one—including the FBI, the academic community, computer users and manufacturers, and the growing colony of computer-security consultants—has accurate data on computer-related crime. What's more, hardly any two experts on the subject can agree on its definition.

For example, a noted attorney writing in the New York Times defined computer-related crime as "a criminal act that has been committed using a computer as the principal tool. It can include theft of money, which might result from entry of fictitious payees into an accounts-payable or payroll system, unauthorized use, such as the use of a computer system by employees running their own service bureau, or just plain sabotage or mischief for the fun of it." Not a bad description.

However, the attorney then used the multi-million-dollar Equity Funding insurance scandal to illustrate the danger of computer-related crime, contradicting his own definition, since Equity Funding was not, at heart, a "computer" fraud. The electronic data processing system was not essential to that crime since less than 20 percent of the fictitious assets of the corporation were ever in machine record form. The computer was used simply to print out the forms essential to the scam; the job could have been accomplished by other, albeit less efficient, means.

In other cases, Wells Fargo officials contend that their alleged losses to a boxing promoter and his ally did not constitute a computer-related crime, and many experts feel that the Security Pacific swindle several years ago does not fit the definition of computer-related crime. In both cases, a serious breach of physical security and poor auditing procedures were involved.

But what about the number of crimes that *are* computer-related? Aren't they growing, if for no other reason than that the number of installed computers is increasing rapidly, especially with the advent of low-cost personal computers? CBEMA does not think so.

One thing that has led to such a distorted view of the "growth" of computer-related crime is the "tip of the iceberg" myth that at least 85 percent, and perhaps 90 percent, of such crimes go unreported for a variety of reasons. Those figures would be laughable if they were not being accepted as gospel by the media, not to mention the FBI, which is the first to admit its data are woefully lacking. This statistic was derived from a 1976 study indicating that 85 percent of violent crimes in Detroit were unreported. By some leap of logic, the report assumed that 85 percent of computer-abuse cases also must go unreported.

The fact is, according to Peter Watkins, a senior consultant for Peat, Marwick and Partners, that computer abuse occurs in approximately 0.02 percent of computer installations, or two cases in every 10,000 installations. Mr Watkins arrives at this incidence rate by dividing what he considers the number of verifiable computer-abuse cases reported annually (75) into the approximate number of computer systems operating (350,000). This is equal to just 1.6 percent of all robberies of US financial institutions in 1978.

He also convincingly challenges annual loss estimates—which range from $300 million to billions—noting that these figures are based on unverifiable cases, including all manner of abuse such as physical destruction, unauthorized use of services, phony computer schools, and so forth. And these figures reflect such notorious but questionable computer-related crime cases as Equity Funding and Wells Fargo, where losses are in the millions. From 1959 to 1979, a worldwide average of $6.96 million was lost from all industries annually owing to financial deception with computers, according to Peat, Marwick. But CBEMA does not endorse even these statistics, though they certainly represent a far more scientific approach to measuring computer-related crime than any other now being cited.

Computer manufacturers do not think that computer-related crime is simply a tempest in a teapot, trumped up by the news media, liberally aided by consultants anxious to land computer-security contracts. We recognize and are working actively to prevent computer abuse; it is in the industry's self-interest to do so. At the same time, however, we attempt to put the problem in perspective so it can be rationally addressed.

Errors and omissions constitute the largest single problem for those concerned with computer security. Very few crimes are committed by data-processing professionals—a combination of economic, technical, and procedural controls make it relatively easy to protect data-processing systems against attack by such professionals. The people who commit crimes by using computers are usually clerks, administrators, and operational people who misuse the resources extended to them to get their jobs done.

Thus, while computer crime is not significantly increasing, it *is* democratizing white-collar crime. This problem should be solved if organizations will hold users of computer systems accountable for their actions, limit users to only the actions needed to perform their assigned work, and keep a record of each transaction.

In short, it is relatively easy to make data-processing systems far safer than manual systems. The three major tools for dealing with information-system security are access control, integrity, and encryption. Major manufacturers can build into the systems safeguards against many security breaches, and they can teach users how to prevent most others. Yet there is no such thing as total security. One cannot protect against every eventuality, nor would it be cost-effective to try.

Therefore, it is important that computer users have adequate legal recourse when their files are criminally violated, and CBEMA has consistently supported state and federal legislation that intelligently addresses prevention of computer-related crime. We were early supporters of the overall intent of the Federal Computer Systems Protection Act (S. 1766) introduced by Senator Abraham Ribicoff. And our Privacy and Security Committee has held discussions with Representative Bill Nelson's staff concerning his recent proposal in the House.

As long as computers are used by humans, there will be abuses and there will be crime. Such occurrences can and are being held to a minimum. Computer crime is not now, never has been, and never will be out of control—unless security is completely ignored. And that is not going to happen.

Data Protection

David Burnham

Cheap computing power makes it possible to keep masses of "transactional information"—records of phone calls, credit card payments, and so on—in huge data bases and to transmit it across the country at low cost. The danger, says New York Times *journalist David Burnham, is that these new computer networks increase the power of big organizations over the individual—and they are wide open to abuse. Taken from chapter 3 of the author's book,* The Rise of the Computer State *(Random House, New York, and Weidenfeld & Nicolson, London, 1983).*

It is a truism that we live in a world radically different from that of our grandparents. One way to measure the great distance we have traveled during this microsecond of human history is to compare the records that documented the life of an American before the turn of the century with the records that document our individual lives today. One hundred years ago, the few records that existed could tell us when a child was born, when a couple were married, when a man or woman died and what the boundaries of the land purchased by a family were. In those days, of course, only a handful of the American people went to college. Social security, income taxes, and life insurance did not exist. Three-quarters of the population was self-employed.

Today, fewer than 5 percent of the American people work for themselves. And of the remaining 95 percent, almost half are employed by large corporations that collect detailed information about the education, health, family, and work habits of their employees. Today, two out of three Americans have life insurance and nine out of ten are covered by health insurance plans. Insurance companies usually collect large amounts of information about their customers—revealing information such as whether they are seeing a psychiatrist, what drugs they use, and whether they have a drinking problem. Today, 60 million students are enrolled in schools and colleges that generally collect detailed personal and financial histories about both the student and his or her parents.

The vast scale of information collected by government agencies, private corporations, and institutions such as hospitals and universities would not be possible without large centralized computers or, alternatively, linked series of smaller computers. It is also true, however, that some of these organizations did in fact collect some of this information before the computer. With armies of meticulous clerks, there were a few industries and a few countries like Germany and Chile that did compile massive handwritten records about the lives of an amazing number of people. The computer, however, has powers well beyond that possessed by human scribes, no matter how numerous, and thus has fundamentally altered the nature of society's records.

The first important change is that the computer mass-produces what has come to be called "transactional information," a new category of information that automatically documents the daily lives of almost every person in the US. Exactly when did you leave your home? Exactly when did she turn on the television? Exactly when did he deposit the check? Exactly when were the calls made from their telephone? How many times have you driven your automobile? In the centuries before the computer, transactional information answering these kinds of questions was almost never collected. And in those very few instances where it was collected, it was not easily available for later inspection.

With the computerized filing systems now available, the larger organizations of our society can easily collect and store this new kind of information. Equally important, they can combine it with automated dossiers containing the traditional kinds of information such as a person's age, place of birth, and the material contained in school and work records.

There is one more important development made possible by the computer: the incredible maze of electronic highways that can move the new and old information about the country in a matter of seconds at an astoundingly low cost. The automatic exchange of information between different data bases was not seriously considered in the first years of the computer age. But as the technology has become more subtle and sophisticated, it gradually is reducing the barriers between these giant repositories, increasing their ability to "talk" with each other.

The contributions of these linked data bases to our daily lives are enormous. The swift granting of lines of credit to a substantial number of American people would not be possible without the computerized data bases maintained by credit reporting companies like TRW and Equifax. The hundreds of millions of checks written each year by tens of millions of Americans could not be processed and cleared without the computerized data bases maintained by the separate banks and the Federal Reserve System. Easy movement about the US would be far harder without the computerized reservation systems of the airlines and car rental agencies. The collection each year of nearly $500 billion in federal taxes from almost 100 million individuals and corporations would be extraordinarily difficult without the computerized data bases of the Internal Revenue Service and the large corporations who each year employ more and more Americans.

There are a variety of reasons why understanding the true significance of all of these changes is very hard. First, there is the fundamental difficulty of putting anecdotal flesh on the bones of the abstract truth that information is power and that organizations increase their power by learning how to swiftly collect and comprehend bits and pieces of information.

This difficulty is greatly multiplied by the sheer force of the tools of the new information age: the machines that can locate a single item in a file of millions in the blink of an eye or that can swiftly develop statistical trends by massing these single items, and the communication links that can shuttle the collected information about the world at almost the speed of light.

When thinking about the impact of these technical achievements, allow your mind to wander beyond the traditionally narrow boundaries of the computer debate. Consider how the technology is altering the power of large organizations. Consider how the technology is affecting our social values such as the notion of checks and balances, the role of work, and the importance of spontaneity. Consider the far-reaching changes it is bringing to the nation's economy.

What does it mean, for example, that the officials and clerks of the US government, each year armed with more and more computers, have collected 4 billion separate records about the people of the US, 17 items for each man, woman, and child in the country? What does it mean that an internal communications network serving just one multinational corporation now links more than 500 computers in over a 100 cities in 18 countries and has been growing at a rate of about one additional computer a week in recent years? What does it mean that 10,000 merchants all over the country are able to obtain a summary fact sheet about any one of 86 million individual Americans in a matter of three or four seconds from a single data base in southern California? What does it mean that a handful of federal agencies, not counting the Pentagon, have at least 3 separate telecommunication networks stretching all over the US?

Two of the world's largest and most complicated systems of linked data bases are controlled by the American Telephone and Telegraph Company and the Federal Bureau of Investigation. AT&T's gigantic network was deliberately developed by the company's scientists, engineers, and businessmen. The FBI system, which seeks to link the computerized data bases operated by a majority of the 50 states, has developed in a more haphazard fashion.

Because the information-collecting and distributing systems of both the telephone company and the FBI began to function long before the birth of the computer, they illustrate the important point that the new tools of information processing usually are extensions of old bureaucracies, not shiny stand-alone machines that can be considered on their independent merit.

A second trait shared by these two very different systems is that neither has been subject to much outside scrutiny during much of their development. AT&T is a private company regulated by 50 state utility bodies of widely varying quality and a federal commission that never had the staff adequately to monitor the company's interstate operations. And very few congressional committees or state legislatures have taken the time or possess the perspective to consider the impact of the gradually growing network linking the FBI to the states.

Computers generate transactional information for many purposes and many organizations. They allow the construction of huge, speedy, and low-cost communication networks to transmit many different kinds of information to thousands of different customers. But the very different information processing systems developed by AT&T and the FBI provide clear examples of how transactional information and mass networks have enhanced the impact of these two pre-computer bureaucracies on all of our lives. First, consider AT&T. Through its millions of miles of cables, microwave highways, and satellite hookups, the American people make 500 million calls a day—four calls, on the average, for each of the nation's 130 million telephones. Thanks to the computerized data bases that are tied into this massive electronic network at a steadily increasing number of junctures, AT&T has become the largest single holder of transactional information in the world. Buried in the computers of the system are records that can be helpful in drawing an amazingly detailed portrait of any single person, group, or corporation that uses the telephone.

The astounding power of these records is not appreciated by the public, the courts, or Congress. But for government and industry investigators, they have become an important tool. A few years ago, for example, the Senate created a special committee to investigate a very sensitive and delicate subject, the relationship between President Carter's brother Billy and the government of Libya. After many months of embarrassed maneuvering, the Democratic Senate committee issued a report on the antics of the brother of the Democratic president. Almost every other page of the committee's 109-page final report contains a footnote to the precise time and day of calls made by Billy Carter and his associates from at least ten different telephones operating in three different states.

The report said that on November 26, 1979, Billy Carter and an associate began driving to Washington from Georgia. Shortly after beginning the trip, the report said, the two men stopped to telephone the Libyan embassy and request a meeting with a high-ranking official. The assertion was supported by a footnote to telephone company records showing that "a five-minute call at 3:43 was charged to Billy Carter's telephone from [a pay telephone in] Jonesboro, an Atlanta suburb." Another footnote said that calls were made from Carter's office telephone in Georgia to the Libyan embassy in Washington "on March 7, March 10, four times on March 11, twice on March 12, three times on March 13, three times on March 14, March 15 and March 17."

The investigations of the special Senate committee were publicly announced and the telephone records that document the report were obtained by a formal legal process. But this sometimes is not the case. One of the top officials in the Nixon White House, for example, claims that, shortly after the automobile accident that claimed the life of Mary Jo Kopechne in Martha's Vineyard, the White House political operatives ordered the FBI to obtain the telephone credit card records of Senator Edward Kennedy. These records, which almost certainly would have revealed who and when Senator Kennedy called immediately after the accident, obviously would have been considered useful to those Nixon advisers who thought Kennedy was a likely opponent in the coming elections. Though reporters for the *New York Times* determined that the records in question disappeared from the files of New England Bell shortly after the accident, they never found documentary evidence confirming the account of a top Nixon lieutenant that they were obtained by the White House.

Both the Billy Carter and Teddy Kennedy cases illustrate why investigators are so interested in transactional information. First, the information can be extraordinarily revealing. Only considering the data that can be collected from a telephone computer, investigators can learn what numbers an individual has called, what time of day and day of the week the calls were made, the length of each conversation, and the number of times an incorrect number was dialed. Considered as a whole, such information can pinpoint the location of an individual at a particular moment, indicate his daily patterns of work and sleep, and even suggest his state of mind. The information also can indicate the friends, associates, business connections, and political activities of the targeted individual.

But there is an even more fundamental consideration at stake. Almost by definition, transactional information is automatically collected and stored in the data bases of the telephone company, the electronic equipment of banks and the computers of two-way interactive television systems. This means that transactional information can be obtained months after the instance when the particular event that is documented by the records actually occurred.

This ability greatly enlarges the scope of any investigator. Before the computer age, it was extremely hard to develop concrete evidence about the activities and whereabouts of an individual unless someone had been assigned to follow him. In most cases, investigators were limited to pursuing the handful of individuals they believed might undertake a forbidden act in the future. Now they can move back in time, easily gathering concrete evidence about any person of interest long after the forbidden act occurred.

The broad broom of transactional information, however, can sweep up much more than the highly revealing computer tracks of an individual citizen. In at least two instances, for example, evidence has recently come to light where this same kind of computerized information was used by AT&T to track the activities of several large corporations and even the ethnic and economic groups living in a single state.

A significant characteristic of the transactional information collected by AT&T and other major computer systems is that its reach is universal. Transactional information is collected and stored about the telephone use and banking habits of everyone who lives in America, rich and poor, ethical and unethical, white and black, Republican and Democrat. Though the information ultimately may cast a revealing light on the activities of a single individual, it is collected about the activities of all.

The system of data bases that gradually are being linked by the computerized network of the FBI does not share this universal quality. Instead, it collects and distributes information about one segment of the population, the millions of Americans who are arrested each year in the US.

The story of the FBI system begins almost a hundred years ago, long before the birth of either the modern computer or the FBI. The occasion was a meeting of reform-minded policemen who decided the time had come to ask Congress to finance an agency in the federal government to maintain a central record of all American criminals. The policemen, who called themselves the National Chiefs of Police Union, argued that the prompt exchange of information about "criminals and criminal classes" made possible by such an agency would improve the ability of their separate departments to control crime.

Nearly nine decades later, in the summer of 1981, a task force appointed by President Reagan's attorney general, William French Smith, acted to bring the original proposal in line with current technology: the panel endorsed a plan to have the FBI make a significant change in its existing communication system to speed up the exchange of the records of millions of people who are arrested each year all over the US. As was suggested by the police chiefs in St Louis, the Reagan administration's task force said the adoption of the proposal would help the government curb serious crime.

There is a good deal of plausibility in the belief of both the police chiefs at the turn of the century and the Reagan administration today that the swift exchange of summary information about an individual's past contacts with the criminal justice system will significantly improve the ability of the police, prosecutors, and judges to catch and punish criminals. Careful examination of the sources of information that lead a policeman to make an arrest or a judge to decide an appropriate sentence, however, suggests that the nearly instantaneous retrieval of criminal history records may be considerably less useful than is generally thought to the business of crime-fighting.

Furthermore, the gradual computerization of criminal history records in most of the states and the development of a single computerized network to link all these state and local data bases raise questions that go to the heart of democratic institutions of the US. So profound are these questions, in fact, that they have provided a handful of critical congressmen, working with communication experts of the Nixon, Ford, and Carter administrations, with the necessary weapons to stall the construction of the federal network for the last decade.

But now, with the blessing of the attorney general's Task Force on Violent Crime, plans to have such a network constructed by the FBI appear to be nearing fruition. Should the FBI get its way, the single most powerful law enforcement agency in the US will control a computerized network linking a system of data bases of almost unimagined size and complexity.

The impact of the information that would be transmitted through the vast maze of telephone lines, data bases, and computer terminals is of even greater concern than the network itself. When FBI officials call for the development of a bureau-controlled network to speed the exchange of criminal history records—rap sheets—most Americans assume the proposal is designed to reduce the chance that they or members of their family will be the victim of a hardened criminal, an experienced mugger, or a drug-crazed rapist.

But the reality of trying to control crime in America is much more ambiguous and complex than is sometimes suggested by advocates of a greatly enlarged computer network for the FBI. For example, only a small portion of the millions of traumatic events logged yearly into the criminal history records involve what are now called career criminals—professional murderers, robbers, and burglars who repeatedly prey on their victims.

The sad truth is that many of the murderers and rapists who terrorize the American people are not marauding strangers. Instead they are wives who have been driven to kill their husbands in the dark heat of the night. They are uncles who have been asked to babysit for their young nieces. They are old acquaintances who fell to fighting at the end of an all-night bout of drinking.

An astoundingly large number of Americans have at one time or another been arrested. Because some individuals have records in more than one jurisdiction, and because millions of those with arrest records have subsequently died, it is not easy to estimate how many people have actually been arrested during the course of their lives. A recent authoritative study done for the Labor Department by Neal Miller, however, concludes that about one out of five living Americans has an arrest record somewhere in the US.

In the past few years, local, state, and federal law enforcement officials throughout the country have successfully argued that society should provide them with statewide and national computerized communication systems that can instantly inform a policeman on the street whether the individual he has just stopped is currently wanted on suspicion of committing a crime. These systems make sense from the standpoint of protecting both the patrolman who is making the stop and society at large from the possibility of further hurtful activities by the person who is being temporarily detained. But the acknowledged value of a hot-line system to inform a patrolman that the person he has stopped *currently* is a suspect for another crime does not necessarily justify a parallel system informing the patrolman that his target was arrested for a crime in a distant city at some possibly *distant period of time*.

The computer panders to the natural human instinct to desire more information about everything. But there are some law enforcement officials who question whether more arrest records are actually going to help them do their work. They contend that the value of the arrest record has been greatly diluted, partly because so many Americans have been arrested for so many significant causes that it is hard to separate the wheat from the chaff. They further believe the records are of questionable value because they frequently do not disclose whether the case was immediately dismissed, whether the defendant was found guilty or innocent, and what sentence, if any, was imposed. Beyond the pragmatic judgment about the utility of providing the patrolman instant access to arrest records, they see a larger ethical and constitutional question: do the American people want their police making arrests for *current* activities partly on basis of *past* behavior?

The idea that a national rap sheet system would make an important contribution to our work here is just a bunch of baloney. Our problem is not to find out who the guy is. Our biggest problem is, once we catch him coming out of a house with the goods, how do we keep him in jail and how do we make sure he stays in jail? If anything, we have over-information-oriented and over-computerized this department.

These critical comments from the police are drawn from a 1981 report to the Office of Technology Assessment as part of a congressional effort to determine the impact—positive and negative—of the FBI's proposed system. This particular report was prepared by Dr Kenneth C. Laudon, a professor at New York's John Jay School of Criminal Justice. During its preparation, Dr. Laudon interviewed over 140 experienced criminal justice officials working in four states and six cities. Dr Laudon found that the prosecutors and judges he talked with were even more doubtful about the utility of the FBI plan than the police.

The complaints of the policemen, prosecutors, and judges interviewed by Dr Laudon went well beyond simply questioning the utility of a summary record that has been lifted out of its original legal and social context. Over and over again, the users complained that the records being moved along the experimental criminal history segment of the FBI's communication network were incomplete and inaccurate. The subjective judgments of the police officials, prosecutors, and judges about the poor quality of the information was supported by a second investigation undertaken by Congress's Office of Technology Assessment.

For this second study, the OTA arranged for Dr Laudon to obtain access to a random sample of the criminal history records that recently had been dispatched to law enforcement and other agencies from five official repositories maintained and operated by three separate states and the FBI. The information in the records from the repositories was then compared with the information in the original records in files of the county courthouses. Procedures were followed that permitted the comparative analysis without disclosing individual names.

The findings are surprising. In North Carolina only 12.2 percent of the summaries were found to be complete, accurate, and unambiguous. In California 18.9 percent were complete, accurate, and unambiguous. In

Minnesota the researchers found almost half the sample—49.5 percent—met the same standards.

The quality of the FBI files, which of course rests on the information submitted by the 50 states, was not noticeably better. Based on a random sample of 400 records dispatched by the Identification Division during a single week in August of 1978, Dr Laudon found that 25.7 percent of the FBI rap sheets met the standards suggested by federal law. Assuming the one-week sample was valid for the entire year, the OTA study suggests that 1.75 million of the 2.35 million records sent by the FBI to criminal justice and other institutions all over the country had various failings.

The quality of the much smaller number of records disseminated by the Computerized Criminal History segment of the FBI's National Crime Information Center was somewhat better than that of the Identification Division rap sheets. In the computerized system, the comparison of the sample summary and the original records disclosed that 45.9 percent were complete, accurate, and unambiguous.

At the request of OTA, Dr Laudon also checked a sample of 400 arrest warrants from the 127,000 contained in the FBI's "hot file" on a single day in August of 1979. Upon comparing the information on the FBI's warrant notices with the information in the local court records, he found that 10.9 percent of the sample already had been cleared or vacated, 4.1 percent showed no record of the warrant at the local agency, and a small additional number of warrants had other problems. Again, assuming the validity of the sample, it appears that on that single day in 1979, 17,340 Americans were subject to false arrest because the FBI computer incorrectly showed they were wanted when the warrants in question had been cleared or vacated.

Official records, of course, were subject to improper use long before the computer became an important part of every major public and private bureaucracy in the US. But an executive in a New York state agency explained how the computer has enlarged the opportunity to abuse.

"Technology may not be the only villain, but it is one villain," the official said. "Before high technology, you could actually control information better; at least, it could not spread very far because it was impractical to transmit it. With the computer system, you can't control it any more, largely because of the automatic interfacing of the system, which makes it difficult for even us to know who's getting our information."

The Office of Technology Assessment asked a team at the Bureau of Governmental Research and Service at the University of South Carolina to try to determine exactly who was now receiving criminal history information and to assess the social impacts of computerizing these histories. According to the South Carolina researchers—Lynne Eickholt Cooper, Mark E. Tompkins, and Donald A. Marchand—the use of criminal records outside the traditional confines of the criminal justice system is enormous and growing.

All applicants for federal positions, all military recruits, many of the hundreds of thousands of citizens working for private contractors who are doing jobs for the federal government, and all new employees of federally chartered and insured banks are among the millions of persons who have long been subject to criminal history record checks for many years, the researcher reported.

In addition, with the recent growth in the number of states and cities requiring licenses and permits for almost any kind of job, the population subject to criminal record checks has further exploded. At the last count, more than 7 million Americans must obtain licenses to earn their living. In California, for example, 47 separate licensing boards, 50 state agencies, and 32 out-of-state agencies have access to the criminal history records stored in the state's computerized record system. In New York the use of criminal records by law enforcement agencies has declined in recent years, while its use by private employers has gone up.

Through a variety of federal and state laws, society has made the collective decision that individuals *convicted* of certain crimes may be properly denied certain privileges. Sometimes such individuals lose the right to vote. Sometimes they may not run for public office. Sometimes they are prohibited from bidding on government contracts.

The us also pays a good deal of lip service to the principle that an arrest or investigation, *without a conviction*, should not be sufficient grounds to deny an individual an honorable place in the job market. In late 1981, President Reagan provided an eloquent testimonial to this principle when asked why he had not asked three officials in his administration to remove themselves from office after they were accused of illegal activities. The three men were Labor secretary Raymond Donovan, cia director William Casey, and National Security Council director Richard Allen.

"I believe in the fairness of the American people, and I believe that in recent years there has been a very dangerous tendency in this country for some to jump to the conclusion that accusation means guilt and conviction," Mr Reagan told reporters. "And I think it is high time we recognize that any individual is innocent until proven guilty of a wrongdoing, and that's what we are going to do."

Mr Reagan's eloquent statement has considerable backing in the laws and regulations of the us. The Civil Service Reform Act of 1978 includes a provision limiting federal suitability checks to conviction records, and only for those crimes that are reasonably related to job performance. Three states have laws explicitly barring employer discrimination against ex-offenders. Many state and local human relations offices have issued rules barring private employers from requesting arrest information and limiting the use of conviction information.

Despite the various restrictions, however, there is considerable evidence that the president's policy for the high officials of his administration frequently does not obtain for the average job hunter. According to one congressional survey, for example, the 50 states handed out a total of 10.1 million criminal history records in 1978. Two million of the records—one out of five—went to private corporations and government agencies that were not part of the criminal history system.

The Labor Department study that concluded that 40 million Americans have an arrest record also estimated that just under 26 million of them were in the job market. With so many people applying for so many different jobs, it is nearly impossible to generalize about the impact of criminal records on their lives. Certainly, the bank is on solid ground when it decides not to hire a convicted bank robber. Certainly, no one would argue about a hospital personnel director who decides not to hire a convicted narcotics dealer.

But there is good evidence that some employers assume that any kind of record, no matter what the offense, no matter what the outcome, is a powerful mark against the individual. More than a decade ago, two sociologists named Richard Schwartz and Jerome Skolnick attempted to measure this kind of bias in an experiment in which a number of employers were shown the employment folders submitted by 100 men who were looking for a menial job in the Catskill area of New York. The applicants were broken down into four groups. The first group had no criminal records. The second group had been arrested for assault and acquitted of the charge with a letter from the judge explaining the presumption of innocence. The third had been arrested for assault and acquitted, but there was no letter from the judge. The fourth group had been convicted.

The employers were asked whether they would be willing to offer the individuals in each group a job. Thirty-six percent said they would hire the men with no record, 24 percent said they would hire the men who had been acquitted and had a letter from the judge, 12 percent said they would hire the men who had been acquitted but had no letter, and 4 percent said they would hire the men who had been convicted.

The findings of this simple study, replicated in a number of subsequent research projects, are a dramatic illustration of the powerful impact of a criminal record, even if the individual under consideration has been acquitted. The implications of these findings—when considered in terms of the number of Americans with records and the accuracy and completeness of these records—are staggering.

Computerized records being rocketed around the country would be a greater help to the police and less damaging to the public if they were accurate and complete. In the pursuit of this worthy goal, after failing to reach a compromise on comprehensive legislation to govern the NCIC, Congress adopted a brief amendment to the Crime Control Act of 1973. The 139-word amendment, which all parties regarded as an interim measure, contained three very simple principles. First, to the maximum extent feasible, criminal history information disseminated with the help of federal funds should contain disposition, as well as arrest data. Second, procedures would be adopted to require that the information is kept current. Third, the information would be used for law enforcement, criminal justice, and "other lawful purposes." As already noted, the exception for other lawful purposes has become something of a gusher.

The 1973 amendment also offered the citizen an apparent remedy. "An individual who believes that criminal information concerning him in an automated system is inaccurate, incomplete, or maintained in violation of this title, shall upon satisfactory verification of his identity, be entitled to review such information and to obtain a copy of it for the purpose of challenge or correction." Almost three years later, the government followed up on the 1973 amendment by issuing a vaguely worded regulation that nominally required the states to develop policies and procedures to ensure the privacy and security of criminal history records. Individual access and review was one of five specific areas the federal government said the states must cover in their regulations.

Ever since the computer became a major force in the administration of large government and business organizations in the mid-1960s, the individual's right to see and correct his own computerized record has been held up as a miracle cure for many of the potential abuses of the computer age. Alan F. Westin, in his pioneering 1967 book *Privacy and Freedom*, was among the first to find wonderful powers in the cure of public access. The principle was subsequently embraced by both the Privacy Act of 1973 and the 1977 report of the Privacy Protection Study Commission.

But the OTA report indicating that more than half of the millions of criminal history records now circulating in the US are incomplete, inaccurate, or ambiguous is compelling evidence that the remedies prescribed by law and regulation are not every effective. And when the congressional researchers looked at the actual record-keeping practices of four states and six urban areas, they found strong evidence that the promises made in the 1973 Crime Control Act amendment are largely an illusion.

The California criminal justice system, for example, contains more than 3 million records, 1.1 million of which are in computers, the balance in manual files. In a recent year, authorized organizations made 5 million separate inquiries of the system. Despite the huge number of transactions, however, the researchers discovered that only 300–400 individuals each year ask to see their records, 80 find something they regard as incorrect, and 40 are actually successful in forcing California to make a correction.

A second approach to assuring accurate records is to require that the record-keepers periodically check their files and dump those that are inaccurate or incomplete. The government, at least in theory, adopted this alternative approach in 1976 when it required that all the states develop plans for regularly auditing and purging their criminal history files.

In 1980 the Office of Technology Assessment sent a questionnaire to the 50 states about how they managed their criminal history records. One question was whether they checked the accuracy of the records in their files. Four out of five of the 49 states answering this question responded that they had never conducted record quality audits. So much for federal regulations.

The astounding finding that only one out of five states has ever sought to audit and purge the information in its criminal history files may explain why so many of the records are inaccurate or incomplete. It also demonstrates how difficult it is for the federal government to force state and local agencies to meet a standard established by Congress.

Computer scientists and manufacturers purport to believe that their machines are neutral. This is true, of course, as long as the technology remains in the showroom. The neutrality evaporates, however, when powerful officials running powerful bureaucracies harness the computers to achieve their collective goals. Often both the goals and methods of achieving them are in the public interest. History tells us, however, that the organizations of fallible men sometimes lose their way.

During the 1960s aud 1970s the leadership of the FBI and a number of police departments throughout the country came to believe that their responsibilities went beyond arresting those who had committed criminal acts. Their job, some policemen thought, extended to trying to channel the political thoughts and lifestyles of the American people along certain narrow paths. Police departments in such cities as Chicago and Los Angeles enormously expanded their surveillance of individuals who did not share conventional police values. The US Army ordered its intelligence agents to attend all sorts of political rallies. The FBI, with the explicit approval of J. Edgar Hoover, undertook a secret program in which it deliberately sought to have critics of the federal government fired from their jobs or otherwise discredited by mailing false and anonymous reports to their neighbors and colleagues.

The FBI's effort to expand its activities beyond the precise boundaries of criminal law did not stop with the anonymous smearing of thousands of Americans who did not happen to share Mr Hoover's view of the universe. From 1971 to 1974 the FBI harnessed the computerized data bases of the National Crime Information Center to pick up the movements of persons who were not criminal suspects.

Under the public relations establishing the NCIC and how it would operate, the FBI declared that only the names of persons who had been formally charged with a crime would be listed in the NCIC computer. In July of 1975, however, John Tunney, then chairman of the Senate Subcommittee on Constitutional Rights, announced the discovery that the Bureau had violated its own regulations by using the NCIC "to keep track of individuals who might be of interest to the FBI for whatever purpose, including possible political reasons."

Three months later, the Justice Department confirmed that from 1971 to 1974 the FBI had instructed its NCIC's computers to sound an alarm any time a local law enforcement agency sent a message to Washington indicating that any one of 4700 individuals was arrested. When it is remembered what a large part of the population have arrest records, and how many persons were being arrested during that particular period for taking part in civil rights and antiwar demonstrations, the automated flagging mechanism can be viewed as a potentially powerful surveillance device.

One indication of how embarrassed the Justice Department was about the FBI's secret operation of a surveillance system outside the limits set down in its own regulation was the way the project was abruptly halted on the precise day that the Senate subcommittee staff made its first inquiry. But seven years later, there are indications that the FBI and other law enforcement agencies of the Reagan administration would like to resurrect the use of the NCIC computers to track persons who are not the subject of a formal arrest warrant.

The first hint of this possible shift in NCIC policy came at a meeting of the NCIC advisory board in June of 1980 in a talk by Kier T. Boyd, an FBI inspector and the deputy assistant director of that part of the bureau that controls the FBI's very large computerized telecommunications system. According to the minutes of the advisory board meeting, Mr Boyd openly advocated the startup of the same kind of surveillance that had been so suddenly halted in 1974.

The official began by acknowledging that the tracking of the early 1970s had run into a wave of adverse reaction because "of the political climate of that time" and that the policy under consideration in the spring of 1981 "smacks of the tracking system which in the past has not been very well received by certain quarters of Congress." In an apparent reference to the growing power of conservatives in the Congress, however, Mr Boyd said he now felt "that the climate is very definitely changing and now we have an opportunity to raise the system to a new level."

In the summer of 1982, the Justice Department approved the request of the Secret Service to use the FBI computer to keep track of persons it suspects may be a danger to officials. But Representative Edwards, the one-time FBI agent, asked the Justice Department not to implement the plan until a thorough public examination of the issues had been completed and possible legislation considered.

Without a law, Mr Edwards said in a letter to Attorney General William French Smith,

what assurances do we have that this system will not evolve into the sort of system maintained by the Secret Service in the 1970s when "dangerousness" and "threat" were interpreted to include political dissent? In 1972, the Secret Service had nearly 50,000 individuals on its lists, including such "threats" as Jane Fonda, Tom Hayden, Ralph Abernathy, Cesar Chavez, Benjamin Spock, and Walter Fauntroy. Among the organizations listed were the NAACP, the Southern Christian Leadership Conference, and the John Birch Society.

It is very hard to challenge a Secret Service request for a tool that it declares will provide improved protection for the President. But it is easy to understand that the principle of restricting the FBI's computerized telecommunications network to tracking criminal suspects is an important one. It also is easy to understand that changes in a powerful instrument of federal surveillance should not be made without public discussion, congressional hearings, and legislative authorization. Once an exception has been made for one category of persons, each succeeding category will be that much easier.

Electronic Politics

Richard M. Neustadt

The new information technology is a double-edged sword. It can be used as a force for good or for evil. Nowhere is this more obvious than in the case of democratic politics, as a former policy adviser to President Carter makes clear in this brief account of campaigning with computers. It first appeared in Howard F. Didsbury (ed.), Communications and the Future *(World Future Society, Bethesda, MD, 1982).*

The first thing Congressman Jim Coyne does when he gets to his Capitol Hill office each morning is read his computer. Coyne is the first congressman to set up an "electronic mailbox" to receive messages from constituents and others who have their own computers. Coyne says he likes his system because it gets him fresh ideas faster than they get to his colleagues.

Coyne's flickering screen is a sign that the notorious "communications revolution" is about to hit politics. A wave of new technology will transform campaigning, political organizing, news coverage, lobbying, and voting. Some of these changes may make campaigning less costly and bring decision-making closer to the people. But the greatest impact may be to fragment our politics, narrowing people's perspectives, shifting more power to special interest groups, and weakening the glue that holds our system together.

The turbulence ahead will be reminiscent of what television did to politics two decades ago. TV turned campaign schedules into a string of "visuals" to draw the cameras—candidates started wading into polluted rivers to deliver position papers on the environment. It started frantic hunts for money to buy precious seconds of air time. Presidents and candidates learned to use TV to talk directly to the people, over the heads of the political parties and the press. Television even changed the kinds of faces, voices, and personalities that win elections.

During the 1980s, television and other media will be reshaped by cable TV, satellite-to-home broadcasting, videotape recorders, personal computers, and electronic publishing. Politicians will have to learn the political implications of these devices, just as they had to learn about television. And all of us ought to start thinking now about how we want politics to work in the new era.

Campaigning on cable

The combination of new technology and deregulation is creating a flood of new television channels. A quarter of US households now subscribes to cable TV; by 1990, the number should be over 50 percent. All the new cable systems have at least 32 channels, and the frantic bidding war for franchises is pushing cable companies to promise as many as 200 channels in some cities. The number of regular radio and TV stations on the air has doubled since 1960. The FCC is about to award hundreds of "low-power television" licenses, creating new outlets in the coverage gaps between existing stations, and it will soon have hundreds of new radio licenses to give away. Direct-broadcast satellites may be operating by 1986, sending as many as 20 channels of television directly from space to low-cost rooftop receivers.

This wave of new channels is creating the economic base for "narrowcasting." The big three networks will survive and will continue to aim at mass audiences, but dozens of new services will spring up to serve smaller audiences with particular interests. Already, the new economics of television has spawned networks dedicated to news, sports, weather, movies, culture, Hispanics, Jews, blacks, senior citizens, children, and even health.

Up to now, politicians used television to send short, universally appealing messages to large audiences. The new television will mean longer messages tailored to smaller audiences. When we watch the narrowcasting networks, we will see campaign ads and news programs showing candidates advocating bilingual education on Spanish channels, defending Social Security on channels aims at the elderly, and playing football on sports channels. Politicians will find the all-news channels particularly fertile ground— viewers with enough appetite for news to watch it continuously will be good prospects to contribute, volunteer, and vote.

The new media's greatest impact may be on congressional and local campaigns that could not afford to advertise on old-style television. For example, a candidate running in Lansing, Michigan, has to pay at least $500 for a 60-second ad on a regular TV station, and most of that money is wasted because the station covers half the state. The same commercial on the local cable system costs $30 and reaches just Lansing residents. Advertising is new to cable—only 10 percent of the systems take local ads today—but it will be universal in a few years.

Cable even offers free time at the local level. Most cable systems have "public access channels," which are available to anyone who wants to use them, first-come, first-served. In many cities, the cable system even supplies the studio and cameras. Few access channels have regular audiences, but they are ideal for a campaign manager to brief volunteers or a political party to run a weekly talk show for its activists.

Most Americans get their news from television. Up to now, TV news has meant mostly 90-second stories, Sunday morning interview shows, and a few "specials" on presidential speeches, election nights, and the like. The new media will turn "specials" into daily fare and change politics in the process.

The cable system in Reading, Pennsylvania, was one of the first to provide live coverage of city council meetings. Viewers got to know who the council members were and what they did, and that changed Reading's politics. Those TV shows helped the current mayor, Karen Miller, attract a following, beat a political machine, and win her seat. "People definitely are watching us," says Mayor Miller.

The Madison, Wisconsin, cable system is another pioneer in local coverage. City councilmen accustomed to anonymity find themselves treated like TV personalities—strangers stop them in the streets to praise or revile their performances.

Most of the cable systems now being built will dedicate a channel to his kind of programming. There is talk in several states about letting cameras into the legislatures. A cable network has carried the US House of Representatives live for the past three years, and the Senate will vote this year on whether to do likewise. Early reports from the House indicate that television has produced some grandstanding but also some improvements. Congressmen have a strong incentive not to look foolish when they speak on the House floor—their opponents can use TV clips in campaign ads.

These channels are creating a novel kind of television news. They bring events directly to people, "gavel-to-gavel," without editing or interpretation. The audiences are small compared to "Laverne and Shirley," but they are huge compared to the numbers who actually show up at city council meetings or House debates.

These channels operate without reporters, but electronic journalism also is growing. Two 24-hour cable news networks are on the air, and two more will start this year. Reporters for these networks will be able to do stories four or five minutes long, increasing substance and subtlety. In Washington and many other cities, newspapers are teaming up with cable systems to produce local news channels. All these services will give us more in-depth reporting and more of what television does best—live coverage of breaking stories.

News coverage on the old media also is improving. CBS is considering expanding its evening news program to an hour. Many local TV stations are increasing their own operations and bringing in more national and international coverage. (The new technology is helping canny political organizations take advantage of the local stations' interest. For example, the Republicans bought half an hour of satellite time to transmit clips of senators praising President Reagan's State of the Union Address to their home-state TV stations.)

Taken together, these trends will provide more information for those who want it. Politics will be more accessible and understandable, and that may get more people involved.

Unfortunately, many people may end up knowing less. Most Americans get their news from television, and the old broadcasters' news programs, superficial though they are, provide a homogeneous information base for the most diverse society in the West. In narrowcasting, by contrast, some channels will provide only news about their particular themes, and many—such as the movie and sports networks—will offer no news at all. Narrowcasting may fragment Americans' perception of events, and the vast menu of entertainment channels may draw some people away from the news altogether. An electorate awash in video information may end up less informed.

One result is that, when the president goes on television in 1990, he may be carried on six networks, but he will not be on two dozen others. For the first time since FDR's fireside chats, the mass audience that presidents have automatically commanded will begin to erode.

Campaigning with computers

Politics requires mountains of paperwork. Candidates and causes need people to make lists, stuff envelopes, and so on. These tasks have helped make housewives and high-school students the indispensable footsoldiers of campaigns. In the last two years, however, computers have become cheap enough to let most campaigns automate their paperwork. In the 1982 election, the most prized volunteers will be those who bring their personal computers to campaign headquarters.

The newly liberated volunteers will be turned loose to contact voters, but computers will play a role there, too. A few campaigns have started using automatic dialing machines that call hundreds of people an hour and play a message when they answer—e.g., a tape of the candidate urging them to vote. The next step will be "voice recognition" machines that "understand" what the person at the other end of the line is saying. Several companies already have built prototypes that can handle callers' airline reservations with startling accuracy. By the end of the decade, we may see electronic canvassing—once people get used to talking to computers.

Computers also will help "target" messages and money to particular groups of voters. A Republican National Committee study last year recommended giving local parties access to the national party's sophisticated data banks, to provide instant information and analysis in the heat of campaigns.

The advent of computers will help well-heeled candidates who can afford the fanciest equipment, but falling prices will let even impoverished candidates use automation. The only certain gainers will be those who move quickly to take advantage of these machines.

The new technologies will help candidates, but they will do even more for interest groups. The US Chamber of Commerce has taken the lead by launching its own television network. "BizNet" will use a satellite to broadcast six hours per day of business news, interviews, and—of course—lobbying messages, directly to receivers placed at corporate headquarters around the country. The Chamber has spent $5 million on studios and will spend several million more each year to run the operation.

BizNet is pioneering a migration that will sweep through political organizations—the shift from paper to electronics. Up to now, most interest groups have relied on mail to raise money and communicate with members, but the cost of mail is going through the roof. Today it costs about $250,000 to print and mail a fundraising pitch or a newsletter to a million people, and that price may double in five years.

In electronics, by contrast, most costs are stable or falling. For the price of a fundraising mailing, a group can now reach over a million people with 100 two-minute ads on a cable network. For the price of a newsletter, an organization can produce up to ten one-hour television shows, and some of the new networks are so hungry for programming that they will give away air time. Such opportunities are unavailable on the old networks, which sell only 30- and 60-second spots—too short to be substantive—and charge tens of thousands of dollars each. With the new channels, an environmental group, for example, can spend a relatively small amount of money to produce a weekly "nature hour" that gets its message to its members and to anyone else who may tune in.

Apart from cost savings, this approach takes advantage of the emotional impact of pictures—in an America reared on TV, video is the way to get people excited. That is why Richard Viguerie, the leading fundraising expert for conservatives causes, says that "cable television will be to the politics of the 1980s what direct mail was to the 1970s."

Only one group has really exploited the new media so far—the churches. In the past few years, several denominations acquired their own UHF-TV stations and created satellite-fed networks that reach millions. Most political organizations are hanging back because of ignorance or penury, but a few savvy groups are moving now, while channels are still available and narrowcasting network time is still cheap. The United Auto Workers, for example, has applied for 24 low-power stations to create the base for its own TV network.

By the end of the decade, trade associations, corporations, unions, and political parties will be turning out oceans of videotape. This "electronic direct mail" will be a key tool of political organizing.

The new technology is giving political organizations another potent tool: the "video teleconference." Last December, for example, Vice President Bush had promised to speak to a group of GOP contributors in California, but he could not spare the time for a full day of traveling. Instead, Bush went to a studio a mile from the White House and talked by satellite. The audience watched him on a screen while he gave his speech and answered their questions. The contributors were happy, Bush saved a day, and the Republican Party saved the cost of an airplane charter.

The National Education Association tried a much bigger teleconference last year. The Association wanted to mobilize 1000 of its organizers across the country for a letter-writing campaign on a budget issue. Instead of flying everyone to Washington, the Association brought them to 49 sites around the country and plotted strategy over a satellite. The conference worked, and it saved a quarter of a million dollars in air fares and hotel bills.

By 1990, similar electronic meetings between congressmen and constituents will be a routine part of lobbying. These sessions will be cheaper and faster to arrange than the traditional flight to Washington, and lobbyists will find it much easier to recruit busy constituents for a one-hour teleconference than a two-day trip. Electronic meetings will not have quite the impact of face-to-face contact, but they will get the message across.

One company already is building a studio within walking distance of the Capitol to exploit this market. Eventually, Congress is likely to wire its own buildings, so that members can hold teleconferences from their offices and committees can hear testimony from out-of-town witnesses.

Electronic polling

The most dramatic changes of all may be in polling. The 1980 Reagan-Carter debate gave us a primitive picture of what lies in store. After the debate, the ABC network asked viewers to call a special telephone number and record their preference. Within hours, an AT&T computer tallied millions of votes and pronounced Reagan the winner.

A two-way cable TV system in Columbus, Ohio, called "QUBE" illustrates the next generation of the technology. Subscribers are given a small black box with several buttons, which is connected to the TV set. The announcer on a television program asks a question and gives the audience four choices, and viewers at home then push the button for the answer they like. Within seconds, the results flash on the screen.

This year, cable companies will install two-way gear in Dallas, Omaha, Pittsburgh, and several other cities. Other companies are developing two-way systems that use the telephone network. By 1990, as much as a quarter of the population may be using these so-called "videotex" systems to get information, bank, and shop from home. All those people will be able to participate in electronic polls.

The new technology may even be used for actual elections. San Diego showed what can happen when it ran a bond referendum by mail last year: the turnout was twice the normal level. Registrar of Voters Rudy Ortiz commented, "In the future, elections may be done by telephonics. We may just be showing the way." If Americans are willing to trust their banking to electronics, voting should be an easy step. This may be a good way to help the disabled and ill to vote, and it might do something for all the healthy people who do not bother to go to the polls.

QUBE is popular—subscribers apparently like talking back to their TV sets. Larry Wangberg, the cable executive who oversees the project, says it is "boosting the citizen interest-level" in politics. Alvin Toffler's book about the information age, *The Third Wave*, predicts that in time this technology will "combine direct citizen participation with 'representation' into a new system of semi-direct democracy." Toffler promises less alienation and better government from this automated politics.

However, electronic polling can be dangerous. First, the results are skewed. The ABC and QUBE polls count only those who bother to respond— hardly a cross-section. Most poor people will not have videotex terminals for a very long time, so the new technology will deny them even the chance to participate. Moreover, the existing two-way systems have no way to tell who is actually pushing the buttons. Many of the "voters" could be four-year-olds.

Second, these polls measure people's instant, emotional reactions to issues. Electronic polling is powerful because it involves many more people than the old-fashioned Gallup and Harris surveys, and it gets answers a week faster. In the process, these polls deny people the chance to think about the issue and omit the old surveys' follow-up questions that measure how deeply people feel. One shudders at the idea of instant referenda on the Panama Canal treaty or what to do about the hostage seizure in Teheran.

Finally, electronic voting offers a perilously convenient way for politicians to bias results or escape hard decisions. "Any fool knows you can control the outcome by how you ask the questions," warns Mayor Miller. "You could have some pretty atrocious acts justified by this." The greatest pitfalls are at the local level, where survey polling has been rare and where it would be all too easy to "go along with the people" on tough zoning or budget decisions.

The press, seduced by novelty, has given heavy coverage to QUBE results. Journalists should be more wary. In a few years, however, millions of people will be pushing buttons, and electronic polls are bound to have an effect.

Fortunately, people have too much common sense—and politicians too much instinct to protect their own power—to turn decisions over to fully automatic "teledemocracy." The subtler but more profound impact of electronic polling and the other new technologies will be on the distribution of power.

The shift of power from the political parties to narrow interest groups has been underway for decades. Radio and TV contributed to this trend by giving candidates a way to talk directly to voters, without having to go through party organizations, and by making candidates more dependent on money.

The new technologies may further dilute the fragile glue of the parties and of public identification with broad ideas. If television shapes the public's image of the world, narrowcasting can splinter that image. As former CBS

News President Richard Salant put it recently, "Instead of a common data base, we'll have smaller and smaller groups knowing more and more about less and less." Meanwhile, politicians and causes will use the new media to bind people more tightly into narrowly focused groups. The art of saying different things to different people is hardly news, but the advent of narrowcasting and electronic direct mail will make it easier to do and harder to catch.

In addition, the new technology will touch off a new scramble for money to buy teleconferencing studios, satellite transmitters, and fancy computers, and only the well-heeled interest groups will have the money to use all the new machines. Few liberal groups can match BizNet's millions.

On the other hand, some of the portents are good. The new technology may increase public participation in politics, through viewing and voting. Live television coverage may make politicians more responsive to the general public, rather than powerful lobbies. It certainly will help politicians speak directly to large numbers of people, with whatever length and depth the audience will tolerate. Low-cost ads on cable, free-access channels, and cheap computers will give impoverished candidates and causes their first chance to buy time on television and use time-saving automation. The parties themselves can use the technology to raise money and organize supporters.

Some of the effects on the balance of power will depend on how the new media are regulated. The Communications Act requires regular TV and radio broadcasters to sell time to candidates at minimum rates, to treat all candidates equally, and to provide balanced coverage of controversial issues. The application of these rules to cable and satellite-to-home broadcasting is thoroughly confused at present, and Congress and the FCC have barely begun to think about the matter.

In 1952, few politicians paid much attention when the first TV cameras were trundled into the national party conventions. Eight years later, television helped elect Kennedy and defeat Nixon. The political impact of the new media will be at least as traumatic. This time around, politicans and the rest of us had better take note and help shape this new world.

Guide to Further Reading

Computer crime

"The Spreading Danger of Computer Crime," *Business Week*, April 20, 1981.
"Computer Security," three-part special, *New Scientist*, July 7, 1983.
"Computer Security: What Can Be Done," *Business Week*, September 26, 1983.
Adrian Norman, *Computer Insecurity* (Chapman & Hall, London, 1983).
Michael Wood, *Introducing Computer Security* (NCC, Manchester, England and John Wiley, New York, 1983).

The *use* of computers for *detecting* crime is discussed in: *An Assessment of Alternatives for a National Computerized Criminal History System* (Office of Technology Assessment, US Congress, Washington, DC, 1982).

Martin D. J. Buss and Lynn M. Salerno, "Common Sense and Computer Security," *Harvard Business Review*, March–April 1984.

Data protection

Deanna C. Nash and David A. Bollier, "Protecting Privacy in the Age of Hometech," *Technology Review*, August–September 1981.
Craig T. Norback, *The Computer Invasion* (Van Nostrand Reinhold, New York, 1981). Really a handbook listing files in 125 US government agencies.
Computer-Based National Information Systems: Technology and Public Policy Issues (Office of Technology Assessment, US Congress, Washington, DC, 1981).
Herbert I. Schiller, *Who Knows: Information in the Age of the Fortune 500* (Ablex, Norwood, NJ, 1981). Argues that the growing privatization of information is a threat to democratic society.
Joseph F. Coates, "The Future of Computer Data Security," *Vital Speeches of the Day*, February 15, 1982.
"Privacy: A Free or Secret State," in Paul Sieghart (ed.), *Microchips With Everything* (Comedia, London, 1982).
G. L. Simons, *Privacy in the Computer Age* (NCC, Manchester, England, 1982).

Politics

James N. Danziger, *Computers and Politics: High Technology in American Local Government* (Columbia University Press, New York, 1981).

Richard M. Neustadt, "Watch Out Politics—Technology Is Coming," *The Washington Post-Sun*, Outlook Section, March 14, 1982.

Andrew Lloyd, "Europe Examines Electronic Democracy," *New Scientist*, June 2, 1983. On Council of Europe report, *New Possibilities for Democratic Participation Resulting From Information and Communication Technology* (Brussels, 1983).

G. Benjamin (ed.), *The Communications Revolution in Politics* (Proceedings of the Academy of Political Science, New York, 1982).

Irving Louis Horowitz, "Printed Words, Computers and Democratic Societies," *The Virginia Quarterly Review*, Autumn 1983.

Ithiel de Sola Pool, *Technologies of Freedom: On Free Speech in an Electronic Age* (Harvard University Press, Cambridge, MA, 1983).

Other problems

Jerry L. Salvaggio, "Social Problems of Information Societies: The US and Japanese Experiences," *Telecommunications Policy*, September 1983. Looks at invasion of privacy, misuse of information, and inequalities of information, etc.

The issue of "hackers"—people who break into computer systems for kicks—is discussed in *Newsweek*, September 5, 1983. A critical review of that article, "Moral Clarity in the Computer Age," appears in *Communications*, October 1983.

Ted Nelson, *Literary Machines* (Nelson, New York, 1981). Looks at the impact of new technology on writers.

Thomas Mandeville, "The Spatial Effects of Information Technology," *Futures*, February 1983. A review of the literature on possible urban/regional problems.

Miles Orvell, "The Screen Revolution," *Technology Review*, February–March, 1982. A look at McLuhan's writings in the light of new developments.

13 Global Issues

Information Technology and the Third World

Juan Rada

Information technology is changing the technological profile of manufacturing and the service industries. The main effect on the less developed countries will be to increase the obsolescence of their industries, services, and development strategies. Juan Rada is with the International Management Institute, Geneva and this paper was first read to the IFAC seminar, Vienna, Austria, March 1983.

Introduction

The effects of information technology on North–South relations, the international division of labor and development, is a subject of growing interest to specialists as well as policy-makers (Mitterand, 1982). I will attempt to review here the reasons for this interest by analyzing aspects of the effects of the technology in some of these areas. In a brief article all aspects cannot be covered, and the intention is to highlight some of the main issues.

Since the beginning of the debate in developed countries about the effects of information technology on their societies and economies, consideration has been given to potential impacts, both positive and negative, on developing countries. In the early stages of discussions and research, the main concern was related to the assessment of economic effects, especially the potential erosion of one of the developing countries' perceived comparative advantages, namely low labor costs (Rada, 1980).

This concern remains central today, but it is qualified by a number of considerations as the understanding of the effects of technology grows. Such considerations include questions of quality, product cycle, change in the function of products, and a better understanding of the behavior and structure of the suppliers of the new type of equipment. In addition, the early concern was especially centered around the erosion of advantages of offshore assembly, especially in those countries with export-oriented economies—which, of course, constitute a rather exceptional group of developing countries.

From the almost exclusive focus on automation in manufacturing and economic effects, new and important elements have been added. From an economic point of view, two of these deserve special attention. The first concerns attempts to understand the effects of automation within developing countries, particularly in those cases where development is geared primarily toward the internal market. This is the case, for instance, with most of the Latin American economies (UNIDO/ECLA, 1982).

The second relates to one aspect of the effects of information technology that, although implicit in early research, was somehow overlooked in terms of its real importance: namely, that it allows the transportability of services to an extent and depth not dreamed of some years ago. This has opened up new questions, especially for developing countries.

Although I shall focus on these two points, the social, cultural, and political effects of automation are of equal, if not of greater, importance. Lately, substantial attention has been paid to these points, particularly in the context of searching for alternative development strategies. This search is not new, but it differs from earlier attempts in its desire to see whether, given the nature of the technology, some entirely new concept of development can be pursued, heavily based on human resources, information, and knowledge-intensive activities: in brief, to determine how developing countries can appropriate information technology in a qualitatively different form, and aim at some sort of leapfrogging.

These views are based on a prospective assessment of the form human activities could take as a result of the changing technological profile of society. This view has been developed out of a mechanical extrapolation of the potential of the technology and is most creatively expressed in works such as Alvin Toffler's *The Third Wave*.

In other words, would it not be better for LDCs to aim at societal models that will truly consider the prospective "information society," rather than to evolve (if possible) through the traditional lines? This view might be tempting for many, but often fails to acknowledge not only the unequal distribution of resources and knowledge, but also the fact that "informatization" is the *consequence* of development and not its cause, although the technology can be used for development purposes.

This is not to negate the fact that development needs to be conceived in a completely different form in order to account for the current technological mutation, which indeed questions the very core of currently pursued development strategies. Such questioning, however, involves developed as well as developing countries, since a global approach will sooner or later be necessary to readdress the direction that current changes seem to be taking.

The starting point in this discussion is to list briefly the main areas of the economic impact of information technology (IT).

IT: the main areas of impact

Effects in production

1 Substitution of mechanical components (e.g. watches)
2 Substitution of electromechanical components (e.g. cash registers and calculators)
3 Substitution of electric and older electronics (e.g. computers)
4 Upgrading of traditional products, creating entirely new capabilities (e.g. word processors)
5 Upgrading of control systems and substitution by electronic ones (e.g. machine tools)
6 New products (e.g. games)

In the case of products, the use of electronics can alter the very function of the product. A good example is cash registers, which, from being only adding machines, have become data entry terminals, potentially part of a system of accounting, control, and ordering. Other products, such as machine tools or word processors, if supplied with the adequate hardware and software, can communicate to other machines, data bases, or computers.

Effects in manufacturing processes

1 Increase in the flexibility, adaptability, and economy of production (e.g. CAD/CAM)
2 Incorporation of skills and functions into equipment (e.g. CNC machines or robots)

In the case of processes, the important developments are essentially two. First there is the programmability, which leads to flexibility, since the same equipment can be reprogrammed to perform a different task rather than changing the machine. This in turn leads to the second important point, which is that there is great resistance to obsolescence and thus an extension of the life-cycle of manufacturing processes. Robots, for example, are very resistant to obsolescence since in most cases it is sufficient to change the program and/or the "hand" to use the equipment for a different operation. This has an important capital-saving effect. In addition, there are savings on downtime for retooling and changing tasks.

Effects in the office

1 Automation of routine clerical work (e.g. data and word processing)
2 Increase in the efficiency and effectiveness of communications, especially in those areas where work is less formalized, such as in professional and managerial areas.

Automation in the office or, more accurately, the use of electronic tools as aids or facilitators of the work is and will have a far-reaching effect on economic activities in general and especially in the production of services. In developed countries the percentage of the labor force working in offices or information activities is constantly growing, in some cases reaching 50 percent. The office sector has traditionally been under-capitalized and its productivity has been low as compared with manufacturing and agriculture.

It is not known how the current process of rapid diffusion of electronic technology in the office, in either clerical or managerial activities, might affect companies' competitiveness, productivity, and the international division of labor.

It is safe to say at this stage that, increasingly, most employees, whether in agriculture or in manufacturing, are concerned with information-processing activities rather than production. In fact, the absolute number of people employed in manufacturing has been decreasing in the US since 1964 and in Europe since the early 1970s. This trend will continue, as it did with agriculture in the past. The decrease in the labor content of agriculture was accompanied by substantial reductions in hours worked and important increases in output.

Effects in services

1   Transportability of services (e.g. remote access to data bases, banks, archives, etc.). This also leads to new services (e.g. Prestel)
2   Increase in self-service (e.g. gas stations, banking, etc.)

These in turn lead to the replacement of human-to-human services by machines. The impact of information technology on services is perhaps the most important in the long run because it is creating entirely new possibilities that are different from past activities. This point will be discussed later.

The re-invention of industry

The first question to ask in terms of the impact of information technology on developing countries is how technology affects developed ones. This question is pertinent owing to the fact that it is in these countries that the technology was first used to increase productivity in manufacturing and services. This problem has been extensively studied, and the main conclusion is that a re-invention of industry will be necessary, that is, a radical change in the technological profile of productive activities.

This change naturally affects countries that have geared their efforts toward industrialization either through import substitution for the internal market or through exports.

The basic hypothesis is that the industrial utilization of IT leads to an erosion of developing countries' comparative advantages and international competitiveness, especially in traditional industries. The main reasons for this are as follows.

*Decrease in the relative importance of labor-intensive manufacturing
and cost of labor*
This is essentially due to the *automation of production*, which tends thereby to
erode the competitiveness of low labor costs. A good illustration is the
comparison of Hong Kong and the US in the manufacturing of electronic
devices. Table 13.1 shows that when the process is manual the difference in

Table 13.1 Manufacturing cost per device (US$)

Process	Hong Kong	US
Manual	0.0248	0.0753
Semi-automatic	0.0183	0.0293
Automatic	0.0163	0.0178

Source: Global Electronics Information News-letter, No. 25, October 1982.

cost is about 1 to 3, decreasing drastically with semi-automatic processes and
automatic ones. While this example refers to components, a similar process is
taking place for systems and consumer products. In the case of TV sets, the
chairman of Electronics Industry of Korea stated that owing to the
automation of assembly and technological change,

the manufacturing costs of a TV set in Korea and that of the US are practically
comparable to each other. Rapid advancement of industrial technology is eliminating
labor-intensive portions of the electronics industry; this tends to make it harder for
Korea to earn enough foreign currency to import expensive new technology. [Kim,
1980].

It should be mentioned that TV sets as a product have changed substantially in
the last few years and are bound to change more in the future when they
become digital and the CRT (cathode-ray tube) is replaced by some other form
of display. In other areas preliminary evidence shows similar trends. For
instance, the expansion of automation in Japan has contributed to a recent
reduction of investment in the Asia/Pacific region involving firms in
electronics, assembly parts and textiles (*Business Asia*, 1982).

In the case of garments, a trend to systems optimization and automation is
clearly underway. Although not yet a "perfect fit," Hoffman and Rush (1982)
conclude that: "Although it has not happened yet, to a great extent there is a
feeling among the large producers that a large share of offshore production
will be brought back [to the developed countries]."

Value-added is pushed out of assembly and into components as integration
increases. This occurs at the product level, while in systems value-added, it is
pushed upwards toward servicing. This process is proper to the industries
where electronics has substituted other components in products (as described
above) and implies that functions previously obtained by assembling pieces
are incorporated in the electronic component itself.

In this category fall, for instance, calculators, telexes, sewing machines, and precision engineering in general. In the case of electronic components the amount of value-added obtained in offshore assembly has been decreasing constantly. The dutiable value of components imported under the US tariff arrangements, that is, the value-added in offshore plants to US products diminished from 57 percent in 1974 to 39 percent in 1978 (USITC, 1979).

The main explanation for this is the increasing value of the parts produced in the US as a result of the growing complexity of devices. This process has continued since: one only needs to see that in 1978 the level of technology was LSI (large-scale integration) rather than the current VSLI (very large-scale integration). As the level of integration of components increases, the value-added obtained in front-end operations also increases. Furthermore, the assembling of chips is being automated and moved, in the case of sophisticated devices, to "clean rooms."

Changes in product cycles
Product cycles in many areas have been considerably condensed while process cycles have increased, owing to the resistance to obsolescence of programmable machines and equipment. Typically, product cycles have been shortened in some industries (e.g. office equipment) from 12 to three or four years. This has led to a concentration of manufacturing investment in capital-intensive flexible manufacturing, and partly explains the erosion of the advantage of developing countries in so-called "mature" or "semi-mature" products.

In other words, the "product cycle" view of international trade needs review, since formerly mature industries or products are being completely revitalized. A case in point is a European company that closed a plant in South America because the short product cycle did not justify the investment. In the past, the payback time was far longer with a more stable technology. At the time of mechanical or electromechanical technology, local manufacturing was justified because of the large amount of value-added obtained in assembly; and, because of the longer cycle, there was a relatively small incidence of amortization and development cost in the final cost of products (Cohen, 1981).

Quality considerations
These are growing in importance as markets become more segmented and competition increases under conditions of low growth. This in itself leads to what has been called the "hands-off" approach in manufacturing or automation, coupled with a change in the skill-mix at the shopfloor. One of the main reasons why Japanese manufacturers in the field of semiconductors make very limited use of offshore facilities in developing countries is precisely the perception that the required level of quality cannot be obtained.

Quality has a cost, and requires an infrastructure and substantial managerial know-how. In some cases the old manufacturing system of assembly is inconsistent with quality requirements and new methods have to be used, notably modular or group-work schemes, with emphasis on a highly multi-skilled labor force. This phenomenon highlights the dilemma for export as well as import substitution strategies.

The elements mentioned above are the tip of the iceberg of current change. Further down the inevitable question is: What about access to technology, its production and application? Three comments are necessary here.

First, almost by definition, advanced and rapidly changing technology is not properly documented and therefore its transfer tends to have peculiarities. In fact, transfer of technology in the area of concern of this paper takes place essentially through three main mechanisms:

1   mobility of personnel, which take with them their own knowledge (this is the so-called "Silicon Valley syndrome"). This accounts for a large part of transfers in the US;
2   second sourcing, which is the agreement between two producers to manufacture fully compatible products. This might or might not entail full exchange of technology. An agreement of this nature implies a partner that can produce at similar technical, economic, and quality standards;
3   cross-licensing agreements, which assumes a mutual exchange of technology.

In brief, transfer tends to take place among established or important producers, and furthermore, the technology is tightly guarded as trade secrets. Many companies in the software area, for instance, do not patent or copyright their products because it entails disclosure of valuable information.

The second comment is that the issue is access not only to a given technology but to the *process of technological change*, because of the dynamism of it. This leads to a number of questions that I shall not discuss in this paper, notably about the innovation environment. The point that I wish to make is that access to the process of technological change in advanced areas (and not only IT) seems to take place essentially, as European companies have discovered, with participation in the equity of companies. The possibility of some developing countries doing this is relatively small, as some exploration has shown, essentially because of the high mobility of the personnel and also because of political considerations. In this respect one should simply mention that in many areas of electronics civilian applications have surpassed military ones, creating an additional obstacle to prospects for transfer of technology.

The third comment refers to production and applications. In terms of production, few LDCs are in a position even to raise the question. Some have implemented policies in this field (i.e. India and Brazil), but their performance cannot be evaluated at this stage, except to say that they at least provide the countries with the capacity to follow the technologies closely. Success will depend largely on the targeting of market segments and technologies (e.g.

uncommitted logic arrays and custom circuits in general). At the systems (e.g. minis, micros, etc.) and software level, the situation is different. The assembly of equipment from components that are bought practically off-the-shelf is taking place in many countries, and this is likely to continue for some time to come. But as the level of integration of components grows, the amount of software incorporated into chips (firmware) will also grow, taking value-added away from the assembly of systems.

To illustrate this further, the trend in microcomputers is to incorporate into the hardware as many "utilities" as possible, such as word processing, Visicalc, and others, in a similar fashion to what happened with the pocket calculator. At the beginning the calculator featured four arithmetic operations; as integration increased, more were added, making the machine more useful and also less expensive. This trend implies that, in the not-too-distant future, the source of value-added will go to systems software, design, and service in a far more pronounced way than today.

The assembly of systems will continue, especially when protected by tariff barriers, incentives of industrial policies, or both (i.e. Brazil). These types of equipment will be used largely in internal markets, and are unlikely to make a dent internationally. In addition, systems in developing countries tend to be far more expensive than in the international marketplace, making less economical their application, especially when labor costs are lower. In one Latin American country the cost of word processors and microcomputers was, respectively, double and triple that of the US. This is explained by a number of reasons, the most important being that suppliers have to cover maintenance, software development, and overheads, selling a rather limited amount of equipment as they operate as "profit-centers." Installation and use of the equipment is also more costly in some cases because of expensive auxiliary installations (i.e. electrical generators), subutilization, or lack of adequate skills, especially managerial ones.

Optimization of systems and office/service automation
One of the most important effects of IT is that it leads to the optimization of business activities as a result of rapid and timely processing of information and the relative ease of communications. I prefer to use the term "service automation" rather than "office automation" because it truly accounts for the nature of the change. The impact of office automation is measurable not at the work-station level (as is done with word processors), but rather at the level of total systems performance, mainly because of the effects on management information systems and managerial effectiveness. To illustrate the point, consider that secretaries account for only 7 percent of the total clerical costs in the US and spend only 20 percent of their time typing.

It is not, then, in the work station that the effects of office automation are to be found, but rather on the synergies, greater numbers of options, faster response, and more informed decisions that are derived from it. This is not to negate important productivity increases at the work station but to treat them as one, and perhaps not the most important, component of current improvement. Research conducted at IMI-Geneva shows precisely this, and furthermore confirms the optimization of systems that takes place within companies utilizing IT.

The effects of these processes on the international division of labor and developing countries is yet unknown, but it is possible to make some tentative hypotheses. First, given the composition of the labor force in the advanced countries and also within manufacturing companies, an improvement in systems performance will further reinforce the advantages derived from automation and product change. For instance, in pharmaceuticals today, typically only about 30 percent of the labor force is employed in production, and the proportion is expected to decrease from current levels by as much as 40 percent by 1990. This implies that manufacturing is decreasing in importance (as measured in total cost) while performances at the systems level and innovation are becoming the key to profit, growth, and survival.

Second, and most important, is the increase in productivity of services, which for the most part are information-processing activities. The transport ability of services is the most important long-term effect of the technology; thus, more efficient production of them reinforces the great advantage that developed countries have in this area. I shall elaborate on this point later.

Changes in skill-mix and conditions for absorption of technology
I mentioned earlier that important changes in skills accompany the product/ process changes. In some cases there are significant skill-saving effects (e.g. CNC machine tools), which can be beneficial to developing countries. In general, the trend seems to be toward higher skills, especially at the systems and design level and not the least in software. Most of the developing countries' labor forces have low skills, or skills of a mechanical nature that in many cases are being substantially altered (for instance, for interface types of work). This again calls for a more active policy on the part of the developing countries in terms of training and education. The absorption of technology is also changing, not only for the reasons stated above about transfer, but also because knowledge tends to be of a more abstract nature.

It is not by chance that much innovation in electronics has taken place around universities (Stanford, Berkeley, MIT, and now Cambridge in the UK). This means that the links between scientific and technological knowledge are becoming tighter, and the neat categories of the past that distinguished invention from innovation are not always tenable.

The need for scientific policies is obvious, especially in areas where these types of knowledge are closely related to technological development (e.g. physics of materials in electronics or genetic engineering), but these policies require a clear focus. It is true that current and future technological progress is based on science, but at the same time innovation does not necessarily require a sophisticated scientific base, as Japan has proven so convincingly during the 1960s and the early 1970s. This is particularly true in relation to process and systems innovation.

In the first case changes tend to be incremental (unless the product changes), and in the second case (systems) changes tend to be of a conceptual and organizational nature. A classical example of this latter type of innovation is self-service, which has boosted productivity in many sectors with little "hardware" investment of R&D work. Credit cards, marketing systems, financial services, leasing and rental operations are other examples. In this respect it would be interesting to compare innovation in banking with the familiar innovation curves in electronics; we would probably find, to our surprise, that they would not differ much.

The skill-mix is changing while new skill requirements are emerging, particularly in software, systems design, and (an almost forgotten one) management and organization. Two different companies or countries with similar skills and other endowments may perform quite differently simply because of differences in their management and organization which lead to varying degrees of technology absorption. Often, skills of different natures can be obtained through training and retraining, but a precondition is the action that creates the *need* for them: that is the answer to the questions what and why, the know-what and know-why. If these two questions are not answered, obtaining know-how will make no difference to performance—on the contrary, it will not even be possible to obtain the proper know-how.

These changes are to an extent already occurring in LDCs, but in some fields the die is cast while in others selectional decisions can still be made. Developing countries can obtain immense benefits from technology if it is applied in the context of a development strategy. For instance, IT is capital-saving in manufacturing and services (lower entry barriers) per unit of output. This leads to the traditional dilemma of technologically induced labor-saving effects, since capital intensity tends to increase.

The main issue for developing countries in terms of employment/technology lies in the field of agriculture rather than computerization in manufacturing or services. The reasons are rather simple. The number of computers currently in use has a practically insignificant effect on the overall volume of employment when they do cause displacement. The two largest users of computers in the developing world are India (about 1000) and Brazil (about 10,000), which in the context of their economies is minimal. The applications to which they are put remain traditional, and they tend to optimize administrative systems that in turn create beneficial effects in the rest of the country's economy.

An illustrative case is the informatization of the postal check system in Algeria, where 176 people were made redundant through voluntary retirement. The general effects can be seen in table 13.2. Examples like this illustrate the trade-off between employment and the use of technology and show how beneficial IT can be when applied in critical bottlenecks, especially

Table 13.2 Computerization of the postal check system in Algeria, 1974–7

	Manual 1974	Computerized 1977
No. of operations	24,360,000	33,620,000
Volume (millions DA)	109.5	210.8
No. of accounts	452,000	709,000
Waiting time at centers before processing of document	15 days	2 days
Payment at cash desk	3–6 hours	2 min.
Saturation ratio	95%	50%
Employment	856	680

Source: Secretariat d'Etat au Plan, Commissariat National a l'Informatique, "L'Informatique en Algerie," Algiers, 1978.

in relation to infrastructure and services. It needs to be done with the normal criteria of appropriateness of technology, that is, selectivity. It is in the fields of administrative services and infrastructure that short-term benefits can be realized rather than in totally new types of applications, which will take a long time to mature and are heavily dependent on equipment performance and characteristics of the human–machine interface.

Capital-saving effects also take place in manufacturing and agricultural applications, together with skill-saving effects. The real challenge in this field is to combine traditional and low technologies with advanced ones, and much needs to be done here.

Developing countries are far more heterogeneous than developed ones, and sometimes this shorthand concept masks tremendous differences. In some countries (e.g. Southeast Asia) companies are combining advanced technologies with lower labor costs, and future developments will depend on the competitive reaction of developing countries' producers. Others are following a policy of technological upgrading in the exporting industries, and even authorization for acquiring foreign computerized equipment depends in some cases on potential export performance. Furthermore, many developing countries have large pools of educated labor in areas most appropriate to current change, such as software, and great potential exists in this field. I insist on the word "potential" since production or export of software is not as easy as it might sound; in some cases import substitution of software might be far more economical than attempting exports.

Possible policies and measures to maximize the benefits of IT for developing countries are not only necessary but urgent. But this should not mask the equally urgent need to search for different development models, South–South cooperation, regionalism, and, most importantly, some sort of social command of technology. Command of technology differs substantially from control in the sense that it maintains the relative autonomy of action and creativity at the technological and scientific level, but provides a direction for the application of that creativity based on fundamental human and social needs. A policy of this nature will emerge only if, at country, regional, and global levels, the priorities of IT are identified as being based on a normative concept of development.

Notwithstanding short-term policies and strategies, the fact remains that the gap between developed and developing countries will increase, as will the gaps *within* developing countries, which is one of the important structural causes of their present state. The gap between countries has been shown in many ways. For our purpose what matters is the situation in IT, which can be seen from tables 13.3 and 13.4.

Table 13.3 Value of data processing equipment* (US $1000)

	1978	%	1983	%	1988	%
Developed countries (US, W. Europe, and Japan)	110	83	180	82	250	80
Other countries (incl. centrally planned economies)	22.5	17	40	18	61	20
Total	132.5	100	220	100	311	100

*Micros are not included.
Source: Diebold, Europe, 1979.

Table 13.4 Worldwide telecommunications equipment market* (US $1000)

	1980	%	1985	%	1990	%
Developed countries	36	90	53.5	89	75.4	86
Developing countries	4	10	6.7	11	12.1	14
Total	40	100	60.2	100	87.5	100

*Includes: telephone, telegraph, telex, data communications, satellite communications, mobile radio and radio telephone, radio paging, and cable TV.
Source: Arthur D. Little Inc.

Three important conclusions can be derived from this purely quantitative check. First, the participation of developing countries in the process of "informatization" is indeed small. Second, the gap is likely to grow by a factor of two during the 1980s. It should be said that telecommunications investments are far more predictable than computer investments, owing to longer planning cycles. If anything, the figures in tables 13.3 and 13.4 are optimistic, and would probably need to be adjusted downwards in the light of the severe financial problems of developing countries that had large telecom and informatic projects (e.g. Nigeria and Mexico).

More precise "informatic indicators," particularly in terms of data and satellite communications are revealed in the data of table 13.5. The gap in transmission and satellites is even larger when one considers the entire telecom market (table 13.4), and it points to a qualitative difference in priorities for investment and the type of emerging infrastructure. The telecommunications infrastructure, especially in data transmission, is the one

Table 13.5 Transmission and satellite markets (us$
million)

	Telegraph, telex & data transmission		Satellite communications	
	1980	1990	1980	1990
Africa	48.4	97.3	3.0	10.5
Latin America	106.5	189.0	14.3	34.9
North America	2481.4	6000.5	122.8	463.7
Europe	733.9	1984.1	59.0	189.2

Source: Arthur D. Little Inc.

that will largely determine the "multiplier effects" of information technology, particularly in terms of knowledge and information-intensive activities.

The third conclusion that can be drawn from tables 13.3–13.5 is that investment in the field of communications and transmission equipment increases almost mechanically the possibility of optimizing systems, increasing office and service productivity, raising the efficiency of production, and furthering conditions for capital-saving effects (e.g. optimization of stocks). This implies that the relative position of developing countries in terms of leapfrogging into the "information age" is even lower than thought if taking traditional indicators of the "industrial age." This is why I stated earlier that IT is the consequence rather than the cause of development, and leapfrogging can be possible only within a global rather than a purely national or regional strategy.

It should be understood that the process of "informatization" of society is one in which greater amounts of knowledge and information are incorporated into goods and services. This also means that knowledge and information activities acquire a dynamism of their own right and become sources of wealth creation and value-added (e.g. design, programming and R&D). As the amount of information and knowledge incorporated into products, processes, and services increases, the relative amount of energy, materials, labor, and capital decreases. Technologies diffuse through society precisely because they are factor-saving, and IT saves simultaneously in all directions while increasing the capacity to create and process information, and therefore contributing to the accumulation of knowledge.

The empirical evidence shows precisely that the current process is one where greater amounts of information and knowledge are going into production, and not, as some might suggest, that we shall live off information exchanges (Gershuny and Miles, 1983; Jonscher, 1983). Indeed, the greater consumers of robots or even computers are those industries such as automobiles or telecommunications that a few years ago were considered "traditional" sectors.

The nature of the technology calls for a more detailed understanding of "knowledge and information-intensive activities." I have chosen here to examine the services (there are other aspects, such as knowledge and information transfers), because perhaps the most far-reaching effect of technology is that it allows the transportability of services: instead of going to a bank or library, we can transport the bank or library to our own terminals in our offices or homes.

Services revisited

The impact of IT on services is vast. It is not possible here to account for all the elements involved, among other reasons because the understanding of services and the data available are limited.

Until recently, interest in the service sector has been due mainly to the tendency for the share of services to rise in total output and employment and the implications of this for the growth of output, employment, and productivity. Two elements make a review necessary: first, the role of services in international trade, and thus on domestic economies, and second, the impact of IT on the production, commercialization, and distribution of services.

Table 13.6　Balance of trade in non-factor services, by selected groups of countries, 1977–80 (US $million)

Selected oil exporters	1977	1978	1979	1980
EEC	9,551	13,400	10,183	10,164
EFTA	2,632	3,920	4,623	4,892
US	3,792	3,793	1,408	3,852
LDCs	−20,270	−25,711	−36,701	−42,528
	−18,987	−23,909	−33,623	−38,803

Source: UNCTAD Secretariat, based on IMF statistics; UNCTAD, "Trade and Development Report," Geneva, 1982.

The first problem one encounters is that of definition, because the service sector is too heterogeneous in terms of its production and consumption. In this context I simply take the current, albeit imperfect, measurement of trade in nonfactor services or service products. This area includes shipping, insurance, banking, other trade-related services and private and public-related services. In this category, the situation of developing countries can be seen in table 13.6.

In areas such as private services, where transportability is increasing, the balance of trade was −$2911 million in 1978 and −$4635 million in 1980. While international trade in goods is either stagnant or decreasing, trade in services has been growing. The case of the UK is illustrative, with private invisible exporters earning a total of £26,472 million in 1981, up 13 percent from 1980, which in itself was a record. A large percentage of the above figure was earnings from investments abroad.

One interesting category is consulting engineers, who exported £214 million in 1976 and £425 million in 1980; this is the type of service that is becoming highly transportable (Committee on Invisible Trade, 1982). These figures give a general view of the type of volume involved. In general, developing countries have high deficits in this field, although the two largest deficits belong to West Germany and Japan. The interest here is to see how IT affects this area, which is a dynamic and expanding one in international trade and one that accounts for the greatest percentage of labor in the developed economies as well as a growing percentage of output.

For the decade 1970–80, as recorded in the balance of payments statistics, world services exports grew at an average annual rate of 18.7 percent. The value of service exports in 1980 was of the order of $350 billion. The growth figure is lower than that for world exports of merchandise, but the latter figures are inflated by the upsurge in oil prices (Office of the United States Trade Representative, 1983).

The application of IT to the transportation services has had four main effects on the economy (besides increases in self-service and replacement of human-to-human services by goods):

1   an increase in the transparency of markets owing to the availability of information; for example, the listing of suppliers in data bases with their prices, or teleshopping that is being applied first to industrial products. This change tends to increase competition since it provides exposure to a wide range of choices;
2   blurring of borders between service providers (i.e. between retailing and banking) and industry and services (i.e. telecommunications and computing);
3   lowering of the barrier to entry in many services: this happens because, once the infrastructure is in place (the data highways), the cost of marketing services is reduced to plugging into networks. For instance, a software producer utilizing Prestel in the UK has theoretically as many retailing points as there are terminals. Furthermore, he/she can transport the product through the telephone lines. In the past, this would have required representatives in many places and an investment in bricks and mortar. Similarly, the use of telebanking allows small banks to compete with large ones, with the added advantage that they have fewer fixed assets to depreciate. The analogy here is the lower barrier to entry into the market for a farmer when a main road passes in front of his land;
4   internationalization of services and specialization in its production. It is this point that I wish to develop further here.

The process of internationalization took place first in mining and agriculture and later in manufacturing. Today it is the turn of services, in accordance with the changes of the structure of the economies. The analogy that could help to clarify the current phenomenon is the effect that the steamboat or the development of railroads had on trade in the last century. The economic transport of bulk cargo destroyed the "natural" barrier to trade internationally and within countries. This was instrumental in the integration of national economies and the development of an international one in tangible goods. This in turn made evident the comparative advantages in the production of goods since distance was no longer an overwhelming impediment to international trade.

The decreasing cost of transmission in telecommunications, the convergence with computers, and the increase in productivity in the production of services is creating similar conditions, but this time in intangibles. Thus, if a European consults a data base in the US he is importing an intangible service, while the US is exporting one. Current change is then making evident the comparative advantages in the production of services, especially those that are knowledge- and information-intensive. One should qualify this statement, by saying that little is known about economies of scale (or scope) in the production of services or the relative factor intensity of their production.

The concentration of knowledge and information-intensive services in developed countries is a fact, and the gap is more pronounced than in manufacturing. The question here is not simply one of import–export but also of job creation, patterns of investment, and production, inasmuch as the latter is becoming also knowledge-intensive. To illustrate this, one can take the example of what the machine-tool industry will look like in the future. Machine tools could be linked via telecommunications to machining data bases, and what will happen to them will be similar to computers: the generation of revenues will be based on software and services rather than hardware. At that point producers will be making "machine packages" rather than discrete pieces of equipment. A manufacturer will buy the entire "package," including the link and servicing. This in itself will change the nature of technology transfer.

Table 13.7   Reference data bases and records: geographical distribution, 1975–9

Area	1975	1977	1979
US			
No. of data bases	177	208	259
No. of records (millions)	46	58	94
Other developed economies			
No. of data bases	124	154	269
No. of records (millions)	6	13	55
Total			
No. of data bases	301	362	528
No. of records (millions)	52	71	149

Source: M. E. Williams, "Data Bases and On-line Statistics for 1979," ASIS Bulletin, December 1980.

Table 13.8   Cost to Canada of import of data processing and computer services

	1975	1978	1980	1985
Cost of imports of services	$155m	—	$560m	$1.5b
Proportion of outside services required	30%	—	41%	52%
Est. job losses	4,400	7,500	11,000	23,000
Data processing jobs represented in losses	—	6%	8%	14%

Source: Consultative Committee on the Implications of Telecommunications for Canadian Sovereignty (Clyne Report), "Telecommunications and Canada," Canadian Government Publishing Center, March 1979.

Tables 13.7 and 13.8 show the concentration and possible consequences of this trend, based on a rather narrow set of criteria, namely, data bases and Canadian imports of data processing services by 400 foreign subsidiaries operating in Canada. Another study (Price Waterhouse Associates, 1981) shows that the trends identified remained valid. (It should be clear that Canada is in a rather exceptional geographical position, and also has special telecommunication arrangements with the us. This means that these types of effects are more pronounced in Canada.)

It is not possible to go into the details of the implications of this emerging phenomenon, especially in the longterm. It should be clear, however, that this is not simply a question of international flows of machine-readable data or transborder data flows, but also a question of industrial and development policy, since the entire production infrastructure is changing in technological profile via an increase in information and knowledge-intensity of products, processes, and services.

In order to respond to current changes, maximize benefits, and minimize negative impacts, we need to assess how the strategic sectors of developing countries are being affected by technological change in the short and long terms. If policies are not thereby adjusted, the negative effects will outweigh potential benefits since development strategies will be built on quicksand. This should be the first step toward the search for regional and global approaches. A proposal of this nature should not impede the application of the technology now, but could give it better direction. It is also in the best interests of developed countries to search for solutions since interdependence is a growing reality.

Conclusions

IT is an expanding reality with far-reaching consequences for the relationships within and between countries. In this context it should be clear that we are at the beginning of the development of IT, not in the middle or at the end.

From the point of view of developing countries, four general conclusions seem to be valid.

1 An erosion of their advantages in low labor costs and "mature industries" is taking place because of changes in products and manufacturing

processes. The final outcome of this trend will depend largely on the LDCs' response at the macro and micro level.

2 Developments of the service sector and in particular of value-added services depend on the industrial base, as it is not an autonomous sector. This means that a deterioration of the industrial base will inevitably affect the creation and development of services. In turn, as the service content of industry increases, the lack of development in services could further affect the possibilities of industrial development.

3 The use of IT in developing countries is limited, and the gap in this field is increasing. With few exceptions, most LDCs have not developed policies to confront the challenge at the different levels required (e.g. skills).

4 A new reality and opportunity is emerging with the internationalization of services. Redressing trends here seem to be a priority, essentially because the situation is still in a state of flux. The lower barrier to entry in services offers some developing countries opportunities that were impossible to imagine with tangible products, among them the possibility of reaching consumers directly.

The final outcome of the effects of IT in developing countries and the global realities will depend largely on the willingness of the actors to approach the problems and opportunities in a global context. The command of technology based on human needs seems to be a priority for what remains of the century.

References

Business Asia (1982), March 12.

Cohen, E. (1981), "Modificaciones provocadas por la microelectronica en el rol de las empresas transnacionales electronicas en los nacionales electronicas en los paises en vias de desarrollo. Analisis de dos casos en el area de maquinas de oficina," Primer Seminario Latinoamericano sobre Microelectronica y Desarrollo, Buenos Aires (mimeo).

Gershuny, J. and Miles, I. (1983), *The New Service Economy* (London).

Hoffman, K. and Rush, H. (1982), "Microelectronics and the Garment Industry: Not Yet a Perfect Fit," in IDS Bulletin, *Comparative Advantage in an Automating World*, vol. 13, no. 2.

Jonscher, C. (1983), "Information Resources and Economic Productivity," *Information Economics and Policy*, vol. 1, no. 1.

Kim, W.H. (1980), "Challenge to US Domination: The Promise of Technology for Newly Industrialized Countries," in *Financial Times Conference: World Electronics Strategies for Success* (London), p. 99.

Mitterrand, F. (1982), *Report to the Summit of Industrialized Countries: Technology, Employment and Growth* (Paris).

Office of the United States Trade Representative (1983), *A US National Study on Trade in Services* (Washington, DC).

Price Waterhouse Associates (1981), "Review of Economic Implications of Canadian Transborder Data Flows", as reported by *Transnational Data Report*, vol. 4, no. 6.

Rada, J. (1980), *The Impact of Microelectronics* (Geneva), chap. 7.

UNIDO/ECLA (1982), "Expert Group Meeting on Implications of Microelectronics for the ECLA Region," Mexico City, June 7–11, 1982. *Conference Proceedings* (Vienna).

USITC (1979), *Competitive Factors Influencing World Trade in Integrated Circuits* (Washington, DC), p. 14.

"Smart" Weapons

William J. Perry and Cynthia A. Roberts

Microelectronics has made possible a new generation of "smart" weapons, precision-guided armaments that incorporate intelligence. The US is ahead in advanced microchip technology, while the Soviet Union—much stronger in conventional weapons—is trying desperately to catch up. The authors are in the Arms Control and Disarmament Program at Stanford University. This article is from Technology Review, *July 1982.*

For over a decade, the Soviet Union has been outproducing the US by more than two to one in almost all categories of conventional weapons—tanks, aircraft, missiles, and ships. Given their high priority on military production and development, together with their growing technological sophistication, the Soviets have been able not only to overwhelm the US in numbers of weapons, but to compete in performance as well. Indeed, impressive advances—from the MIG-27 to the Typhoon submarine—typify the high-performance systems now entering the Soviet armed forces.

The Soviets have achieved this numerical advantage not by building simple, cheap equipment as is popularly supposed: they lead the US because they spend twice as much on military equipment. The large increases in defense spending proposed by the Reagan adminstration will not by themselves redress this disparity, because Moscow's larger outlays for military equipment depend not on a larger Soviet defense budget, but on allocations within the defense budget.

The Soviets have an advantage because they allocate less than one-fourth of their military budget to labor costs. By comparison, more than half the US defense budget goes to salaries for the volunteer force, retirement pay, and related labor costs. As a result, even with equal defense budgets, Moscow can afford to spend twice as much on weapons procurement. During the 1970s the Soviets spent $350 billion (1982 dollars) more than the US on military equipment, R&D, and facilities, even though the overall defense budgets were about equal during the first half of the decade.

Thus, the Soviets do not have to trade off quantity for quality; they are able to pursue both. There is no simple way for the US to respond to this challenge. As a relatively affluent society with attendant high labor costs, the US is essentially caught in a vise: it cannot simply double the quantity of weapons it produces, even at the expense of quality, because larger quantities of equipment require increases in labor that only aggravate the basic cost problem.

Nor can the US any longer assume that the quality of its weapons systems will ensure the needed performance edge. Ten years ago, American fighter airplanes were so superior to their Soviet counterparts that the US could compete successfully despite a two-to-one numerical disadvantage. But present Soviet tactical aircraft are characterized by improved maneuverability, extended range, and upgraded avionics. The US cannot compete with this performance level simply by increasing the complexity of its airplanes. American aircraft are already designed near the "knee of the curve," and further performance improvements will require significant increases in cost and complexity.

Rather than attempt to compete with the Soviets by matching them weapon for weapon, the US must exploit its particular strengths to the fullest by using its technological lead to "finesse" Soviet numerical superiority. This "offset strategy" should be based on the selective application of technology where it leads to high-leverage increases in military effectiveness. Such a strategy is not rooted in a naive faith in the magic power of technology, nor does it overlook the importance of reducing key numerical disparities.

Of course, the US is not alone in confronting the Soviet challenge—it has allies whose aggregate industrial strength is equivalent to its own. But the alliance must be truly coherent for this potential advantage to be realized. In particular, the United States must make available to its allies the same technological advantages that are incorporated in its own weapons. This will require a much higher degree of cooperation than has been evident to date.

Systemic differences

To understand how best to apply the US technological advantages, we must first appreciate the basic differences between the US and Soviet processes of acquiring weapons. First, according to CIA estimates, the Soviet Union spends about twice as much on defense research and development as the US. Second, the Soviet educational system graduates many more engineers than the US. In 1980 alone, the Soviet Union conferred degrees on 300,000 engineers, while the US figure was roughly 60,000. Third, the Soviet Union is intensely developing approximately twice as many different weapons systems as the US. And finally, the Soviet system of design bureaus provides a continuity to weapons development that has almost no parallel in the US. Not only do chief designers and industrial ministers tend to have long tenure, but the number of workers and the level of activity at major Soviet R&D installations grow at a relatively constant rate. In contrast, only rarely does an American defense contractor work on one program for 20 or more years.

While these differences are real enough, taken by themselves they can be misleading. Although the Soviet Union does invest twice as much as the US in defense technology, this military effort is the country's major area of research and development. In the US, Europe, and Japan, on the other hand, there is a vast technological infrastructure not funded by defense but exploited by it. No real counterpart to this commercial high-technology sector exists in the Soviet Union.

The field of microelectronics is probably the best example of this difference. The US semiconductor industry derives less than 10 percent of its sales from the Defense Department. Thus, nearly all of that industry's R&D is supported by profits from commercial sales. Yet the West has the greatest advantage in defense systems involving microelectronic components and the computers derived from semiconductors. The microprocessor, which plays a key role in the new generation of precision-guided munitions, was essentially a commercial development. This area of significant American advantage is not apparent from a comparison of US and Soviet defense expenditures.

That the Soviets have been able to narrow the American lead in certain key technologies, including navigation, optics, and propulsion, is indicative of the intensity of Soviet efforts and a powerful reminder that the US cannot take its technological advantage for granted. Indeed, the successful application of the offset strategy depends on significant real growth in defense R&D as well as an improvement in the US ability to make timely applications of commercial technology to its weapons systems.

A "body count" of US vs Soviet engineers is also misleading because it is not indicative of the relative ability to create new technology. Soviet engineers are neither trained nor encouraged to be innovative, and many are placed in jobs that are filled in the US by technicians. American engineers, on the other hand, are substantially more creative, and this creativity is clearly reflected in the enormous US lead in technical innovations. It is no accident that the Soviets themselves worry about the inefficient use of their many engineers as well as their general lack of creativity. Some Soviet analysts have even urged that changes be made in the education of Soviet engineers.

The fact that the Soviet Union is building more weapons systems than the US is also a mixed blessing for Soviet leaders. Large-scale series production of related systems, such as tanks and missiles, reduces uncertainty and may lead to decreased unit costs. It has also led to a corresponding Soviet advantage in the age of weapons in the field. The Soviets often field one-and-a-half to two generations of equipment while the West fields one generation. But building such enormous quantities of military equipment can also result in waste and inefficiency. The Soviet military procurement system apparently does not include a way to cancel programs once they are started.

For example, the Soviets have developed at least 11 different intercontinental ballistic missiles (22 counting modifications). All were produced and deployed, except for two, the ss-10 and ss-16, which appear to have been technical failures. Such redundancy in design and production is bound to absorb defense resources—both roubles and engineers—without yielding corresponding benefits.

Finally, the fact that Soviet design bureaus and factories maintain not only a large effort but also continuity *is* reason for concern. For example, sustained high rates of Soviet tank production permit evolutionary improvements in design and technology. Moreover, during the last few years Soviet military production facilities have been constructed at the highest level of the past two decades. Yet the USSR's 135 major assembly plants and over 3500 individual factories do not compensate for Soviet barriers to successful innovation, which are rooted in the centralized system of planning and supply and administrative control of prices and bonus incentives. Indeed, the failure of certain high-technology industries (such as computers) to keep pace with technical developments in the West has continually restricted Soviet military options.

The development of the cruise missile clearly reflects the shortcomings of the Soviet approach. Because the Soviet air-launched cruise missile is an evolutionary modification of jet aircraft, it is very heavy, has a short range, and is relatively inaccurate. In contrast, the American system of competition among contractors paved the way for development of a strategic cruise missile that is a fivefold improvement over the Soviet version in range, weight, and accuracy. It will probably be at least five years until a comparable cruise missile becomes operational in the Soviet Union.

Thus, the nature of each weapons system is a determining factor in the ability of the Soviet Union to compete with the US in system performance. Notably, the Soviets excel in the development of military systems where improvements in effectiveness can be achieved through progressive modifications. And their true metier is an ability to create innovative designs for weapons constructed with off-the-shelf components, as evidenced by the ZSU-23/4 anti-aircraft gun and the BMP armored personnel carrier. But the Soviet record for technological innovation, such as those required by the cruise missile program, is less impressive.

The bottom line is that the US leads the Soviet Union by approximately five or more years in the technologies of greatest significance to defense. Although the Soviets have closed or are closing the gap in some areas, this is not the case for the most important technologies such as microelectronics. As a result, the US has the opportunity to use technology to offset the Soviet effort. Equally important, the US can exploit technology to produce defense equipment that is easier to operate and maintain and less expensive to procure.

Tanks versus smart weapons

East–West competition in armored forces shows how technology can be used to offset a numerical disadvantage. The Soviet Union currently fields more than 40,000 battle tanks, compared with just over 10,000 for the US. Moreover, the balance in tank inventories will deteriorate further when the Soviet Union produces its newest tank, the T-80, at a probable rate of 2000 units per year, or twice the number of tanks the US hopes to produce during the same period.

The offset strategy does not call for the US to procure 30,000 additional tanks or try to create a tank many times more effective than the Soviet tank. According to Lanchester in his famous treatise on military combat, there is a square-law relationship between quality and quantity. Thus, American tanks would have to be 16 times as effective as Soviet tanks to offset a four-to-one Soviet quantitative advantage. Considering that Soviet tanks are already quite impressive, this is not a realistic option. Even if the US proceeds with planned production of the M-1 tank, as we believe it should, America will still be outnumbered by four to one in tank forces. The alternative is to deploy the most effective and cost-efficient mix of armor and anti-armor weapons in numbers sufficient to meet the numerically superior threat.

Critical to this approach is the US deployment of anti-tank guided missiles that exploit the technology of precision-guided munitions (PGMs). Popularly referred to as "smart" bombs, PGMs are conventional munitions that can be precisely guided to the target after being launched. Such weapons literally "zero in" on either fixed or moving targets such as tanks.

The vast array of anti-armor weapons under advanced development today has come a long way from the unreliable devices used in World War II. Major advances in smart weapons now permit accuracies that approach a "one-shot, one-kill" capability. Moreover, contemporary precision-guided munitions may be deployed in a variety of ways. They can be fired from airplanes and helicopters or by individual soldiers—they are the portable modern bazooka. Precision guidance can also be incorporated in artillery projectiles, so that a 155 mm artillery round can make a direct hit on a moving tank at a range of 10 miles. By coupling the large quantity of US self-propelled artillery with the so-called cannon-launched guided projectile—the Copperhead—designers can convert an anti-personnel weapon into a very effective anti-tank weapon.

To ensure that this offset strategy is viable, the US must maintain a vigorous effort on all three generations of precision-guided munitions. First-generation PGMs developed in the mid-1960s are now in service. However, these accurate, relatively inexpensive weapons have two major drawbacks: their accuracy is affected by weather conditions, and the operator is vulnerable to counterfire. For example, after firing the TOW (tube-launched, optically tracked, wire-guided) missile, a gunner still has to keep the cross hairs of the sight on the target. Electrical signals transmitted through the wire automatically correct any deviation between the sight path and the track of the missile.

Second-generation laser-guided PGMs, such as the Copperhead and the Hellfire, were developed during the 1970s and are now entering NATO inventories. Although these anti-tank weapons are much more lethal and easier to operate, they are still weather-dependent, and their operators are vulnerable. Yet first-generation and second-generation PGMs will be the mainstay of NATO forces through the 1980s. As of 1980, Warsaw Pact countries had twice the number of anti-tank missiles in central Europe as NATO, so deployment of these weapons should continue until third-generation systems are well into production.

In contrast to the precision-guided munitions now deployed, third-generation PGMs will be "fire and forget," greatly reducing operator vulnerability and weather limitations and giving NATO forces a competitive edge in ground combat. The technologies now under development for this generation include millimeter-wave radar and infrared sensors configured in "focal-plane arrays." These are clusters of approximately 1000 individual sensors, each on a tiny chip. The resolution of these sensors is sharp enough to distinguish a vehicle the size of a tank from one much smaller or larger, thus permitting selective targeting. These technologies will allow third-generation PGMs to operate effectively in adverse weather conditions and with less susceptibility to enemy fire and other countermeasures.

Significant advances in the field of miniaturization in the past four years have not only made these weapons more efficient but have also opened up completely new possibilities. Sophisticated sensors and computers that a decade ago would have filled a large room are now small and rugged enough to fit into an artillery shell such as the Copperhead. Another example is the US Tank Breaker, a portable weapon being developed with focal-plane arrays and infrared guidance that will allow a soldier to engage a main battle tank.

Still more ambitious is the Assault Breaker, a long-range system designed to interdict rear-echelon enemy armour. Each warhead contains 30–40 submunitions guided independently to multiple targets. Thus, one such Assault Breaker could engage a whole company of tanks. These technological developments will provide true "fire and forget" capability and make an enormous difference in overcoming the vast numerical inferiority of US tank forces.

Stealth and electronic countermeasures

Another technology that can neutralize the Soviet threat lies in the US ability to penetrate Soviet air defenses. The Soviets' lead in air defense is even greater than their advantage in tanks. Whereas the US maintains no continental surface-to-air missiles, the Soviet Union has deployed more than 10,000 surface-to-air missiles for strategic air defense alone. (This figure does not include surface-to-air missiles deployed with Soviet combat forces.) Overall, the Soviet Union has invested more than $100 billion in strategic air defense.

The most practical US approach for dealing with this discrepancy is not to invest $100 billion in air defense but to develop ways of defeating Soviet air defenses. Two technologies, stealth (or "low observables") and electronic counter-measures together constitute a formidable response.

An airplane is considered "low-observable" if its characteristics detectable by radar can be reduced tenfold. For example, reducing the cross-section of a fighter bomber, as it appears to a radar operator, from ten square meters to one square meter decreases the range within which an opposing air defense system can operate by about a factor of two. Consequently, the air defense system has only half as much time to engage the low-flying, fast-moving airplane, making the job more difficult. And *very* low observable aircraft are virtually impossible to detect and intercept.

Although public discussion of stealth technology has focused on the strategic bomber, "low-observables" technology will be incorporated to some extent in all future military airplanes and missile systems. The first stealth vehicle, the air-launched cruise missile (ALCM), will become operational this year. The radar cross-section of the ALCM is one-thousandth that of the B-52 strategic bomber, and thus can defeat existing Soviet air defenses. The next generation of stealth programs will apply this technology to tactical airplanes, strategic bombers, and the ALCM follow-on.

A special synergism is created by combining stealth and electronic countermeasures (ECM) such as radar jamming, which reduces the ability of an air defense system to track its target. For example, if the cross-section of a military aircraft is reduced by a factor of 100, the power required to jam enemy radar may be reduced by the same factor. An ECM system requiring large, complex, high-power tubes, for instance, may be replaced with a system that is entirely solid-state. Stealth therefore gives an airplane the same level of ECM effectiveness while reducing the complexity, cost, and weight of the ECM components as well as increasing reliability.

That stealth makes the task of radar jamming less demanding does not mean that the US should allow its state-of-the-art ECM capability to lag. Advancing ECM technology is imperative to cope with the rapidly expanding threat from new Soviet fighters, long-range interceptors equipped with a "look-down, shoot-down" capability, airborne warning and control systems, and missiles using monopulse radars, which pose particularly difficult problems from an ECM system.

Avoiding dogfights

The third category of weapons development, tactical aircraft, is the most controversial and misunderstood example of technology's contribution to US defense preparedness. Proponents of the "simpler-but-more" school argue, as James Fallows did in his recent book, *National Defense*, that the US should abandon its costly pursuit of technological perfectionism and instead build large quantities of simpler and less expensive airplanes.

Advocates of the "superior-quality" approach, on the other hand, emphasize that, by capitalizing on its technological superiority, the US can continue developing and deploying sophisticated aircraft and avionics systems superior to those of the Soviet Union. But extreme proponents of this school generally underestimate the importance of numbers of weapons. And by succumbing to the "technological imperative," these analysts are often reluctant to advocate procurement of systems that fall short of perfection, thus slowing deployment even further. Framing the debate in "quantity-versus-quality" terms is therefore misleading.

The crux of the problem turns on the steady improvement in Soviet tactical aircraft, which have progressed from being a predominately defensive force to one with improved offensive capabilities. In fact, the latest generation of Soviet tactical aircraft is more complex than its American counterparts. Soviet fighters today, including the MIG-27 Flogger and the Su-24 Fencer, have improved maneuverability, extended range, and upgraded avionics and weapons. Moreover, the Fencer, like the MIG-25M Super Foxbat, carries a weapons officer in the two-seat cockpit. The most striking change in Soviet aircraft design is the fact that all but one of the latest generation of Soviet fighters have variable-geometry wings. This adds both complexity and cost to the airplanes and makes them more difficult to maintain.

In the latest American airplanes, including the F-15, F-16, and F-18, the incremental improvement in performance afforded by variable-geometry wings was abandoned in favor of less costly and more reliable aircraft, an important consideration generally overlooked by critics of American military designers. This example also illustrates the underlying assumption of the offset strategy: that technology should be applied only when it yields significantly superior war-fighting capability.

In line with this strategy, the US may derive important lessons from the simulated combat exercises held at Nellis Air Force Base in 1977. These exercises demonstrated that in close-range, "dogfight" situations, numbers of aircraft are crucial. Incremental increases in aircraft performance simply do not offset a four-to-one difference in numbers. Therefore, in addition to procuring more airplanes, the US should develop weapons and strategies to minimize the risk of having to engage in dogfights when outnumbered. With superior tactical intelligence, US pilots can avoid situations where they are outnumbered. With the means to attack multiple targets at long ranges, pilots can reduce the numerical disadvantage before they reach dogfight range.

Although US pilots will inevitably encounter some dogfights, improved intelligence and air-to-air missiles can act as powerful equalizers. The key to this approach is the transfer of technological sophistication from the airplane to its weapons. Thus, F-16s procured in large quantities may be preferable to fewer numbers of the more formidable F-15 if the F-16s are equipped with improved air-to-air missiles.

Two technologies are critical to the ability of the US to maintain a performance edge in air combat. First, the US can achieve an advantage in the avionics equipment that gives pilots "situation awareness," informing them of the type and location of all aircraft in their battle space. This information can be displayed electronically and updated every few seconds. Such situation awareness requires a combination of advanced systems, including an airborne warning and control system (AWACS), navigational satellites, and digital data communications systems that relay the data from collection centers to individual pilots.

A second advantage will be provided by the next generation of air-to-air missiles, particularly the Advanced Medium-range Air-to-Air Missile (AMRAAM), which marks a revolutionary advance from present technology. With advanced terminal guidance and signal-processing techniques, AMRAAM has a high probability of making a direct hit on low-flying targets despite the presence of sophisticated electronic countermeasures. Moreover, its longer range enables pilots to intercept simultaneously three or four enemy airplanes beyond visual range. Consequently, AMRAAM allows pilots to engage multiple enemy fighters before the fighters engage them. This missile is much more impressive than the Sidewinder missile, which functions like an advanced machine gun and is primarily useful in one-on-one encounters. Thus, even if US tactical fighter forces were outnumbered in a particular theater, superior situation awareness would enable American pilots to avoid being outnumbered in their own locality. And AMRAAM allows them to operate in stand-off engagements instead of dogfights.

Sea cruise

The fourth weapons category, naval developments, is particularly interesting because of the Reagan administration's proposal to move from a 450-ship navy to a 600-ship navy to offset the Soviet numerical edge in naval forces. We agree with the administration's objective of maintaining US maritime superiority. But a more appropriate way of dealing with the Soviet surface navy would be to emphasize improved surveillance and cruise missiles.

Ocean surveillance implies the continuous tracking of Soviet surface ships in broad ocean areas. Presently, the US is capable of monitoring Soviet vessels in moderate-sized areas using airborne radars such as AWACS and E-2C aircraft. This capability could be enhanced significantly by the use of both infrared and radar surveillance satellites.

Once a ship is located, the best means of attack is not with another ship but with a cruise missile, as demonstrated by the Argentinian use of the Exocet missile against the British frigate *Sheffield* in the Falkland Islands dispute. Indeed, the cruise missile has rendered direct ship-to-ship combat obsolete. The Tomahawk antiship cruise missile, currently in production, has five times the range of its predecessor and will make a vital contribution to the US Navy's sea-control mission.

American cruise missiles are vastly superior to their Soviet counterparts. Soviet cruise missiles resemble unmanned jet airplanes, each weighing more than 20,000 pounds. For example, the Soviet AS-3 air-launched cruise missile is approximately the same size and weight as the US F-5 tactical fighter. In sharp contrast, American cruise missiles are designed to make maximum use of miniaturized electronics and very efficient minijet engines and weigh about 3000 pounds. Thus, the US can put 20 cruise missiles on a B-52 bomber, whereas the Soviets can put only 2 of their cruise missiles on a Backfire bomber. This enormous military and cost advantage stems directly from superior technology.

The formidable fleet of Soviet submarines poses a serious threat to America's surface navy and sea lines of communications. The Soviet two-to-one quantitative advantage in submarines is partially offset by the fact that US submarines have substantially superior performance. Even more important is the superior stealth of US submarines, which are significantly quieter than their Soviet counterparts. This allows the US to detect Soviet submarines at several times the ranges at which the Soviet Union can detect American submarines. Finally, more advanced US detection technology, particularly long-range acoustic systems and refined signal-processing and data-integration techniques, greatly increases America's advantage.

More technology, less complexity

Most of these defense technologies are based on advances in the fields of microelectronics and computer technology, where the US enjoys approximately a five-year lead over the Soviet Union. Determined to maintain this edge, the Department of Defense has focused its research and development efforts on very high speed integrated circuits (VHSIC). Over the next four to five years, the Pentagon will invest approximately $300 million in VHSIC, and this does not include a comparable investment by private corporations. The purpose of this program is to advance the date by which advanced microelectronics will be incorporated into weapons systems. The circuitry will make weapons more reliable, since the number of connections among integrated circuits will be reduced. The program will also result in major savings in circuitry cost, weight, size, and power.

Although the Soviet Union has built devices with integrated circuits, Soviet shortcomings in software and manufacturing techniques are a major impediment to establishing a large-scale computer industry. Furthermore, the Soviets are trying to close the gap in microelectronics and computers by "reverse-engineering" Western technology. To maintain the US lead in this crucial area, the Defense Department has tightened controls on the transfer of critical technology and "keystone" equipment to Soviet-bloc countries. These controls have not been completely effective: circuits have been shipped out of the country illegally and are often available from sources outside the US. However, as long as the Soviets persist in their efforts to copy US technology, and the US continues to develop new technology, the Soviets will remain in a "tailchase" mode.

Although technological innovations such as VHSIC are often the key to finessing Soviet quantitative superiority, it is becoming fashionable to deride advanced American technology as being too expensive, difficult to operate and maintain, and unreliable. However, these critics display a lack of insight into the potential of modern technology: they identify as a problem what is America's best hope for a solution.

Many of the problems they describe in operating and maintaining military hardware can be traced to complex electromechanical devices used in equipment prior to the microelectronics revolution. Anyone who has operated a hand-held, solid-state calculator realizes that the difference between it and the desk calculator it replaced is not simply in performance—ease of operation and maintenance, reliability, and durability have all been significantly increased. Defense procurement programs are gradually replacing older, complex electromechanical systems with modern, solid-state microelectronics equipment.

This same technology is also being used to make various types of simulators that are revolutionizing the training of pilots and other equipment operators. Computer simulators recreate, at a reasonable cost, the fast-paced combat conditions that are often too dangerous and expensive to duplicate in reality. These simulated two-sided combat scenarios, similar to video games, will lead to vast improvements in operator proficiency and help reduce what Clauswitz has called the "friction of war."

The notion that such technology increases equipment cost has no basis in reality—complexity, not technology, is the culprit. In fact, almost every category of consumer purchases *except* electronics has seen a tenfold increase in price in the last few decades. In 1950, a black-and-white television set listed for about $500; a better set today may be purchased for less than $100. Even a more sophisticated color set runs about $500. Similarly, the cost of a calculator has been reduced from $1000 for the 1960s model to $10 for today's solid-state model. And computers have gone from a few million dollars to a few tens of thousands of dollars for equivalent capability. Thus, the selective application of electronics is the key to limiting defense costs.

The B-1 bomber illustrates the need to distinguish between military effectiveness and technological sophistication. The original B-1 was designed with a supersonic capability that added greatly to its cost and complexity but contributed only marginally to its improved survivability. In fact, dropping the supersonic requirement not only decreased cost and complexity but probably increased net survivability. It allowed the size of the engine air-intake system to be reduced, which in turn led to a reduction in its radar cross-section, making the aircraft more difficult to detect. Similarly, most commercial airlines have elected to forgo the supersonic option because the benefit of reduced flight time does not offset the liability of increased costs and reduced passenger loads. Thus, wide-body subsonic transports are a commercial success, while the supersonic Concorde has proven to be uneconomical.

The 1980s will be a dangerous period for the US because of the serious challenges in weapons development posed by the Soviet Union. The US is capable of meeting these challenges if it exploits American technology intelligently and energetically. Soviet leaders also recognize the crucial importance of technology. Brezhnev himself has said, "In the competition between the two world-opposed forces, science and technology will play a critical role, and this makes further advances in science and technology of decisive significance." But he was resorting to wishful thinking: such a strategy requires ongoing access to technological innovations, an area where the US excels.

There is no reason to be complacent about the US technological lead—the Soviets are driving hard to close it. But the US has a fundamental advantage in technological as opposed to numerical competition, and should exploit it fully.

Long Waves of Economic Development

Chris Freeman

The depression of the 1980s is part of a long-term cycle first identified by Kondratiev. Information technology will provide the engine of renewed economic growth, but new post-Keynesian policies are also needed to help get us out of the mess. Professor Freeman is with the Science Policy Research Unit at Sussex University, England and this article is taken from an address given to the annual meeting of the British Association for the Advancement of Science, August 1983, and published in the Symposium's proceedings, New Technology and the Future of Work and Skills *(Frances Pinter, London, 1984).*

Introduction

Forecasts of the future level of unemployment are almost uniformly pessimistic. Even those forecasts that assume a sustained recovery of production, investment, and international trade over the next few years are generally cautious about any major reduction in unemployment. Many assume a further deterioration, and the notion that an unemployment rate of 10 percent of the labor force is somehow inevitable has become quite widespread.

Yet it seems only yesterday that almost every government in the OECD area was committed to full employment as a primary objective of government policy. Unemployment levels of 2 percent or even 1 percent were commonplace in most European countries. In the 1950s and 1960s, it was often assumed that Keynesian policies had solved the prewar problem of persistent large-scale unemployment.

Keynes himself was not quite so optimistic about the prospects for unemployment in postwar Britain. He wrote to Beveridge that there was "no harm in trying" to reach the level of 3 percent employment, which had been adopted as a working definition of "full employment."[1] Beveridge himself only came to adopt the 3 percent definition at quite a late stage during the course of World War II.[2] In an earlier Report on social insurance, he had actually assumed an average postwar rate of 8 percent. Many other economists were far more pessimistic, and official government projections had at one time assumed a level of 10 percent unemployment for a decade ahead.

Before World War I, on the other hand, as in the 1950s and 1960s, the climate of opinion was far more optimistic. During the *Belle Époque*, the great reforming Liberal governments adopted the commonly held belief that improvements in the flow of information, through the introduction of labor exchanges and other social reforms, would remove the main imperfections in the labor market. Unemployment was indeed at that time much lower than during the interwar period.

Thus, during the course of the twentieth century, we have had two long swings in the conventional wisdom about unemployment: from a relatively optimistic view at the beginning of the century to deep pessimism during the 1930s, then once more to over-optimism in the 1950s and again to deep pessimism in the 1980s. It seems therefore that the beliefs of economists, and of the governments that they advised, were heavily influenced by the experience of the previous decade. Those involved in the business of long-term forecasting will recognize this as a familiar syndrome.

Some eminent economists have explicitly assumed that growth rates will remain depressed for a long time. They frequently also did this in the 1930s and 1940s. "It is my considered guess," said Paul Samuelson, "that the final quarter of the 20th Century will fall far short of the third quarter in its achieved rate of economic progress. The dark horoscope of my old teacher Joseph Schumpeter may have particular relevance here."[3]

Samuelson's reference to Schumpeter serves to remind us that, rather than simply extrapolate the experience of recent years, it may make better sense to try and understand the long-term fluctuations in the behavior of the economic system. Indeed, this may help to explain the long-term changes in the opinions and theories of the economists themselves.

Such half-century-long cycles are of course associated primarily with the name of the Russian economist, Nikolai Kondratiev. Although many other economists, as diverse as Jevons, Pareto, and van Gelderen, had already discussed the phenomenon before World War I, it was Kondratiev whose name was given to these cycles. During the 1920s, when he was the director of the Institute of Applied Economics Research in Moscow, he made an intensive effort to study and explain them, before he perished in Siberia in the 1930s. It was Joseph Schumpeter who popularized the notion of Kondratiev cycles in the English-speaking world, and who put the main emphasis on the spread of major new technologies as the most important characteristic of such long cycles.

Great controversy still surrounds these cycles or, as many people prefer to call them, "long waves" in economic development. Many economists doubt their very existence, and some historians adopt them only as a convenient form of historical periodization. The statistical debate about long cycles in the eighteenth and nineteenth centuries is likely to continue indefinitely. We have experience of only four long cycles during the industrial epoch, and the first of these was almost entirely a British phenomenon. As Kondratiev himself discovered, it is extremely difficult to reconstruct long-term time series of production, investment, employment, trade, prices, and interest rates over two centuries.

For the purpose of this paper, it is unnecessary to adopt a position of belief or disbelief in relation to this continuing controversy. It is sufficient to take the agnostic, pragmatic position of van der Zwan[4] that we should endeavor to learn something from the experience of the severe structural crises that many (but not all) countries have experienced in the 1930s and 1980s, and that some also experienced in the 1880s. In particular, we should endeavor to understand the long swings in the labor market, which almost all industrial countries have experienced during the course of this century. These long-term swings have also been associated with very big changes in the rate of immigration, in the female participation rate in the labor force, and in many other important social trends.

Such long-term fluctuations cannot be explained simply in terms of conventional business cycle theory, but require an additional dimension of analysis. This involves the rise of new technologies, the rise and decline of entire industries, major infrastructural investments, changes in the international location of industry and technological leadership and other related structural changes. It was primarily Schumpeter and Kondratiev who introduced these sorts of topic into the debate. Only if they are taken into account will it be possible to develop satisfactory counter-cyclical policies capable of coping with the deeper structural crisis that we are now experiencing.

The basic theme of this paper is that the depression of the 1980s cannot be dealt with as though it was just another slightly more severe recession, like those experienced in the 1950s and 1960s. Nor can it be left simply to a process of spontaneous recovery. The development of a viable strategy to cope with the scale of the unemployment problem of the 1980s requires a combination of the insights of Keynes with those of Schumpeter and Kondratiev.

Kondratiev and Schumpeter

In his earlier work, Kondratiev confined himself to a descriptive account of long waves and to assembling statistical data that might demonstrate their existence.[5] It was several years later that he attempted to offer his own explanations for them. He did, however, from the outset reject certain explanations based on the impact of exogenous events, such as "(1) changes in techniques, (2) wars and revolutions, (3) the assimilation of new countries into the world economy and (4) fluctuations in gold production."

In criticizing those economists who had suggested that these phenomena might explain the long waves, he said: "Their weakness lies in the fact that they reverse the connections and take the consequence to be the cause, or see an accident where we have really to deal with a law governing the events."

The "laws" that governed the long wave movement of the economy, he later suggested,[6] were the replacement cycles of long-lasting infrastructural investments and the availability of loan capital for these very large investments:

The production of the kind of capital goods in question necessitates a vast outlay of capital over a relatively long time-span. The occurrence of such periods of increased production of capital goods, i.e. periods of long ascending waves, is hence dependent on a series of preconditions. These preconditions are: (1) a high intensity of saving activity; (2) a relatively abundant and cheap supply of loan capital; (3) its accumulation in the hands of powerful enterprises and centres of finance; (4) a low level of commodity prices, which acts as a stimulant to savings activity and long term capital investments. The pressure of these preconditions creates a situation which will lead sooner or later to an increase in the production of the kind of basic capital goods mentioned above and hence to the emergence of a long ascendant economic wave.

If this theory is taken to be simply an "echo theory" based on the need to *replace* certain long-lived capital goods every half-century or so, then it cannot be justified empirically. The lifetime of such assets is extremely varied. While there is some evidence in the US for a "building cycle," or "Kuznets cycle," as it is normally called, this is shorter than the Kondratiev cycle, and is not necessarily related to replacement investment. But if Kondratiev's argument is modified to take account of the periodic need for *new types* of infrastructural investment—railways, communication systems, power generation and distribution networks, oil and petrol storage and distribution networks, highways, and so forth—then it becomes far more plausible.

Kondratiev expressed the opinion that an "especially large number of important discoveries and inventions in the technique of production and communication" are made during the recession of a long wave, but suggested that these inventions "are usually applied on a large scale only at the beginning of the next long upswing."[7] Although he was probably wrong about the timing of discoveries and inventions and failed to make the crucial distinctions between inventions and innovations, there is here the germ of a theory that could link the diffusion of new technologies to the movement of the long wave. It was, however, Schumpeter and not Kondratiev who developed this aspect of long wave theory.

The basic justification for attempting to relate the long waves to technical change is that the process of diffusion of any major new technology is a matter of decades and not of months or years. As Rosenberg has pointed out, the diffusion process is seldom one of simple replication of carbon copies of a new product.[8] It almost always involves a *cluster* of new inventions and innovations, affecting processes, components, subsystems, materials, and management systems, as well as the products themselves. The introduction of railways, of electric power, of the internal combustion engine, or of the computer are all examples of major economic and social transformations involving entirely new industries, new types of capital goods, components, materials, new skills at all levels, new management attitudes and systems, new education and training systems, new occupational and industrial classifications, new design and development systems, new legislation, and new forms of finance, company organization, and ownership. For Schumpeter the long waves in economic life were a succession of technological transformations of the economic system.[9] These necessitated deep structural change—a process that he called "creative destruction."

In Schumpeter's theory, the initiative of entrepreneurs, drawing upon the earlier discoveries and inventions of scientists and inventors, creates entirely new opportunities for investment, growth, and employment. The profits made from these innovations are then the decisive impulses for new surges of growth, acting as a signal to swarms of imitators. The fact that one or a few innovators can make exceptionally large profits, which they sometimes do, does not of course mean that all the imitators necessarily do so. When the bandwagon starts rolling some people fall off and profits are gradually "competed away" until recession sets in.

Whether or not the diffusion of new technologies and the associated rapid growth of new leading sectors of the economy offers a plausible explanation of long waves in economic development depends crucially on whether some of these innovations are so large in their impact as to cause major perturbations in the entire system. This might occur, as Kondratiev suggested, because some of them, such as railways or electric power, were so big and required such a long period for their construction that they could impart an upward thrust to the entire economy.

In principle, a big upswing might also occur because of the clustering of a series of basic innovations at certain periods. Following Kondratiev's suggestion, Gerhard Mensch has maintained that such clustering did indeed take place during the depression of the 1880s and 1930s,[10] but John Clark, Luc Soete, and myself have argued that the empirical evidence does not support this view.[11] It rather supports the notion that "new technological systems" often have their origin several decades before their widespread adoption in the economy, not in the immediately preceding depression. The upward thrust to economic growth comes not from the first innovations, but from a pattern of change associated with diffusion investments.

Carlota Perez has suggested that the big boom periods of expansion occur when there is a "good match" between the new technological "paradigm" or "style" of a long wave and the socio-institutional climate.[12] Depressions, in her view, represent periods of mis-match between the emerging new technological paradigms (already quite well advanced during the previous long wave) and the institutional framework. The widespread generalization of the new technological paradigms is possible only after social institutions have adapted to the requirements of the new technology.

The "productivity paradox," which has been the subject of widespread comment, provides strong support for this view. During the 1950s and 1960s there was generally a strong association between the growth of output of particular branches of industry and the labor productivity growth of each branch. This relationship was generally known as "Verdoorn's Law" and was often explained in terms of exploitation of economics of scale in the fastest-growing branches of the economy, such as petrochemicals, oil refining, vehicles, and so forth. Both the overall growth and the labor productivity growth depended on the good match that then existed between the low-cost energy-intensive technological paradigm and the favorable institutional framework within which it was exploited.

During the 1970s and 1980s labor productivity growth has slowed down markedly, especially in the US. Yet almost every engineer and scientist would agree that the *technological* potential for productivity increases has never been greater. Labor productivity has continued to increase very rapidly in the electronics industry itself, and the potential applications of microelectronic technology are innumerable, extending to every branch of manufacturing and the services. Productivity slowdown therefore can in no way be explained by the slowdown of technology, but must on the contrary be explained by some degree of incompatibility or "mis-match" between the new technological paradigm and the institutional or social framework. The recent labor productivity gains during a period of depression in the UK are due not so much to the "Verdoorn" effect as to the "Verdun" effect—the wholesale closure or scrapping of the less efficient older vintages of capital equipment that exist in every industry.

Keynes

It is often said that Keynes was deeply rooted in the neoclassical tradition of economics, and this is no doubt true. Nevertheless, even in his earliest writings, it is possible to trace his awareness of the limitations of the self-regulating market mechanism. Moggridge points out that already in 1913, in his book on *Indian Currency and Finance,* he insisted on "the essential fragility of the economic order which others took to be natural and automatic and emphasized the need for conscious management." He cites the following passage:

The time may not be far distant when Europe . . . will find it possible to regulate her standard of value on a more rational and stable basis. It is not likely that we shall leave permanently the most intimate adjustments of our economic organism at the mercy of a lucky prospector, a new chemical process, or a change of ideas in Asia.[13]

This already foreshadows his more general onslaught on *"laissez-faire"* in the 1920s.

The world is *not* so governed from above that private and social interest always coincide. It is *not* so managed here below that in practice they coincide. It is *not* a correct deduction from the Principles of Economics that enlightened self-interest generally *is* enlightened; more often individuals seeking separately to promote their ends are too ignorant or too weak to attain even these.[14]

In 1934 his broadcast was even more explicit:

On the one side are those who believe that the existing economic system is, in the long run, a self-adjusting mechanism, though with creaks and groans and jerks and interrupted by the time lags, outside interference and mistakes . . . on the other side of the gulf are those who reject the idea that the existing economic system is, in any significant sense, self-adjusting . . . The strength of the self-adjusting school depends on its having behind it almost the whole body of organized economic thinking and doctrine of the last hundred years. This is a formidable power . . . For it lies behind the education and the habitual modes of thought, not only of economists, but of bankers and businessmen and civil servants and politicians of all parties . . . thus if the heretics on the other side of the gulf are to demolish the forces of 19th century orthodoxy . . . they must attack them in their citadel. No successful attack has yet been made . . . I range myself with the heretics.[15]

This broadcast foreshadowed the publication of Keynes' *General Theory of Employment, Interest and Money* in 1936[16] which, at least temporarily, was indeed a fairly successful attack on the "citadel," and which argued that "the duty of ordering the current volume of investment cannot safely be left in private hands" and advocated the "socialization of investment." By this he meant of course, not public ownership or socialism, but public responsibility for the overall level of investment and employment. He insisted that, if private decisions to invest were inadequate to overcome a depression, then it was the responsibility of government to compensate for this deficiency.

Yet, in the *General Theory* he did not look at the question of investment directed toward the more rapid and effective introduction of *new* technologies, which is a far more effective stimulus to "animal spirits" than all the wonders of ancient Egypt or medieval Europe. Since he rejected the orthodox notion that a reduction in wages would stimulate a recovery from depression, it is surprising that he did not take up the role of new technologies in raising the marginal efficiency of capital. This is all the more astonishing in view of the fact that he had clearly made the link with Schumpeterian theory six years earlier in his *Treatise on Money*. He said:

In the case of fixed capital, it is easy to understand why fluctuations should occur in the rate of investment. Entrepreneurs are induced to embark on the production of fixed capital or deterred from doing so by their expectations of the profit to be made. Apart from the many minor reasons why these should fluctuate in a changing world, Professor Schumpeter's explanation of the major movements may be unreservedly accepted. He points to:

The innovations made from time to time by the relatively small numbers of exceptionally energetic businessmen—their practical application of scientific discoveries and mechanical inventions, their development of new forms of industrial and commercial organisation, their introduction of unfamiliar products, their conquests of new markets, exploitation of new resources, shifting of trade routes, and the like. Change of this sort, when made on a large scale, alter the data on which the mass of routine businessmen have based their plans. But when a few highly endowed individuals have achieved their success, their example makes the way easier for a crowd of imitators. So, once started a wave of innovation gains momentum.

It is only necessary to add to this that the pace at which the innovating entrepreneurs will be able to carry their projects into execution at a cost in interest which is not deterrent to them will depend on the degree of complaisance of those responsible for the banking system. Thus, while the stimulus to a credit inflation comes from outside the banking system, it remains a monetary phenomenon in the sense that it only occurs if the monetary machine is allowed to respond to the stimulus.[17]

This passage is remarkable for its unequivocal acceptance of Schumpeter's explanation of the major surges of investment in capitalist societies. It is all the more surprising that neither Keynes nor the Keynesians followed up this recognition of the crucial role of technical innovation. In fact, in the *General Theory* Keynes regressed to a position of neglect of technology when he introduced the largely artificial concept of a secular decline in the marginal efficiency of capital wholly unrelated to the actual changes in techniques or in the capital stock. Schumpeter was therefore justified in one of the main points of his critique of the *General Theory*:

It limits applicability of this analysis to a few years at most—perhaps the duration of the "40 months cycle"—and in terms of phenomena, to the factors that *would* govern the greater or the smaller utilisation of an industrial apparatus *if* the latter remains unchanged. All the phenomena incident to the creation and change in this apparatus, that is to say, the phenomena that dominate the capitalist process, are thus excluded from consideration.[18]

For the Keynesians it became a matter of relative indifference *which* were the new technologies and the fast-growing industries. They ignored the problem of long-term swings in the direction of technical change from a *job-generating* to a *job-displacing* emphasis or vice-versa. It was simply assumed that this could take care of itself so long as aggregate demand was sustained at a high enough level. From our standpoint it matters very much what is the direction of technical change, *which* are the important new technological systems, because the employment effects, the government policies, private and public R&D and investment strategies, and the institutional changes that are required to advance them may be very different.

There is no intention here to belittle the work of Keynes. He was by far the greatest British economist of the twentieth century and his achievement dwarfs that of his detractors today. But this cannot change the fact that Keynesian policies were and are deficient with respect to long-term changes in technology, structural change in the economy, and the specifics of infrastructural investment. It is true, as Keynes remarked, in the long run we are all dead, but this does not mean that long-term problems can be ignored.

This weakness of Keynesian theory and Keynesian policies was masked by the events of World War II. Rearmament was rightly acclaimed as a vivid demonstration that full employment could be achieved fairly quickly by massive public procurement and investment. What was overlooked was the dramatic way in which new technologies were promoted and diffused in the forcing house of war. The technical, social, and institutional changes made already *during* World War II meant that the problems of achieving a good "match," in the Perez sense, were already largely resolved in the 1940s and the conversion to a rapidly expanding peacetime economy was relatively smooth.

If war could have been avoided, it would have required a far more drastic application of policies for institutional change to achieve those social changes. This was evident from the relative failure of Roosevelt's New Deal to take the US out of depression or get anywhere near to full employment throughout the 1930s. Today, every sane person must wish to avoid tackling the problems of long-term structural and social change via the path of rearmament and war, but this means that the issue of new technological systems and relevant infrastructural investment must be faced directly and not swept under the carpet or allowed to go by default, in the hope of a spontaneous solution that may never come.

The SPRU program

In our own program of research on technical change and employment at the Science Policy Research Unit (SPRU), we have therefore attempted to take into account the long-term effects of technical change discussed above. The theoretical framework for the research has been described in our book, *Unemployment and Technical Innovation: A Study of Long Waves in Economic Development.*[19] Within this generally Schumpeterian framework, we are now doing empirical research to test the validity of this mode of analysis and to indicate the main lines of a post-Keynesian approach to full employment.

The main elements of such an approach would ideally be as follows.

1 We would need to identify the major new technological systems that might plausibly be supposed to be the engines of any new upswing in the British and the world economy over the next 10–20 years. We should make some assessment of the probable direction of technical change resulting from these technologies in each of the main sectors of the economy, the potential for entirely new industries and services, and the changing skill and capital requirements.

2 We would need to identify the changing long-term pattern of relationships between investment and employment in each of the main sectors of the economy. It would be important to avoid any assumption of a fixed, constant

capital–output ratio, and essential to take into account the heritage of past capital stock, as well as the employment likely to be generated by future vintages of investment, embodying more advanced technology.

3 We would need to assess various possible "scenarios" for the future evolution of investment and employment; to identify possible patterns of future investment, both public and private, that, in combination with changes in working hours and other social arrangements might lead the economy back to a new type of "full employment" in the 1980s. This would also involve identifying the main social and institutional barriers to a more rapid diffusion of new technological systems.

4 Finally, we would need to consider the development of an international expansionary framework, within which national attempts to tackle unemployment would stand some chance of success. This involves going beyond the essentially Keynesian framework that stood the world in good stead for the quarter-century after World War II, and tackling the problems of Third World indebtedness, of the severe and persistent disequilibrium in international payments associated with uneven technological development, and of international access to new technologies.

In common with most other researchers, our findings point to the overwhelming importance of microelectronics and the associated information technologies. This new technological paradigm affects every other branch of the economy, in terms of both its present and future employment and skill requirements, and its future market prospects. It has resulted in a drastic fall in *costs*, as well as vastly improved technical performance both within the electronic industry and outside it. This combination is relatively rare in the history of technology, but it means that microelectronics satisfies all the requirements for a Schumpeterian revolution in the economic system.

By contrast, the newer biotechnologies, although they certainly also have enormous future potential, have not yet reached the point where their macroeconomic impact could be great enough to carry the entire economy forward in the next decade or two. This illustrates the importance of the Mensch debate. It is the diffusion of the innovations of the 1950s, 1960s, and 1970s, rather than the first innovations of the 1980s, that must provide the main impetus for a new economic upswing. The new biotechnologies will provide very important auxiliary growth areas and will ultimately revolutionize agriculture, the food industry, and the chemical industry; but the main elements of the new technological paradigm for the fifth Kondratiev cannot come from this source.

Still less can nuclear technology play this role. Its applications are extremely limited. Its capital cost is astronomical, so that a large program would severely aggravate any capital shortage problems. Its cost advantages are even now dubious, and there are strong environmental, social, and political arguments for limiting its diffusion, especially in the case of the fast breeder reactor.

There are of course innumerable other types of technical change that affect particular processes and products, many of them important for individual industries. But our research, like that of similar work elsewhere, points unambiguously to the conclusion that the dominant new technological paradigm is associated with the combination of microelectronics, computers, telecommunications, and information technologies. This new paradigm may be loosely described as the "information revolution."

What are the implications of this revolution for investment and employment? We must emphasize at the outset that we have attached great importance in our research to the changing relationship between capital investment, output and employment. In most sectors of the economy we see the relationship between physical capital and employment as "clay" rather than "putty." In the course of the program we have attempted to develop a "vintage model" of the changing capital stock in each of the 40 main sectors distinguished in the Cambridge model of the British economy. This vintage model enables us to get some feeling for the potential future growth of employment in each sector on various alternative assumptions about the pattern of demand, future investment, technical change, and international technological competition.

From the evidence of past long waves of development, we assumed that there will be some tendency for the employment-generating effects of new technologies to be strongest in the early period of growth.[20] At this stage there is little standardization of components or subsystems, and a specialized industry supplying capital goods has not yet developed. However, as growth accelerates, economies of scale, specialization, and standardization lead to a more capital-intensive pattern, and in the downswing of a Kondratiev cycle the pressure to reduce costs and combat the erosion of profit margins generates an increasingly job-displacing pattern of technical change.

Therefore we did *not* postulate (as many growth models do) an unchanging capital–output ratio. On the contrary, we expected that, as an industry grew and matured, there would be a tendency for capital productivity to decline. This tendency we expected to be associated with a diminishing increment of employment associated with each new vintage of investment. If these tendencies were general, then they could induce stagnation in the economy generally, via two mechanisms:

1 diminishing marginal efficiency of capital leading to a slowdown of investment: this was the problem that Keynes described in the *General Theory*, but whereas he saw it as a secular trend, we see it as a cyclical problem, and not a permanent one;
2 capital shortage problems in the sense that the rate of investment needed to generate a return to full employment could be so vast as to strain the resources of the capital market, even if the prospects of profit were good.

Both of these problems could be exacerbated by weaknesses in international technological competition, such as those experienced by the UK in the past 30 years.

If they were not overcome or counteracted by social policies, such tendencies might generate conditions leading to a weak recovery and a prolonged period of mass unemployment as in the 1930s. In our empirical work, in relation to each of the major sectors of manufacturing and services, we have found confirmation of these tendencies. Marginal capital productivity *has* been falling in most sectors of manufacturing for a fairly long time, and international comparisons indicate that this is by no means simply a UK problem, but a feature of the world economy.

In the past experience of long waves, such tendencies in mature industries and sectors have been overcome on the one hand by structural change—i.e. a shift in the pattern of demand toward more labor-intensive sectors, with lower-than-average investment needs per unit of output—and on the other hand by major technical changes in various sectors of the economy, which brought about a rise in capital productivity, reversing the previous trend of the downswing of the long wave.

We certainly would not underestimate the importance of the remaining labor-intensive areas of employment in advanced economies. On the contrary, we would stress that there are certain types of services that are in principle rather labor-intensive, in the sense that direct human contact is the very essence of the service or activity. These include many types of caring and personal service, such as child care, psychiatric care, counselling services, many types of health care, and much education and training. They also include many types of creative work, including artistic, scientific, and craft activities.

The provision of these services has depended historically on a combination of growing public provision (usually on a non-profit basis) and voluntary part-time provision. The growth of such services is, we would maintain, one of the hallmarks of a civilized society; but for that growth to be sustained, and employment growth to continue in this area, it is essential to sustain high rates of productivity increase in the rest of the economy.

The impetus to a major upswing of the economy could come in principle, as it has come in the past, from a major new technological paradigm. Carlota Perez has pointed out that each of the major new technological paradigms has involved a drastic reduction in *costs*, so that it became managerial and engineering "common sense" to take advantage of the new technology in product and process design and in relation to all types of business decision-making.[21] This occurred with respect to the fall in transport costs and in steel costs during the nineteenth century, and the fall in energy costs (especially oil) in the twentieth century. It is happening now with information technology and microelectronics.

It is essential, however, not to underestimate the vast scope of institutional change that is needed. It will involve enormous changes in the pattern of skills of the workforce and therefore in the education and training systems; in management and labor attitudes, in the pattern of industrial relations and worker participation; in working arrangements; in the pattern of consumer demand; in the conceptual framework of economists, accountants, and governments, and in social, political and legislative priorities.

Perhaps the most important point is that "intangible" capital investment must now be recognized in its own right as *more* important than the transitory physical capital investment, which is today the main focus of attention for most managements, accountants, and economists. For a long time firms in the computer industry have devoted more resources to R&D, education and training, information services, design, and software development, than to physical capital investment.[22] This balance will now be tilted even more toward intangible investment as the information systems available to firms, government departments, and other institutions are becoming their most critical resource.

This applies both to formal information systems and data banks and to the costs of developing, educating, and training the people involved. The scale of the change in skills and in occupations is still underestimated. Even in a period of high unemployment, there are still persistent skill shortages for certain types of labor, particularly in relation to electronic engineering, software design, and systems analysis. In addition, there are many types of skilled people whose level of training and qualification is inadequate for the new types of work that they are being called upon to perform, or ought to be able to perform.

The information revolution also affects the whole climate and conduct of industrial relations. It makes possible the most sophisticated centralized type of "Big Brother" time and motion study, whether of sales workers in a supermarket or of miners operating an underground shearer-loader. But it also make possible a very high degree of decentralization, local responsibility, and initiative. Rosenbrock's work has shown the importance of designing production and management systems that leave a high degree of autonomy, responsibility, and skill with the workforce; otherwise the alienation effect of centralization and deskilling could generate social tensions even greater than those on the Tayloristic assembly line.

Computerization will also facilitate the introduction of much greater flexibility in working hours. In many occupations flexi-time has already been introduced, and the scope for this will be increased. Part-time work and work-sharing are likely to become more widespread. But here again, this social change can take a variety of different forms. It *could* be introduced as a means of depressing wage levels and of reducing social insurance benefits, if part-time workers are treated as a lower-grade type of labor. Or it could be introduced mainly as a response to the desire of many workers—male as well as female—at various stages in their working lives to work part-time, so that they can spend more time with young children, on pursuing education and training, or on other activities.

The institutional framework for the fifth Kondratiev cycle is only now being shaped. Most of our institutions and ideologies are still geared to the old technological paradigm of the fourth Kondratiev. Only through social and political debate and conflict shall we determine how we reshape our institutions and our way of life to match the potential of the new technology. The new pattern of employment that will emerge should be of a kind that encourages flexibility in hours of work and continuing education, and ensures to everyone who is seeking paid employment the opportunity to work in socially useful activity. This should mean a renewed commitment to the goal of "full employment," but in a new social context. It is our task in the 1980s to find the path to that goal, as the Keynesians did in very different circumstances in the 1930s and 1940s.

Notes

1 T. W. Hutchinson, *Keynes v the Keynesians*, Institute of Economic Affairs, London, 1977.
2 W. H. Beveridge, *Social Insurance and Allied Services*, Cmd 6404, HMSO, London, 1942.
3 P. A. Samuelson, "The World's Economy at Century's End," *Japan Economic Journal*, March 10, 1981, p. 20.
4 A. Van der Zwan, "On the Assessment of the Kondratiev Cycle and Related Issues," Centre for Research in Business Economics. Erasmus University, Rotterdam, 1979.
5 N. Kondratiev, "The Major Economic Cycles," *Voprosy Konjunktury*, vol. 1, pp. 28–79 (Moscow), 1925.
6 E. Mandel, *Late Capitalism (Der Spätkapitalismus)* Frankfurt, 1972 (English edition, New Left Books, London, 1975, p. 135).
7 Kondratiev, "Major Economic Cycles."
8 N. Rosenberg, *Perspectives on Technology*, Cambridge University Press, England, 1976.
9 J. A. Schumpeter, *Business Cycles*, 2 vols, McGraw Hill, New York, 1939.
10 G. Mensch, *Das technologische Patt: Innovationen überwinden Depression*, Umschau, Frankfurt, 1975.
11 C. Freeman, J. A. Clark, and L. Soete, *Unemployment and Technical Innovation: A Study of Long Waves in Economic Development*, Frances Pinter, London, 1982.
12 Carlota Perez, "Structural Change and the Assimilation of New Technologies in the Economic and Social Systems: A Contribution to the Current Debate on Kondratiev Cycles," paper given at Royal College of Art Seminar on April 14, 1983, London; to be published.
13 D. E. Moggridge, *Keynes*, Fontana, London, 1976, p. 43.
14 J. M. Keynes, "The End of Laissez-Faire," in *Essays in Persuasion*, Macmillan, London, 1931, p. 312.
15 J. Eatwell, *Whatever Happened to Britain?*, Duckworth/BBC, London, 1982, p. 43.
16 J. M. Keynes, *The General Theory of Employment, Interest and Money*, Harcourt Brace, New York, 1936.
17 J. M. Keynes, *Treatise on Money*, vol. 2, 1930, p. 85–6.

18 J. A. Schumpeter, *Ten Great Economists*, Allen and Unwin, London, 1952, p. 283.
19 Freeman, Clark and Soete, *Unemployment and Technical Innovation*.
20 Ibid.
21 Perez, "Structural Change."
22 C. Freeman, *Economics of Industrial Innovation*, Frances Pinter, London, 1982, p. 4.

Guide to Further Reading

Colin Norman, *The God That Limps: Science and Technology in the Eighties* (W. W. Norton, New York, 1981). Interesting discussion of the global issues arising out of the information technology revolution from the Worldwatch Institute.

Third World

Juan Rada, *The Impact of Microelectronics* (ILO, Geneva, 1980).

Juan Rada, "The Microelectronics Revolution: Implications for the Third World," *Development Dialogue*, 1981, no. 2 (Uppsala, Sweden).

Juan Rada, "A Third World Perspective," in Paul Sieghart (ed.), *Microchips With Everything* (Comedia, London, 1982).

Kurt Hoffman and Howard Rush, "The Microelectronics Industry and the Third World," *Futures*, August 1980.

Kurt Hoffman and Howard Rush, *Microelectronics and Clothing: The Impact of Technical Change on a Global Industry* (ILO, Geneva, 1983).

Raphael Kaplinsky (ed.), *Comparative Advantage in an Automating World*, Bulletin of the Institute of Development Studies, University of Sussex, England, March 1982. Includes pieces by Rada, Hoffman and Rush, etc.

T. Sagafi-Nejad, *Controlling International Technology Transfer: Issues, Perspectives and Policy Implications* (Pergamon, Elmsford, NY, 1981).

Frank Feather and Rashmi Mayur, "Communications for Global Development: Closing the Information Gap," in Howard F. Didsbury (ed.), *Communications and the Future* (World Future Society, Bethesda, MD, 1982).

Rita Cruise O'Brien (ed.), *Information, Economy and Power: The North–South Dimension* (Westview, Boulder, CO, 1983).

Marilyn Carr and George McRobie, "Mass Production or Production for the Masses?" *Mazingira: The International Journal for Environment and Development*, vol. 7 no. 2, 1983.

"Smart" weapons

James Fallows, "America's High-Tech Weaponry," *The Atlantic*, May 1981.
Wolfgang K. H. Panofsky, "Science, Technology and the Arms Race," *Physics Today*, June 1981.
Paul F. Walker, "Precision-Guided Weapons," *Scientific American*, August 1981.
"Battle of the Microchips," *Time*, May 17, 1982.
Frank Barnaby, "Microelectronics in War," in Günter Friedrichs and Adam Schaff, *Microelectronics and Society* (Pergamon Press, Oxford, England and Elmsford, NY, 1982). Very interesting overview.
Paul F. Walker, "Smart Weapons in Naval Warfare," *Scientific American*, May 1983. Account of the use of precision-guided weapons such as the Exocet missile in the Falklands conflict.

Space wars

"Killer Electronic Weaponry," *Business Week*, September 20, 1982.
James Canan, *War in Space* (Harper & Row, New York, 1982).
David Baker, *The Shape of Wars to Come* (Stein & Day, New York, 1982).
Bhupendra Jasani, "How Satellites Promote the Arms Race," *New Scientist*, November 11, 1982.
Marie Cunningham and Mariana Fitzpatrick, *Future Fire* (Warner, New York, 1983).
David Fishlock, "The Beam in the Pentagon's Eye," *Financial Times*, March 31, 1983; and "Star Wars Weapon Research," *Financial Times*, November 2, 1983.

Space

"The Future in Space," special issue of *New Scientist*, October 1, 1981.
Technology Review, October 1981; also a special issue on space and the implications of the shuttle.
Michiel Schwarz and Paul Stares (eds), special issue of *Futures*, October 1982, on all aspects of space exploration.
David Marsh, "Future Factories May Orbit in Space," *Financial Times*, March 14, 1983.
Christopher Joyce, "Countdown Begins for Space Factories," *New Scientist*, June 2, 1983.
"The US Returns to Space," *Business Week* Special Report, June 20, 1983.
Peter Smolders, "Soviets Unveil Space Station Plans," *New Scientist*, June 30, 1983.
David Fishlock, "Industrial Parks That Spin in Space," *Financial Times*, September 15, 1983.
Ellis Rubenstein (ed.), "Space 25," special issue of *IEEE Spectrum*, September 1983.
John Noble Wilford, "Big Business in Space: NASA's Next Mission," *New York Times Magazine*, September 18, 1983.
Peter Marsh, "Scientists Gain a Foothold in Space," *New Scientist*, September 22, 1983.
Peter Marsh, "Robot Repairers in Space," *Financial Times*, December 13, 1983; also December 23, 1983, January 3, 1984, and April 6, 1984.
"Perfect Spheres: The First Product From Space," *Business Week*, January 30, 1984.
Christopher Joyce and Ian Anderson, "Reagan Votes For Space Station," *New Scientist*, February 2, 1984.

Long waves

Special issue of *Futures*, August 1981, plus carry-over of articles into October 1981 issue.
Peter Hall, "The Geography of the Fifth Kondratieff Cycle," *New Society*, March 26, 1981.
H. D. Haustein and E. Neuwirth, "Long Waves in World Industrial Production," *Technological Forecasting and Social Change*, September 1982.
W. W. Rostow, "Technology and Unemployment in the Western World," *Challenge*, March–April 1983. Like Freeman, brings together Kondratiev waves and the new information technology.
Samuel Brittan, "The Myth of the Kondratieff," *Financial Times*, April 17, 1983.
Carlota Perez, "Structural Change and the Assimiliation of New Technologies in the Economic and Social System," *Futures*, October 1983.

The informal economy

Carl P. Simon and Ann D. Witte, *Beating the System: the Underground Economy* (Auburn House, Boston, MA, 1981).
"The Underground Economy's Hidden Force," *Business Week*, April 5, 1982.
Rolf G. Heinze and Thomas Olk, "Development of the Informal Economy," *Futures*, June 1982.
Jonathan Gershuny and Ian Miles, *The New Service Economy* (Frances Pinter, London, 1983).
Jonathan Gershuny, *Social Innovation and the Division of Labour* (Oxford University Press, Oxford, England and New York, 1983).
James S. Henry, *The Underground Economy* (W. W. Norton, New York, 1984).

14 Parameters of the Post-industrial Society

Computopia

Yoneji Masuda

The emerging information society will be completely different from industrial society, argues Masuda. Indeed, we can look forward to "Computopia" on Earth, if only we understand and direct the underlying social forces. Masuda was author of the Japanese Plan for an Information Society: A National Goal Toward the Year 2000, *published as early as 1971. This reading is taken from his most recent book,* The Information Society as Post-Industrial Society *(World Future Society, Bethesda, MD, 1981 and 1983), and provides us with a rare glimpse of Japanese thinking about the future.*

What is the image of the information society?

1 The information society will be a new type of human society, completely different from the present industrial society. Unlike the vague term "post-industrial society," the term "information society" as used here will describe in concrete terms the characteristics and the structure of this future society. The basis for this assertion is that *the production of information values and not material values will be the driving force* behind the formation and development of society. Past systems of innovational technology have always been concerned with material productive power, but the future information society must be built within a completely new framework, with a thorough analysis of the system of computer-communications technology that determines the fundamental nature of the information society.

2 The developmental pattern of industrial society is the societal model from which we can predict the overall composition of the information society. Here is another bold "historical hypothesis": *the past developmental pattern of human society can be used as a historical analogical model for future society.* Putting the components of the information society together piece by piece by using this historical analogy is an extremely effective way for building the fundamental framework of the information society.

The overall composition of the information society

Table 14.1 presents the overall framework of the information society based upon these two premises. This table presents the overall composition of the information society based on a historical analogy from industrial society. Let me explain each of the major items.

1 The prime innovative technology at the core of development in industrial society was the steam engine, and its major function was to substitute for and amplify the physical labor of man. In the information society, "computer technology" will be the innovational technology that will constitute the developmental core, and its fundamental function will be to *substitute for and amplify the mental labor of man.*

  2 In industrial society, the motive power revolution resulting from the invention of the steam engine rapidly increased material productive power, and made possible the mass production of goods and services and the rapid transportation of goods. In the information society, "an information revolution" resulting from development of the computer will rapidly expand information productive power, and make possible *the mass production of cognitive, systematized information, technology, and knowledge.*

3 In industrial society, the modern factory, consisting of machines and equipment, became the societal symbol and was the production center for goods. In the information society the *information utility* (a computer-based public infrastructure) consisting of information networks and data banks will replace the factory as the *societal symbol*, and become the production and distribution center for information goods.

4 Markets in industrial society expanded as a result of the discovery of new continents and the acquisition of colonies. The increase in consumption purchasing power was the main factor in expansion of the market. In the information society, "the knowledge frontier" *will become the potential market*, and the increase in the possibilities of problem-solving and the development of opportunities in a society that is constantly and dynamically developing will be the primary factor behind the expansion of the information market.

  5 In industrial society, the leading industries in economic development are machinery and chemicals, and the total structure comprises primary, secondary, and tertiary industries. In the information society the leading industries will be the *intellectual industries,* the core of which will be the knowledge industries. *Information-related industries* will be newly added as *the quarternary group* to the industrial structure of primary, secondary, and tertiary. This structure will consist of a matrix of information-related industries on the vertical axis, and health, housing, and similar industries on the horizontal axis.

6 The economic structure of industrial society is characterized by (1) a sales-oriented commodity economy; (2) specialization of production-utilizing divisions of labor; (3) complete division of production and consumption between enterprise and household. In the information society (1) informa-

Table 14.1 Pattern comparison of industrial society and the information society

	Industrial society	Information society
Innovational technology		
Core	Steam engine (power)	Computer (memory, computation, control)
Basic function	Replacement, amplification of physical labor	Replacement, amplification of mental labor
Productive power	Material productive power (increase in per capita production)	Information productive power (increase in optimal action–selection capabilities)
Socioeconomic structure		
Products	Useful goods and services	Information, technology, knowledge
Production center	Modern factory (machinery, equipment)	Information utility (information networks, data banks)
Market	New world, colonies, consumer purchasing power	Increase in knowledge frontiers, information space
Leading industries	Manufacturing industries (machinery industry, chemical industry)	Intellectual industries, (information industry, knowledge industry)
Industrial structure	Primary, secondary, tertiary industries	Matrix industrial structure (primary, secondary, tertiary, quaternary/systems industries)
Economic structure	Commodity economy (division of labor, separation of production and consumption)	Synergetic economy (joint production and shared utilization)
Socio-economic principle	Law of price (equilibrium of supply and demand)	Law of goals (principle of synergetic feedforward)
Socio-economic subject	Enterprise (private enterprise, public enterprise, third sector)	Voluntary communities (local and informational communities)
Socio-economic system	Private ownership of capital, free competition, profit maximization	Infrastructure, principle of synergy, precedence of social benefit
Form of society	Class society (centralized power, classes, control)	Functional society (multicenter, function, autonomy)
National goal	GNW (gross national welfare)	GNS (gross national satisfaction)
Form of government	Parliamentary democracy	Participatory democracy
Force of social change	Labor movements, strikes	Citizens' movements, litigation
Social problems	Unemployment, war, fascism	Future shock, terror, invasion of privacy
Most advanced stage	High mass consumption	High mass knowledge creation
Values		
Value standards	Material values (satisfaction of physiological needs)	Time-value (satisfaction of goal achievement needs)
Ethical standards	Fundamental human rights, humanity	Self-discipline, social contribution
Spirit of the times	Renaissance (human liberation)	Globalism (symbiosis of man and nature)

tion, the axis of socioeconomic development, will be produced by the information utility; (2) self-production of information by users will increase; information will accumulate; (3) this accumulated information will expand through synergetic production and shared utilization; and (4) the economy will change structurally from an exchange economy to a *synergetic economy*.

7 In industrial society the law of price, the universal socioeconomic principle, is the invisible hand that maintains the equilibrium of supply and demand, and the economy and society as a whole develop within this economic order. In the information society the *goal principle* (a goal and means principle) will be the fundamental principle of society, and the synergetic feedforward, which apportions functions in order to achieve a common goal, will work to maintain the order of society.

8 In industrial society, the most important subject of social activity is the enterprise, the economic group. There are three areas: private enterprise, public enterprise, and a third sector of government ownership and private management. In the information society the most important subject of social activity will be the *voluntary community*, a socioeconomic group that can be broadly divided into local communities and informational communities.

9 In industrial society the socioeconomic system is a system of private enterprise characterized by private ownership of capital, free competition, and the maximization of profits. In the information society, the socioeconomic system will be a voluntary civil society characterized by the superiority of its infrastructure, as a type of both public capital and knowledge-oriented human capital, and by a fundamental framework that embodies the *principle of synergy and social benefit*.

10 Industrial society is a society of centralized power and hierarchical classes. The information society, however, will be a multi-centered and complementary voluntary society. It will be horizontally functional, maintaining social order by *autonomous and complementary functions of a voluntary civil society*.

11 The goal of industrial society is to establish a Gross National Welfare Society, aiming to become a cradle-to-grave high welfare society. The information society will aim for the *realization of time value* (value that designs and actualizes future time) for each human being. The goal of society will be for everyone to enjoy a worthwhile life in the pursuit of greater future possibilities.

12 The political system of industrial society is a parliamentary system and majority rule. In the information society the political system will become a *participatory democracy*. It will be the politics of participation by citizens; the politics of autonomous management by citizens, based on agreement, participation, and synergy that take in the opinions of minorities.

13 In industrial society, labor unions exist as a force for social change, and labor movements expand by the use of labor disputes as their weapon. In the information society, *citizen movements* will be the force behind social change; their weapons will be litigation and participatory movements.

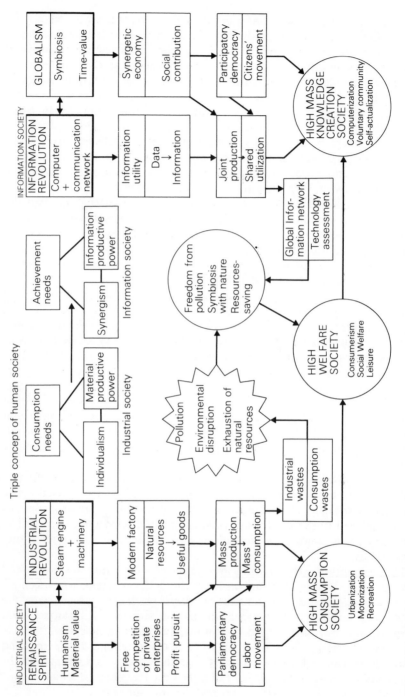

Figure 14.1 The transformation process from industrial society to information society.

14 In industrial society there are three main types of social problems: recession-induced unemployment, wars resulting from international conflict, and the dictatorship of fascism. The problems of the information society will be future shocks caused by the inability of people to respond smoothly to rapid societal transformation, acts of individual and group terrorists such as hijackings, *invasions of individual privacy*, and the crisis of a *controlled society*.

15 The most advanced stage of industrial society is a high mass-consumption stage, centering on durable goods, as evidenced by motorization (the diffusion of the automobile). The most advanced stage of the information society will be *the high mass knowledge creation society* in which computerization will make it possible for each person to create knowledge and to go on to self-fulfillment.

16 In industrial society, the materialistic values of satisfying physiological and physical needs are the universal standards of social values; but in the information society, seeking the *satisfaction of achieved goals* will become the universal standard of values.

17 Finally, the spirit of industrial society has been the renaissance spirit of human liberation, which ethically means respect for fundamental rights and emphasis on the dignity of the individual, and a spirit of brotherly love to rectify inequalities.

The spirit of the information society will be the *spirit of globalism*, a symbiosis in which man and nature can live together in harmony, consisting ethically of *strict self-discipline and social contribution*.

A vision of Computopia

Looking back over the history of human society, we see that, as the traditional society of the Middle Ages was drawing to a close, the curtain was rising on the new industrial society. Thomas More, Robert Owen, Saint Simon, Adam Smith, and other prophets arose with a variety of visions portraying the emerging society. The one that is of special interest to me is Adam Smith's vision of a *universal opulent society*, which he sets out in *The Wealth of Nations*. Smith's universal affluent society conceives the condition of plenty for the people, economic conditions that should free the people from dependence and subordination, and enable them to exercise true independence of spirit in autonomous actions.

Smith presented *The Wealth of Nations* to the world in 1776. Strangely, James Watt's first steam engine was completed in the same year, but although the Industrial Revolution was under way, Smith's grand vision of a universal society of plenty was still far off when he died in 1790. His vision seems to be half-realized two centuries later, as society reaches Rostow's High Mass Consumption stage. The High Mass Consumption stage means that the material side of Smith's vision, of people having wealth in plenty, is partially accomplished, at least in the advanced countries. The wider vision he had, of individual independence and autonomy that would follow, has clearly not been realized, because the axis around which the mass production and consumption of industrial goods turns in industrial society comprises machines and power. Capital investments are necessarily immense, with the result that the concentration of capital and corresponding centralized power are the dominating factors. This is the fundamental structure of all industrial societies, something that transcends the question of a society being capitalistic or socialistic.

Industrial societies are characterized by centralized government supported by a massive military and administrative bureaucracy; and in capitalist states supranational enterprises have been added that make the modern state dependent on the trinity of industry, the military, and the government bureaucracy. In industrial societies the individual has freedom to take social action in three ways. A person is able to participate indirectly in government policy by voting in elections once every few years. He or she has the freedom of using income (received as compensation for subsistence labor) to purchase food and other articles necessary to sustain life, which implies freedom to use time on weekends and holidays as one likes. This freedom of selection, however, is freedom only in a limited sense, quite removed from the voluntary action selection that Adam Smith envisioned.

As the twenty-first century approaches, however, the possibilities of a universally opulent society being realized have appeared in the sense that Smith envisioned it, and the information society that will emerge from the computer communications revolution will be a society that actually moves toward a universal society of plenty.

The most important point I would make is that the information society will *function around the axis of information values rather than material values*, cognitive and action-selective information. In addition, the information utility, the core organization for the production of information, will have the fundamental character of an infrastructure, and knowledge capital will predominate over material capital in the structure of the economy.

Thus, if industrial society is a society in which people have affluent material consumption, the information society will be a society in which *the cognitive creativity of individuals flourishes throughout*. And if the highest stage of industrial society is the high mass consumption society, then the highest stage of the information society will be the *global futurization society*, a vision that greatly expands and develops Smith's vision of a universal opulent society; this is what I mean by "Computopia." This global futurization society will be a society in which everyone pursues the possibilities of his or her own future, actualizing his or her own self-futurization needs by acting in a goal-oriented way. It will be global, in which *multi-centered voluntary communities of citizens participating voluntarily in shared goals and ideas flourish simultaneously throughout the world.*

Computopia is a wholly new long-term vision for the twenty-first century, bearing within it the following concepts.

Pursuit and realization of time value
My first vision of Computopia is that it will be a society in which each individual pursues and realizes time value. In Japan, the advanced welfare society is often talked about, and people are now calling for a shift of emphasis from rapid economic growth to stable growth, stressing social welfare and human worth, sometimes expressed as a shift from a GNP society to a GNW society, i.e. gross national welfare. The current idea of an advanced welfare society, however, tends to place the emphasis on the importance of living in a green environment where the sun shines. Obviously, in seeking to escape from the pollution and congestion of cities, and from the threat of a controlled society, this concept is significant, as indicative of our times. Yet it does not embrace a dynamic vision of the future, which I feel is its greatest weakness. The disappearance of pollution and congestion or even escape from the cities will not alone bring satisfaction. The need for self-realization is of a very high dimension that must be actively satisfied. The future society, as I see it, will be a society in which each individual is able to pursue and satisfy the need for self-fulfillment.

The self-realization I refer to is nothing less than the need to realize time value, and time value, of course, involves painting one's own design on the invisible canvas of one's future, and then setting out to create it. Such self-fulfillment will not be limited merely to individuals all pursuing their own self-realization aims, but will expand to include mini-groups, local societies, and functional communities.

Freedom of decision and equality of opportunity
The concepts of freedom and equality grew out of the Puritan Revolution (1649–60), which occurred in England around the end of the Middle Ages. Initially, the ideas of freedom from absolute authority and legal equality underlay these concepts, backed by the theories of social contract and individual consent as the basis of political authority, theories that maintain that freedom and equality are natural rights for all people. These two ideas provided the theoretical base for the formation of modern civil society.

As the capit..list economic system came into being, freedom and equality developed conceptually to include "freedom to work at something of one's own choice," "equality of ownership," "freedom to select an occupation," and "industrial equality," more commonly referred to as free competition.

The information society will offer new concepts of freedom and equality, embodying freedom of decision and equality of opportunity.

As I have said, the information society will be a society in which each individual pursues and realizes time value. In this type of society the freedom that an individual will want most will be *freedom to determine voluntarily the direction of time value realization in the use of available future time.* Call it "freedom of decision." Freedom of decision is the freedom of decision-making for selection of goal-oriented action, and refers to the right of each individual to determine voluntarily how to use future time in achieving a goal. This will be the most fundamental human right in the future information society.

"Equality of opportunity" is *the right all individuals must have, meaning that the conditions and opportunities for achieving the goals they have set for themselves must be available to them.* This will guarantee that all individuals have complete equality in all opportunities for education, and the opportunity to utilize such opportunities for action selection. Guaranteed equality of opportunity will, for the first time, assure that the people will share equally the maximum opportunities for realizing time value.

Flourishing diverse voluntary communities
A society composed of highly educated people with a strong sense of community has long been a dream of mankind, and several attempts have been made to bring it into being. Recently, communes have been formed by young groups, and a number of cooperative communities have been formed in Japan. One was the *Yamagishi-kai*, formed after the war. The rapid growth of information-productive power built around the computer will see some big advances and developments beyond the ideas and attempts of the past. There will be enhanced independence of the individual, made possible for the first time by the high level of the information-productive power of the information society. The development of information-productive power will liberate people by reducing dependence on subsistence labor, with rapidly increasing material productive power as the result of automation, thus increasing the amount of free time one can use. There will also be an expanded ability to solve problems and pursue new possibilities, and then to bring such possibilities into reality; that is to say, it will expand one's ability for futurization.

The development of this information-productive power will offer the individual more independence than can be enjoyed now.

Another point to be noted is the autonomous expansion of creativity that will follow. The keynote of Utopian societies in the past has been the establishment of communal life through the common ownership of the means of production, based more or less on the prototype of primitive communism. This type of society has inevitably operated with a relatively low level of productive power; but the future information society will ensure more active voluntary communities, because humans will be liberated from dependence on subsistence labor, and because of the expanded possibilities for future time-value realization.

As a consequence, Utopian societies will move on from being merely cooperative societies, where most time must still be given to sustaining existence, to become dynamic and creative voluntary communities. It is people with common goals who will form the new voluntary communities, communities that will always be carried on by voluntary activity and the creative participation of individuals; individual futurization and group futurization will be harmoniously co-ordinated with societal futurization. In the mature information society of the future, nature communities, non-smoking communities, energy conservation communities, and many other new types of voluntary communities will prosper side by side.

Interdependent synergistic societies

A synergistic society is one that develops as individuals and groups cooperate in complementary efforts to achieve the common goals set by the society as a whole. The functioning societal principle is *synergism*, a new principle to replace the free competition of the current capitalistic society.

In the future information society, information utilities, whose structure of production is characterized by self-multiplication and synergy, will take the place of the present large factories, and become the societal symbol of the information society. These information utilities will be the centers of productive power, yielding time value that will be the common goal of voluntary communities, because of the *self-multiplication* that characterized production in the information utility. Unlike material goods, information does not disappear by being consumed; and even more important, the value of information can be amplified indefinitely by constant additions of new information to the existing information. People will thus continue to utilize information which they and others have created even after it has been used; and, at macro-level, the most effective way to increase the production and utilization of information will be for people to *work together to make and share societal information*. This economic rationality means that the information utility itself will become part of the infrastructure. It will be the force behind the productive power that gives birth to socioeconomic values, and corresponding new socioeconomic laws and systems will come into being as a matter of course. *Synergistic feedforward* will function as the new societal principle to establish and develop social order, with the resulting societies becoming voluntary communities.

Functional societies free of overruling power

The history of the rule of man over man is long, continuing right into the present, simply changing form from absolute domination by an aristocracy linked with religion in feudal society to economic domination of enterprises in capitalist society, and to political domination by the bureaucracy in both socialist and capitalist society. The future information society, however, will become a *classless society*, free of overruling power, the core of society being voluntary communities. This will begin as informational and local communities which a limited number of people steadily develop and expand.

A voluntary community is a society in which the independence of the individual harmonizes with the order of the group, and the social structure is a multi-centered structure characterized by mutual cohesion. By "multi-centered" I mean that *every individual and group in a voluntary community is independent, and becomes a center.* "Mutual cohesion" means that *both individuals and groups that constitute the centers share a mutual attraction to form a social group.* Behind this mutual attraction lies the common goal, the spirit of synergy, with the ethics of self-imposed restraints. In other words, as individuals pursue their own time value, they work synergetically as a group to achieve a shared goal, and all exercise self-restraint so that there will be no interference with the social activities of others. This social structure is the overall control system of a voluntary community.

In the political system, democracy based on participation of the citizens will be the general mode of policy-making, rather than the indirect democracy of the parliamentary system. The technological base to support this participatory democracy will consist of (1) information networks made possible by the development of computer-communications technology; (2) simulation of policy models; and (3) feedback loops of individual opinions; with the result that policy-making will change from policy-making based on majority versus minority rule to policy-making based on the balance of gain and loss to individuals in the spectrum of their areas of concern, both in the present and in future time. In policy-making by this means, the feedback and accumulation of opinions will be repeated many times until agreement is reached, to insure the impartial balance of merits and demerits of the policy decision as it affects individuals and groups with conflicting interests.

The present bureaucratic administrative organization will be converted into a *voluntary management system of the citizens.* Only a small staff of specialists will be needed to carry out administrative duties, officers who are really professionals responsible for the administrative functions. The bureaucratic organization of a privileged class will disappear. In this voluntary civil society, ruling, coercion, and control over others will cease. Society will be *synergistically functional,* the ideal form that the information society should take.

Computopia: can it become a reality?

Can these visions of Computopia be turned into reality? We cannot escape the need to choose, before it is established, either "Computopia" or an "Automated State." These inescapable alternatives present two sharply contrasting bright and dark pictures of the future information society. If we choose the former, the door to a society filled with boundless possibilities will open; but if the latter, our future society will become a horrible and forbidding age.

As far as present indications go, we can say that there is a *considerable danger that we may move toward a controlled society.*

This is seen in the following tendencies.

During the first 15–20 years of their availability, computers were used mainly by the military and other governmental organizations and large private institutions. Medium and small enterprises and individuals were generally barred from using computer-communications technology, since large-scale computers at the early stage of automation were extremely costly. This situation caused a significant delay in democratic applications of computers. Initially, they were used mainly for automatic control and labor-saving purposes, rather than "problem solving" applications. The development of automatic control of separate systems to integrated real-time control systems covering broad areas is increasing the danger of a controlled society.

The utilization of computers for major scientific and technological applications, such as space development, has led us to neglect the need for coexistence with nature, while our impact on nature has grown immeasurably. The development of "big" science and technology has operated in such a way as to further increase the imbalance between human and nature systems.

If computerization continues in this direction, the possibility of a controlled society increases alarmingly.

However, I believe and predict that the catastrophic course to an "Automated State" will be avoided, and that our choice will be to follow the path to "Computopia." I give two logical reasons for my confidence.

The first theoretical basis is that *the computer as innovational technology is an ultimate science*. By "ultimate science" I mean a science that will bring immeasurable benefits to humanity if wisely used, but which would lead to destruction if used wrongly. Nuclear energy, for example, can be an extremely useful source of energy, but it could kill the greater part of the human race in an instant. The computer may, in one sense, be more important, as an ultimate science, than atomic energy.

If computers were to be used exclusively for automation, a controlled society, the alienation of mankind, and social decadence would become a reality. But if used fully for the creation of knowledge, a high mass-knowledge-creation society will emerge in which all people will feel their lives to be worth living. Further, an on-line, real-time system of computers connected to terminals with communication lines would turn society into a thoroughly managed society if utilized in a centralized way, but if their utilization is decentralized and open to all persons, it will lead to creation of a high mass-knowledge-creation society. Similarly, if data banks were to be utilized by a small group of people in power to serve their political purposes, a country would become a police state; but if used for health control and career development, every person can be saved from the sufferings of disease, and be able to develop full potentialities, opening up new future opportunities and possibilities.

The computer thus confronts us with these alternatives: an "Automated State" or a "Computopia." So it is not the *forecasting* of the state of a future information society, but our *own choice* that is decisive. There is only one choice for us—the road to Computopia. We cannot allow the computer, an ultimate science, to be used for the destruction of the spiritual life of mankind.

The second theoretical basis of my confidence is that *the information society will come about through a systematic, orderly transformation.* Information-productive power will develop rapidly to replace material productive power, a development that will bring about a qualitative conceptual change in production, from production of material goods to the production of far-reaching systems that include everything from production systems for material goods (such as automated factories), to social systems (wired cities, self-education systems), to political systems (direct citizen participation systems), and even to ecological systems.

Obviously, information-productive power centering on the computer communications network will be the powerful thrust to bring about societal systems innovations. New social and economic systems will be created continuously, and society as a whole will undergo dynamic changes: not the drastic social changes of the past, typified by the power struggles of ruling classes, wars between nation-states, and the political revolutions of mass revolt, but a *systematic, orderly transformation.* As old socioeconomic systems gradually become ineffectual and unable to meet the needs of the times, they will atrophy, and new, responsive socioeconomic systems will take their place, in the way that a metamorphosis takes place with an organism, the useless parts of the body atrophying and other parts developing in response to the new demands.

Moreover, this systematic transformation of the societal structure will be brought about by citizen action, *changing means- and goal-oriented modes of action into cause-and-effect modes of action.* I have pointed out that human modes of action will become goal-oriented in the information society: these modes of goal-oriented action will evolve to the point where they function as a goal principle, to become the principle of social action. When this happens, social action will be logical, means-oriented action for the pursuit of common goals. So we can replace the term "goal–means-oriented action" with the term "cause–effect relationship," following the idea of Max Weber, who changed this concept of goal–means relationship into a concept of cause–effect relationship. In the information society, the social actions of citizens in general will become goal–means relationships that operate as cause–effect relationships.

The rebirth of theological synergism of man and the supreme being

The final goal of Computopia is the *rebirth of theological synergism* of man and the supreme being, or if one prefers it, the *ultimate life force*—expressions that have meaning to both those of religious faith and the irreligious. This can be called the ultimate goal of Computopia. The relation existing between man and nature was the beginning of civilization. For many thousands of years man was completely encompassed by the systems of nature, which he had to obey or be destroyed by them. Five or six thousand years ago, man succeeded in harnessing these systems of nature in a limited way to increase agricultural production, and the first civilizations were built. This marked the beginning of man's conquest of nature. But with the Industrial Revolution the conquest of nature meant the destruction of nature, and now nature's retaliation has begun, the sequel to man's relation with nature that turned into destruction.

Now, a new relationship is beginning. At last, man and nature have begun to act together in a new ecological sense, on a global scale, in synergistic society. At the base of this conversion of human society into an ecological system is the awareness of the limitations of scientific technology. It means awareness that scientific technology is simply the application of scientific principles, and that these cannot be changed by man; nor can he create new principles to work and live by. It is also a new awareness of the commonality of man's destiny, in that there is no place where man can live except on this earth, which first gave him life; from this very awareness is emerging the idea of a synergistic society where man and nature must exist in true symbiosis.

This is the assertive, dynamic idea that man can live and work together with nature, not in a spirit of resignation that says man can only live within the framework of natural systems; but man and nature will work together as one. Put another way, man approaches the universal supra life, with man and god acting as one.

"God" does not refer to a god in the remote heavens; it refers to nature with which we live our daily lives. The scientific laws that we have already identified and are aware of are simply manifestations of the activity of this supreme power. The ultimate ideal of the global futurization society will be for man's actions to be in harmony with nature in building a synergistic world.

This synergism is a modern rebirth of the theological synergism which teaches that "spiritual rebirth depends upon the cooperation of the will of man and the grace of God," however it may be expressed. It aims to build an earthly, not a heavenly, synergistic society of god and man.

When we open the book of history, we see that, when man brought about the accumulation of wealth and an increase in productive power, various choices had to be made. The Greeks built magnificent temples to Apollo and carved beautiful statues of Venus. The Egyptians built gigantic pyramids for their Pharaohs, and the Romans turned the brutalities of the Colosseum into a religious rite. The Chinese built the Great Wall to keep out the barbarians. Now man has made the fires of heaven his own, and left footprints on the craters of the moon.

We are moving toward the twenty-first century with the very great goal of building a Computopia on earth, the historical monument of which will be only several chips one inch square in a small box. But that box will store many historical records, including the record of how 4 billion world citizens overcame the energy crisis and the population explosion, achieved the abolition of nuclear weapons and complete disarmament, conquered illiteracy, and created a rich symbiosis of god and man, without the compulsion of power or law, but by the voluntary cooperation of the citizens to put into practice their common global aims.

Accordingly, the civilization to be built as we approach the twenty-first century will not be a material civilization symbolized by huge constructions, but will be virtually *an invisible civilization*. Precisely, it should be called an "information civilization." *Homo sapiens*, who stood at the dawn of the first material civilization at the end of the last glacial age, is now standing at the threshold of the second, the information civilization after ten thousand years.

A New Industrial Revolution?

Howard Rosenbrock et al.

In contrast to the euphoria of Masuda, the authors of this piece delve into history with a look at the British Industrial Revolution of 1780–1830. They conclude firmly that we are not witnessing a social revolution of equivalent magnitude, because the new information technology is not yet bringing about a new way of living. Rosenbrock is Professor of Control Engineering at the University of Manchester Institute of Science and Technology, Manchester, England. From New Technology: Society, Employment and Skill, *Council for Science and Society, London, 1981.*

This is a contribution to what is now a widespread discussion on the future development of technology and its interaction with society: on the ways in which society will influence the development of technology, and technology in turn will influence society. It is a discussion generally marked by uncertainties and hesitation. Great changes are predicted, but there are wide differences of view about what they will be, the rate at which they will occur, and the extent to which they will be beneficial.

The mood of this debate can be contrasted with a similar debate in England in early Victorian times. Then, in some quarters, there was an equally strong conviction of rapid future change. There was also great criticism of some of the aspects of early industrialization: loss of independence of the workers, bad working conditions, truck shops, child labor, etc; and the threat to Christian decencies, to deference, and to order. But these were seen predominantly as defects which could be eliminated, and the underlying change was generally seen as beneficial. Men believed in "progress," which would alleviate human toil, eliminate poverty, improve communications and transport, and make available to many what had been available only to a few. The steam engine, gas-lighting, the achromatic microscope, the railway, the electric telegraph, steamships, photography, new production machinery: all of these were seen as contributing to a future in which increasing knowledge would give increasing power over nature, and increasing wealth.

The mood today seems to be different. We have unparalleled knowledge and power over nature, yet this faces us with moral dilemmas and responsibilities for which we are ill-prepared. We can if we choose keep alive the victims of brain damage, or cruelly deformed children who once would have died: reason and humanity are both confused by such choices, and the exercise of the power we have attained, in either direction, leaves us tainted. Our technology has made it easy for us to burn up in a few decades the oil produced in many millions of years, or to destroy the last forests, or drive to extinction an increasing range of living creatures. We are uneasily aware of the judgment that later generations will pass on our response to these tempting opportunities. Buried in our minds are the pictures of Dachau and Auschwitz and Hiroshima and Gulag, which make a simple belief in progress no longer possible, while the doctrine of "Mutually Assured Destruction" has accustomed us to the sober rational calculus of the unthinkable.

In many ways we have the feeling of living at the end of an era, rather than the beginning of a new one. Both those who are optimistic about the future and those who are pessimistic agree in predicting rapid change. Few believe that the world in 20 years' time will be very similar to the world as it is today, yet the feeling is of moving away from what we have, rather than moving toward a welcoming future.

The Victorian situation led toward the danger of complacency. Our own offers a choice of two: on the one hand, despondency, and on the other, a millennial optimism, both of which can be seen at the present time.

Industrial revolutions

The future, it is widely believed, is going to see a period of very rapid change in technology, to such an extent that there is much talk of another industrial revolution. It is therefore useful to look at the first industrial revolution, say from 1780 to 1830, to see what it was about that period that led it to be regarded as revolutionary. It was not just the change in technology, but the fact that the new technology brought into being a new way of living. In this it was comparable to the transition in Stone Age times from hunting to agriculture.

The Industrial Revolution was chiefly an English phenomenon, partly Scottish and Welsh, and hardly touched other countries until a later period. It is therefore convenient to speak only of England. In 1780 England was still an agricultural country, with nearly 80 percent of the population living in the countryside. Outside London, the largest towns were Birmingham, Bristol, and Liverpool, each with about 50,000 inhabitants. Relative to earlier times, the eighteenth century was prosperous, and agriculture had been greatly improved. The enclosures of land were accelerating, driving people into the towns. Yet, for the majority, life went on in a way that can be traced back with only the slowest changes for many centuries. Country beliefs and traditions often had at their heart the rituals of pre-christian Stone Age farmers, from whom there was a continuous descent.

By 1830, though much of this still existed, it no longer seemed the way of the future. The great manufacturing towns had sprung up—Manchester based upon cotton with 182,000 people at the 1831 census, Leeds based upon wool with 123,000, and Birmingham upon engineering with 144,000, while the port of Liverpool had 202,000. There were in all 14 English towns in 1831 with a population of 50,000 or more. Life in these rapidly growing towns is largely hidden from us, but in them the old links with the countryside and its traditions were broken, and the village community was replaced by something much looser. The push from the land and from the decaying handicrafts was matched by a pull toward the towns and the new mills. An intelligent and ambitious young man could leave the countryside and live in relative prosperity as a mill-hand. A young woman would go to the mill in preference to "service" because of the greater freedom.

Conversely, those who became ill and unfit for work could no longer rely on the same care by the community. Orphans could be sent from their parishes to a kind of slavery as apprentices in the mills, while in the workhouses that were soon to be set up, husband and wife and children were separated. Above all, a new spirit of inhumanity arose, based upon a calculation of "economic necessity."

The Industrial Revolution had its heroic craftsmen in the millwrights and steam engine makers and their like. But traditional craftsmanship declined, and with it the design of manufactured articles. Around 1820 there was a sharp drop in the standard of book production. The quality of furniture and of household goods of many kinds deteriorated, and the old, easy, natural sense of seemliness and fitness in design was lost.

The worst of these developments produced in time their remedies. Factory legislation regulated safety and hours of work. Trade unions grew slowly to equalize in some degree the bargaining power of employer and employed. Much later, there began a slow and faltering attempt toward better design.

Toward the end of the nineteenth century, England was unrecognizably different from its condition in 1780: no longer based upon the countryside, but centered on the towns and on manufacturing, and past the peak of its international pre-eminence as a manufacturing country. Heavy physical labor was no longer the almost universal lot, though some perhaps in factory conditions worked more unremittingly than in the past. Food was plentiful and varied, and not yet as debased as it was later to become. Education to a certain level had become universal. A new sense of community, born in adversity, was felt in many working-class districts: later often to be destroyed by rebuilding or increased mobility. There was a feeling that life had become easier; almost a feeling of having climbed out of a pit: though, indeed, one can ask whether this was not the same pit that had opened in 1780.

How one assesses the gains and losses is not very significant. It is rather as if one were to compare the hunter with the farmer, and to assess the loss of freedom and satisfaction in the chase against the labor, but also the settled community life, of agriculture. In both cases one culture replaced the other, and to all appearances irreversibly. Not the least of the reasons for the irreversibility, in both cases, is that the new organization could support a much larger population than the old. Once the population had increased to the extent that the change permitted, a return to the earlier conditions was impossible.

Such a diversion into the past may not seem relevant to the next 20 years but it is intended to make this point: if the next 20 years are to produce something that we are justified in calling an "industrial revolution," it will have to be something as all-pervasive and far-reaching as we have described. It will have to be not just a change in our technology, but a change in the whole manner of our life. Judged by this criterion, none of the other "industrial revolutions" that have been suggested at various times qualifies for the title—whether based on the coming of railways, or electricity, or the automobile or atomic energy.

The coming changes will be based on the computer, on telecommunications, and on the power that they will give us over the manipulation of information. Will these changes be as profound as those that began around 1780? If so, the hunting and pastoral phases of man's development will have lasted some millions of years; the agricultural phase will have lasted some thousands; and the industrial phase will have lasted just about two hundred.

Among many opinions that have been expressed, it is possible to distinguish three major kinds of reply, or, rather, three extremes between which most others lie. The first answers no: the magnitude of the changes and their speed have been exaggerated, and what we shall see is just one more stage in the development of industry, comparable to other developments that have gone before. This line of thought can find support from the widespread expectation in the 1950s that automation would bring about rapid and fundamental changes in society: expectations that have not, so far, been fulfilled.

The second school of thought answers yes: we are moving out of the stage of industrial society into the stage of post-industrial society. We shall in the future need no more people to produce the goods we use than we need at present to grow the food we eat—say, 5 percent of the population for each. The remaining 90 percent will work in service industries, many of them concerned with the storage, manipulation, and dissemination of knowledge; though much that is referred to as "knowledge" is merely facts. In such a society we shall probably divorce the two functions of work, producing and earning. Leisure will become widespread, and work will be a minor and occasional diversion. This theme gives an infinite scope for imaginative elaboration, which can easily incorporate much of William Morris's "News from Nowhere."

The third school also answers yes, but predicts a total breakdown of society. There will be 5 million unemployed in Britain, and the resulting unrest will lead to the breakdown of government and some form of dictatorship, either of the right or the left. Out of this, through a revolutionary development, will arise the kind of society described in the preceding paragraph. Russia and the East European countries are usually assumed, on no apparent evidence, to be immune from these difficulties, being able to move smoothly from their present shortage, or at least apparent shortage, of labor to the leisure of the highly automated and computerized society.

Types of industry

Any detailed consideration of the effects of new developments in technology has to accept that these may be very different in different areas of industry. We have made a threefold division in the following way.

First there are the process industries, such as oil, chemicals, glass, paper, cement, iron and steel making (but not finishing), some parts of food production, etc. It is difficult to define these precisely, but they are most often continuously operating, and deal with flows of material. They are highly automated, and often already use computers in their control systems. The operating labor force is usually very small, to the extent that its further reduction is often not a serious economic consideration, and may not be possible for safety reasons: a certain minimum staff is needed to deal with accidents to men or machines. Maintenance, planning, and office work connected with their operation may, on the other hand, employ large numbers. Utilities such as gas, electricity, and water can also be put in this category.

In the operation of such plants we do not expect the new developments in microcomputers and communications to make a rapid qualitative change. The industries will probably make early and extensive use of the new technology, but chiefly to do rather better what is already being done. There will probably be greater integration of their operation, and a great deal of automatic data-gathering. In the longer term, some new processes might become feasible through better control, but this is not an imminent or a widespread possibility. Other developments, such as new processes based on biotechnology, will probably have greater effect.

At the other extreme is office work of various kinds—correspondence, ordering, invoicing, etc. Here the capital investment has in the past been relatively low, so that new equipment does not have to bear the cost of making older equipment obsolete. A relatively small investment can bring great increases in productivity. The equipment can be developed and installed rapidly, so it can use the latest developments in technology. In this area, which includes word processors and, later, the electronic office and electronic mail, we expect changes to be more rapid and extensive.

In between is the area of engineering production, where numerically controlled (NC) machine tools are beginning to spread fairly rapidly, and robots are beginning to be used in small numbers. Computer-aided design and manufacture are at an early stage, with great scope for development. In this area we expect an intermediate rate of change.

Technology and society

Although, looking backward, we can select a date around 1780 as the beginning of a rapid change which is almost a discontinuity, contemporaries did not have our acute awareness of a break with the past. Even in 1830, the new was still embedded in a matrix of the old. Moreover, all the developments that we associate with the industrial revolution can be traced backwards in the eighteenth and seventeenth centuries, and in some respects at least as far as the monasteries of the Middle Ages. Newcomen's steam engine was in use from 1705, iron was made with coke by Abraham Darby in 1709, and throughout the eighteenth century there was a series of inventions and improvements in cotton spinning and weaving.

In the same way, if we are about to see a great acceleration in our ability to handle information, to control and automate machinery, and to eliminate human intervention, then the roots of this change lie far back in the past. The period around 1830 was one of special awareness, and Charles Babbage had already grasped in principle the path that was to be followed. Machinery would be made "self-acting" so that human skill and effort were no longer needed. Or if human aid was still needed, the work would be made as simple as possible. More and more, work would be broken down into its elements, which could be done by unskilled workers under supervision, and later by machine. This applied not only to manual tasks, but also to such apparently intellectual work as the calculation of mathematical tables. Babbage's "analytical engine" was an attempt to carry this process a full century forward, and to produce in the mid-nineteenth century what was achieved only in the mid-twentieth.

The later developments of this theme, by F. W. Taylor, the Gilbreths, Henry Ford, and many others, are well-known. Throughout, the development of more automatic machinery has been accompanied by fragmentation of the jobs associated with them, removal of initiative and skill from the worker, and pacing of work by the machine.

The skills removed from the workplace have been concentrated in planning and supervisory organizations and have undergone a process of development and change. New occupations and new skills have arisen, which are less widespread than those they replaced. Some of these new skills and new occupations are of a professional type and status, and they have swelled the professional class, which has also grown in other ways.

The computer now appears to be offering the power to continue this process of deskilling at the lower levels, and to extend it also to higher levels.

The spray-painting of a car can be taught to a robot, which then indefinitely replicates the movements of the man whose job it was. The diagnostic skills of a physician can be incorporated, in a few weeks, into a computer, which can then match his accuracy closely. By incorporating the skills of several physicians, the computer may perform better than any one of them; and it may in time be able to learn from its own experience. Not all of this can be done at present, but none of it seems impossible in the near future. In the same way, skilled engineering designers may be largely replaced by computer systems.

If we look at past experience, it seems likely that possibilities of this kind, if they can be realized profitably with the computer, will be implemented despite any protests by those concerned. A hard-pressed hospital administration will believe that only by using computers can it offer a satisfactory service to its patients. An engineering company will see itself becoming uncompetitive unless it incorporates its design procedures in a computer.

To follow such a path of increasing automation usually requires an additional expenditure on capital equipment. Profitability then depends upon a reduction of employment for a given output, or at least the substitution of less-skilled, and so cheaper labor for more highly skilled. Both courses reduce the demands that are made on human ability, and a classical economic argument sees this as the creation of new opportunity. The human resources set free are available for other needs of society, or to increase the production of goods. Moreover, an economic mechanism will automatically ensure that this opportunity is fully used.

Yet the experience of the last 50 years does little to establish confidence in this self-regulating mechanism. The demoralizing unemployment of the 1930s ended only with the beginning of World War II, and it is not clear that the depression would have ended without the war. The 1970s, against expectations, saw a renewed increase of unemployment. During the whole period, a large proportion of those employed have done work below their capability. What is striking is that very great effort is expended upon the creation of the opportunity that unemployment or under-employment represents, and, in comparison, almost none upon using that opportunity.

A new industrial revolution?

We do not see microelectronics and computers and communication systems as bringing a new departure comparable to that which occurred in England between 1780 and 1830. Then, a stable society based upon agriculture and the countryside gave way to economic expansion, to great cities, and to a new expectation of continual "progress." Rather than a break with the past of that kind, we should see the effect of new technology in the next 20 years as an intensification of existing tendencies, and their extension to new areas.

To say that the next 20 years will see not a new beginning, but an intensification of old tendencies, is not to diminish the importance of the changes that will occur. These are likely to be faster and more extensive than anything in recent experience. They will for the first time impinge severely on white-collar and some professional work, as well as intensifying the pressure upon blue-collar jobs.

We also suggest that we are not likely to enter upon an era when work becomes largely unnecessary, a mere diversion from leisure. Our prosperity is based upon 200 years and more of increasingly rapacious exploitation of the earth, and it is not likely that we can continue on this path much longer. Energy and raw materials will become scarcer and more expensive under the pressure of a rapidly increasing world population and an increasing demand from poorer countries, while our impact upon the environment will have to be curbed. The new technology may serve rather to alleviate increasing difficulties than to lead us to an age of plenty.

If these judgments are correct, they go some way to explain the public mood on which we commented above. The problems we shall face will be not wholly new ones, but those with which we are familiar, though perhaps intensified and extended. There is likely, for example, to be a progression away from blue-collar jobs to white-collar or service occupations, but these could well take on more of the character that factory work has had in the past, if the tendency is not resisted.

Coming to terms with the past

We have emphasized the inhumanity and injustice that accompanied the early stages of the Industrial Revolution. These were not new things in England, but they had not previously been justified and defended by the prevailing philosophy. So powerful were the effects of this philosophy that to those who looked down from a higher level in society, the suffering became invisible, or, if not invisible, then transparent; and their view was not arrested by it but looked through it at what they took to be economic verities beyond.

Even so sympathetic an observer as Mrs Gaskell could not trust the clear evidence of her eyes. She describes, in a time of economic hardship, the suffering of the weavers in Manchester, and the contrasting prosperity of the mill-owners:

it is a bewildering thing to the poor weaver to see his employer removing from house to house, each grander than the last . . . while all the time the weaver . . . is struggling on for bread. . . . Large houses are occupied, while spinners' and weavers' cottages stand empty. . . . Carriages still roll along the streets, concerts are still crowded . . . while the workman loiters away all his unemployed time in watching these things, and thinking of the pale uncomplaining wife at home, and the wailing children asking for enough food. . . . The contrast is too great. Why should he alone suffer from bad times?"

Yet, she continues immediately, "I know that this is not really the case; and I know what is the truth in such matters. . . .'

We have not yet as a nation fully acknowledged the wrong that was done at that time, and we have not fully rejected it. Until we do so, we are unlikely to recover our energy and self-confidence. And until we do so, fresh shoots of the same philosophy will continually arise from its underground roots.

Productivity

If there is one aim that is common to all industrial nations, it must be to increase labor productivity. This is held out as the cure for inflation and for an adverse balance of payments. It is regarded as a promise of increased employment following from increased competitiveness, and as the only way toward an increase in material prosperity.

We suggest that these conclusions are less firm than they seem, and will bear a deeper analysis than they have received. To begin with, at least three different ideas are confused under the one name, "productivity." First, there is the idea of benefit to mankind. There is certain essential work that has to be accomplished for survival and convenience; a reduction in the effort needed for this work frees time and energy for other desirable activities. It is the appeal of this idea that underlies the other two, though they do not necessarily offer the same rewards.

Second, there is the idea of productivity of a company or commercial enterprise, which allows it to reduce its costs and improve its competitive position. This idea is not the same as the first, because an increase in output obtained by a proportional increase in effort still seems like an increase in productivity to the enterprise. So does a subdivision of jobs, which reduces the wage bill without reducing the total effort demanded.

Third, there is the idea of productivity of a nation, in which men and women are either employed or unemployed. An increase in productivity that leads to an increase in unemployment can be thought of as an opportunity: the unemployed effort is available for other desirable purposes. But it becomes a benefit only when it is used for those purposes, and though great effort is devoted to creating the opportunity, much less is devoted to using it.

Because of the differences between these interpretations of productivity, the efforts of an enterprise can fail to bring benefits to society, and may instead bring disadvantages. They can result in one part of the population working long hours under high pressure at uninteresting jobs, while another part is unemployed. An improved competitive situation of the enterprise may also fail to be reflected in an equivalent improvement of national competitiveness. If it is achieved by an increase in unemployment, it throws a burden upon the nation which largely cancels the benefit.

The chief agent in pursuing an increase of productivity is the individual enterprise. What it is actually pursuing is not the simple benefit contained in the first of the three interpretations, nor the direct improvement of national competitiveness. If the aims of the enterprise could be more closely aligned with the other two objectives, its activity and initiative could more directly contribute to the aims of society.

Economic determinism

A number of economic arguments are often deployed to show that things which would otherwise be desirable cannot be achieved. For example, it would be desirable to develop a different kind of technology from the one that we have: a technology that would set out not to eliminate skill and initiative, but rather to collaborate with them in increasing productivity. But then it is said that the technology that we have has evolved by competition in the marketplace, and is therefore economically superior to any alternative. To develop an alternative would impose a cost, reduce our competitive advantage, and lead to a reduction in wealth.

This conclusion follows only if a particular set of assumptions is made. By means of an elementary example, it has been shown how an equally plausible set of assumptions leads to an exactly opposite conclusion. That is, the pursuit of a short-term advantage, enforced by market action, can lead to a technology that is economically inferior to its alternatives.

More generally, it is suggested that the development of economic theory suffers from a number of difficulties. Controlled experiments are usually impossible, and the effect of one variable cannot be studied in isolation from others. The economic system is continually changing, and there is a question whether we can learn about it more quickly than it changes. Moreover, within an economic system there are many groups pursuing their own interests, so that a deterministic model is not appropriate. Something like the theory of games is needed, which so far has not developed to the necessary degree.

For these reasons, and also because of the historical experience, it is suggested that undue reliance on economic theory is not justified. A more direct, and probably safer, course is suggested instead: this is, that so far as possible, the conditions in which the major interest groups operate should be changed, in such a way that their perceived advantage aligns itself more closely with the aims of society.

Technological determinism

A related but different kind of determinism is often suggested in technology, and it is easy to see how it arises. The historical development—of steam engines, spinning and weaving machinery, machine tools, electrical power, computers, factory organization, and the rest—has led to a quite astonishing increase in productivity and material wealth. We have no experience of any alternative way in which technology could have been developed. We therefore are easily led to believe that the historical path was the only one that could have been pursued.

In particular, we may be led to believe that the development of technology must inevitably lead to the elimination of skill in the jobs that it directly affects. The conclusion that we suggest is a different one. The new technologies of microelectronics and computers and communications can be used to reinforce and extend the historical process of subordinating men and women to machines, and of eliminating their initiative and control in their work. They can also be used to reverse this process, to develop a technology that is subordinate to human skill and cooperates with it. Which of these outcomes will ensue depends upon the struggles of those concerned. What is important is to believe in the possibility of the second outcome, and to believe that the full use of human abilities is a higher and more productive goal than the perfection of machines.

Specific areas in which it appears clearly possible to develop a technology that preserves and collaborates with human skill include:

1 computer-aided design, where the process of deskilling is under way, but probably not irreversible;
2 NC machine tools and FMS systems, where for many industrial products the reintegration of the machinist's job would most likely offer economic benefits;
3 many applications of "expert systems," where development is still at an early age.

There are probably many more, but because the effort devoted to this possibility is so much less than that given to the alternative, they are difficult to identify.

On the other hand, where fragmentation of jobs has been carried to a very high degree over a very long period, the problem of developing a valid alternative may be much greater. The Volvo plant at Kalmar, and some of the work in the German "Humanisation of Work" program, show at least the beginnings of a path heading in the desired direction.

Changing the economic structure

If great effort is applied persistently by many people toward some not wholly impossible goal, there is likely to be a degree of success. Such effort has been applied toward increasing the productivity of labor according to one special definition, and it has succeeded astonishingly. No such persistent and tenacious effort has been devoted to making full use of human abilities. It has been assumed that full employment will follow from the working of the economy, either unaided or with some government control. The provision of jobs that use to the full the human abilities of those who fill them has not been a major objective: rather the reverse, since effort has been applied to reducing the skill and control of the worker.

We suggest that this order of priorities is wrong; that consistent effort toward the full use of human ability would embrace the subordinate aim of increasing productivity. It would also strive toward full employment and toward engaging the initiative and increasing the skill of workers.

To make these aims the intense and active concern of those engaged in the economic life of the country requires a change in their perceived incentives, which could arise through changes in public opinion, through trade union action, or through changes in the law. Without suggesting that they offer a complete solution, we propose that the following would repay study in this connection:

1   the operation of large Japanese companies, in which a commitment to continued employment for a large part of the workforce appears to give the company an incentive to use the ability and initiative of these workers to the full. It also removes from workers much of the fear that new technology will lead to their unemployment. We recognize that the Japanese system has its own difficulties, though lower economic efficiency is not among them;

2   the Scandinavian experience, in which trade unions have highly trained members who participate in the planning and introduction of new technology—for example the data processing shop stewards in some Norwegian plants;

3   the successful and self-sustaining development of workers' cooperatives at Mondragon in Spain, which in principle should ensure that aims similar to those we have mentioned are pursued;

4   the Lucas Aerospace Shop Stewards' "Alternative Corporate Plan," which clearly demonstrates the reserves of untapped initiative that were latent in the company.

Interdisciplinary research

The central complex of questions that have been considered in this report, relating to technology and work and society, is approached from different directions by different disciplines. These include engineering, computer science, industrial psychology and sociology, ergonomics, technology assessment, management science, economics, system science, social and economic history, political science, and probably others.

What hardly exists is a study of this complex in which the different disciplines, with their different assumptions and methods, are brought into contact with one another. Until this is done, the results obtained within a single discipline are likely to be highly misleading. If it could be done, the benefit to the disciplines themselves, and to the value of their results, should be considerable. We suggest that this is a task that might be particularly appropriate for the recently formed Technical Change Center.

A similar point arises, on a somewhat narrower front, in the development of new technology. Decisions are made, during the research stage of new technologies and of new equipment, that will later constrain all efforts to design the jobs in connection with them. This research phase should therefore attract the interest of social scientists and of trade unions. Yet this involvement hardly exists except in a few small and recent examples.

Education

If the changes that have been suggested above could be brought about, they would have profound implications for all levels of education. If skills were valued and preserved in work, education could link in a much more fruitful way with later experience. As at present, a range of skills, both physical and mental, would be acquired in the process of education. What would be more important would be to develop an ability to acquire further skills. Just as the scientific and technical knowledge learned in one's youth needs to be constantly updated, so one should regard skill as something that will continually need to change and develop.

In a similar way, the interdisciplinary work suggested in the previous section, if it were successful, could lead to deep changes in the education of technologists. At present, when elements of social sciences are introduced into technological courses, they are usually taught on the basis of an outlook and assumptions that are very different from those of technology. Any integration has usually to be provided by the student himself. A true integration of ideas around the theme of technology and society would find a much more ready acceptance.

Some Questions for the Information Society

Michael Marien

In this survey, the author takes a critical look at the notion of the "information society" and various versions of it. He then poses some tough questions which deserve careful consideration by all of us. Marien is the editor of Future Survey, *a monthly abstract of material on futures-related topics, and this essay concludes* De Informatiemaatschappij *(Natuur en Techniek, Maastricht, Netherlands, 1983). It also appeared in the* World Future Society Bulletin, *September–October 1983.*

The cluster of technologies described as the "communications revolution" has resulted in the new social condition presently called the Information Society. This article seeks to provide an overview and some sense of the range of future possibilities by providing an interrelated set of broad questions, as well as some brief and tentative answers. It is hoped that this initial inventory of questions will stimulate a better list, continuously updated and addressed in depth as new conditions generate new questions.

In addition to urging better questions, this article also urges better answers. Most of the questions raised here can only be roughly and tentatively answered, even when applied to present conditions. When considering the future, there are of course no firm answers, but only rough estimates, probabilities, and speculations. We should not be ashamed of such fuzziness and imperfection, nor should we turn away from questions about the future. As noted by French political philosopher Bertrand de Jouvenel (1967), it is natural and necessary to have visions of the future.

But three types of preconception should be avoided. The first, and most widespread, is the uncritical, euphoric stance that is expressed by commercial interests, which invariably emphasize only the positive attributes of new technology. This same preconception is frequently found in the narrow world view of the technician (who cannot imagine any negative consequences), and in the Utopian passion of wanting to help people with this or that technology.

A second type of over simplified view is the opposite of the first: the hypercritical, pessimistic stance that perceives all modern technology as a human disaster, or focuses solely on growing corporate or government control of information systems. This pessimistic stance can bring forth some important truths (e.g., see Schiller, 1981; Woodward, 1980), but it seldom offers guidance for positively shaping an Information Society.

The third over-simplified view acknowledges both of these positions, and concludes that there are opportunities for good and evil, centralization and decentralization, freedom and oppression, wealth and poverty. But this balanced view is often expressed superficially, merely concluding, for example, that we must choose between Computopia or a Big Brother society. The reality, however, is likely to be complex and ambiguous, requiring many critical choices over time and incorporating elements of simultaneous euphoria and gloom that fluctuate in their balance.

Because ambiguity and uncertainty are highly probable, it is important that we continually ask the right questions, supply the best possible answers, and share these answers across national boundaries. The following questions illustrate the type of concerns that ought to be addressed, and the type of answers that might follow.

Will we live in an information society?

The first question is too often assumed as a given. It is now a fact that in the industrially developed nations of the world the bulk of the labor force is engaged in some manner of producing or disseminating information. Yet, ironically, these very nations are *under*developed insofar as being information societies. Moreover, this dominant characteristic of the society and its labor force will not necessarily continue into the future.

Fashions, values, or insights may change, or society itself may change in such a way that "information society" is clearly inappropriate as a societal label. The term "information society" has only been used for about a decade, superseding the less specific label "service society," and the even more ambiguous "post-industrial society." Information Society was apparently first used in Japan in the late 1960s (Kohyama, 1968), and was the focus of the *Plan for an Information Society* (Masuda, 1981). Forerunners to "information society" include the terms "age of cybernation" (used widely in various forms during the 1960s), "electronic age," and "age of information" (both proposed by Marshall McLuhan in 1964); "knowledge society," described by Peter Drucker in 1969, and the ungainly "technetronic society" suggested by Zbigniew Brzezinski in 1970.

But societal labels come and go, and hundreds have been proposed in recent years. Why should "information society" necessarily last to 2000, or even to 1990? A new variant might come into usage, such as "telematic society" (proposed by Nora and Minc, 1980, and adopted by James Martin, 1982), or even "the age of Infoglut," which focuses on the pervasive condition of information overload (Marien, 1982). Unemployment caused by the automation of office work and other informational services may be extensive and, if not compensated by an equal number of new jobs in the information sector, could result in a labor force no longer dominated by information-related occupations. The major activity of society would then be some other occupation, or even involuntary idleness—the lack of any occupation—a condition that already characterizes some Third World nations.

It should also be acknowledged that there are other concurrent technological revolutions that could have an even greater impact on society than the new information technologies. The biological revolution could retard the aging process and lengthen human life spans so that we become, essentially, a "Society of Immortals" (albeit a crowded one). Advances in solar cell technology could bring a "soft path" revolution in energy use. The spread of armaments in general, and nuclear weapons in particular, could lead to a cataclysmic detonation of weapons, either through accident or design. (Either instance, ironically, could be seen as a profound failure of communication.) In fact, the EMP or electromagnetic pulse, of a single high-altitude nuclear blast could burn out a nation's electrical systems (Broad, 1983). In the wake of such a grim scenario, "information society" would seem in retrospect to be the ultimate illusion of modern man.

Although the nuclear threat is growing, most experts still think it unlikely that we will experience a nuclear Armageddon. Nevertheless, there is an increasing possibility of serious disruption by war or terrorism (O'Heffernan, et al. 1983). If we survive, it is likely that we will have an information society of some sort, although we might not call it that. An intelligent approach to assessing the future would acknowledge the potential for changing images. We must consider the many ways in which we may not have an information society, as well as the many ways in which it may be realized, if we hope to shape such a society to the greatest benefit of the greatest number of people.

Will we experience a communications revolution?

Similar to the first question, a "revolution" in communications is widely assumed. But this, too, should be posed as a question. The phrase "revolution" has been used promiscuously in recent years, even among those whose scientific training would seemingly inhibit rash statements about technical and social developments. Consider, for example, the proclamations in the late 1960s about the Green Revolution—new crops that would solve the world's hunger problems. Similar assertions were also made at that time about communications. In 1970, Isaac Asimov, the prolific author of science fiction and popularized science books, announced a Fourth Revolution of electronic communications (following speech, writing, and the printing press), which when truly established would bring worldwide electronic literacy, the library of mankind available to any person at any time, a personal immediacy to justify the sense of a global village, lessened differences among people, and cities spreading out and disappearing. Shortly thereafter, in 1972, the Carnegie Commission on Higher Education also proclaimed a Fourth Revolution brought on by electronics. A decade later, there are still intimations of such a revolution in higher education and the entire world. But one can hardly say that it has taken place.

Remarkable developments have occurred in the cluster of technologies comprising the "communications revolution." These include not only the notable reductions in computer size and cost, combined with dramatically enhanced capability, but also the expanded use of satellites, cable television, home information services, and the many applications of microchips. And the string of inventions has by no means run its course; in the words of Adam Osborne (1979), an innovative designer and manufacturer of computers, these new technologies are "running wild."

It is impossible to forecast the ultimate configuration of this rapidly evolving cluster of complementary and competing technologies, or to assess their multiple impacts on human life. The best that can be done is to engage in systematic technology forecasting and assessment. Developments in technology can to some degree be anticipated, especially by procedures of collective thinking such as the Delphi method, which assembles expert forecasts, refining and revising them with two or more rounds of questioning (e.g., Pelton, 1981). The potential impacts of individual technologies can be assessed (e.g., Nilles, 1982, on the personal computer, Tydeman et al., 1982, on teletext and videotext, and Wise et al., 1980, on microcomputers). The French government has sponsored a study of the impacts of computerization on society (Nora and Minc, 1980), and the Science Council of Canada (1982) has expressed concern about planning for an information society. Wilson Dizard (1982) has described what the US should do to understand and control the potentially dehumanizing and anti-democratic effects of this "massive technocratic drive."

Unfortunately, no effort has been made to collect all of these forecasts, assessments, speculations, and warnings to determine what is known and not known, identify areas of agreement and disagreement, and establish the range of proven policies that might be pursued. Ironically, in the midst of an inchoate revolution in communications technology, this relatively simple act of communication between researchers and responsible policy-makers has not occurred.

A cautious approach to the "communications revolution" would be to withold such a label for the present. In contrast to the changes wrought by other technological revolutions in the past 200 years—railroads, the telegraph, telephones, electric power, automobiles, radio, and television—the new cluster of communications technologies has yet to shape the lives of most people profoundly. Still, these new technologies are likely to be immensely influential in the next few decades, if nuclear weapons have not obliterated much or all of humanity, and if a worldwide economic collapse (which would probably retard the development of the information society) is avoided. The development and dissemination of new technologies will probably not be as rapid or as widespread as many enthusiasts today believe. But the "revolution" will proceed, for there is little or no public opposition, and governments at best have only been able to influence—not control—it. The nature of the revolution—how it affects the way we communicate and our lives in general—remains problematic.

Will we communicate better?

The question of whether we will communicate better—and even whether we are communicating better today than we did in the recent past—appears obvious and simple. But it is a complex question with deep significance, and little or no attention has been paid to it. The failure to consider this question may stem from the trained incapacity of communications experts to consider big questions, or from the twin assumptions that new technologies of communication will necessarily improve communication, and that all attempts to communicate are realized. Rather, it is important to recognize that, while error-free communication is an ideal, non-communication in modern society appears to be widespread (Marien, 1982). Examples include failed communications (important messages not sent or received), flawed communications (wrong messages sent as a result of unintentional error or intentional lying or distortion), miscommunication (messages not understood or believed, or resulting in an unintended effect), and junk communication (trivial messages that are received, but are of no importance).

The new communication technologies will greatly multiply capacities for storage, transmission, and manipulation of information. But will they improve human communication, or inadvertently make it more difficult? To sketch some tentative answers to this question, it is useful to briefly examine a few prospects in eight general situations in which people communicate.

Work

Major changes in the workplace are likely, with many jobs eliminated and new jobs created. Robots in factories will presumably perform many tasks that are dangerous or boring. Automation of office work will displace a large part of the female labor force (Menzies, 1981), but, presumably, will also enable better communication within and among organizations, and allow some degree of decentralization of workplaces to rural areas and individual homes (as suggested by the romantic image of the "electronic cottage," proposed by Alvin Toffler).

Commerce
Relationships between buyers and sellers may be improved with the advent of teleshopping (the display of wares on the home video screen), which would give consumers better information on alternative products. Credit cards activated by thumbprints should be a further advance toward a cashless society.

Health
Computerized communications already offer physicians better access to medical knowledge, and the computerization of personal medical histories (with proper safeguards, one hopes) can supply valuable patient information. Individuals will also have much better access to medical information for their own self-care, with new devices to monitor body processes, worn on the wrist or implanted in the body. Microprocessor implants, for example, could detect the first sign of malignant cells being generated. New developments in microelectronics promise at least some degree of hearing for the deaf and sight for the blind, and computers might even act as psychotherapists (Evans, 1980).

Entertainment
Many people will surely have more electronic options for their pleasure, including videogames and simulated experiences, 100-channel cable television, cheap collector dishes enabling access to increasingly sophisticated communication satellites, and various videodisk and videocassette recordings. Some or all of these will be accompanied by greatly improved presentation in the home, such as wall-size television displays, improved high-definition pictures, and stereo sound. It remains to be seen, however, whether this will result in an abundance of high-quality options or merely a multiplication of banality, a vibrant free market between entertainers and audiences or cultural monopoly and control by governments and information conglomerates.

Education
It is doubtful that any information utility will be a completely free service, as Utopians imagine. Nevertheless, the potential for electronic access to the world's knowledge and for computer programs of instruction, multiple cable television channels devoted to education, instructional videotapes and more, offer enticing possibilities for an education revolution, both within and outside of schools and colleges, that would affect the learning of both children and adults. But the general caveat for all aspects of the communication revolution—not as much, nor as soon—is especially applicable to education. Educational institutions serve many functions other than learning, and they are difficult to change. Extensive self-directed adult learning is possible, but serious utilization of information abundance must compete with

the many enticements of non-serious entertainment. Indeed, the mind-deadening influence of television is a major explanation for the steady decline in test scores among American high school students over the past 20 years. Sober analysis of who is learning what as a result of the new technologies will be required. It may well be found that information technology will further widen existing divisions between the rich and poor, and create a new generation gap between the computer-literate young and their print-literate elders.

Politics

Equal caution should be applied to predictions of enhancements in political communications. The new communications technologies offer many promising ways to make societies more democratic: e.g., electronic plebiscites and opinion polls, teleconferencing with representatives, cable television channels devoted to legislative proceedings, and better voter information on candidates. But these possibilities will not necessarily be realized, and could be more than offset by a dossier society utilizing a centralized data bank, improved surveillance capabilities, new lie detector technology, and narrow-casting of political messages enabling candidates to say different things to different groups of voters. Automatic language translators might facilitate intercultural communication and lessen the tensions of world society, but Western and particularly American culture may very well increase its dominance over the Third World, leading to more angry but unheeded calls for a New World Information Order (MacBride et al., 1980; Smith, 1980).

Intergroup relations

Both within and between nations, the new abundance of information, and its further fragmentation to meet the needs and interests of myriad racial, ethnic, religious, intellectual, political, commercial, and leisure interests, is unlikely to facilitate intergroup harmony and sharing. Mass broadcasting lacked diversity and generally aimed at the lowest common denominator; but it did at least provide a shared experience. The de-massification of the media enhances the virtue of greater variety, but the dark side of this trend is a chaos of specialized interests. And to the degree that one finds ample entertainment and education in the home, people will spend more time in their living rooms, and less time with their neighbors or in public places.

Families

Relations between men and women, husbands and wives, and parents and children may be enhanced by the imminent spread of mobile telephones and the possible development of two-way wrist video devices (popularized for many years in the "Dick Tracy" comic strip). But the multitude of specialized entertainments may nevertheless serve to strain these intimate relations. Another important factor affecting communications within families will be the amount of leisure time available, which may or may not be improved by productivity increases at work and decentralization of work-places (thus reducing commuting time or enabling more work at home). Involuntary leisure forced by unemployment often places a major strain on family relationships.

To summarize this cursory survey of basic communication nexuses, it is difficult to say whether human communication has been improving or will improve in the future. It appears likely that the new communications technologies will produce overall improvements in work, commerce, health, and entertainment. These technologies also offer considerable potential for improving education and politics, but we should be very cautious in making forecasts and assessments. Communications between groups and within families could improve, but are perhaps more likely to worsen. In all these areas, considerable monitoring will be required to provide a reasonably comprehensive answer to the question of whether we are communicating better.

Will our lives be better?

Will the new information technologies lead to improvements in the quality of life? One must immediately ask, improvements for whom? There will surely be winners and losers, within and among nations. Economic abundance may result from the information revolution; Paul Hawken, for example, postulates that we are in the midst of a tumultuous transition from the mass economy of the industrial age to the "informative economy," where information increasingly replaces mass as a factor in production, resulting in better goods and services. But this does not necessarily lead to more equal sharing, or to enhancement of the *median* benefit. Will all nations benefit from an expanded economic pie (as some argue), or will rich nations increase their share of a relatively limited economic pie (as argued by others)? Will illiterate peasants of the Third World benefit in any way from the new abundance of information? Even within the richer nations, the unemployed and underemployed may not benefit, unless economies expand enough to fund a more generous welfare state.

One must also ask how "quality of life" is measured, which is no simple matter. There may be a growing maldistribution of work, with the current worldwide problem of surplus, unemployable populations aggravated by further advances in automation. The promise of many technologies, and information technology in particular, is that they will be labor-saving, allowing more people to have more leisure. But what is to be done with those who have involuntary leisure forced upon them? And among the employed, voluntary leisure in a technological age can prove illusory, as pointed out by the Swedish economist, Steffan Linder (1970). Many professionals today find their workloads increasing, rather than decreasing. New ways to process information may only add to the chaos, much as superhighways have inadvertently led to more traffic congestion.

Besides economic indicators of quality of life, one must also assess political and human indicators. Will information technologies increase or decrease the threat of nuclear war? Can we maintain and expand democratic forms of governance (Wicklein, 1981)? Will non-democratic governments be strengthened by the new technologies (as now seems likely)? Will we spend more time in pleasant and productive interaction with people, or will our interaction with machines increase to the point where computers become our most important teachers and our best friends?

Who has to learn what?

Every society requires a minimum standard of competence for its members. The information society will demand a new standard of literacy, well beyond merely knowing how to use a computer. To survive in any civilized fashion, we must learn to cope with what Alvin Toffler refers to as our "blip culture" of immense diversity, contradictory fragmented images, and shattered consensus. The new literacy will require the ability to distinguish between knowledge and mere information, and to seek out wisdom amid abundant knowledge (Work, 1982). Lifelong learning has been extolled for many years, but the new economic literacy will require an ability to learn, and a willingness to retrain for new occupations. Finally, the new civic literacy must include an understanding of global affairs, for the new technologies of communications, combined with such problems as pollution, access to ocean resources, monetary chaos, and the arms race, are accelerating the process of globalization.

But, in contrast to these needs, what do we in fact know? Victor Ferkiss (1969) asserts that technological man does not yet exist, and neoprimitive man continues to be trapped in a technological environment, in which things—not human beings—are in control. In introducing his notion of "meaning lag," Canadian sociologist Orrin Klapp (1982) warns that meaning formation (the limited human capacity for processing information) is slow and inefficient, compared with the speed and amount of information now accumulating in society.

We must recognize that we live in an Age of Ignorance, where the learning needs of all age groups are outracing their attainments. Our nation is indeed at risk. However, the most important learning needs are not among children, but among adults—especially our political, intellectual, scientific, corporate, and religious leaders—the decision-makers who will be shaping the information society over the next two decades. Their decisions, for better or worse, will largely determine whether the information society is humane, just, productive, free, participatory, and safe, or whether it is a society characterized by greater inequalities, more centralization, accelerating dangers, and further alienation.

Such a reorientation of educational priorities is unlikely, though—at least during the next few years. We still fail to recognize our widespread ignorance and the need to focus on adult learning because the academic degrees that many of us hold convince us that we are well educated as individuals and as societies. We wish to appear knowledgeable and sophisticated, and this image is reinforced by the sophisticated technologies at our individual and collective command—tools such as computers and automobiles that many are able to use, although comparatively few really understand them very well. The citizens of the industrialized countries live in a society that is *developed* in manufacturing era terms—but *overdeveloped* when assessed by humane and ecological measures, and *underdeveloped* as an information society. Our society is underdeveloped because of the gap between our present abilities and our need to learn new skills and shape new worldviews. This "ignorance gap" appears to be growing.

As with the previous questions about the information society, this question also points to the need for empirical research—in this instance, research into what people in all age groups in all nations know. And some global standard of information society literacy must emerge from this sobering assessment.

Who will address these questions?

The questions raised here are presumably of central importance to all individuals and nations. They could be considered as a matter of national security, deserving the funding equivalent that is now invested in a single battleship or aircraft, let alone the modern arsenals that every nation seems compelled to have. Worldwide, an estimated $650 billion was spent on military preparations during 1982 (Palme, 1982). It is doubtful whether even 0.01 percent of this amount ($65 million) has been devoted worldwide to considering questions about the information society. Annual spending on this aspect of national security is probably less than 0.001 percent ($6.5 million). Whatever the figure, our societal investment priorities appear to be grossly distorted.

Several hundred books have been published in recent years, dealing with some aspect of the information society. Yet there is little communication among those who address some aspect of this subject. This is largely due to our obsolete industrial era colleges and universities, which encourage attention to small and "manageable" questions, technical questions that result in "hard" answers, and questions that conform to the configurations of the established disciplines and professions.

The fragmentation of perspectives increasingly found in the wider society is reflected in the subject of communications itself, which is studied by the professions of journalism, education, and information science (formerly library science), and such cross-disciplinary areas as computer science, management science, behavioral science, language and area studies, and future studies. Adding to this intellectual tumult, researchers in the social sciences often specialize in the economics, politics, and sociology of information and communications. Occasional government studies attempt to provide some overview, but little or no effort has been made by governments, foundations, research institutes, or leading universities to try systematically to overcome the rampant bureaucratization of knowledge in general and thinking about communications in particular.

People in all nations need the best and most up-to-date answers to the most important questions about the information society. The "national security" benefits of stimulating inquiry, refining and synthesizing the plethora of observations and policy proposals, and encouraging public understanding and dialogue would certainly be great. The cost, in contrast to these benefits, would be miniscule.

Ideally, we need schemes such as the World Brain, proposed by H. G. Wells in 1938, to bring together the scattered mental wealth of the human race and make it universally accessible. Such a reorientation of education and information has been updated by the concept of WISE, the World Information Synthesis & Encyclopedia (Kochen, 1975). Similarly, to deal intelligently with questions of the future, we need an on-going surmising forum, as advocated by Bertrand de Jouvenel, to bring together and debate the many "futuribles" about what might happen and what ought to happen.

Will anything of this sort take place? Appeals to higher and wider vision, such as this essay, are easy to make. Creating and sustaining the institutions needed to promote such vision will be far more difficult. Even with the help of such coordinating and synthesizing organizations, the evolution to an information society will not be an easy one, but only will be made a little less turbulent than it would otherwise be. Continuing the status quo of informational chaos greatly heightens the likelihood that we will not realize a viable and humane outcome. Most countries have avoided choices, with the expectation of muddling through (Nanus, 1982). In such an event, the information society will arrive stillborn—if it arrives at all—owing to our failure wisely to generate and employ information.

References

Asimov, Isaac (1970), "The Fourth Revolution," *Saturday Review*, 24, October, 17–20.

Broad, William J. (1983), "The Chaos Factor," *Science* 83, 4:1, Jan–Feb 1983, 40–9; also "Nuclear Pulse," *Science*, vol. 212, May 29/June 5/June 12, 1981, 1009–1012, 1116–1120, and 1248–1251.

Brzezinski, Zbigniew (1970), *Between Two Ages: America's Role in the Technetronic Era*. New York: Viking.

Carnegie Commission on Higher Education (1972), *The Fourth Revolution: Instructional Technology in Higher Education*. New York: McGraw-Hill.

De Jouvenel, Bertrand (1967), *The Art of Conjecture*. New York: Basic Books.

Dizard, Wilson P. (1982), *The Coming Information Age: An Overview of Technology, Economics, and Politics*. New York: Longman.

Drucker, Peter F. (1969), *The Age of Discontinuity: Guidelines to Our Changing Society*. New York: Harper & Row.

Evans, Christopher (1980), *The Micro Millennium*. New York: Viking (published in Britain in 1979 as *The Mighty Micro*.)

Ferkiss, Victor (1969), *Technological Man: The Myth and the Reality*. New York: George Braziller.

Forester, Tom (ed.) (1980), *The Microelectronics Revolution*. Cambridge MA: MIT Press, 1980.

Hawken, Paul (1983), *The Next Economy*. New York: Holt, Rinehart and Winston.

Klapp, Orrin E. (1982), "Meaning Lag in the Information Society," *Journal of Communication*, 32:2, Spring, 56–66.

Kochen, Manfred (ed.) (1975), *Information for Action: From Knowledge to Wisdom*. New York: Academic Press.

Kohyama, Kenichi (1968), "Introduction to Information Society Theory," *Chuo Koron*, Winter 1968; cited by Yoneji Masuda in Yoshihiro (ed.), *Changing Value Patterns and their Impact on Economic Structure*. Tokyo: University of Tokyo Press, 1982, p. 174.

Linder, Steffan Burenstam (1970), *The Harried Leisure Class*, New York: Columbia University Press.

MacBride, Sean, et al. (1980), *Many Voices One World: Communication and Society, Today and Tomorrow*. New York: Unipub (final report of the UNESCO International Commission for the Study of Communication Problems).

Marien, Michael (1982), "Non-Communication and the Future," in Howard Didsbury Jr (ed.). *Communications and the Future*. Bethesda, MD: *World Future Society*.

Martin, James (1981), *Telematic Society: A Challenge for Tomorrow*. Englewood Cliffs. NJ. Prentice, Hall (first published as *The Wired Society* in 1978).

Masuda, Yoneji (1981), *The Information Society as Post-Industrial Society*. Bethesda, MD: *World Future Society*.

McLuhan, Marshall (1964), *Understanding Media: The Extensions of Man*. New York: McGraw-Hill.

Menzies, Heather (1981), *Women and the Chip: Case Studies of the Effects of Informatics on Employment in Canada*. Toronto: Institute for Research on Public Policy.

Nanus, Burt (1982), "Developing Strategies for the Information Society," *The Information Society Journal*, 1:4, 339–56.

Nilles, Jack M. (1982), *Exploring the World of the Personal Computer*. Englewood Cliffs, NJ, Prentice-Hall.

Nora, Simon, and Minc, Alain (1980), *The Computerization of Society: A Report to the President of France*. Cambridge, MA: MIT Press (first published in France in 1978).

O'Heffernan, Patrick, Lovins, Amory B. and Lovins, L. Hunter (1983), *The First Nuclear World War*. New York: William Morrow.

Osborne, Adam (1979), *Running Wild: The Next Industrial Revolution*, Berkeley, California: Osborne/McGraw-Hill.

Palme, Olaf (1982), "Military Spending: The Economic and Social Consequences," *Challenge*, 25:4, September–October, 4–21 (chapter 4 of the report of the Independent Commission on Disarmament and Security Issues, *Common Security: A Blueprint for Survival*. New York: Simon & Schuster).

Pelton, Joseph N. (1981), "The Future of Telecommunications: A Delphi Survey," *Journal of Communication*, 31:1, Winter, 177–89.

Schiller, Herbert I. (1981), *Who Knows: Information in the Age of The Fortune 500.* Norwood, NJ: Ablex.

Science Council of Canada (1982), *Planning Now for an Information Society: Tomorrow Is Too Late.* Hull, Quebec: Canadian Government Publishing Centre (report no. 33).

Smith, Anthony (1980), *The Geopolitics of Information: How Western Culture Dominates the World.* New York: Oxford University Press.

Toffler, Alvin (1982), *The Third Wave.* New York: William Morrow.

Tydeman, John, et al. (1982), *Teletext and Videotext in the United States: Market Potential Technology, Public Policy Issues.* New York: McGraw-Hill.

Wells, H.G. (1938), *World Brain.* New York: Doubleday, Doran.

Wicklein, John (1981), *Electronic Nightmare: The New Communications and Freedom.* New York: Viking.

Wise, Kensall D., et al. (1981), *Microcomputers: A Technology Forecast and Assessment to the Year 2000.* New York: Wiley-Interscience.

Woodward, Kathleen (ed.) (1980), *The Myths of Information: Technology and Post-Industrial Culture.* Madison, Wisconsin: Coda Press.

Work, William (1982), "Communication Education for the Twenty-First Century," *Communication Quarterly*, 30:4, Fall, 265–9.

Guide to Further Reading

The information society

Alvin Toffler, *The Third Wave* (William Morrow, New York; Collins, London, 1980).

Edward Cornish, "The Coming of an Information Society," and Graham T. T. Moliter, "The Information Society: The Path to Post-industrial Growth," *The Futurist*, April 1981.

Tarja Cronberg and Inga-Lisa Sangregorio, "More of the Same: The Impact of Information Technology on Domestic Life in Japan," *Development Dialogue*, no. 2, 1981, Uppsala, Sweden.

Edward Ploman, "The Communications Revolution," *Intermedia*, September 1981.

"The Information Society," special issue of *Journal of Communication*, Winter 1981.

"The Information Society: Plug in or Drop Out," *New Scientist* special issue, December 9, 1982.

Wilson P. Dizard, *The Coming Information Age* (Longman, New York, 1982).

Frederick Williams, *The Communications Revolution* (Sage, Beverly Hills, 1982; Mentor, London, 1983). Popular overview with a "middle ground" approach.

Barry Jones, *Sleepers, Wake! Technology and the Future of Work* (Wheatsheaf, Brighton, England, 1982). Lively contribution from Australia.

"The Information Age," special issue of *Political Quarterly*, vol. 54, no. 2, 1983.

Gordon Thompson, *Memo from Mercury*, Occasional Paper no. 10 (Institute for Research on Public Policy, Ottawa, Canada).

Tom Stonier, *The Wealth of Information: A Profile of the Post-industrial Economy* (Thames Methuen, London, 1983).

Critical perspectives

Kathleen Woodward (ed.), *The Myths of Information: Technology and Postindustrial Culture* (Coda and Indiana University Press, Bloomington, IN, 1980).

M. Lehman (ed.), *Communication Technologies and Information Flow* (Pergamon, Elmsford, NY, 1981).

H. P. Gassman, "Is There a Fourth Economic Sector?", *OECD Observer*, November 1981.

Jennifer Daryl Slack and Fred Fejes, *The Ideology of the Information Age* (Ablex, Norwood, NJ, 1981).

Vincent Mosco, *Pushbutton Fantasies: Critical Perspectives on Videotex and Information Technology* (Ablex, Norwood, NJ, 1982).

Orrin E. Klapp, "Meaning Lag in the Information Society," *Journal of Communication*, Spring 1982.

Michael Marien, "Non-Communication and the Future," in Howard F. Didsbury (ed.), *Communications and the Future* (World Future Society, Bethesda, MD 1982).

Kimon Valaskakis, "Post-industrial or Re-industrial: Two Interpretations of the Information Revolution," *World Future Society Bulletin*, July–August 1982.

Planning Now for an Information Society: Tomorrow Is Too Late, Report no. 33 (Science Council of Canada, Hull, Quebec, 1982).

Burt Nanus, "Developing Strategies for the Information Society," *Information Society Journal*, vol. 1, no. 4, 1982. US *laissez-faire* or Japanese interventionism?

Ithiel de Sola Pool, "Tracking the Flow of Information," *Science*, August 12, 1983.

Raphael Kaplinsky, *Automation: The Technology and Society* (Longman, Harlow, England, 1984). Marxist "indictment of contemporary capitalism"—argues that automation is the response to a crisis, not its cause.

Donald N. Michael, "Too Much of a Good Thing?"
Vital Speeches of the Day, November 1983.

General Guide to Further Reading

Introductory material

W. H. Mayall, *The Challenge of the Chip* (Science Museum/HMSO, London, 1980). Illustrated.

Eric Morgan, *Microprocessors: A Short Introduction* (Department of Industry/HMSO, London, 1980).

Robin Bradbeer, Peter De Bono and Peter Laurie, *The Computer Book: An Introduction to Computers and Computing (BBC, London, 1982)*. Based on the 1982 BBC-TV series. Very good.

John Shelley, *Microfuture* (Pitman, London, 1981).

Peter Zorkoczy, *Information Technology: An Introduction* (Pitman, London, 1982).

Gordon Pask and Susan Curran, *Micro Man: Computers and the Evolution of Consciousness* (Macmillan, London, and Century, New York, 1982).

Michael Crichton, *Electronic Life* (Knopf, New York, 1983).

The history of computers is dealt with in:

Joel N. Shurkin, *Engines of the Mind* (W. W. Norton, New York, 1984).

"The Birth of Computers," *New Scientist*, September 15, 1983.

"The Computer Moves In," *Time* cover story, January 13, 1983.

Social impact studies

Michael L. Dertouzos and Joel Moses (eds), *The Computer Age: A Twenty-Year View* (MIT Press, Cambridge, MA, 1979).

Simon Nora and Alain Minc, *The Computerization of Society* (MIT Press, Cambridge, MA, 1980).

Juan Rada, *The Impact of Microelectronics* (ILO, Geneva, Switzerland, 1980).

H. Dominic Covvey and Neil Harding McAlister, *Computer Consciousness—Surviving the Automated 80s* (Addison-Wesley, Reading, MA, 1980) and *Computer Choices: Beware of Conspicuous Computing* (Addison-Wesley, Reading, MA, 1982).

Alan Burns, *The Microchip—Appropriate or Inappropriate Technology?* (Ellis Horwood/ John Wiley, Chichester, England, 1981).

Nils Bjorn-Anderson, Michael Earl, Olav Holst, and Enid Mumford (eds), *Information Society: For Richer, For Poorer* (Elsevier, Amsterdam–North Holland, 1982).

Liam Bannon, Ursula Barry, Olav Holst (eds), *Information Technology: Impact on the Way of Life* (Tycooly, Dublin, Eire, 1982).

Howard F. Didsbury (ed.), *Communications and the Future: Prospects, Promises and Problems* (World Future Society, Bethesda, MD, 1982).

Paul Sieghart (ed.), *Microchips With Everything: The Consequences of Information Technology* (Comedia, London, 1982).

Gunter Friedrichs and Adam Schaff, *Microelectronics and Society: For Better or For Worse, A Report to the Club of Rome* (Pergamon Press, Elmsford, NY, 1982).

John Eaton and Jeremy Smithers, *This is IT* (Philip Allan, Deddington, England, 1982).

P. K. Marstrand (ed.), *New Technology and The Future of Work and Skills* (Frances Pinter, London and Dover, NH, 1984).

Other sources

In the US, a series of Office of Technology Assessment (OTA) reports have covered various aspects of the information technology revolution (the most relevant are mentioned in the Guides to Further Reading at the end of each chapter).

While in the UK, a series of ACARD (Advisory Council for Applied Research and Development) reports have covered such topics as robots, industrial innovation and CAD/CAM. In addition, there were:

New Information Technologies, 27th Report of the House of Lords' Select Committee on the European Communities (HMSO, London, 1981).

New Information Technologies and Social Change, Report of the Standing Committee on Employment (European Commission, Brussels, 1981).

Making a Business of Information: A Survey of New Opportunities, Report of the Information Technology Advisory Panel (HMSO, London, 1983).

Crisis Facing UK Information Technology, National Economic Development Office, London, August 1984.

The National Computing Centre, Manchester, has a whole range of publications on microelectronics, communications, expert systems, and automation, while LAMSAC (Local Authorities Management Services and Computer Committee) has also published a great deal of material on computer applications.

Obviously, it makes sense to monitor the science magazines like *Science, Scientific American, New Scientist, Technology Review*, and *High Technology* for news of the latest technical developments, while general magazines like *Time* and *Newsweek* are good at spotting social trends arising out of these developments. Business-orientated magazines like *Fortune, Business Week*, the *Harvard Business Review*, the *Economist*, and the business press in general are equally perceptive with their frank assessments of the prospects for various social innovations. After all, it is businessmen who make them work, by trying to make them pay.

Finally, a couple of dictionaries:

A. J. Meadows, M. Gordon, and A. Singleton, *Dictionary of New Information Technology* (Century, New York, 1982).

Dennis Longley and Michael Shain, *Dictionary of Information Technology* (Macmillan, London, 1982).

Index